MW00711051

STUDENT SOLUTIONS MANUAL

# PRECALCULUS
# MATHEMATICS

## FIFTH EDITION

# STUDENT SOLUTIONS MANUAL

## LAUREL TECHNICAL SERVICES

# PRECALCULUS MATHEMATICS

## FIFTH EDITION

## MAX SOBEL

## NORBERT LERNER

PRENTICE HALL, UPPER SADDLE RIVER, NJ 07458

Production Editor: *Carole Suraci*
Production Supervisor: *Joan Eurell*
Acquisitions Editor: *Melissa Acuna*
Supplement Acquisitions Editor: *Audra Walsh*
Production Coordinator: *Ben Smith*

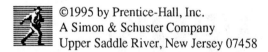 ©1995 by Prentice-Hall, Inc.
A Simon & Schuster Company
Upper Saddle River, New Jersey 07458

All rights reserved. No part of this book may be
reproduced, in any form or by any means,
without permission in writing from the publisher.

Printed in the United States of America

10   9   8   7   6   5   4   3   2


ISBN: 0-13-159626-8


Prentice-Hall International (UK) Limited, *London*
Prentice-Hall of Australia Pty. Limited, *Sydney*
Prentice-Hall Canada Inc., *Toronto*
Prentice-Hall Hispanoamericana, S.A., *Mexico*
Prentice-Hall of India Private Limited, *New Delhi*
Prentice-Hall of Japan, Inc., *Tokyo*
Simon & Schuster Asia Pte. Ltd., *Singapore*
Editora Prentice-Hall do Brasil, Ltda., *Rio de Janeiro*

# Contents

# Chapter 1: Fundamentals of Algebra

## Exercises 1.1

1. $-15$ is a negative *integer* which may be written as the fraction $\frac{-15}{1}$. Therefore, $-15$ is also a *rational number* in the set of *real numbers*.
   Answer: (c), (d), (f).

3. $\sqrt{51} = 7.1414284 \ldots$, a non-repeating, non-terminating decimal. Therefore, $\sqrt{51}$ is an *irrational number* in the set of *real numbers*.
   Answer: (e), (f).

5. $\frac{16}{2} = \frac{8}{1} = 8$, a *natural number*, *whole number*, *integer*, and *rational number* in the set of *real numbers*.
   Answer: (a), (b), (c), (d), (f).

7. $0$ is *not* a counting, or natural, number, but it is a *whole number*, *integer*, and *rational number* $\left(\frac{0}{a},\ a \neq 0\right)$ in the set of *real numbers*.
   Answer: (b), (c), (d), (f).

9. $\sqrt{12} = 3.4641016\ldots$, a non-repeating, non-terminating decimal. Therefore, $\sqrt{12}$ is an *irrational number* in the set of *real numbers*.
   Answer: (e), (f).

11. $\{N : N < 5\} = \{1, 2, 3, 4\}$

13. $\{W : 2 < W < 7\} = \{3, 4, 5, 6\}$

15. $\{I : -3 < I < 0\} = \{-2, -1\}$

17. $\{I : I < 1\} = \{\ldots, -3, -2, -1, 0\}$

19. $\{W : W \notin I\} = \{\ \ \}$, because the set $W$ of *whole numbers* is a complete subset of the set $I$ of *integers*.

21. True.

23. False. Given $a, b \in I$. If $a < b$, then $\frac{a}{b} \notin I$.

25. False. Given $a, b \in I$. $a - b \neq b - a$; rather, $a - b = -(b - a)$.

27. True.

29. True.

31. $\sqrt[3]{5}$ and $7$ are both real numbers, so the statement "$\sqrt[3]{5} + 7$ is a real number" illustrates the *Closure Property of Addition*.

33. Since $(-5) + 5 = 0$, $(-5)$ and $5$ are the *additive inverses* of one another. The property illustrated is the *Inverse Property for Addition*.

35. $(5 \times 7) \times 8 = (7 \times 5) \times 8$ illustrates the *Commutative Property for Multiplication*, since $5 \times 7 = 7 \times 5$.

37. $\frac{1}{4} + \frac{1}{2} = \frac{1}{2} + \frac{1}{4}$ illustrates the *Commutative Property for Addition*.

39. $-13 + 0 = -13$ illustrates the *Identity Property for Addition*.

41. $\frac{1}{2} + \left(-\frac{1}{2}\right) = 0$ illustrates the *Inverse Property for Addition*.

43. $0\left(\sqrt{2} + \sqrt{3}\right) = 0$ illustrates the *Multiplication Property of Zero*.

45. $(3 + 9)(7) = (3)(7) + (9)(7)$ illustrates the *Distributive Property*.

47. $7 + n = 3 + 7$ is a true statement if $n = 3$.

49. $(3 + 7) + n = 3 + (7 + 5)$ is a true statement if $n = 5$.

**51.** $5(8 + n) = (5 \times 8) + (5 \times 7)$ is a true statement if $n = 7$.

**53.** If $5(x - 2) = 0$, then $x - 2 = 0$ because according to the *Zero Product Property*, if $ab = 0$, either $a = 0$ or $b = 0$. Since $5 \neq 0$, then $x - 2 = 0$, which implies that $x = 2$.

**55.** No. It is true that $2^4 = (2^2)^2 = 4^2$, but it is not generally true that $a^N = N^a$. For example, $2^3 = 8 \neq 9 = 3^2$.

**57.** **(a)** Let $n = 0.454545...$
Then $100n = 45.454545...$
$\phantom{Then 100n} - n = \phantom{4}0.454545...$
$$99n = 45 \Rightarrow n = \frac{45}{99} = \frac{5}{11}$$

**(b)** Let $n = 0.373737...$
Then $100n = 37.373737...$
$\phantom{Then 100n} - n = \phantom{3}0.373737...$
$$99n = 37 \Rightarrow n = \frac{37}{99}$$

**(c)** Let $n = 0.234234234...$
Then $1000n = 234.234234234...$
$\phantom{Then 1000n} - n = \phantom{23}0.234234234...$
$$999n = 234 \Rightarrow n = \frac{234}{999} = \frac{26}{111}$$

**59.** If $\dfrac{0}{0}$ were a value $x$ which could be determined, then $\dfrac{0}{0} = x$ such that $x$ would be a unique real number. Therefore, $0 \cdot x = 0$ would be a unique real number. But any real number $x$ would satisfy this new equation. Therefore, $x$ cannot be uniquely determined.

**61.**

| $\dfrac{2}{3}$ | $\dfrac{8}{5}$ | $\dfrac{18}{13}$ | $\dfrac{44}{31}$ | $\dfrac{106}{75}$ | $\dfrac{256}{181}$ | |
|---|---|---|---|---|---|---|
| 0.667 | 1.6 | 1.385 | 1.419 | 1.413 | 1.414 | |

| $\dfrac{1}{2}$ | $\dfrac{5}{3}$ | $\dfrac{11}{8}$ | $\dfrac{27}{19}$ | $\dfrac{65}{46}$ | $\dfrac{157}{111}$ | |
|---|---|---|---|---|---|---|
| 0.5 | 1.667 | 1.375 | 1.421 | 1.413 | 1.414 | |

| $\dfrac{1}{4}$ | $\dfrac{9}{5}$ | $\dfrac{19}{14}$ | $\dfrac{47}{33}$ | $\dfrac{113}{80}$ | $\dfrac{273}{193}$ | $\dfrac{659}{466}$ |
|---|---|---|---|---|---|---|
| 0.25 | 1.8 | 1.357 | 1.424 | 1.413 | 1.415 | 1.414 |

The sequence of fractions converges to $\sqrt{2} \approx 1.414$.

**63.** At the point on the number line with coordinate "2", construct a segment of length 1 unit perpendicular to the number line. The endpoint of this segment should be connected to the point on the number line which coincides with the value "0". The length of this connected segment is $\sqrt{5}$; if this distance is used as the radius of the circle with center "0", the value $\sqrt{5}$ may be transferred to the number line.

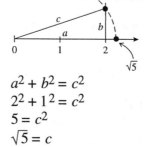

$a^2 + b^2 = c^2$
$2^2 + 1^2 = c^2$
$5 = c^2$
$\sqrt{5} = c$

## Exercises 1.2

**1.** $3x - 2 = 10$
$3x = 12$
$x = 4$

**3.** $-2x + 1 = 9$
$-2x = 9 - 1 = 8$
$-2x \div (-2) = 8 \div (-2)$
$x = -4$

**5.** $-3x - 5 = 7$
$-3x = 7 + 5 = 12$
$-3x \div (-3) = 12 \div (-3)$
$x = -4$

**7.** $2x - 1 = -17$
$2x = -17 + 1 = -16$
$\frac{1}{2}(2x) = \frac{1}{2}(-16)$
$x = -8$

**9.** $2(x + 1) = 11$
$2x + 2 = 11$
$2x = 11 - 2 = 9$
$\frac{1}{2}(2x) = \frac{1}{2}(9)$
$x = \frac{9}{2}$

**11.** $3x + 7 = 2x - 2$
$\underline{-2x \qquad -2x}$
$x + 7 = \qquad -2$
$x \quad = \quad -9$

**13.** $\frac{1}{2}x + 7 = 2x - 3$
$\underline{-\frac{1}{2}x \qquad -\frac{1}{2}x}$
$7 = \frac{3}{2}x - 3$
$10 = \frac{3}{2}x$
$\frac{2}{3}(10) = \frac{2}{3}\left(\frac{3}{2}x\right)$
$\frac{20}{3} = x$

**15.** $\frac{4}{3}x - 7 = \frac{1}{3}x + 8$
$\underline{-\frac{1}{3}x \qquad -\frac{1}{3}x}$
$\frac{3}{3}x - 7 = 8$
$\frac{3}{3}x \quad = 15$
$x \quad = 15$

**17.** $\frac{3}{5}(x - 5) = x + 1$
$\frac{3}{5}x - 3 = x + 1$
$\underline{-\frac{3}{5}x \qquad -\frac{3}{5}x}$
$-3 = \frac{2}{5}x + 1$
$-4 = \frac{2}{5}x$
$\frac{5}{2}(-4) = \frac{5}{2}\left(\frac{2}{5}x\right)$
$-10 = x$

**19.** $\frac{7}{2}x + 5 + \frac{1}{2}x = \frac{5}{2}x - 6$
$\frac{8}{2}x + 5 = \frac{5}{2}x - 6$
$\frac{3}{2}x = -11$
$x = -\frac{22}{3}$

**21.** $-3(x + 2) + 1 = x - 25$
$-3x - 6 + 1 = x - 25$
$-3x - 5 = x - 25$
$20 = 4x$
$5 = x$

**23.** $1 - 12x = 7(1 - 2x)$
$1 - 12x = 7 - 14x$
$2x = 6$
$x = 3$

**25.** $x + 2\left(\frac{1}{6}x + 2\right) = \frac{6}{5}x + 16$
$x + \frac{1}{3}x + 4 = \frac{6}{5}x + 16$
$\frac{4}{3}x + 4 = \frac{6}{5}x + 16$
$\frac{20}{15}x + 4 = \frac{18}{15}x + 16$
$\frac{2}{15}x = 12$

*Chapter 1: Fundamentals of Algebra*

$$x = (12)\left(\frac{15}{2}\right)$$
$$x = 90$$

27. $A = \frac{1}{2}h(b_1 + b_2)$
$$2A = h(b_1 + b_2)$$
$$\frac{2A}{b_1 + b_2} = h$$

29. $V = \pi r^2 h$
$$\frac{V}{\pi r^2} = h$$

31. $I = Prt$
$$P = \frac{I}{rt}$$
If $I = \$220$, $r = 5\frac{1}{2}\%$, and $t = 2$, then
$$P = \frac{220}{(0.055)(2)} = \frac{110}{0.055} = \$2,000.$$

33. $V = \frac{1}{3}\pi h^2(3R - h)$
$$\frac{3V}{\pi h^2} = 3R - h$$
$$\frac{3V}{\pi h^2} + h = 3R$$
$$\frac{V}{\pi h^2} + \frac{h}{3} = R$$

35.

Let $w$ = rectangle's width in inches
$w + 4$ = rectangle's length in inches
$$2w + 2(w + 4) = 56$$
$$2w + 2w + 8 = 56$$
$$4w = 48$$
$$w = 12$$
width = 12 inches;
length = $12 + 4 = 16$ inches

37. Let $x$ = number of 10-cent stamps
purchased = 10
$\frac{1}{2}x$ = number of 25-cent stamps purchased
= 5
$x + 3$ = number of 30-cent stamps
purchased = 13
$$10x + 25\left(\frac{1}{2}x\right) + 30(x + 3) = 615$$
$$10x + \frac{25}{2}x + 30x + 90 = 615$$
$$\frac{105}{2}x = 525$$
$$105x = 1050$$
$$x = 10$$

39. Let $t$ = time in hours each car travels.
$$45t + 55t = 350$$
$$100t = 350$$
$$t = 3.5, \text{ or } 3\frac{1}{2} \text{ hours}$$

41. Let $x$ = first of three consecutive odd
integers, $x + 2$ = second of three
consecutive odd integers, $x + 4$ = third of
three consecutive odd integers.
$$x + (x + 2) + (x + 4) = 180$$
$$3x + 6 = 180$$
$$3x = 174$$
$$x = 58$$
But 58, which is the only possible solution
to this equation, is not odd but even.
Therefore, the consecutive integers
totaling 180 must be even.

43. Let $w$ = rectangle's width in inches,
$3w - 1$ = rectangle's length in inches.
$$(3w - 1) + 6 = 2(w + 5)$$
$$3w + 5 = 2w + 10$$
$$w = 5 \text{ inches}$$
The length is $3(5) - 1 = 15 - 1 = 14$
inches.

45. Let $x$ = the first of three consecutive odd
integers, $x + 2$ = the next of three

consecutive odd integers, $x + 4 =$ the last
of three consecutive odd integers
$3x + 6 = 237$
$3x = 231$
$x = 77 \Rightarrow$ The 3 consecutive odd integers
are 77, 79, 81.

47. Let $c =$ time traveled on the county road,
$5c =$ time traveled on the highway.
$35c + 48(5c) = 27.5$
$35c + 240c = 27.5$
$275c = 27.5$
$c = 0.1 = \dfrac{1}{10}$ hour = 6 minutes
Therefore, highway time = 30 minutes.
Total time of the trip = 36 minutes.

49. Let $P =$ amount invested at 9% annual
interest rate, $P + 2700 =$ amount invested
at 12% annual interest rate.
$1794 = 0.09P + 0.12(P + 2700)$
$1794 = 0.09P + 0.12P + 324$
$1470 = 0.21P$
$7000 = P$
If \$7,000 is invested at 9%, then \$9,700 is
invested at 12%.

51. Let $B =$ base cost of the car.
$B + 0.06B = 9010$
$1.06B = 9010$
$B = 8500$
The base cost of the car was \$8,500.

## Challenge

The first time the boats meet, the first has
traveled 700 feet and the second has traveled
$d - 700$ feet, where $d$ is the width of the river.
When they meet the second time, the first boat
has traveled $(d - 700) + 400 = d - 300$ more feet,
and the second boat has traveled
$700 + (d - 400) = d + 300$ more feet. Since the
ratios of distance traveled are equal,
$$\dfrac{d - 700}{700} = \dfrac{d + 300}{d - 300}$$
$(d - 700)(d - 300) = 700(d + 300)$
$d^2 - 1000d + 210{,}000 = 700d + 210{,}000$
$d^2 = 1700d$
$d = 1700$
The width of the river is 1700 feet.

1. False. $0 < 2$ and 0 is not negative.

3. True.

5. False. $0 < 1$, but $1 = 1^2$.

7. True.

9. $[-5, 2]$

11. $[-6, 0)$

13. $(-10, 10)$

15. $(-\infty, 5)$

17. $[-2, +\infty)$

19. $(-\infty, -1]$

21. $(-3, -1)$

23. $[-3, -1)$

25. $[0, 5]$

27. $(-\infty, 0]$

29. $-1 \le x < 3$; $[-1, 3)$

31. $-1 < x < 3$; $(-1, 3)$

33. $x < 1$; $(-\infty, 1)$

35. $x + 5 > 17$
$x > 12$ 　　　　　　　$\{x \mid x > 12\}$

37. $x - 7 \ge -3$
$x \ge 4$ 　　　　　　　$\{x \mid x \ge 4\}$

39. $5x - 4 < 6 + 4x$
$x - 4 < 6$
$x < 10$ 　　　　　　　$\{x \mid x < 10\}$

**41.** $3x > -21$
$x > -7$ $\quad\quad$ $\{x \mid x > -7\}$

**43.** $-5x < 50$
$x > -10$ $\quad\quad$ $\{x \mid x > -10\}$

**45.** $5x - 3 \le 22$
$5x \le 25$
$x \le 5$ $\quad\quad$ $\{x \mid x \le 5\}$

**47.** $2x + 7 \le 5 - 6x$
$8x + 7 \le 5$
$8x \le -2$
$x \le -\dfrac{1}{4}$
$\left\{ x \mid x \le -\dfrac{1}{4} \right\}$

**49.** $-5x + 5 < -3x + 1$
$5 < 2x + 1$
$4 < 2x$
$2 < x$
$\{x \mid x > 2\}$

**51.** $2(x + 1) < x - 1$
$2x + 2 < x - 1$
$x + 2 < -1$
$x < -3$
$\{x \mid x < -3\}$

**53.** $\dfrac{1}{2}x - 5 > \dfrac{1}{4}x + 3$
$\dfrac{1}{4}x > 8$
$x > 32$
$\{x \mid x > 32\}$

**55.** $-\dfrac{3}{5}x - 6 < -\dfrac{2}{5}x + 7$
$-6 < \dfrac{1}{5}x + 7$
$-13 < \dfrac{1}{5}x$
$-65 < x$
$\{x \mid x > -65\}$

**57.** $\dfrac{1}{x} < 0$
$x < 0$
$\{x \mid x < 0\}$

**59.** $3x + 5 \ne 8$
$3x \ne 3$
$x \ne 1$
$\{x \mid x \ne 1\}$

**61.** $3x - 2 \not< 1$
$3x \not< 3$
$x \not< 1 \Rightarrow x \ge 1$
$\{x \mid x \ge 1\}$

**63.** $3(x - 1) \not> 5(x + 2)$
$3x - 3 \not> 5x + 10$
$-3 \not> 2x + 10$
$-13 \not> 2x$
$-\dfrac{13}{2} \not> x \Rightarrow -\dfrac{13}{2} \le x$
$\left\{ x \mid x \ge -\dfrac{13}{2} \right\}$

**65.** $\dfrac{5}{3}x \not< 2x - 1$
$0 \not< \dfrac{1}{3}x - 1$
$1 \not< \dfrac{1}{3}x$
$3 \not< x$
$\{x \mid x \le 3\}$

**67.** $2x + 3 < 11$ $\quad\quad$ $\{x \mid x < 4\}$
$2x < 8$
$x < 4$

**69.** $\dfrac{1}{2}x + 2 \le 1$ $\quad\quad$ $\{x \mid x \le -2\}$
$\dfrac{1}{2}x \le -1$
$x \le -2$

**71.** $2(x + 1) < 3(x + 2)$ $\quad\quad$ $\{x \mid x > -4\}$
$2x + 2 < 3x + 6$

$$2 < x + 6$$
$$-4 < x$$

**73.** Let $x =$ an integer.
$3x - 5 = 5$ less than 3 times the integer
$$34 < x + 3x - 5 < 54$$
$$34 < 4x - 5 < 54$$
$$39 < 4x < 59$$
$$\frac{39}{4} < x < \frac{59}{4}$$
Solution Pairs:

| $x$ | $3x - 5$ |
|------|----------|
| 10 | 25 |
| 11 | 28 |
| 12 | 31 |
| 13 | 34 |
| 14 | 37 |

**75.** Let $x =$ fourth test grade.
$$86 \leq \frac{1}{5}(85 + 86 + 93 + 2x) < 90$$
$$430 \leq 264 + 2x < 450$$
$$166 \leq 2x < 186$$
$$83 \leq x < 93$$
The score must be at least 83 but less than 93.

**77.** $25 < C < 30$
$$\frac{9}{5}(25) < \frac{9}{5}C < \left(\frac{9}{5}\right)30$$
$$\frac{9}{5}(25) + 32 < \frac{9}{5}C + 32 < \left(\frac{9}{5}\right)30 + 32$$
$$45 + 32 < F < 54 + 32$$
$$77° < F < 86°$$

**79.** $l + g \leq 110$ and $l \geq 2; \ w \geq 2; \ h \geq 2.$
   **(a)** If $l = 42$, $42 + g \leq 110$ and
      $g \leq 68 \Rightarrow 2w + 2h \leq 68.$
      Since $w$ and $h$ must each be at least 2 inches, $g$ must be at least 8 inches. Therefore, $8 \leq g \leq 68.$
   **(b)** If $l = 42$, $2w + 2h \leq 68.$
      If $w = 18$, $36 + 2h \leq 68.$
$$2h \leq 32$$
$$h \leq 16$$
      Allowable values for $h$ (in inches) are $2 \leq h \leq 16.$

**81.** Let $x =$ earnings of 1 part-timer,
$x =$ earnings of next part-timer,
$x - 12 =$ earnings of last part-timer.
$$210 \leq 3x - 12 \leq 252$$
$$222 \leq 3x \leq 264$$
$$74 \leq x \leq 88 \Rightarrow 62 \leq x - 12 \leq 76$$
The first 2 part-timers earn from \$74 to \$88 weekly, while the third part-timer earns from \$62 to \$76 weekly.

**Exercises 1.4**

**1.** $\left|-\frac{1}{3}\right| \overset{?}{=} \frac{1}{3}$
$$\left|-\frac{1}{3}\right| = \frac{1}{3} \Rightarrow -\left|-\frac{1}{3}\right| = -\frac{1}{3}$$
$$-\frac{1}{3} \neq \frac{1}{3}$$
False.

**3.** $\left|-\frac{1}{2}\right| \overset{?}{=} 2$
$$\left|-\frac{1}{2}\right| = \frac{1}{2}$$
$$\frac{1}{2} \neq 2$$
False.

**5.** $\left|\dfrac{x}{y}\right| \overset{?}{=} |x| \cdot \dfrac{1}{|y|}$
$$\left|\frac{x}{y}\right| = \frac{|x|}{|y|} = |x| \cdot \frac{1}{|y|}$$
True.

**7.** $\|x\| \overset{?}{=} |x|$
If $x \geq 0$, $|x| = x$ and $\|x\| = |x| = x.$
If $x < 0$, $|x| = -x$ and $\|x\| = |-x| = |x| = -x.$
True.

**9.** $|a| - |b| \overset{?}{=} a - b$
If $a \geq 0$ and $b < 0$, then $|a| - |b|$
$= a - (-b) = a + b \neq a - b$
False.

**11. (a)** If $x_1 = 3$ and $x_2 = 9$, the coordinate of the midpoint of

$$\overline{x_1 x_2} = \frac{3+9}{2} = \frac{12}{2} = 6.$$

**(b)** If $x_1 = -8$ and $x_2 = -2$, the coordinate of the midpoint of

$$\overline{x_1 x_2} = \frac{-8-2}{2} = \frac{-10}{2} = -5.$$

**(c)** If $x_1 = -12$ and $x_2 = 0$, the coordinate of the midpoint of

$$\overline{x_1 x_2} = \frac{-12+0}{2} = \frac{-12}{2} = -6.$$

**(d)** If $x_1 = -5$ and $x_2 = 8$, the coordinate of the midpoint of $\overline{x_1 x_2} = \frac{-5+8}{2} = \frac{3}{2}$.

**13.** $|x| = \dfrac{1}{2} \Rightarrow x = \dfrac{1}{2}$ or $x = -\dfrac{1}{2}$

$$\left\{ x \,|\, x = \pm\frac{1}{2} \right\}$$

**15.** $|x - 1| = 3 \Rightarrow x - 1 = 3$ or $x - 1 = -3$
$x = 4$ or $x = -2$
$\{x \,|\, x = -2, 4\}$

**17.** $|2x - 3| = 7 \Rightarrow 2x - 3 = 7$ or $2x - 3 = -7$
$2x = 10$ or $2x = -4$
$x = 5$ or $x = -2$
$\{x \,|\, x = -2, 5\}$

**19.** $|4 - x| = 3 \Rightarrow 4 - x = 3$ or $4 - x = -3$
$-x = -1$ or $-x = -7$
$x = 1$ or $x = 7$
$\{x \,|\, x = 1, 7\}$

**21.** $|3x + 4| = 16 \Rightarrow 3x + 4 = 16$
or $3x + 4 = -16$
$3x = 12$ or $3x = -20$
$x = 4$ or $x = \dfrac{-20}{3}$

$$\left\{ x \,|\, x = \frac{-20}{3},\, 4 \right\}$$

**23.** $\dfrac{|x|}{x} = 1$

If $x > 0$, $\dfrac{|x|}{x} = \dfrac{x}{x} = 1$
for all real values $x$.

If $x = 0$, $\dfrac{|x|}{x} = \dfrac{0}{0}$
which is <u>undefined</u>.

If $x < 0$, $\dfrac{|x|}{x} = \dfrac{-x}{x} = -1$
for all real values $x$.

Therefore $\dfrac{|x|}{x} = 1$ for all real values $x > 0$.
$\{x \,|\, x > 0\}$

**25.** $|x + 1| = 3 \Rightarrow x + 1 = 3$ or $x + 1 = -3$
$x = 2$ or $x = -4$

**27.** $|x - 1| \geq 3 \Rightarrow x - 1 \geq 3$ or $x - 1 \leq -3$
$x \geq 4$ or $x \leq -2$

**29.** $|x + 2| \leq 3 \Rightarrow x + 2 \leq 3$ and $x + 2 \geq -3$
$x \leq 1$ and $x \geq -5$

**31.** $|-x| = 5 \Rightarrow -x = 5$ or $-x = -5$
$x = -5$ or $x = 5$

**33.** $|x| \geq 5 \Rightarrow x \geq 5$ or $x \leq -5$

**35.** $|x - 5| \leq 3 \Rightarrow x - 5 \leq 3$ and $x - 5 \geq -3$
$x \leq 8$ and $x \geq 2$

**37.** $|x - 3| < 0.1 \Rightarrow x - 3 < 0.1$ and $x - 3 > -0.1$
$x < 3.1$ and $x > 2.9$

**39.** $|2x - 1| < 7 \Rightarrow 2x - 1 < 7$ and $2x - 1 > -7$
$2x < 8$ and $2x > -6$

$x < 4$ and $x > -3$

$$-5\,-4\,-3\,-2\,-1\;0\;1\;2\;3\;4\;5$$

**41.** $|4 - x| < 2 \Rightarrow -2 \le 4 - x < 2$
$$-6 < -x < -2$$
$$6 > x > 2$$

$$-3\,-2\,-1\;0\;1\;2\;3\;4\;5\;6\;7$$

**43.** $|x - 4| \not> 1 \Rightarrow |x - 4| \le 1 \Rightarrow -1 \le x - 4 \le 1$
$$3 \le x \le 5$$

$$-5\,-4\,-3\,-2\,-1\;0\;1\;2\;3\;4\;5$$

**45.** $\dfrac{1}{|x - 3|} > 0$

$\dfrac{1}{|x - 3|} > 0$ for all values of $x$ such that

$|x - 3| > 0$.

$|x - 3| > 0 \Rightarrow x - 3 > 0$ or $x - 3 < 0$

$x > 3$ or $x < 3$

$$-5\,-4\,-3\,-2\,-1\;0\;1\;2\;3\;4\;5$$

**47.** **(a)** $|x - y| \ge \big||x| - |y|\big|$
Case(i): $x > 0, y > 0$
$$|8 - 6| \ge \big||8| - |6|\big|?$$
$$|2| \ge |8 - 6|?$$
$$2 \ge 2? \qquad \text{True}$$
Case (ii): $x > 0, y < 0$
$$|8 - (-6)| \ge \big||8| - |-6|\big|?$$
$$|14| \ge |8 - 6|?$$
$$14 \ge 2? \qquad \text{True}$$
Case(iii): $x < 0, y > 0$
$$|-8 - 6| \ge \big||-8| - |6|\big|?$$
$$|-14| \ge |8 - 6|?$$
$$14 \ge 2? \qquad \text{True}$$
Case (iv): $x < 0, y < 0$
$$|-8 - (-6)| \ge \big||-8| - |-6|\big|?$$
$$|-2| \ge |8 - 6|?$$
$$2 \ge 2? \qquad \text{True}$$
**(b)** $|x + y| \le |x| + |y|$
Case(i): $x > 0, y > 0$
$$|3 + 5| \le |3| + |5|$$
$$|8| \le 3 + 5?$$
$$8 \le 8? \qquad \text{True}$$

Case (ii): $x > 0, y < 0$
$$|3 + (-5)| \le |3| + |-5|?$$
$$|-2| \le 3 + 5?$$
$$2 \le 8? \qquad \text{True}$$
Case(iii): $x < 0, y > 0$
$$|-3 + 5| \le |-3| + |5|?$$
$$|2| \le 3 + 5?$$
$$2 \le 8? \qquad \text{True}$$
Case (iv): $x < 0, y < 0$
$$|-3 + (-5)| \le |-3| + |-5|?$$
$$|-8| \le 3 + 5?$$
$$8 \le 8? \qquad \text{True}$$

## Critical Thinking

**1.** The set of real numbers is not associative with respect to division. For example,
$(1 \div 2) \div 2 = 0.5 \div 2 = 0.25$ but
$1 \div (2 \div 2) = 1 \div 1 = 1$
So, $(1 \div 2) \div 2 \ne 1 \div (2 \div 2)$

**3.** In order to show that a statement claimed to be true is in fact false, one example is all that is necessary. This one example, which demonstrates the statement is false, is called a counterexample.

**5.** $|x - (-2)| < 3$ or $|x + 2| < 3$

**7.** If $a < b$ and $c < d$, it is not true that $a - c < b - d$. For example, $1 < 2$ and $2 < 3$, but $1 - 2 \not< 2 - 3$ since $-1 \not< -1$.

## Exercises 1.5

**1.** $3^4 \cdot 3^2 = 3^8?$          False
$3^4 \cdot 3^2 = 3^{4+2} = 3^6$

**3.** $2^5 \cdot 2^2 = 4^7?$          False
$2^5 \cdot 2^2 = 2^{5+2} = 2^7$

**5.** $\dfrac{10^4}{5^4} = 2^4?$          True

$\dfrac{10^4}{5^4} = \dfrac{(2 \cdot 5)^4}{5^4} = \dfrac{2^4 \cdot 5^4}{5^4} = 2^4$

*Chapter 1: Fundamentals of Algebra*   

**7.** $(-27)^0 = 1$?  True

**9.** $3^4 + 3^4 = 3^8$?  False
$3^4 + 3^4 = 2(3^4)$

**11.** $(a+b)^0 = a + 1$?  False
$(a+b)^0 = 1$

**13.** $\dfrac{1}{2^{-3}} = -2^3$?  False
$\dfrac{1}{2^{-3}} = 2^3$

**15.** $\dfrac{2^{-5}}{2^3} = 2^{-2}$?  False
$\dfrac{2^{-5}}{2^3} = 2^{-5-3} = 2^{-8}$

**17.** $2^0 + 2^1 + 2^2 = 1 + 2 + 4 = 7$

**19.** $\left(\dfrac{2}{3}\right)^0 + \left(\dfrac{2}{3}\right)^1 = 1 + \dfrac{2}{3} = \dfrac{5}{3}$

**21.** $\left(\dfrac{1}{2}\right)^4 (-2)^4 = \left(\dfrac{1}{16}\right)\left(\dfrac{16}{1}\right) = 1$

**23.** $\dfrac{(-2)^5}{(-2)^3} = (-2)^{5-3} = (-2)^2 = 4$

**25.** $\dfrac{2^3 \cdot 3^4 \cdot 4^5}{2^2 \cdot 3^3 \cdot 4^4} = 2^{3-2} \cdot 3^{4-3} \cdot 4^{5-4}$
$= 2 \cdot 3 \cdot 4 = 24$

**27.** $\dfrac{2^{10}}{4^3} = \dfrac{2^{10}}{(2^2)^3} = \dfrac{2^{10}}{2^6} = 2^4 = 16$

**29.** $\left(\dfrac{2}{3}\right)^{-2} + \left(\dfrac{2}{3}\right)^{-1} = \left(\dfrac{3}{2}\right)^2 + \dfrac{3}{2}$
$= \dfrac{9}{4} + \dfrac{3}{2} = \dfrac{9}{4} + \dfrac{6}{4} = \dfrac{15}{4}$

**31.** $200{,}000{,}000 = 2 \times 10^8$

**33.** $0.00037 = 3.7 \times 10^{-4}$

**35.** $0.0000000555 = 5.55 \times 10^{-8}$

**37.** $7.89 \times 10^{-4} = 0.000789$

**39.** $2.25 \times 10^5 = 225{,}000$

**41.** $(x^{-3})^2 = x^{-6} = \dfrac{1}{x^6}$

**43.** $x^3 \cdot x^9 = x^{12}$

**45.** $(2a)^3 (3a)^2 = 8a^3 \cdot 9a^2 = 72a^5$

**47.** $(-2a^2 b^0)^4 = (-2)^4 a^8 \cdot 1^4 = 16a^8$

**49.** $\dfrac{(x^2 y)^4}{(xy)^2} = \dfrac{x^8 y^4}{x^2 y^2} = x^6 y^2$

**51.** $\left(\dfrac{x^3}{y^2}\right)^4 \left(\dfrac{-y}{x^2}\right)^2 = \dfrac{x^{12}}{y^8} \cdot \dfrac{y^2}{x^4}$
$= \dfrac{x^{12} y^2}{x^4 y^8} = \dfrac{x^8}{y^6}$

**53.** $\dfrac{x^{-2} y^3}{x^3 y^{-4}} = x^{-5} y^7 = \dfrac{y^7}{x^5}$

**55.** $\dfrac{5x^0 y^{-2}}{x^{-1} y^{-2}} = \dfrac{5 \cdot 1}{x^{-1}} \cdot \dfrac{1}{1} = \dfrac{5}{x^{-1}} = 5x$

**57.** $\dfrac{(-3a)^{-2}}{a^{-2} b^{-2}} = \dfrac{(-3)^{-2} a^{-2}}{a^{-2} b^{-2}} = \dfrac{\left(-\frac{1}{3}\right)^2}{b^{-2}}$
$= \dfrac{\frac{1}{9}}{b^{-2}} = \dfrac{1}{9} b^2 = \dfrac{b^2}{9}$

**59.** $\dfrac{(a+b)^{-2}}{(a+b)^{-8}} = (a+b)^{-2-(-8)} = (a+b)^6$

**61.** $\dfrac{-12x^{-9}y^{10}}{4x^{-12}y^7} = -3x^{-9-(-12)}y^{10-7} = -3x^3y^3$

**63.** $\dfrac{(a+3b)^{-12}}{(a+3b)^{10}} = (a+3b)^{-12-10}$

$= (a+3b)^{-22} = \dfrac{1}{(a+3b)^{22}}$

**65.** $x^{-2} + y^{-2} = \dfrac{1}{x^2} + \dfrac{1}{y^2}$

**67.** $\left(\dfrac{x^{-2}}{y^3}\right)^{-1} = \dfrac{x^2}{y^{-3}} = x^2 \cdot \dfrac{1}{y^{-3}} = x^2y^3$

**69.** $-2(4-5x)^{-3}(-5) = (-2)(-5)(4-5x)^{-3}$

$= \dfrac{10}{(4-5x)^3}$

**71.** $2^x \cdot 2^3 = 2^{12} \Rightarrow \quad x + 3 = 12$

$\quad\quad\quad\quad\quad\quad\quad\quad\quad x = 9$

**73.** $2^x \cdot 2^x = 2^{16} \Rightarrow \quad x + x = 16$

$\quad\quad\quad\quad\quad\quad\quad\quad\quad 2x = 16$

$\quad\quad\quad\quad\quad\quad\quad\quad\quad\quad x = 8$

**75.** $\dfrac{2^x}{2^2} = 2^{-5} \Rightarrow x - 2 = -5$

$\quad\quad\quad\quad\quad\quad\quad\quad x = -3$

**77.**

| Hour | Amount Remaining |
|------|------------------|
| 0 | $\left(\dfrac{1}{2}\right)^0$ 640 grams |
| 1 | $\left(\dfrac{1}{2}\right)^1$ 640 grams |
| 2 | $\left(\dfrac{1}{2}\right)^2$ 640 grams |
| $n$ | $\left(\dfrac{1}{2}\right)^n$ 640 grams |

After 7 hours, $\left(\dfrac{1}{2}\right)^7 (640)$ grams remain.

$\left(\dfrac{1}{2}\right)^7 (640) = \left(\dfrac{1}{128}\right)(640) = 5$ grams

After $n$ hours, $\left(\dfrac{1}{2}\right)^n (640)$ grams remain

$= \dfrac{640}{2^n}$

**79.**

| Cut # | Amount Remaining |
|-------|------------------|
| 0 | $243\left(\dfrac{2}{3}\right)^0$ feet   (If $\dfrac{1}{3}$ is cut off each time, $\dfrac{2}{3}$ remains) |
| 1 | $243\left(\dfrac{2}{3}\right)^1$ feet |
| 2 | $243\left(\dfrac{2}{3}\right)^2$ feet |
| $n$ | $243\left(\dfrac{2}{3}\right)^n$ feet |

After 5 cuts, $(243)\left(\dfrac{2}{3}\right)^5$ feet remain.

$(243)\left(\dfrac{2}{3}\right)^5 = 243\left(\dfrac{32}{243}\right) = 32$ feet remain.

After $n$ cuts, $(243)\left(\dfrac{2}{3}\right)^n$ feet remain.

**81.** $A = P(1 + i)^n$.
If $P = \$1000$, $i = 10\%$, $n = 3$:
$A = 1000(1 + 0.10)^3$
$= 1000(1.1)^3$
$= 1000(1.331)$
$= \$1,331$

**83.** Is "raising to a power" commutative?
If so, 3 *raised to the power of* 4 = 4 *raised to the power of* 3.
In other words, $3^4 = 4^3$ but $81 \neq 64$. So "raising to a power" is *not* a commutative operation.

Is raising to a power" associative?
If so, (2 *raised to the power of* 3)
*raised to the power of* 4
= 2 *raised to the power of*
(3 *raised to the power of* 4).

In other words, $(2^3)^4 = 2^{(3^4)}$
but $2^{12} \neq 2^{81}$.
So "raising to a power " is *not* an
associative operation.

**85.** $D = rt \Rightarrow t = \dfrac{D}{r}$

$$t = \frac{93{,}000{,}000 \text{ miles}}{186{,}000 \text{ miles per second}} = \frac{9.3 \times 10^7}{1.86 \times 10^5}$$

$$= 5 \times 10^2 \text{ seconds} = 500 \text{ seconds}$$

## Challenge

You will snap your fingers once in 0 minutes,
twice in 1 minute, three times in 3 minutes, etc.
So, you will snap your fingers $n$ times in $2^{n-1} - 1$
minutes. There are $(365)(24)(60) = 525{,}600$
minutes in one year. Since $2^{19} = 524{,}288$, you
will only be able to snap your fingers 20 times in
one year.

## Exercises 1.6

**1.** $81^{-1/2} = \dfrac{1}{81^{1/2}} = \dfrac{1}{9}$

**3.** $(64)^{-2/3} = (64^{1/3})^{-2} = (4)^{-2} = \dfrac{1}{4^2} = \dfrac{1}{16}$

**5.** $(-125)^{2/3} = (-125^{1/3})^2 = (-5)^2 = 25$

**7.** $\sqrt[3]{9} \cdot \sqrt[3]{-3} = \sqrt[3]{-27} = -3$

**9.** $\dfrac{\sqrt[3]{-3}}{\sqrt[3]{-24}} = \sqrt[3]{\dfrac{-3}{-24}} = \sqrt[3]{\dfrac{1}{8}} = \dfrac{1}{2}$

**11.** $\dfrac{\sqrt{9}}{27^{-1/3}} = (9)^{1/2}(27)^{1/3} = 3 \cdot 3 = 9$

**13.** $\sqrt[3]{(-125)(-1000)} = (-125)^{1/3}(-1000)^{1/3}$
$= (-5)(-10) = 50$

**15.** $\sqrt[3]{\sqrt[3]{-512}} = \left[(-512)^{1/3}\right]^{1/3} = (-8)^{1/3} = -2$

**17.** $\sqrt{144 + 25} = (169)^{1/2} = 13$

**19.** $\left(\dfrac{1}{8} + \dfrac{1}{27}\right)^{1/3} = \left(\dfrac{27}{216} + \dfrac{8}{216}\right)^{1/3} = \left(\dfrac{35}{216}\right)^{1/3}$

$\dfrac{(35)^{1/3}}{6} = \dfrac{\sqrt[3]{35}}{6}$

**21.** $\left(\dfrac{16}{81}\right)^{3/4} + \left(\dfrac{256}{625}\right)^{1/4} = \left(\dfrac{16^{1/4}}{81^{1/4}}\right)^3$

$+\left(\dfrac{256^{1/4}}{625^{1/4}}\right) = \left(\dfrac{2}{3}\right)^3 + \dfrac{4}{5}$

$= \dfrac{8}{27} + \dfrac{4}{5} = \dfrac{40 + 108}{135} = \dfrac{148}{135}$

**23.** $\left(-\dfrac{125}{8}\right)^{1/3} - \left(\dfrac{1}{64}\right)^{1/3} = \dfrac{(-125)^{1/3}}{(8)^{1/3}} - \dfrac{1}{4}$

$= \dfrac{-5}{2} - \dfrac{1}{4} = \dfrac{-10}{4} - \dfrac{1}{4} = \dfrac{-11}{4}$

**25.** $\sqrt{2} + \sqrt{18} = \sqrt{2} + \sqrt{9 \cdot 2}$
$= \sqrt{2} + 3\sqrt{2} = 4\sqrt{2}$

**27.** $\sqrt{6} \cdot \sqrt{12} = \sqrt{6} \cdot \sqrt{6 \cdot 2} = \sqrt{6} \cdot \sqrt{6} \cdot \sqrt{2}$
$= 6\sqrt{2}$

**29.** $2\sqrt{5} + 3\sqrt{125} = 2\sqrt{5} + 3\sqrt{25 \cdot 5}$
$= 2\sqrt{5} + 15\sqrt{5} = 17\sqrt{5}$

**31.** $2\sqrt{200} - 5\sqrt{8} = 2\sqrt{100 \cdot 2} - 5\sqrt{4 \cdot 2}$
$= 20\sqrt{2} - 10\sqrt{2} = 10\sqrt{2}$

**33.** $\sqrt[3]{128} + \sqrt[3]{16} = \sqrt[3]{64 \cdot 2} = \sqrt[3]{8 \cdot 2}$
$= 4\sqrt[3]{2} + 2\sqrt[3]{2} = 6\sqrt[3]{2}$

**35.** $\dfrac{8}{\sqrt{2}} + 2\sqrt{50} = \dfrac{8}{\sqrt{2}} + 10\sqrt{2}$

$= \dfrac{8\sqrt{2}}{2} + 10\sqrt{2} = 14\sqrt{2}$

**37.** $3\sqrt{8x^2} - \sqrt{50x^2} = 3|x|\sqrt{8} - 5|x|\sqrt{2}$

$= 6|x|\sqrt{2} - 5|x|\sqrt{2} = |x|\sqrt{2}$

**39.** $3\sqrt{10} + 4\sqrt{90} - 5\sqrt{40}$

$= 3\sqrt{10} + 12\sqrt{10} - 10\sqrt{10} = 5\sqrt{10}$

**41.** $\dfrac{10}{\sqrt{5}} + 3\sqrt{45} - 2\sqrt{20} = \dfrac{10\sqrt{5}}{5} + 9\sqrt{5} - 4\sqrt{5}$

$= 2\sqrt{5} + 9\sqrt{5} - 4\sqrt{5} = 7\sqrt{5}$

**43.** $3\sqrt{9x^2} + 2\sqrt{16x^2} - \sqrt{25x^2}$

$= 9|x| + 8|x| - 5|x| = 12|x|$

**45.** $\sqrt{x^2 y} + \sqrt{8x^2 y} + \sqrt{200x^2 y}$

$|x|\sqrt{y} + 2|x|\sqrt{2y} + 10|x|\sqrt{2y}$

$= |x|\sqrt{y} + 12|x|\sqrt{2y}$

**47.** $\dfrac{8x}{\sqrt{2}} = \dfrac{8x\sqrt{2}}{2} = 4x\sqrt{2}$

**49.** $\dfrac{1}{\sqrt{18}} \cdot \dfrac{\sqrt{18}}{\sqrt{18}} = \dfrac{\sqrt{18}}{18} = \dfrac{3\sqrt{2}}{18} = \dfrac{\sqrt{2}}{6}$

**51.** $\dfrac{24}{\sqrt{3x^2}} \cdot \dfrac{\sqrt{3x^2}}{\sqrt{3x^2}} = \dfrac{24|x|\sqrt{3}}{3x^2} = \dfrac{8|x|\sqrt{3}}{|x| \cdot |x|} = \dfrac{8\sqrt{3}}{|x|}$

**53.** $\dfrac{8}{\sqrt[3]{2}} \cdot \dfrac{\sqrt[3]{2}}{\sqrt[3]{2}} \cdot \dfrac{\sqrt[3]{2}}{\sqrt[3]{2}} = \dfrac{8(2)^{2/3}}{2} = 4\left(2^2\right)^{1/3} = 4\sqrt[3]{4}$

**55.** $\left(a^{-1/2} b^{1/3}\right)\left(a^{1/2} b^{-1/3}\right)$

$= a^{-1/2+1/2} b^{1/3+(-1/3)} = a^0 b^0 = 1$

**57.** $\left(\dfrac{64a^6}{b^{-9}}\right)^{2/3} = \dfrac{64^{2/3} a^{6 \cdot 2/3}}{b^{-9 \cdot 2/3}} = \dfrac{\left[(64)^{1/3}\right]^2 a^4}{b^{-6}}$

$= \dfrac{16a^4}{b^{-6}} = 16a^4 b^6$

**59.** $\left(\dfrac{a^{-2} b^3}{a^4 b^{-3}}\right)^{-1/2}\left(\dfrac{a^4 b^{-5}}{ab}\right)^{-1/3}$

$= \left(a^{-6} b^6\right)^{-1/2}\left(a^3 b^{-6}\right)^{-1/3}$

$= \left(a^3 b^{-3}\right)\left(a^{-1} b^2\right) = a^2 b^{-1} = \dfrac{a^2}{b}$

**61.** $\dfrac{1}{2}(3x^2 + 2)^{-1/2} \cdot 6x = \dfrac{6x}{2} \cdot \dfrac{1}{(3x^2 + 2)^{1/2}}$

$= \dfrac{3x}{(3x^2 + 2)^{1/2}}$

**63.** $\dfrac{1}{2}(x^2 + 4x)^{-1/2}(2x + 4)$

$= \dfrac{2x + 4}{2} \cdot \dfrac{1}{(x^2 + 4x)^{1/2}}$

$= \dfrac{x + 2}{(x^2 + 4x)^{1/2}}$

**65. (a)**

$l = 20$ cm

$w = 15$ cm

$d = \sqrt{l^2 + w^2}$

$d = \sqrt{20^2 + 15^2} = \sqrt{400 + 225}$

$= \sqrt{625} = 25$

Therefore, $d = 25$ cm

**(b)**

$l = 16$ cm

$w = 10$ cm

$d = \sqrt{l^2 + w^2}$

$d = \sqrt{256 + 100} = \sqrt{356} \doteq 18.9$ cm

**67.** 
$$\sqrt[3]{\frac{32}{x^2}} - \frac{2\sqrt[3]{x}}{\sqrt[3]{2x^3}} = \frac{(32)^{1/3}}{x^{2/3}} - \frac{2x^{1/3}}{2^{1/3}x}$$

$$= \frac{2^{5/3}}{x^{2/3}} - \frac{2x^{1/3}}{2^{1/3}x}$$

$$= \left(\frac{2^{1/3}x^{1/3}}{2^{1/3}x^{1/3}}\right)\left(\frac{2^{5/3}}{x^{2/3}}\right) - \frac{2x^{1/3}}{2^{1/3}x}$$

$$= \frac{2^{6/3}x^{1/3} - 2x^{1/3}}{2^{1/3}x} = \frac{2^2 x^{1/3} - 2x^{1/3}}{2^{1/3}x}$$

$$= \frac{4x^{1/3} - 2x^{1/3}}{2^{1/3}x}$$

**69.** 
$$\sqrt[3]{\frac{x^{3n+1}y^n}{x^{3n+4}y^{4n}}} = \sqrt[3]{\frac{x^{-3}}{y^{3n}}} = \sqrt[3]{\frac{1}{x^3 y^{3n}}} = \frac{1}{xy^n}$$

**71.** 
$$\sqrt{\frac{x^n}{x^{n-2}}} = \sqrt{x^2} = x, \text{ assuming } x \text{ represents}$$
a positive real number.

**73.** $x = \sqrt[n]{a}$ and $y = \sqrt[n]{b} \Rightarrow x^n = a$ and $y^n = b$

$$\frac{a}{b} = \frac{x^n}{y^n} = \left(\frac{x}{y}\right)^n$$

$$\sqrt[n]{\frac{a}{b}} = \frac{x}{y} \text{ but } \frac{x}{y} = \frac{\sqrt[n]{a}}{\sqrt[n]{b}}.$$

Therefore, $\sqrt[n]{\dfrac{a}{b}} = \dfrac{\sqrt[n]{a}}{\sqrt[n]{b}}.$

**75.** 
$$|xy| = \sqrt{(xy)^2} = \sqrt{x^2 y^2}$$
$$= \sqrt{x^2} \cdot \sqrt{y^2} = |x| \cdot |y|$$

## Exercises 1.7

**1.**
$$
\begin{array}{r}
3x^2 + 5x - 2 \\
+\ 5x^2 - 7x + 9 \\
\hline
8x^2 - 2x + 7
\end{array}
$$

**3.**
$$
\begin{array}{r}
3x^3 - 7x^2 + 8x + 12 \\
+\ x^3 - 2x^2 + 8x - 9 \\
\hline
4x^3 - 9x^2 + 16x + 3
\end{array}
$$

**5.**
$$
\begin{array}{r}
4x^2 + 9x - 17 \\
+\ 2x^3 - 3x^2 + 2x - 11 \\
\hline
2x^3 + x^2 + 11x - 28
\end{array}
$$

**7.**
$$
\begin{array}{r}
3x^3 - 2x^2 - 8x + 9 \\
-(2x^3 + 5x^2 + 2x + 1) \\
\hline
x^3 - 7x^2 - 10x + 8
\end{array}
$$

**9.**
$$
\begin{array}{r}
4x^3 + x^2 - 2x - 13 \\
-(2x^2 + 3x + 9) \\
\hline
4x^3 - x^2 - 5x - 22
\end{array}
$$

**11.** $5x + (1 - 2x) = 3x + 1$

**13.** $(y + 2) + (2y + 1) + (3y + 3)$
$= (y + 2y + 3y) + (2 + 1 + 3)$
$= 6y + 6$

**15.** $(x^3 + 3x^2 + 3x + 1) - (x^2 + 2x + 1)$
$= x^3 + (3x^2 - x^2) + (3x - 2x) + (1 - 1)$
$= x^3 + 2x^2 + x$

**17.** $5y - [y - (3y + 8)] = 5y - [y - 3y - 8]$
$= 5y - [-2y - 8] = 5y + 2y + 8 = 7y + 8$

**19.** $(5x - 2xy + x^2y^2) - (2x + xy - x^2y^2)$
$= (x^2y^2 + x^2y^2) + (-2xy - xy) + (5x - 2x)$
$= 2x^2y^2 - 3xy + 3x$

**21.** $2x^2(2x + 1 - 10x^2) = 2x^2(-10x^2 + 2x + 1)$
$= -20x^4 + 4x^3 + 2x^2$

**23.** $(x + 1)(x + 1) = (x + 1)^2 = x^2 + 2x + 1$

**25.** $(4x - 2)(x + 7) = 4x^2 + 28x - 2x - 14$
$= 4x^2 + 26x - 14$

**27.** $(-2x+3)(3x+6) = -6x^2 - 12x + 9x + 18$
$= -6x^2 - 3x + 18$

**29.** $(-2x-3)(3x-6) = -6x^2 + 12x - 9x + 18$
$= -6x^2 + 3x + 18$

**31.** $\left(\dfrac{2}{3}x+6\right)\left(\dfrac{2}{3}x+6\right) = \left(\dfrac{2}{3}x+6\right)^2$
$= \dfrac{4}{9}x^2 + 8x + 36$

**33.** $(7-3x)(4x-9) = (-3x+7)(4x-9)$
$= -12x^2 + 27x + 28x - 63$
$= -12x^2 + 55x - 63$

**35.** $\left(\dfrac{1}{5}x - \dfrac{1}{4}\right)\left(\dfrac{1}{5}x - \dfrac{1}{4}\right) = \left(\dfrac{1}{5}x - \dfrac{1}{4}\right)^2$
$= \dfrac{1}{25}x^2 - \dfrac{1}{10}x + \dfrac{1}{16}$

**37.** $\left(\sqrt{x} - 10\right)\left(\sqrt{x} + 10\right) = \left(\sqrt{x}\right)^2 - (10)^2$
$= x - 100$

**39.** $\left(\sqrt{x} + 2\right)\left(\sqrt{x} - 2\right) = \left(\sqrt{x}\right)^2 - (2)^2$
$= x - 4$

**41.** $(x^2 + x + 9)(x^2 - 3x - 4)$
$= x^2(x^2 - 3x - 4) + x(x^2 - 3x - 4)$
$\quad + 9(x^2 - 3x - 4)$
$= x^4 - 3x^3 - 4x^2 + x^3 - 3x^2 - 4x + 9x^2$
$\quad - 27x - 36$
$= x^4 - 2x^3 + 2x^2 - 31x - 36$

**43.** $(x-2)(x^2 + 2x + 4)$
$= x(x^2 + 2x + 4) - 2(x^2 + 2x + 4)$
$= x^3 + 2x^2 + 4x - 2x^2 - 4x - 8$
$= x^3 - 8$

**45.** $(x-2)(x^4 + 2x^3 + 4x^2 + 8x + 16)$
$= x(x^4 + 2x^3 + 4x^2 + 8x + 16)$
$\quad -2(x^4 + 2x^3 + 4x^2 + 8x + 16)$
$= x^5 + 2x^4 + 4x^3 + 8x^2 + 16x$
$\quad \dfrac{-2x^4 - 4x^3 - 8x^2 - 16x - 32}{}$
$= x^5 \qquad\qquad\qquad\qquad -32$
$= x^5 - 32$

**47.** $(x^{2n} + 1)(x^{2n} - 2) = x^{4n} - 2x^{2n} + x^{2n} - 2$
$= x^{4n} - x^{2n} - 2$

**49.** $3x(1-x)(1-x) = 3x(1-x)^2$
$= 3x(1 - 2x + x^2) = 3x - 6x^2 + 3x^3$
$= 3x^3 - 6x^2 + 3x$

**51.** $(2x+1)(3x-2)(3-x)$
$= (6x^2 - x - 2)(3-x)$
$= 18x^2 - 3x - 6 - 6x^3 + x^2 + 2x$
$= -6x^3 + 19x^2 - x - 6$

**53.** $(x^3 - 2x + 1)(2x) + (x^2 - 2)(3x^2 - 2)$
$= 2x^4 - 4x^2 + 2x + (3x^4 - 8x^2 + 4)$
$= 5x^4 - 12x^2 + 2x + 4$

**55.** $(x^4 - 3x^2 + 5)(2x+3) + (x^2 + 3x)(4x^3 - 6x)$
$= (2x^5 - 6x^3 + 10x + 3x^4 - 9x^2 + 15)$
$\quad + (4x^5 - 6x^3 + 12x^4 - 18x^2)$
$= 6x^5 + 15x^4 - 12x^3 - 27x^2 + 10x + 15$

**57.** $(x-1)^3 = (x-1)(x-1)^2$
$= (x-1)(x^2 - 2x + 1)$
$= x^3 - 2x^2 + x - x^2 + 2x - 1$
$= x^3 - 3x^2 + 3x - 1$

**59.** $(a+b)^4 = (a+b)^2(a+b)^2$
$= (a^2 + 2ab + b^2)(a^2 + 2ab + b^2)$
$= a^4 + 2a^3b + a^2b^2 + 2a^3b + 4a^2b^2 + 2ab^3$
$\quad + a^2b^2 + 2ab^3 + b^4$
$= a^4 + 4a^3b + 6a^2b^2 + 4ab^3 + b^4$

**61.** $(2x+3)^3 = (2x+3)(2x+3)^2$
$= (2x+3)(4x^2 + 12x + 9)$
$= 8x^3 + 24x^2 + 18x + 12x^2 + 36x + 27$
$= 8x^3 + 36x^2 + 54x + 27$

**63.**
$$\left(\frac{1}{3}x+3\right)^3 = \left(\frac{1}{3}x+3\right)\left(\frac{1}{3}x+3\right)^2$$
$$= \left(\frac{1}{3}x+3\right)\left(\frac{1}{9}x^2+2x+9\right)$$
$$= \frac{1}{27}x^3 + \frac{2}{3}x^2 + 3x + \frac{1}{3}x^2 + 6x + 27$$
$$= \frac{1}{27}x^3 + x^2 + 9x + 27$$

**65.**
$$\frac{12}{\sqrt{5}-\sqrt{3}}\left(\frac{\sqrt{5}+\sqrt{3}}{\sqrt{5}+\sqrt{3}}\right) = \frac{12\left(\sqrt{5}+\sqrt{3}\right)}{5-3}$$
$$= \frac{12\left(\sqrt{5}+\sqrt{3}\right)}{2} = 6\left(\sqrt{5}+\sqrt{3}\right)$$

**67.**
$$\frac{14}{\sqrt{2}-3}\left(\frac{\sqrt{2}+3}{\sqrt{2}+3}\right) = \frac{14\left(\sqrt{2}+3\right)}{2-9}$$
$$= \frac{14\left(\sqrt{2}+3\right)}{-7} = -2\left(\sqrt{2}+3\right) = -2\sqrt{2}-6$$

**69.**
$$\frac{\sqrt{x}+\sqrt{y}}{\sqrt{x}-\sqrt{y}}\left(\frac{\sqrt{x}+\sqrt{y}}{\sqrt{x}+\sqrt{y}}\right) = \frac{x+2\sqrt{xy}+y}{x-y}$$

**71.**
$$\frac{\sqrt{5}+3}{\sqrt{5}}\left(\frac{\sqrt{5}-3}{\sqrt{5}-3}\right) = \frac{5-9}{5-3\sqrt{5}} = \frac{-4}{5-3\sqrt{5}}$$

**73.**
$$\frac{\sqrt{x}+\sqrt{y}}{\sqrt{x}-\sqrt{y}}\left(\frac{\sqrt{x}-\sqrt{y}}{\sqrt{x}-\sqrt{y}}\right) = \frac{x-y}{x-2\sqrt{xy}+y}$$

**75.**
$$\frac{\sqrt{4+h}-2}{h}\cdot\left(\frac{\sqrt{4+h}+2}{\sqrt{4+h}+2}\right) = \frac{4+h-4}{h\left(\sqrt{4+h}+2\right)}$$
$$= \frac{h}{h\left(\sqrt{4+h}+2\right)} = \frac{1}{\sqrt{4+h}+2}$$

**Challenge Problem**

$$P^3 = 2^{18} + 2^{12}\cdot 3^5 + 2^6\cdot 3^9 + 3^{12}$$
$$= (2^6)^3 + 3\cdot(2^6)^2\cdot 3^4 + 3\cdot 2^6(3^4)^2 + (3^4)^3$$
$$= (2^6 + 3^4)^3$$
So, $P = 2^6 + 3^4 = 64 + 81 = 145$

**Exercises 1.8**

**1.** $4x^2 - 9 = (2x+3)(2x-3)$

**3.** $a^2 - 121b^2 = (a+11b)(a-11b)$

**5.** $x^3 + 64 = x^3 + 4^3 = (x+4)(x^2-4x+16)$

**7.** $125x^3 - 64 = (5x)^3 - 4^3$
$= (5x-4)(25x^2+20x+16)$

**9.** $8x^3 + 343y^3 = (2x)^3 + (7y)^3$
$= (2x+7y)(4x^2-14xy+49y^2)$

**11.** $3 - 4x^2 = (\sqrt{3})^2 - (2x)^2$
$= (\sqrt{3}+2x)(\sqrt{3}-2x)$

**13.** $x - 36 = \left(\sqrt{x}\right)^2 - 6^2 = \left(\sqrt{x}+6\right)\left(\sqrt{x}-6\right)$

**15.** $8 - 3x = \left(2\sqrt{2}\right)^2 - \left(\sqrt{3x}\right)^2$
$= \left(2\sqrt{2}+\sqrt{3x}\right)\left(2\sqrt{2}-\sqrt{3x}\right)$

**17.** $7 + a^3 = \left(\sqrt[3]{7}\right)^3 + a^3$
$= \left(\sqrt[3]{7}+a\right)\left(\sqrt[3]{49}-\sqrt[3]{7}a+a^2\right)$

**19.** $27x + 1 = \left(3\sqrt[3]{x}\right)^3 + 1^3$
$= \left(3\sqrt[3]{x}+1\right)\left(9\sqrt[3]{x^2}-3\sqrt[3]{x}+1\right)$

**21.** $3x - 4 = \left(\sqrt[3]{3x}\right)^3 - \left(\sqrt[3]{4}\right)^3$
$= \left(\sqrt[3]{3x}-\sqrt[3]{4}\right)\left(\sqrt[3]{9x^2}+\sqrt[3]{4}\sqrt[3]{3x}+\sqrt[3]{16}\right)$
$= \left(\sqrt[3]{3x}-\sqrt[3]{4}\right)\left(\sqrt[3]{9x^2}+\sqrt[3]{12x}+\sqrt[3]{16}\right)$

**23.** $x^2 - y - x + xy = (x^2-x)+(xy-y)$
$= x(x-1)+y(x-1) = (x+y)(x-1)$

**25.** $-y - x + 1 + xy = xy - x + (-y+1)$
$= x(y-1)-1(y-1) = (x-1)(y-1)$

**27.** $2 - y^2 + 2x - xy^2$

$= (2x + 2) + (-xy^2 - y^2)$

$= 2(x + 1) - y^2(x + 1) = (2 - y^2)(x + 1)$

or $= (x + 1)(-y^2 + 2)$

**29.** $7x^3 + 7h^3 = 7(x^3 + h^3)$

$= 7(x + h)(x^2 - hx + h^2)$

**31.** $a^8 - b^8 = (a^4 + b^4)(a^4 - b^4)$

$= (a^4 + b^4)(a^2 + b^2)(a^2 - b^2)$

$= (a^4 + b^4)(a^2 + b^2)(a + b)(a - b)$

**33.** $a^5 - 32 = a^5 - 2^5$

$= (a - 2)(a^4 + 2a^3 + 4a^2 + 8a + 16)$

**35.** $a^3x - b^3y + b^3x - a^3y$

$= (a^3x - a^3y) + (b^3x - b^3y)$

$= a^3(x - y) + b^3(x - y) = (a^3 + b^3)(x - y)$

$= (a + b)(a^2 - ab + b^2)(x - y)$

$= (x - y)(a + b)(a^2 - ab + b^2)$

**37.** $x^5 - 16xy^4 - 2x^4y + 32y^5$

$= (x^5 - 2x^4y) + (32y^5 - 16xy^4)$

$= x^4(x - 2y) + 16y^4(2y - x)$

$= x^4(x - 2y) - 16y^4(x - 2y)$

$= (x^4 - 16y^4)(x - 2y)$

$= (x - 2y)(x^2 + 4y^2)(x^2 - 4y^2)$

$= (x - 2y)(x^2 + 4y^2)(x + 2y)(x - 2y)$

$= (x - 2y)^2(x^2 + 4y^2)(x + 2y)$

**39.** $20a^2 - 9a + 1 = (5a - 1)(4a - 1)$

**41.** $9x^2 + 6x + 1 = (3x + 1)(3x + 1) = (3x + 1)^2$

**43.** $14x^2 + 37x + 5 = (2x + 5)(7x + 1)$

**45.** $8x^2 - 9x + 1 = (8x - 1)(x - 1)$

**47.** $8x^2 - 16x + 6 = 2(4x^2 - 8x + 3)$

$= 2(2x - 3)(2x - 1)$

**49.** $12a^2 - 25a + 12 = (4a - 3)(3a - 4)$

**51.** $4x^2 + 4x - 3 = (2x - 1)(2x + 3)$

**53.** $24a^2 + 25ab + 6b^2 = (8a + 3b)(3a + 2b)$

**55.** $a^2 - 2a + 2$, Not factorable with integer coefficients because neither $(2) + (1)$ nor $(-2) + (-1) = -2$.

**57.** $6x^2 + 2x - 20 = 2(3x^2 + x - 10)$

$= 2(3x - 5)(x + 2)$

**59.** $2b^2 + 12b + 16 = 2(b^2 + 6b + 8)$

$= 2(b + 4)(b + 2)$

**61.** $a^3b - 2a^2b^2 + ab^3 = ab(a^2 - 2ab + b^2)$

$= ab(a - b)^2$

**63.** $16x^2 - 24x + 8 = 8(2x^2 - 3x + 1)$

$= 8(2x - 1)(x - 1)$

**65.** $25a^2 + 50ab + 25b^2 = 25(a^2 + 2ab + b^2)$

$= 25(a + b)^2$

**67.** $a^6 - 2a^3 + 1 = (a^3 - 1)^2 = (a^3 - 1)(a^3 - 1)$

$= (a - 1)(a^2 + a + 1)(a - 1)(a^2 + a + 1)$

$= (a - 1)^2(a^2 + a + 1)^2$

**69.** $6x^5y - 3x^3y^2 - 30xy^3$

$= 3xy(2x^4 - x^2y - 10y^2)$

$= 3xy(2x^2 - 5y)(x^2 + 2y)$

**71.** $(x+2)^3(2)+(2x+1)(3)(x+2)^2$

$= (x+2)^2[(x+2)(2)+(2x+1)(3)]$

$= (x+2)^2[2x+4+6x+3]$

$= (x+2)^2(8x+7)$

**73.** $(x^3+1)^3(2x)+(x^2-1)(3)(x^3+1)^2(3x^2)$

$= (x^3+1)^2[(x^3+1)(2x)$

$\quad +(x^2-1)(3)(3x^2)]$

$= (x^3+1)^2[2x^4+2x+9x^4-9x^2]$

$= (x^3+1)^2(11x^4-9x^2+2x)$

$= (x^3+1)^2(x)(11x^3-9x+2)$

**75. (a)** $x^5+32=x^5-(-2)^5$

$= [x-(-2)][x^4+(-2)x^3+(-2)^2x^2$

$\quad +(-2)^3x^1+(-2)^4]$

$= (x+2)(x^4-2x^3+4x^2-8x+16)$

**(b)** $128x^8+xy^7=x(128x^7+y^7)$

$= x[(2x)^7+y^7]=x[(2y)^7-(-y)^7]$

$= x[2x-(-y)][2^6x^6+2^5x^5(-y)^1$

$\quad +2^4x^4(-y)^2+2^3x^3(-y)^3$

$\quad +2^2x^2(-y)^4+2x(-y)^5+(-y)^6]$

$= x(2x+y)(64x^6-32x^5y+16x^4y^2$

$\quad -8x^3y^3+4x^2y^4-2xy^5+y^6)$

**77.** From (b) in Example 3,

$x^6-1=(x-1)(x+1)(x^4+x^2+1)$ and
from (c) in Example 3,

$x^6-1=(x-1)(x^2+x+1)$

$\quad \cdot (x^2-x+1)(x+1)$

So $(x-1)(x+1)(x^4+x^2+1)$

$= (x-1)(x^2+x+1)(x^2-x+1)(x+1)$

$\Rightarrow x^4+x^2+1=(x^2+x+1)(x^2-x+1)$

**79.** $5(4x^2+4x+1)^4(8x+4)$

$= 5[(2x+1)^2]^4(4)(2x+1)=20(2x+1)^9$

**81.** $y=x\sqrt{x^2+2} \Rightarrow y^2=[x(x^2+2)^{1/2}]^2$

$\qquad y^2=x^2(x^2+2)$

$\qquad y^2=x^4+2x^2$

$\qquad y^2+1=x^4+2x^2+1$

$\qquad 1+y^2=(x^2+1)^2$

$\qquad \sqrt{1+y^2}=x^2+1$

**83.** The area of the entire large circle $=\pi R^2$.

The area of each small circle $=\pi a^2$.
The area of the shaded region *excludes* all
9 small circles, so its area is

$\pi R^2-9\pi a^2=\pi(R^2-9a^2)$

$= \pi(R+3a)(R-3a)$

If $R=15.7$ and $a=3.1$, the shaded region's
area would be

$\pi[15.7+3(3.1)][15.7-3(3.1)]$

$= \pi(15.7+9.3)(15.7-9.3)$

$= \pi(25)(6.4)=\pi(160)$

$= 160\pi$

**Challenge Problem**

$x^3-y^3+xy^2-x^2y-x+y$

$= (x-y)(x^2+xy+y^2)-xy(x-y)-1(x-y)$

$= (x-y)(x^2+xy+y^2-xy-1)$

$= (x-y)(x^2+y^2-1)$

**Exercises 1.9**

**1.** $\dfrac{5}{7}-\dfrac{2}{3}=\dfrac{3}{4}$ ? $\dfrac{5}{7}-\dfrac{2}{3}=\dfrac{15-14}{21}=\dfrac{1}{21}$
False.

**3.** $\dfrac{3ax-5b}{6}=\dfrac{ax-5b}{2}$?

$\dfrac{3ax-5b}{6}=\dfrac{3ax}{6}-\dfrac{5b}{6}=\dfrac{ax}{2}-\dfrac{5b}{6}$
False.

**5.** $x^{-1}+y^{-1}=\dfrac{y+x}{xy}$? $\dfrac{1}{x}+\dfrac{1}{y}=\dfrac{y+x}{xy}$
True.

**7.** $\dfrac{8xy}{12yz} = \dfrac{2x}{3z}$

**9.** $\dfrac{45x^3 + 15x^2}{15x^2} = \dfrac{45x^3}{15x^2} + \dfrac{15x^2}{15x^2} = 3x + 1$

**11.** $\dfrac{12x^3 + 8x^2 + 4x}{4x} = \dfrac{4x(3x^2 + 2x + 1)}{4x}$
$= 3x^2 + 2x + 1$

**13.** $\dfrac{a^2b^2 + ab^2 - a^2b^3}{ab^2} = \dfrac{ab^2(a + 1 - ab)}{ab^2}$
$= a - ab + 1$

**15.** $\dfrac{6a^2x^2 - 8a^4x^6}{2a^2x^2} = \dfrac{2a^2x^2(3 - 4a^2x^4)}{2a^2x^2}$
$= 3 - 4a^2x^4$

**17.** $\dfrac{x^2 - 5x}{5 - x} = \dfrac{x(x - 5)}{-(x - 5)} = -x$

**19.** $\dfrac{n + 1}{n^2 + 1}$ cannot be simplified, as both numerator and denominator are "prime" polynomials.

**21.** $\dfrac{3x^2 + 3x - 6}{2x^2 + 6x + 4} = \dfrac{3(x^2 + x - 2)}{2(x^2 + 3x + 2)}$
$= \dfrac{3(x + 2)(x - 1)}{2(x + 2)(x + 1)} = \dfrac{3(x - 1)}{2(x + 1)}$

**23.** $\dfrac{4x^2 + 12x + 9}{4x^2 - 9} = \dfrac{(2x + 3)^2}{(2x + 3)(2x - 3)} = \dfrac{2x + 3}{2x - 3}$

**25.** $\dfrac{a^2 - 16b^2}{a^3 + 64b^3} = \dfrac{(a + 4b)(a - 4b)}{(a + 4b)(a^2 - 4ab + 16b^2)}$
$= \dfrac{a - 4b}{a^2 - 4ab + 16b^2}$

**27.** $\dfrac{2x^2}{y} \cdot \dfrac{y^2}{x^3} = \dfrac{2y}{x}$

**29.** $\dfrac{2a}{3} \cdot \dfrac{3}{a^2} \cdot \dfrac{1}{a} = \dfrac{2}{a^2}$

**31.** $\dfrac{3x}{2y} - \dfrac{x}{2y} = \dfrac{3x - x}{2y} = \dfrac{2x}{2y} = \dfrac{x}{y}$

**33.** $\dfrac{a - 2b}{2} - \dfrac{3a + b}{3} = \dfrac{3a - 6b - 6a - 2b}{6}$
$= \dfrac{-3a - 8b}{6}$

**35.** $\dfrac{x - 1}{3} \cdot \dfrac{x^2 + 1}{x^2 - 1} = \dfrac{(x - 1)(x^2 + 1)}{3(x - 1)(x + 1)} = \dfrac{x^2 + 1}{3(x + 1)}$

**37.** $\dfrac{1 - x}{2 + x} \div \dfrac{x^2 - x}{x^2 + 2x} = \dfrac{1 - x}{2 + x} \cdot \dfrac{x(x + 2)}{x(x - 1)}$
$= \dfrac{-1(x - 1)(x)(x + 2)}{(x + 2)(x)(x - 1)} = -1$

**39.** $\dfrac{2}{x} - y = \dfrac{2 - xy}{x}$

**41.** $\dfrac{3y}{y + 1} + \dfrac{2y}{y - 1} = \dfrac{3y(y - 1) + 2y(y + 1)}{(y + 1)(y - 1)}$
$= \dfrac{3y^2 - 3y + 2y^2 + 2y}{(y + 1)(y - 1)} = \dfrac{5y^2 - y}{(y + 1)(y - 1)}$
$= \dfrac{y(5y - 1)}{(y + 1)(y - 1)}$

**43.** $\dfrac{2x^2}{x^2 + x} + \dfrac{x}{x + 1} = \dfrac{2x^2}{x(x + 1)} + \left(\dfrac{x}{x}\right)\left(\dfrac{x}{x + 1}\right) = \dfrac{2x^2 + x^2}{x(x + 1)} = \dfrac{3x^2}{x(x + 1)} = \dfrac{3x}{x + 1}$

**45.** $\dfrac{5}{x^2 - 4} - \dfrac{3 - x}{4 - x^2} = \dfrac{5}{x^2 - 4} - \dfrac{x - 3}{x^2 - 4} = \dfrac{8 - x}{x^2 - 4}$

**47.** $\dfrac{2x}{x^2-9}+\dfrac{x}{x^2+6x+9}-\dfrac{3}{x+3}=\dfrac{2x}{(x+3)(x-3)}+\dfrac{x}{(x+3)(x+3)}-\dfrac{3}{x+3}$

$=\dfrac{2x(x+3)+x(x-3)-3(x+3)(x-3)}{(x+3)^2(x-3)}=\dfrac{2x^2+6x+x^2-3x-3x^2+27}{(x+3)^2(x-3)}$

$=\dfrac{3x+27}{(x+3)^2(x-3)}=\dfrac{3(x+9)}{(x+3)^2(x-3)}$

**49.** $\dfrac{x+3}{5-x}-\dfrac{x-5}{x+5}+\dfrac{2x^2+30}{x^2-25}=\dfrac{-x-3}{x-5}-\dfrac{x-5}{x+5}+\dfrac{2x^2+30}{(x-5)(x+5)}$

$=\dfrac{(-x-3)(x+5)-[(x-5)(x-5)]+2x^2+30}{(x-5)(x+5)}=\dfrac{-x^2-8x-15-x^2+10x-25+2x^2+30}{(x-5)(x+5)}$

$=\dfrac{2x-10}{(x-5)(x+5)}=\dfrac{2(x-5)}{(x-5)(x+5)}=\dfrac{2}{x+5}$

**51.** $\dfrac{x^3+x^2-12x}{x^2-3x}\cdot\dfrac{3x^2-10x+3}{3x^2+11x-4}=\dfrac{x(x^2+x-12)}{x(x-3)}\cdot\dfrac{(3x-1)(x-3)}{(3x-1)(x+4)}=\dfrac{(x+4)(x-3)}{x+4}=x-3$

**53.** $\dfrac{n^3-8}{n+2}\cdot\dfrac{2n^2+8}{n^3-4n}\cdot\dfrac{n^3+2n^2}{n^3+2n^2+4n}=\dfrac{(n-2)(n^2+2n+4)(2)(n^2+4)(n^2)(n+2)}{(n+2)(n)(n^2-4)(n)(n^2+2n+4)}=\dfrac{2(n^2+4)}{n+2}$

**55.**

$$
\begin{array}{r}
x^2-\phantom{0}5x+2 \\
x+3\overline{\smash{\big)}\,x^3-2x^2-13x+6} \\
\underline{-(x^3+3x^2)\phantom{-13x+6}} \\
-5x^2-13x\phantom{+6} \\
\underline{-(-5x^2-15x)\phantom{+6}} \\
2x+6 \\
\underline{2x+6} \\
0
\end{array}
$$

$x^2-5x+2,\ r=0$

**57.**

$$
\begin{array}{r}
x^2\phantom{{}-x^2+7} \\
x-1\overline{\smash{\big)}\,x^3-x^2+7} \\
\underline{-(x^3-x^2)\phantom{+7}} \\
7
\end{array}
$$

$x^2,\ r=7$

**59.**

$$\begin{array}{r} x^2 + 7x + 7 \\ x - 2 \overline{\smash{\big)}\, x^3 + 5x^2 - 7x + 8} \\ \underline{-(x^3 - 2x^2)} \\ 7x^2 - 7x \\ \underline{-(7x^2 - 14x)} \\ 7x + 8 \\ \underline{-(7x - 14)} \\ 22 \end{array}$$

$x^2 + 7x + 7,\ r = 22$

**61.**

$$\begin{array}{r} 4x + 3 \\ x^2 - 2x \overline{\smash{\big)}\, 4x^3 - 5x^2 + x - 7} \\ \underline{-(4x^3 - 8x^2)} \\ 3x^2 + x \\ \underline{-(3x^2 - 6x)} \\ 7x - 7 \end{array}$$

$4x + 3,\ r = 7x - 7$

**63.**

$$\begin{array}{r} x + 2 \\ x^2 - 3x + 5 \overline{\smash{\big)}\, x^3 - x^2 - x + 10} \\ \underline{-(x^3 - 3x^2 + 5x)} \\ 2x^2 - 6x + 10 \\ \underline{-(2x^2 - 6x + 10)} \\ 0 \end{array}$$

$x + 2,\ r = 0$

**65.** $\dfrac{\dfrac{5}{x^2 - 4}}{\dfrac{10}{x - 2}} = \dfrac{5(x - 2)}{10(x^2 - 4)} = \dfrac{5(x - 2)}{10(x - 2)(x + 2)} = \dfrac{1}{2(x + 2)}$

**67.** $\dfrac{\dfrac{1}{4+h} - \dfrac{1}{4}}{\dfrac{h}{1}} = \dfrac{\dfrac{4-(4+h)}{4(4+h)}}{\dfrac{h}{1}} = \dfrac{\dfrac{-h}{4(4+h)}}{\dfrac{h}{1}} = \dfrac{-h}{4h(4+h)} = \dfrac{-1}{4(4+h)}$

**69.** $\dfrac{\dfrac{1}{x+3} - \dfrac{1}{3}}{\dfrac{x}{1}} = \dfrac{\dfrac{3-(x+3)}{3(x+3)}}{\dfrac{x}{1}} = \dfrac{\dfrac{-x}{3(x+3)}}{\dfrac{x}{1}} = \dfrac{-x}{3x(x+3)} = \dfrac{-1}{3(x+3)}$

**71.** $$\dfrac{\frac{1}{x^2}-\frac{1}{16}}{\frac{x+4}{1}}=\dfrac{\frac{16-x^2}{16x^2}}{\frac{x+4}{1}}=\dfrac{16-x^2}{16x^2(x+4)}=\dfrac{-1(x^2-16)}{16x^2(x+4)}=\dfrac{-(x+4)(x-4)}{16x^2(x+4)}=\dfrac{-(x-4)}{16x^2}=\dfrac{4-x}{16x^2}$$

**73.** $$\dfrac{\frac{x^{-1}-y^{-1}}{1}}{\frac{1}{x^2}-\frac{1}{y^2}}=\dfrac{\frac{1}{x}-\frac{1}{y}}{\frac{1}{x^2}-\frac{1}{y^2}}=\dfrac{\frac{y-x}{xy}}{\frac{y^2-x^2}{x^2y^2}}=\dfrac{(y-x)(x^2y^2)}{(y^2-x^2)(xy)}=\dfrac{xy}{y+x}$$

**75.** $$\dfrac{(1+x^2)(-2x)-(1-x^2)(2x)}{(1+x^2)^2}=\dfrac{-2x+(-2x^3)-2x+2x^3}{(1+x^2)^2}=\dfrac{-4x}{(1+x^2)^2}$$

**77.** $$\dfrac{x^2(4-2x)-(4x-x^2)(2x)}{x^4}=\dfrac{4x^2-2x^3-8x^2+2x^3}{x^4}=\dfrac{-4x^2}{x^4}=\dfrac{-4}{x^2}$$

**79.** $$\dfrac{a^{-1}-b^{-1}}{a-b}=\dfrac{\frac{1}{a}-\frac{1}{b}}{a-b}=\dfrac{\frac{b-a}{ab}}{\frac{a-b}{1}}=\dfrac{b-a}{ab(a-b)}=\dfrac{-(a-b)}{ab(a-b)}=\dfrac{-1}{ab}$$

**81.** $$\dfrac{x^{-2}-y^{-2}}{xy}=\dfrac{\frac{1}{x^2}-\frac{1}{y^2}}{\frac{xy}{1}}=\dfrac{\frac{y^2-x^2}{x^2y^2}}{\frac{xy}{1}}=\dfrac{y^2-x^2}{x^3y^3}=\dfrac{(y+x)(y-x)}{x^3y^3}$$

**83.** **(a)** $$\dfrac{\frac{A\cdot D}{B}+\frac{C}{1}}{\frac{D}{1}}=\dfrac{\frac{AD+BC}{B}}{\frac{D}{1}}=\dfrac{AD+BC}{BD}=\dfrac{AD}{BD}+\dfrac{BC}{BD}=\dfrac{A}{B}+\dfrac{C}{D}$$

**(b)** $$\left[\dfrac{\left(\frac{AB}{D}+\frac{C}{1}\right)D}{\frac{F}{1}}+E\right]F=\left[\dfrac{\frac{AB+CD}{1}}{\frac{F}{1}}+E\right]F=\left(\dfrac{AB+CD}{F}+\dfrac{E}{1}\right)F=AB+CD+EF$$

**85.** $$\dfrac{2}{\frac{1}{60}+\frac{1}{40}}=\dfrac{2}{\frac{40+60}{2400}}=\dfrac{\frac{2}{1}}{\frac{100}{2400}}=\dfrac{2\cdot2400}{100}=48\text{ mph}$$

A simplified form:

$$\dfrac{\frac{2}{1}}{\frac{1}{s_1}+\frac{1}{s_2}}=\dfrac{\frac{2}{1}}{\frac{s_2+s_1}{s_1s_2}}=\dfrac{2s_1s_2}{s_1+s_2}$$

**87.** Given that $y^3-x^3=8$ :

$$\dfrac{2xy^2-2x^2y\left(\frac{x^2}{y^2}\right)}{y^4}=\dfrac{\frac{2xy^2}{1}-\frac{2x^4y}{y^2}}{\frac{y^4}{1}}$$

$$=\dfrac{\frac{2xy^4-2x^4y}{y^2}}{\frac{y^4}{1}}=\dfrac{2xy^4-2x^4y}{y^6}$$

$$=\dfrac{2xy(y^3-x^3)}{y^6}=\dfrac{2x(8)}{y^5}=\dfrac{16x}{y^5}$$

**89.** Given that $y = \dfrac{x^2}{8} - \dfrac{2}{x^2}$:

$$y^2 = \left(\frac{x^2}{8} - \frac{2}{x^2}\right)^2 = \frac{x^4}{64} - \frac{1}{2} + \frac{4}{x^4}$$

$$y^2 + 1 = \frac{x^4}{64} + \frac{1}{2} + \frac{4}{x^4} = \left(\frac{x^2}{8} + \frac{2}{x^2}\right)^2$$

$$\sqrt{y^2 + 1} = \sqrt{\left(\frac{x^2}{8} + \frac{2}{x^2}\right)^2} = \frac{x^2}{8} + \frac{2}{x^2}$$

## Challenge Problem

The problem with the story is that $\dfrac{1}{2} + \dfrac{1}{3} + \dfrac{1}{9} \neq 1$, so the man didn't divide up all the horses.

## Critical Thinking

**1.** If $3\left(\dfrac{x+1}{x-1}\right) = \dfrac{3x+1}{3x-1}$, then it would be true for all values of $x$, and in particular, $x = 0$.

But, $3\left(\dfrac{0+1}{0-1}\right) = 3(-1) = -3$ and

$\dfrac{3 \cdot 0 + 1}{3 \cdot 0 - 1} = -1$. So, $3\left(\dfrac{x+1}{x-1}\right) \neq \dfrac{3x+1}{3x-1}$.

**3.** $(a+b)^2 \geq 4ab$
To prove this is true, start with
$$(a-b)^2 \geq 0$$
$$a^2 - 2ab + b^2 \geq 0$$
$$a^2 + b^2 \geq 2ab$$
$$a^2 + 2ab + b^2 \geq 4ab$$
$$(a+b)^2 \geq 4ab$$

## Exercises 1.10

**1.** True

**3.** True

**5.** False

**7.** $5 + \sqrt{-4} = 5 + \sqrt{4} \cdot \sqrt{-1} = 5 + \sqrt{4}\,i = 5 + 2i$

**9.** $-5 = -5 + 0i$

**11.** $\sqrt{-16} = \sqrt{16} \cdot \sqrt{-1} = 4i$

**13.** $\sqrt{-144} = \sqrt{144} \cdot \sqrt{-1} = 12i$

**15.** $\sqrt{-\dfrac{9}{16}} = \sqrt{\dfrac{9}{16}} \cdot \sqrt{-1} = \dfrac{3}{4}i$

**17.** $-\sqrt{-5} = -\sqrt{5} \cdot i = -\sqrt{5}\,i$

**19.** $\sqrt{-9} \cdot \sqrt{-81} = 3i \cdot 9i = 27i^2 = 27(-1) = -27$

**21.** $\sqrt{-3} \cdot \sqrt{-2} = \sqrt{3}i \cdot \sqrt{2}i = \sqrt{6}i^2 = -\sqrt{6}$

**23.** $(-3i^2)(5i) = -15i^3 = 15i$

**25.** $\sqrt{-9} + \sqrt{-81} = 3i + 9i = 12i$

**27.** $\sqrt{-8} + \sqrt{-18} = 2\sqrt{2}i + 3\sqrt{2}i$
$= 5\sqrt{2}i = 5i\sqrt{2}$

**29.** $\sqrt{-9} - \sqrt{-3} = 3i - \sqrt{3}i = \left(3 - \sqrt{3}\right)i$

**31.** $(7 + 5i) + (3 + 2i) = 10 + 7i$

**33.** $(8 + 2i) - (3 + 5i) = 5 - 3i$

**35.** $\left(7 + \sqrt{-16}\right) + \left(3 - \sqrt{-4}\right)$
$= (7 + 4i) + (3 - 2i)$
$= 10 + 2i$

**37.** $2i(3 + 5i) = 6i + 10i^2 = -10 + 6i$

**39.** $(3 + 2i)(2 + 3i) = 6 + 9i + 4i + 6i^2$
$= 6 + 13i - 6 = 13i$

**41.** $(5 - 2i)(3 + 4i) = 15 + 20i - 6i - 8i^2$
$= 15 + 14i + 8 = 23 + 14i$

**43.** $\dfrac{3 + 5i}{i} = \dfrac{3}{i} + \dfrac{5}{1} = \dfrac{3i}{i^2} + \dfrac{5}{1} = 5 - 3i$

**45.** $\dfrac{5+3i}{2+i}\left(\dfrac{2-i}{2-i}\right)=\dfrac{10+i-3i^2}{4-i^2}=\dfrac{13+i}{5}$

$=\dfrac{13}{5}+\dfrac{1}{5}i$

**47.** $\dfrac{3-i}{3+i}\left(\dfrac{3-i}{3-i}\right)=\dfrac{9-6i+i^2}{9-i^2}=\dfrac{9-6i-1}{10}$

$=\dfrac{8-6i}{10}=\dfrac{4}{5}-\dfrac{3}{5}i$

**49.** $3i^3=-3i$

**51.** $2i^7=(2)(1)(i^3)=-2i$

**53.** $-4i^{18}=(-4)(i^{16})(i^2)=(-4)(1)(-1)=4$

**55.** $(3+2i)^{-1}=\dfrac{1}{3+2i}\left(\dfrac{3-2i}{3-2i}\right)=\dfrac{3-2i}{9-4i^2}$

$=\dfrac{3-2i}{13}=\dfrac{3}{13}-\dfrac{2}{13}i$

**57.** $\sqrt{(-4)(-9)}=\sqrt{-4}\cdot\sqrt{-9}$ ?

$\sqrt{36}=\sqrt{4i}\cdot\sqrt{9i}$ ?

$6=2i\cdot 3i$ ?

$6=6i^2$ ?

$6=-6$ ?

No.

**59.** $\dfrac{a+bi}{c+di}\left(\dfrac{c-di}{c-di}\right)=\dfrac{ac-adi+bci-bdi^2}{c^2-d^2i^2}$

$=\dfrac{ac-adi+bci+bd}{c^2+d^2}$

$=\dfrac{ac+bd}{c^2+d^2}+\dfrac{bc-ad}{c^2+d^2}i$

**61.** $[(3+i)(3-i)](4+3i)=[9-i^2](4+3i)$

$=10(4+3i)=40+30i$

and $(3+i)[(3-i)(4+3i)]$

$=(3+i)[12+5i-3i^2]=(3+i)[15+5i]$

$=45+30i+5i^2=45+30i-5=40+30i$

**63.** $(5+4i)+2(2-3i)-i(1-5i)$

$=(5+4i)+(4-6i)-(i-5i^2)$

$=5+4i+4-6i-i-5=4-3i$

**65.** $\dfrac{(2+i)^2(3-i)}{2+3i}=\dfrac{(4+4i+i^2)(3-i)}{2+3i}$

$=\dfrac{(3+4i)(3-i)}{2+3i}=\dfrac{9+9i-4i^2}{2+3i}$

$=\dfrac{13+9i}{2+3i}\left(\dfrac{2-3i}{2-3i}\right)=\dfrac{26-21i-27i^2}{4-9i^2}$

$=\dfrac{53-21i}{13}=\dfrac{53}{13}-\dfrac{21}{13}i$

**67.** $x^2+1=x^2-(-1)=x^2-i^2=(x+i)(x-i)$

**69.** $3x^2+75=3(x^2+25)=3[x^2-(-25)]$

$=3[x^2-(5i)^2]=3(x+5i)(x-5i)$

**71.** $i\cdot i^2\cdot i^3\cdot\ldots\cdot i^{98}\cdot i^{99}\cdot i^{100}$

$=(i\cdot i^{100})\cdot(i^2\cdot i^{99})\cdot(i^3\cdot i^{98})\cdot\ldots\cdot(i^{50}\cdot i^{51})$

$=(i^{101})\cdot(i^{101})\cdot(i^{101})\cdot\ldots\cdot(i^{101})$

$=(i^{101})^{50}$

$=i^{5050}=(i^{5048})(i^2)=(1)(i^2)$

$=(1)(-1)=-1$

## Chapter 1 Review Exercises

**1.** A rational number is any number expressed in the form $\dfrac{a}{b}$, where $a$ is any integer and $b$ is an integer other than zero.

**3.** (a) $-13$ is an integer, rational number, and real number.

(b) $\sqrt{13}$ is an irrational number and real number.

(c) $\dfrac{3}{5}$ is a rational number and real number.

(d) $7$ is a natural number, whole number, integer, rational number, and real number.

**(e)** $\sqrt{36} = 6$, which belongs to the same subset as (d) above.

**5.** Are real numbers commutative with respect to subtraction? No.
$5 - 3 = 2 \neq -2 = 3 - 5$
Are real numbers associative with respect to subtraction? No.
$(5 - 4) - 3 = 1 - 3 = -2$
$\neq 4 = 5 - 1 = 5 - (4 - 3)$

**7.** The identity element for addition is 0 (zero).
The identity element for multiplication is 1 (one).

**9.** $15 + (8 + n) = (15 + 8) + 9$
$n = 9$

**11.** $2.5(8 + n) = (2.5 \times 8) + (2.5 \times 10)$
$n = 10$

**13.** If $a = b$, then $a + c = b + c$ and $a - c = b - c$.

**15.** **(a)** $3(x - 1) = x + 2$
$3x - 3 = x + 2$
$2x = 5$
$x = \dfrac{5}{2}$
**(b)** $5(x + 1) = 2(x - 2)$
$5x + 5 = 2x - 4$
$3x = -9$
$x = -3$

**17.** Let $w$ = rectangle's width in centimeters, $2w + 1$ = rectangle's length in centimeters.
$P = 2w + 2(2w + 1) = 32$
$2w + 4w + 2 = 32$
$6w = 30$
$w = 5$ centimeters
Length $= 2(5) + (1) = 11$ centimeters

**19.** $I = Prt \Rightarrow r = \dfrac{I}{Pt}$
$r = \dfrac{405}{(4500)(2)}$
$r = \dfrac{405}{9000}$
$r = 0.045 = 4.5\%$

**21.** $a < b$ if and only if, for any real numbers $a$ and $b$, $b - a > 0$.

**23.** For all real numbers $a$, $b$, and $c$:
If $a < b$ and $c$ is positive, $ac < bc$.
If $a < b$ and $c$ is negative, $ac > bc$.

**25.** $5x - 2 \leq 3x + 6$
$2x \leq 8$
$x \leq 4$
$\{x \mid x \leq 4\}$

**27.** **(a)** $(-5, 2)$
**(b)** $[-5, 2]$
**(c)** $(-5, 2]$
**(d)** $[-5, 2)$

**29.** If $a < b$ and $b < c$, then $a < c$.

**31.** For any real numbers $a$ and $b$, only one of the following is true: $a < b$, $a = b$, or $a > b$.

**33.** $|a| = \begin{cases} a, & \text{if } a \geq 0 \\ -a, & \text{if } a < 0 \end{cases}$

**35.** $|a| < k \Rightarrow -k < a < k$

**37.** $\dfrac{|x - 2|}{x - 2} = 1$

$\dfrac{|x - 2|}{x - 2}$ is not defined at $x = 2$.

If $x > 2$, $|x - 2| = x - 2 \Rightarrow \dfrac{x - 2}{x - 2} = 1$ ✓

If $x < 2$, $|x - 2| = -(x - 2) \Rightarrow \dfrac{-(x - 2)}{x - 2} = -1$

$\{x \mid x > 2\}$

**39.** **(a)** $|x| < 5 \Rightarrow -5 < x < 5$
**(b)** $|x + 2| > 1 \Rightarrow x + 2 > 1$ or $x + 2 < -1$
$x > -1$ or $x < -3$

**41.** $|x + y| = |x| + |y|$ when $x > 0$ *and* $y > 0$ *or* when $x < 0$ *and* $y < 0$. The equality is not true when $x > 0$ *and* $y < 0$ *or* when $x < 0$ *and* $y > 0$.

**43.** $-5^2 = -25$ and $(-5)^2 = (-5)(-5) = 25$

**45.** (a) $3x^2 \cdot x^4 = 3x^6$

(b) $\dfrac{-8^3}{(-4)^2} = \dfrac{-512}{16} = -32$

(c) $10^3 \left(\dfrac{1}{5}\right)^3 = \dfrac{1000}{125} = 8$

(d) $\dfrac{x^{12}y^0}{(2x)^2(2y)^{-1}} = \dfrac{x^{12}}{4x^2 2^{-1}y^{-1}}$

$= \dfrac{x^{12}(2y)}{4x^2} = \dfrac{x^{10}y}{2}$

**47.** $3^{-2} \cdot 3^x = 3^5 \Rightarrow -2 + x = 5$

$\qquad\qquad\qquad\qquad x = 7$

**49.** $\dfrac{3^x}{3^2} = 3^5 \Rightarrow \begin{array}{l} x - 2 = 5 \\ x \quad\;\; = 7 \end{array}$

**51.** (a) $3.25 \times 10^5 = 325{,}000$

(b) $2.5 \times 10^{-6} = 0.0000025$

**53.** $\sqrt[n]{a}$, if $a > 0$, is the positive number $x$ such that $x^n = a$.

**55.** (a) $(x + y)^{1/5} = \sqrt[5]{x + y}$ \qquad True

(b) $\sqrt{x^{16}} = x^4$ ? \qquad\qquad\quad False

$(x^{16})^{1/2} = x^8 \neq x^4$

**57.** (a) $\sqrt{3a} \cdot \sqrt{5b} = \sqrt{15ab}$

(b) $\dfrac{\sqrt[3]{-54x^5}}{\sqrt[3]{2x^2}} = \dfrac{(-54)^{1/3}x^{5/3}}{(2)^{1/3}x^{2/3}}$

$= \dfrac{(-27 \cdot 2)^{1/3}x}{2^{1/3}} = \dfrac{-27^{1/3} \cdot 2^{1/3}x}{2^{1/3}}$

$= -3x$

(c) $\sqrt[5]{16x} \cdot \sqrt[5]{-2x^4} = \sqrt[5]{-32x^5} = -2x$

**59.** $\left(\dfrac{-64a^{-3}}{b^6}\right)^{2/3} = \dfrac{\left[(-64)^{1/3}\right]^2 a^{-2}}{b^4} = \dfrac{16}{a^2 b^4}$

**61.** $\sqrt[3]{16x^3} + 2\sqrt[3]{-54x^3} = 2x\sqrt[3]{2} + (-6)x\sqrt[3]{2}$

$= -4x\sqrt[3]{2}$

**63.** $\dfrac{6}{\sqrt{12}} + 2\sqrt{3} - 3\sqrt{75} = \dfrac{6\sqrt{12}}{12} + 2\sqrt{3} - 15\sqrt{3}$

$= \sqrt{3} + 2\sqrt{3} - 15\sqrt{3} = -12\sqrt{3}$

**65.** $\sqrt[m]{\sqrt[n]{a}} = \left[a^{1/n}\right]^{1/m} = a^{1/mn} = \sqrt[mn]{a}$

$\sqrt{\sqrt[3]{64}} = \left[64^{1/3}\right]^{1/2} = 64^{1/6} = \sqrt[6]{64} = 2$

**67.** $a_n x^n + a_{n-1}x^{n-1} + a_{n-2}x^{n-2} + \dots$

$+ a_2 x^2 + a_1 x + a_0$

**69.** $(3x^3 - 8x^2 + 2x - 5) + (x^2 - 7x + 1)$

$= 3x^3 - 7x^2 - 5x - 4$

**71.** $2x^2(3x^3 - 2x^2 + x - 5)$

$= 6x^5 - 4x^4 + 2x^3 - 10x^2$

**73.** $(2x + 1)(3x - 5) = 6x^2 - 7x - 5$

**75.** $(2x^3 - 5x + 1)(x^2 - 3x + 2)$

$= 2x^5 - 6x^4 + 4x^3 - 5x^3 + 15x^2 - 10x$

$\qquad\qquad + x^2 - 3x + 2$

$= 2x^5 - 6x^4 - x^3 + 16x^2 - 13x + 2$

**77.** $\dfrac{6}{\sqrt{5} - \sqrt{2}} \left(\dfrac{\sqrt{5} + \sqrt{2}}{\sqrt{5} + \sqrt{2}}\right) = \dfrac{6(\sqrt{5} + \sqrt{2})}{5 - 2}$

$= 2(\sqrt{5} + \sqrt{2})$

**79.** $5x^6 + 25x^4 - 15x^2 = 5x^2(x^4 + 5x^2 - 3)$

**81.** $27 - 8x^3 = 3^3 - (2x)^3$

$= (3 - 2x)(9 + 6x + 4x^2)$

**83.** $a^4 - b^4 = (a^2 + b^2)(a^2 - b^2)$
$= (a^2 + b^2)(a + b)(a - b)$

**85.** $15 + ax^2 - 3x - 5ax$
$= (15 - 5ax) + (ax^2 - 3x)$
$= 5(3 - ax) + x(ax - 3)$
$= 5(3 - ax) - x(3 - ax)$
$= (5 - x)(3 - ax)$ or $(ax - 3)(x - 5)$

**87.** $x^2 - 2xy + y^2 = (x - y)^2$

**89.** $6x^2 + x - 1 = (3x - 1)(2x + 1)$

**91.** $x^2 - 3 = x^2 - \left(\sqrt{3}\right)^2 = \left(x + \sqrt{3}\right)\left(x - \sqrt{3}\right)$

**93.** $a^n - b^n = (a - b)(a^{n-1}b^0 + a^{n-2}b^1$
$+ a^{n-3}b^2 + \ldots + a^2 b^{n-3} + a^1 b^{n-2} + a^0 b^{n-1})$

**95.** A "rational expression" is any fraction with a polynomial numerator and a non-zero polynomial denominator;
e.g., $\dfrac{3x + y}{2x}$, $x \neq 0$.

**97.** $\dfrac{2a - 3b}{3b - 2a} = \dfrac{-1(3b - 2a)}{3b - 2a} = -1$

**99.** $\dfrac{x^2 + 2x}{x - 3} \cdot \dfrac{3 + 2x - x^2}{x}$
$= \dfrac{x(x + 2)(3 - x)(1 + x)}{(x - 3)x}$
$= -1(x + 2)(x + 1)$

**101.** $\dfrac{4}{x^2 - x} + \dfrac{3}{1 - x^2} = \dfrac{4}{x(x - 1)} - \dfrac{3}{(x + 1)(x - 1)}$
$= \dfrac{4x + 4 - 3x}{x(x + 1)(x - 1)} = \dfrac{x + 4}{x(x + 1)(x - 1)}$

**103.**

$$2x - 1 \overline{\smash{\big)}\, 2x^3 + x^2 - 5x + 2} \qquad x^2 + x - 2$$

$$\begin{array}{r} x^2 + x - 2 \\ \underline{-(2x^3 - x^2)} \\ 2x^2 - 5x \\ \underline{-(2x^2 - x)} \\ -4x + 2 \\ \underline{-(-4x + 2)} \\ 0 \end{array}$$

Check:
$(2x - 1)(x^2 + x - 2)$
$= 2x^3 + 2x^2 - 4x - x^2 - x + 2$
$= 2x^3 + x^2 - 5x + 2$ ✓

**105.** $\dfrac{\frac{1}{2+h} - \frac{1}{2}}{\frac{h}{1}} = \dfrac{\frac{2 - (2 + h)}{2(2 + h)}}{\frac{h}{1}} = \dfrac{-h}{2h(2 + h)}$
$= \dfrac{-1}{2(2 + h)}$

**107.** $i = \sqrt{-1}$

**109.** (a) $\sqrt{-9} + \sqrt{-36} = 3i + 6i = 9i$
(b) $\sqrt{-9} \cdot \sqrt{-36} = 3i \cdot 6i = 18i^2 = -18$

**111.** $\dfrac{3 + 2i}{2 + i}\left(\dfrac{2 - i}{2 - i}\right) = \dfrac{6 + i - 2i^2}{4 - i^2} = \dfrac{8 + i}{5}$
$= \dfrac{8}{5} + \dfrac{1}{5}i$

**113.** $(9 - 5i) - (2 + 3i) = (9 - 2) + (-5i - 3i)$
$= 7 - 8i$

**115.** $5i^{-3} = \dfrac{5}{i^3} = \dfrac{5}{-i} = \dfrac{5}{-i}\left(\dfrac{i}{i}\right) = \dfrac{-5i}{i^2} = 5i$

**117.** $(a + bi)(c + di) = ac + adi + bci + bdi^2$
$= (ac - bd) + (ad + bc)i$

## Chapter 1 Test: Standard Answer

**1.** (a) False
(b) True
(c) False
(d) True

**(e)** False
**(f)** False
**(g)** True
**(h)** False

**2.** $\dfrac{-7+x}{2} = -2 \Rightarrow -7 + x = -4$

$$x = 3$$

**3.** $\dfrac{2}{3}(x-3)+1 = 2x+3$

$\dfrac{2}{3}x - 2 + 1 = 2x + 3$

$\dfrac{2}{3}x - 1 = 2x + 3$

$-4 = \dfrac{4}{3}x$

$-3 = x$

**4.** Let $w$ = the rectangle's width in inches,
$2w + 5$ = the rectangle's length in inches.
$P = 2w + 2(2w + 5) = 52$
$2w + 4w + 10 = 52$
$6w = 42$
$w = 7$ inches;
the length = $2(7) + 5 = 19$ inches

**5.** Let $t$ = time in hours traveled by the first car
$55(t) + 45(t - 1) = 200$
$55t + 45t - 45 = 200 \Rightarrow 100t = 245$
$t = 2.45$ hours

$\dfrac{45}{100} = \dfrac{x}{60}$

$100x = 2700$
$x = 27$ minutes
At 2:27 pm, the cars will be 200 miles apart.

**6.** **(a)** $375,000,000 = 3.75 \times 10^8$
**(b)** $0.0000318 = 3.18 \times 10^{-5}$

**7.** $\dfrac{|x+2|}{x+2} = -1$

If $x = -2$, $\dfrac{|x+2|}{x+2}$ is undefined.

If $x > -2$, $\dfrac{|x+2|}{x+2} = \dfrac{x+2}{x+2} = 1$

If $x < -2$, $\dfrac{|x+2|}{x+2} = \dfrac{-(x+2)}{x+2} = -1$

$\{x \mid x < -2\}$

**8.** $|x + 2| < 1 \Rightarrow -1 < x + 2 < 1$
$-3 < x < -1$
$\{x \mid -3 < x < -1\}$

**9.** **(a)** $2(5x - 1) < x$
$10x - 2 < x$
$9x < 2$
$x < \dfrac{2}{9}$

**(b)** $|2x - 1| \geq 3 \Rightarrow$
$2x - 1 \geq 3$ or $2x - 1 \leq -3$
$2x \geq 4$ or $2x \leq -2$
$x \geq 2$ or $x \leq -1$

**10.** **(a)** $\dfrac{x^3(-x)^2}{x^5} = \dfrac{x^3 x^2}{x^5} = \dfrac{x^5}{x^5} = 1 \neq x$
False.

**(b)** $\left(\dfrac{3}{2+a}\right)^{-1} = \left(\dfrac{2+a}{3}\right) = \dfrac{2}{3} + \dfrac{a}{3}$
True.

**(c)** $(-27)^{-1/3} = \dfrac{1}{(-27)^{1/3}} = \dfrac{1}{-3} \neq 3$
False.

**(d)** $(x+y)^{3/5} = \left(\sqrt[5]{x+y}\right)^3 \neq \left(\sqrt[3]{x+y}\right)^5$
False.

**(e)** $\sqrt{9x^2} = 3|x|$
True.

**(f)** $(8+a^3)^{1/3} \neq 8^{1/3} + (a^3)^{1/3} = 2 + a$
False.

**11.** **(a)** $\dfrac{(2x^3 y^{-2})^2}{x^{-2}y^3} = \dfrac{4x^6 y^{-4}}{x^{-2}y^3}$

$= 4x^8 y^{-7} = \dfrac{4x^8}{y^7}$

**(b)** $\dfrac{(3x^2y^{-3})^{-1}}{(2x^{-2}y^2)^{-2}} = \dfrac{3^{-1}x^{-2}y^3}{2^{-2}x^4y^{-4}}$

$= \dfrac{2^2y^3y^4}{3^1x^2x^4} = \dfrac{4y^7}{3x^6}$

**12. (a)** $\sqrt{8}\cdot\sqrt{6} = \sqrt{48} = 2\sqrt{12} = 4\sqrt{3}$

**(b)** $\dfrac{\sqrt{360}}{2\sqrt{2}} = \dfrac{6\sqrt{10}}{2\sqrt{2}} = \dfrac{3\sqrt{20}}{2} = \dfrac{6\sqrt{5}}{2} = 3\sqrt{5}$

**(c)** $\dfrac{\sqrt[3]{-243x^8}}{\sqrt[3]{3x^2}} = \dfrac{\sqrt[3]{-27x^6}\cdot\sqrt[3]{9x^2}}{\sqrt[3]{3x^2}}$

$= \dfrac{-3x^2\sqrt[3]{3^2\,x^2}}{\sqrt[3]{3x^2}} = \dfrac{-3x^2(3)^{2/3}x^{2/3}}{3^{1/3}x^{2/3}}$

$= -3x^2(3)^{1/3} = -3x^2\sqrt[3]{3}$

**13. (a)** $\sqrt{50} + 3\sqrt{18} - 2\sqrt{8}$

$= 5\sqrt{2} + 9\sqrt{2} - 4\sqrt{2} = 10\sqrt{2}$

**(b)** $\dfrac{12}{\sqrt{3}} + 2\sqrt{3} = \dfrac{12\sqrt{3}}{3} + 2\sqrt{3}$

$= 4\sqrt{3} + 2\sqrt{3} = 6\sqrt{3}$

**14.** $5\sqrt{3x^2} + 2\sqrt{27x^2} - 3\sqrt{48x^2}$

$= 5|x|\sqrt{3} + 6|x|\sqrt{3} - 12|x|\sqrt{3} = -|x|\sqrt{3}$

**15.** $(x^2 + 3x)(3x^2) + (x^3 - 1)(2x + 3)$

$= 3x^4 + 9x^3 + 2x^4 + 3x^3 - 2x - 3$

$= 5x^4 + 12x^3 - 2x - 3$

**16.**

$$
\begin{array}{r}
3x^2 - x + 2 \\
2x+1\overline{\smash{\big)}\,6x^3 + x^2 + 3x + 2} \\
\underline{-(6x^3 + 3x^2)} \\
-2x^2 + 3x \\
\underline{-(-2x^2 - x)} \\
4x + 2 \\
\underline{-(4x + 2)} \\
0
\end{array}
$$

$3x^2 - x + 2$

**17.** $64 - 27b^3 = 4^3 - (3b)^3$

$= (4 - 3b)(16 + 12b + 9b^2)$

**18.** $6x^2 - 7x - 3 = (3x + 1)(2x - 3)$

**19.** $2x^2 - 6xy - 3y^3 + xy^2$

$= (2x^2 + xy^2) - (6xy + 3y^3)$

$= x(2x + y^2) - 3y(2x + y^2)$

$= (x - 3y)(2x + y^2)$

**20.** $\dfrac{x^2 - 9}{x^3 + 4x^2 + 4x}\cdot\dfrac{2x^2 + 4x}{x^2 + 2x - 15}$

$= \dfrac{(x+3)(x-3)(2x)(x+2)}{(x)(x+2)(x+2)(x+5)(x-3)}$

$= \dfrac{2(x+3)}{(x+2)(x+5)}$

**21.** $\dfrac{x^3 + 8}{x^2 - 4x - 12} \div \dfrac{x^3 - 2x^2 + 4x}{x^3 - 6x^2}$

$= \dfrac{(x+2)(x^2 - 2x + 4)(x^2)(x-6)}{(x+2)(x-6)(x)(x^2 - 2x + 4)}$

$= \dfrac{x^2}{x} = x$

**22.** $\dfrac{\frac{1}{x^2} - \frac{1}{49}}{\frac{x-7}{1}} = \dfrac{\frac{49 - x^2}{49x^2}}{\frac{x-7}{1}} = \dfrac{\frac{-1(x^2 - 49)}{49x^2}}{\frac{x-7}{1}}$

$= \dfrac{-1(x-7)(x+7)}{49x^2(x-7)} = \dfrac{-1(x+7)}{49x^2} = -\dfrac{x+7}{49x^2}$

**23.** $\dfrac{1}{x+3} - \dfrac{2}{x^2 - 9} + \dfrac{x}{2x^2 + x - 15}$

$= \dfrac{1}{x+3} - \dfrac{2}{(x+3)(x-3)} + \dfrac{x}{(2x-5)(x+3)}$

$= \dfrac{(2x-5)(x-3) - 2(2x-5) + x(x-3)}{(x+3)(x-3)(2x-5)}$

$= \dfrac{2x^2 - 11x + 15 - 4x + 10 + x^2 - 3x}{(x+3)(x-3)(2x-5)}$

$= \dfrac{3x^2 - 18x + 25}{(x+3)(x-3)(2x-5)}$

**24.** $(3 + 7i)(5 - 4i) = 15 - 12i + 35i - 28i^2$

$= (15 + 28) + 23i = 43 + 23i$

**25.** $\dfrac{3+7i}{5-4i}\left(\dfrac{5+4i}{5+4i}\right) = \dfrac{15+12i+35i+28i^2}{25-16i^2}$

$= \dfrac{-13+47i}{41} = \dfrac{-13}{41} + \dfrac{47}{41}i$

## Chapter 1 Test: Multiple Choice

**1.** I and II are both true, but III is false — the rational numbers are just one subset of the real numbers represented by the number line.
The answer is (c)

**2.** The answer is (c)

**3.** The answer is (d)

**4.** $|2 - 7| = -(2 - 7) = 5$
The answer is (b)

**5.** The answer is (a)

**6.** The answer is (d)

**7.** $-3(x + 1) < 2x + 2$
$-3x - 3 < 2x + 2$
$-5x < 5$
$x > -1$
The answer is (b)

**8.** Let $x$ represent the width and $3x - 2$ represent the length.
$(3x - 2) + 5 = 5(x - 1)$
$3x + 3 = 5x - 5$
The answer is (c)

**9.** A negative exponent indicates the reciprocal of the expression to which the negative exponent applies.
The answer is (b)

**10.** The answer is (a)

**11.** $\dfrac{8}{\sqrt{5}-1}\left(\dfrac{\sqrt{5}+1}{\sqrt{5}+1}\right) = \dfrac{8(\sqrt{5}+1)}{5-1}$

$= \dfrac{8(\sqrt{5}+1)}{4} = 2(\sqrt{5}+1)$
The answer is (c)

**12.** I is false — $2\sqrt{x+1} = \sqrt{4x+4}$ and III is false. II is a true statement.
The answer is (b)

**13.** $x^2(x - 1) - 2x(x - 1) + (x - 1)$
$= (x - 1)(x^2 - 2x + 1)$
$= (x - 1)(x - 1)^2 = (x - 1)^3$
The answer is (d)

**14.** $x^3 - y^3 = (x - y)(x^2 + xy + y^2)$,
so I is false.
$x^2 + y^2$ is not factorable, so II is false.
$(2x - 3y)^2 = 4x^2 - 12xy + 9y^2$,
so III is false.
The answer is (e)

**15.** The answer is (d)

**16.**

(b) would exclude $3 \le x < 7$, all of which are within five units of 2.
(c) would exclude $-3 < x \le 3$, all of which are within five units of 2.
(d) excludes *all* values within five units of 2.
Consider (a).
$|x - 2| < 5 \Rightarrow x - 2 < 5$ and $x - 2 > -5$
$x < 7$ and $x > -3$
This describes the real numbers isolated in the above picture.
The answer is (a)

**17.** The answer is (b)

**18.**

$|x - 2| \le 4 \Rightarrow x - 2 \le 4$ and $x - 2 \ge -4$
$x \le 6$ and $x \ge -2$
The answer is (c)

**19.** The answer is (c)

**20.** I is true — $\dfrac{1}{i} \cdot \dfrac{i}{i} = \dfrac{i}{i^2} = -i$.

II is true — *any* expression raised to a power of 0 is 1.

III is true —

$$\frac{1+i}{1-i}\left(\frac{1+i}{1+i}\right) = \frac{1+2i+i^2}{1-i^2}$$

$$= \frac{1+2i-1}{1+1} = \frac{2i}{2} = i$$

The answer is (d)

# Chapter 2: Linear and Quadratic Functions with Applications

**Exercises 2.1**

**1.** If $f(x) = -x^2 + 3$, does $f(3) = -6$?　　True
$f(3) = -3^2 + 3 = -9 + 3 = -6$

**3.** If $f(x) = -x^2 + 3$, does $3f(2) = -33$?　False
$f(2) = -2^2 + 3 = -4 + 3 = -1$
$3f(2) = 3(-1) = -3$

**5.** If $f(x) = -x^2 + 3$,
does $f(3) - f(2) = -5$?　　　　True
$f(3) = -3^2 + 3 = -9 + 3 = -6$ and
$f(2) = -2^2 + 3 = -4 + 3 = -1$
$f(3) - f(2) = -6 - (-1) = -6 + 1 = -5$

**7.** If $f(x) = -x^2 + 3$,
does $f(x) - f(4) = -(x-4)^2 + 3$?　False
$f(4) = -4^2 + 3 = -16 + 3 = -13$
$f(x) - f(4) = -x^2 + 3 - (-13) = -x^2 + 16$

**9.** If $f(x) = -x^2 + 3$,
does $f(4 + h) = -h^2 - 8h - 13$?　　True
$f(4 + h) = -(4 + h)^2 + 3$
$= -16 - 8h - h^2 + 3 = -h^2 - 8h - 13$

**11.** $y = x^3$ does define $y$ to be a function of $x$. The domain is the set of all real numbers.

**13.** $y = \dfrac{1}{\sqrt{x}} = x^{-1/2}$ does define $y$ to be a function of $x$. The domain is the set of all real numbers $x > 0$.

**15.** $y^2 = 2x$ does *not* define $y$ to be a function of $x$. For example, if $x = 8$, $y$ could equal 4 or $-4$.

**17.** $y = \dfrac{1}{x+1}$ does define $y$ to be a function of $x$. The domain is the set of all real numbers $x$ except $x \neq -1$.

**19.** $y = \dfrac{1}{1 \pm x}$ does not define $y$ to be a function of $x$. For example, if $x = 5$, $y = \dfrac{1}{6}$ and $-\dfrac{1}{4}$.

**21.** $f(x) = 2x - 1 \Rightarrow$
(a) $f(-1) = 2(-1) - 1 = -3$
(b) $f(0) = 2(0) - 1 = -1$
(c) $f\left(\dfrac{1}{2}\right) = 2\left(\dfrac{1}{2}\right) - 1 = 0$

**23.** $f(x) = x^2$
(a) $f(-1) = (-1)^2 = 1$
(b) $f(0) = (0)^2 = 0$
(c) $f\left(\dfrac{1}{2}\right) = \left(\dfrac{1}{2}\right)^2 = \dfrac{1}{4}$

**25.** $f(x) = x^3 - 1$
(a) $f(-1) = (-1)^3 - 1 = -1 - 1 = -2$
(b) $f(0) = (0)^3 - 1 = 0 - 1 = -1$
(c) $f\left(\dfrac{1}{2}\right) = \left(\dfrac{1}{2}\right)^3 - 1 = \dfrac{1}{8} - \dfrac{8}{8} = -\dfrac{7}{8}$

**27.** $f(x) = x^4 + x^2$
(a) $f(-1) = (-1)^4 + (-1)^2 = 1 + 1 = 2$
(b) $f(0) = 0^4 + 0^2 = 0 + 0 = 0$
(c) $f\left(\dfrac{1}{2}\right) = \left(\dfrac{1}{2}\right)^4 + \left(\dfrac{1}{2}\right)^2 = \dfrac{1}{16} + \dfrac{4}{16} = \dfrac{5}{16}$

**29.** $f(x) = \dfrac{1}{x-1}$
(a) $f(-1) = \dfrac{1}{-1-1} = \dfrac{1}{-2} = -\dfrac{1}{2}$
(b) $f(0) = \dfrac{1}{0-1} = \dfrac{1}{-1} = -1$
(c) $f\left(\dfrac{1}{2}\right) = \dfrac{1}{\frac{1}{2} - 1} = \dfrac{1}{-\frac{1}{2}} = -2$

**31.** $f(x) = \dfrac{1}{\sqrt[3]{x}}$

**(a)** $f(-1) = \dfrac{1}{(-1)^{1/3}} = \dfrac{1}{-1} = -1$

**(b)** $f(0) = \dfrac{1}{(0)^{1/3}} = \dfrac{1}{0}$ undefined at $x = 0$

**(c)** $f\left(\dfrac{1}{2}\right) = \dfrac{1}{\left(\frac{1}{2}\right)^{1/3}} = \sqrt[3]{2}$

**33.** $g(x) = x^2 - 2x + 1$

**(a)** $g(10) = 10^2 - 2(10) + 1$
$= 100 - 20 + 1 = 81$

**(b)** $5g(2) = 5[2^2 - 2(2) + 1] = 5[1] = 5$

**(c)** $g\left(\dfrac{1}{2}\right) + g\left(\dfrac{1}{3}\right)$

$= \left[\left(\dfrac{1}{2}\right)^2 - 2\left(\dfrac{1}{2}\right) + 1\right]$

$+ \left[\left(\dfrac{1}{3}\right)^2 - 2\left(\dfrac{1}{3}\right) + 1\right]$

$= \dfrac{1}{4} - 1 + 1 + \dfrac{1}{9} - \dfrac{2}{3} + 1$

$= \dfrac{9}{36} + \dfrac{4}{36} + \dfrac{12}{36} = \dfrac{25}{36}$

**(d)** $g\left(\dfrac{1}{2} + \dfrac{1}{3}\right) = g\left(\dfrac{5}{6}\right)$

$= \left[\left(\dfrac{5}{6}\right)^2 - 2\left(\dfrac{5}{6}\right) + 1\right]$

$= \dfrac{25}{36} - \dfrac{10}{6} + 1$

$= \dfrac{25}{36} - \dfrac{24}{36} = \dfrac{1}{36}$

**35.** $h(x) = x^2 + 2x$

**(a)** $3h(2) = 3[2^2 + 2(2)] = 3(8) = 24$

**(b)** $h(6) = 6^2 + 2(6) = 36 + 12 = 48$
Therefore, $3h(2) \ne h(3 \cdot 2)$

**37.** $y = \begin{cases} 2x - 1 \text{ if } x \le -2 \\ 1 - 2x \text{ if } x > 2 \end{cases}$

**(a)** If $x = -5$,
$y = 2(-5) - 1 = -10 - 1 = -11$

**(b)** If $x = -2$, $y = 2(-2) - 1 = -4 - 1 = -5$

**(c)** If $x = 0$, $y$ is not defined

**(d)** If $x = 2$, $y$ is not defined

**(e)** If $x = 5$, $y = 1 - 2(5) = 1 - 10 = -9$

**39.** Find $\dfrac{f(x) - f(3)}{x - 3}$ if $f(x) = x^2$

$\dfrac{x^2 - 9}{x - 3} = \dfrac{(x+3)(x-3)}{x-3} = x + 3, x \ne 3$

**41.** Find $\dfrac{f(x) - f(3)}{x - 3}$ if $f(x) = \dfrac{1}{x}$

$\dfrac{\frac{1}{x} - \frac{1}{3}}{\frac{x-3}{1}} = \dfrac{\frac{3-x}{3x}}{\frac{x-3}{1}} = \dfrac{-1(x-3)}{(x-3)(3x)} = \dfrac{-1}{3x}$

$x \ne 3, 0$

**43.** Find $\dfrac{f(x) - f(3)}{x - 3}$ if $f(x) = 2x + 1$

$\dfrac{(2x+1) - 7}{x - 3} = \dfrac{2x - 6}{x - 3} = \dfrac{2(x-3)}{x-3} = 2$

**45.** Find $\dfrac{f(2+h) - f(2)}{h}$ if $f(x) = x$

$\dfrac{(2+h) - 2}{h} = \dfrac{h}{h} = 1$

**47.** Find $\dfrac{f(2+h) - f(2)}{h}$ if $f(x) = -x^2$

$\dfrac{-(2+h)^2 - (-2^2)}{h} = \dfrac{-(4 + 4h + h^2) + 4}{h}$

$= \dfrac{-4 - 4h - h^2 + 4}{h} = \dfrac{-h^2 - 4h}{h}$

$= -h - 4$

**49.** Find $\dfrac{f(2+h) - f(2)}{h}$ if $f(x) = \dfrac{1}{x^2}$

$\dfrac{\frac{1}{(2+h)^2} - \frac{1}{4}}{h} = \dfrac{\frac{1}{4+4h+h^2} - \frac{1}{4}}{\frac{h}{1}}$

$$= \frac{\frac{4-(4+4h+h^2)}{4(4+4h+h^2)}}{\frac{h}{1}} = \frac{-h^2-4h}{4h(h^2+4h+4)}$$

$$= \frac{-h-4}{4(h^2+4h+4)} = -\frac{h+4}{4(h+2)^2}$$

**51.** $h(t) = -16t^2 + v_0t$. If $v_0 = 128$ ft/second,
$h(t) = -16t^2 + 128t$.
At time "0", $h(0) = -16(0)^2 + 128(0) = 0$
(The object hasn't been thrown yet.)
At time 8 seconds,
$h(8) = -16(8)^2 + 128(8)$
$= -1024 + 1024 = 0$
(The object has returned to the ground.)

**53. (a)** $\frac{x}{h} = \frac{2}{5} \Rightarrow 2h = 5x$

$h = \frac{5}{2}x \Rightarrow h(x) = \frac{5}{2}x$

**(b)** $A = \frac{1}{2}(5h) = \frac{5}{2}h = \frac{5}{2}\left(\frac{5}{2}x\right) = \frac{25}{4}x$

$\frac{25}{4}x = A(x)$

**55.**

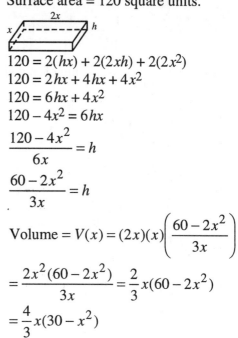

$\frac{s}{s+x} = \frac{6}{24} = \frac{1}{4}$

$4s = s + x$

$3s = x$

$s = \frac{1}{3}x \Rightarrow s(x) = \frac{1}{3}x$

**57.** $V(x) = (50-2x)(50-2x)(x)$
$= (x)(50-2x)^2$ centimeters$^3$.
[Volume $= L \cdot W \cdot H$. A length of $x$ is cut
from each corner of each side of the piece
of tin, thus the length and width are each
$(50-2x)$ centimeters. The height ($H$) is
the length $x$ cut from each side's corner.]

**59.** Width $= x$, length $= 2x$, altitude $= h$.
Surface area $= 120$ square units.

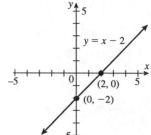

$120 = 2(hx) + 2(2xh) + 2(2x^2)$
$120 = 2hx + 4hx + 4x^2$
$120 = 6hx + 4x^2$
$120 - 4x^2 = 6hx$

$\frac{120-4x^2}{6x} = h$

$\frac{60-2x^2}{3x} = h$

Volume $= V(x) = (2x)(x)\left(\frac{60-2x^2}{3x}\right)$

$= \frac{2x^2(60-2x^2)}{3x} = \frac{2}{3}x(60-2x^2)$

$= \frac{4}{3}x(30-x^2)$

**Challenge Problem**

**1.** $f(x) = \sqrt{x^2+x-6}$
The domain is where $x^2 + x - 6 \geq 0$ or
$(x+3)(x-2) \geq 0$. This occurs when
$x+3 \geq 0$ and $x-2 \geq 0$ or $x+3 \leq 0$ and
$x-2 \leq 0$. In the first case, $x \geq 2$. In the
second case, $x \leq -3$. The domain is
$x \geq 2$ or $x \leq -3$.

**Exercises 2.2**

**1.** $y = x - 2$

| $x$ | $-3$ | $-2$ | $-1$ | $0$ | $1$ | $2$ |
|---|---|---|---|---|---|---|
| $y$ | $-5$ | $-4$ | $-3$ | $-2$ | $-1$ | $0$ |

**3.** $y = 2x - 4$

| $x$ | $-2$ | $-1$ | $0$ | $1$ | $2$ |
|---|---|---|---|---|---|
| $y$ | $-8$ | $-6$ | $-4$ | $-2$ | $0$ |

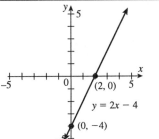

**5.** $x + 2y = 4$

$x$-intercept:  $x + 2(0) = 4$
$(4, 0)$        $x + 0 = 4$
                $x = 4$

$y$-intercept:  $0 + 2y = 4$
$(0, 2)$        $y = 2$

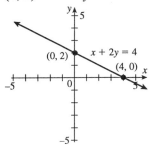

**7.** $x - 2y = 4$

$x$-intercept:  $x - 2(0) = 4$
$(4, 0)$        $x - 0 = 4$
                $x = 4$

$y$-intercept:  $0 - 2y = 4$
$(0, -2)$       $y = -2$

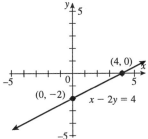

**9.** $2x - 3y = 6$

$x$-intercept:  $2x - 3(0) = 6$
$(3, 0)$        $2x = 6$
                $x = 3$

$y$-intercept:  $2(0) - 3y = 6$
$(0, -2)$       $-3y = 6$
                $y = -2$

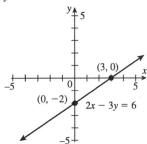

**11.** $y = -3x - 9$

$x$-intercept:  $0 = -3x - 9$
$(-3, 0)$       $9 = -3x$
                $-3 = x$

$y$-intercept:  $y = -3(0) - 9$
$(0, -9)$       $y = -9$

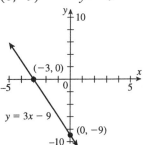

**13.** $y = 2x - 1$

$x$-intercept:  $0 = 2x - 1$
$\left( \dfrac{1}{2}, 0 \right)$        $1 = 2x$

                $\dfrac{1}{2} = x$

$y$-intercept:  $y = 2(0) - 1$
$(0, -1)$       $y = -1$

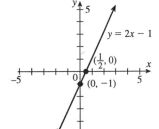

**15.**

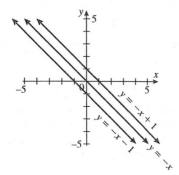

$y = -x + 1$
$y = -x$
$y = -x - 1$

**17.** $y = \frac{1}{2}x; -6 \le x \le 6$

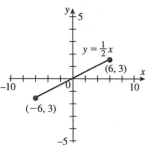

$y = \frac{1}{2}x$
$(6, 3)$
$(-6, 3)$

**19.** $y = 3x - 5; 1 \le x \le 4$

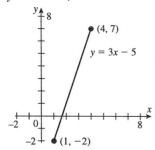

$(4, 7)$
$y = 3x - 5$
$(1, -2)$

**21. (a)** $A(-2, 3), C(2, 1)$

$$\text{Slope} = \frac{3 - 1}{-2 - 2} = \frac{2}{-4} = -\frac{1}{2}$$

**(b)** $B(0, 2), D(4, 0)$

$$\text{Slope} = \frac{2 - 0}{0 - 4} = \frac{2}{-4} = -\frac{1}{2}$$

**(c)** $C(2, 1), D(4, 0)$

$$\text{Slope} = \frac{1 - 0}{2 - 4} = \frac{1}{-2} = -\frac{1}{2}$$

**(d)** $A(-2, 3), E(6, -1)$

$$\text{Slope} = \frac{3 - (-1)}{-2 - 6} = \frac{4}{-8} = -\frac{1}{2}$$

**(e)** $B(0, 2), E(6, -1)$

$$\text{Slope} = \frac{2 - (-1)}{0 - 6} = \frac{3}{-6} = -\frac{1}{2}$$

**(f)** $C(2, 1), E(6, -1)$

$$\text{Slope} = \frac{1 - (-1)}{2 - 6} = \frac{2}{-4} = -\frac{1}{2}$$

**23.** $A(4, 3), B(-5, 2)$

$$\text{Slope } \overline{AB} = \frac{3 - 2}{4 - (-5)} = \frac{1}{9}$$

**25.** $A(6, -7), B(106, -7)$

$$\text{Slope } \overline{AB} = \frac{-7 - (-7)}{6 - 106} = \frac{0}{-100} = 0$$

**27.** $A\left(2, -\frac{3}{4}\right), B\left(-\frac{1}{3}, \frac{2}{3}\right)$

$$\text{Slope } AB = \frac{-\frac{9}{12} - \frac{8}{12}}{\frac{6}{3} - \left(-\frac{1}{3}\right)} = \frac{-\frac{17}{12}}{\frac{7}{3}}$$

$$= \frac{(-17)(3)}{(12)(7)} = \frac{-17}{28}$$

**29.** $A(0, 2); m = \frac{3}{4}$

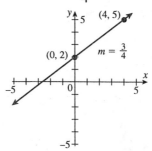

$(4, 5)$
$(0, 2)$
$m = \frac{3}{4}$

**31.** $(-3, 4); m = -\frac{1}{4}$

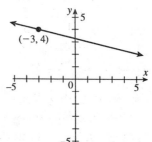

$(-3, 4)$

**33.** $\left(-2, \dfrac{3}{2}\right)$; $m = 0$

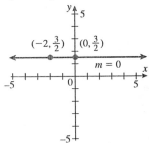

**35.**

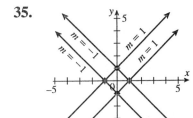

**37.** $A(6, -5)$, $B(8, -8)$

$\overline{AB}$ has slope $m = \dfrac{3}{-2} = -\dfrac{3}{2}$

$C(-3, 12)$, $D(1, 6)$

$\overline{CD}$ has slope $m_2 = \dfrac{6}{-4} = -\dfrac{3}{2}$

$\overline{AB}$ and $\overline{CD}$ are parallel because their slopes are equal.

**39.** $P(5, 11)$; $Q(-7, 16)$; $R(-12, 4)$; $S(0, -1)$

The slope of $\overline{PQ} = \dfrac{16 - 11}{-7 - 5} = -\dfrac{5}{12}.$

The slope of $\overline{QR} = \dfrac{16 - 4}{-7 - (-12)} = \dfrac{12}{5}.$

The slope of $\overline{RS} = \dfrac{-1 - 4}{0 - (-12)} = -\dfrac{5}{12}.$

The slope of $\overline{SP} = \dfrac{11 - (-1)}{5 - 0} = \dfrac{12}{5}.$

$\overline{PQ}$ is perpendicular to both $\overline{QR}$ and $\overline{SP}$ because the products of their slopes $= -1$. Similarly, $\overline{RS}$ is perpendicular to both $\overline{QR}$ and $\overline{SP}$.

The slope of diagonal $\overline{PR} = \dfrac{11 - 4}{5 - (-12)} =$

$\dfrac{7}{17}$, while the slope of diagonal

$\overline{QS} = \dfrac{16 - (-1)}{-7 - 0} = -\dfrac{17}{7}.$ $\overline{PR}$ is

perpendicular to $\overline{QS}$ because the product of their slopes equals $-1$.

**41.** The slope of any horizontal line is 0, which has no reciprocal. Further, the slope of any vertical line is not defined, so a comparison of slopes of horizontal and vertical lines can't be made.

**43.** $A(-1, 1)$, $B\left(1, \dfrac{1}{2}\right)$, $C\left(1, \dfrac{1}{2}\right)$, $D(7, t)$

If $\overline{AB} \perp \overline{CD}$, then $\dfrac{1 - \frac{1}{2}}{-1 - 1} \cdot \dfrac{t - \frac{1}{2}}{7 - 1} = -1$

$\dfrac{\frac{1}{2}\left(t - \frac{1}{2}\right)}{-2(6)} = -1$

$\dfrac{1}{2}\left(t - \dfrac{1}{2}\right) = 12$

$t - \dfrac{1}{2} = 24$

$t = 24\dfrac{1}{2} = \dfrac{49}{2}$

**45.**

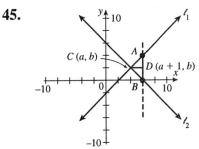

(a) Slope of $l_1 = m_1 = \dfrac{DA}{CD} = \dfrac{DA}{1} = DA$

(b) Slope of $l_2 = m_2 = \dfrac{BD}{DC} = \dfrac{BD}{1} = BD$, $m_2 < 0$.

(c) $\dfrac{BD}{CD} = \dfrac{CD}{AD}$

$\dfrac{BD}{1} = \dfrac{1}{AD}$ or $\dfrac{-m_2}{1} = \dfrac{1}{m_1}$

$-m_2 m_1 = 1$

$m_2 m_1 = -1$

**47.** $D = 5000 - 25x$
$S = 10x + 275$  When does $S = D$?
$10x + 275 = 5000 - 25x$
$35x = 4725$
$x = \$135$
When $x = \$135$,
$D = 5000 - 25(135) = 5000 - 3375$
$= 1625$ radios

**49.** The area of $\triangle OPQ = \frac{1}{2}(OP)(PQ)$

$OP = x$ and $\dfrac{OP}{OS} = \dfrac{PQ}{10} \Rightarrow \dfrac{x}{4} = \dfrac{PQ}{10}$

$4PQ = 10x$

$PQ = \dfrac{10}{4}x = \dfrac{5}{2}x$

Area of $\triangle OPQ = \dfrac{1}{2}(x)\left(\dfrac{5}{2}x\right) = \dfrac{5}{4}x^2$

**51.**

The sum of the areas:

$A(x) = \dfrac{x^2}{16} + \dfrac{(50 - x)^2}{4\pi}$

Each side of the square with perimeter $x$

has length $\dfrac{x}{4}$

The area of the square formed

$= \left(\dfrac{x}{4}\right)^2 = \dfrac{x^2}{16}.$

The circle formed by $PB$ has
circumference $50 - x$. So

$2\pi r = 50 - x$

$r = \dfrac{50 - x}{2\pi}$

The area of that circle

$= \pi \left(\dfrac{50 - x}{2\pi}\right)^2 = \dfrac{(50 - x)^2}{4\pi}$

## Exercises 2.3

**1.** $y = 2x + 3$

**3.** $y = x + 1$

**5.** $y = 5$

**7.** $y = \dfrac{1}{2}x + 3$

**9.** $y = \dfrac{1}{4}x - 2$

**11.** $y - 3 = x - 2$

**13.** $y - 3 = 4(x + 2)$

**15.** $y - 5 = 0$

**17.** $y - 1 = \dfrac{1}{2}(x - 2)$

**19.** $y = 5x$

**21.** $y + \sqrt{2} = 1(x - \sqrt{2}) = x - \sqrt{2}$

**23.** $3x + y = 4$  Slope $= m = -3$
$y = -3x + 4$  $y$-intercept $= (0, b) = (0, 4)$

**25.** $6x - 3y = 1$
$-3y = -6x + 1$
$y = 2x - \dfrac{1}{3}$
Slope $= m = 2$
$y$-intercept $= (0, b) = \left(0, -\dfrac{1}{3}\right)$

**27.** $3y - 5 = 0$  Slope $= m = 0$
$3y = 5$  $y$-intercept $= \left(0, \dfrac{5}{3}\right)$
$y = \dfrac{5}{3}$

**29.** $4x - 3y - 7 = 0$  Slope $= \dfrac{4}{3}$
$-3y = -4x + 7$  $y$-intercept $= \left(0, -\dfrac{7}{3}\right)$
$y = \dfrac{4}{3}x - \dfrac{7}{3}$

**31.** $\dfrac{1}{4}x - \dfrac{1}{2}y = 1$  Slope $= \dfrac{1}{2}$
$-\dfrac{1}{2}y = -\dfrac{1}{4}x + 1$  $y$-intercept $= (0, -2)$
$y = \dfrac{1}{2}x - 2$

**33.** $(2, 3), (3, 2)$  Slope $= m = \dfrac{3-2}{2-3} = \dfrac{1}{-1} = -1$

$y - 3 = -1(x - 2)$
$y - 3 = -x + 2$
$x + y = 5$

**35.** $(3, 0), (0, -3)$
Slope $= m = \dfrac{0 - (-3)}{3 - 0} = \dfrac{3}{3} = 1$
$y - 0 = 1(x - 3)$
$y = x - 3$
$x - y = 3$

**37.** $(-1, -13), (-8, 1)$
Slope $= m = \dfrac{-13 - 1}{-1 - (-8)} = \dfrac{-14}{7} = -2$
$y - 1 = -2(x + 8)$
$y - 1 = -2x - 16$
$2x + y = -15$

**39.** $(10, 27), (12, 27)$ makes a horizontal line.
Slope $= m = 0$
$y - 27 = 0(x - 10)$
$y - 27 = 0$
$y = 27$

**41.** The line parallel to the $x$-axis containing $(5, -7)$ has equation $y = -7$. The line parallel to the $y$-axis containing $(5, -7)$ has equation $x = 5$.

**43.** A line parallel to $2x + 3y = 6$ will have the same slope as $2x + 3y = 6$.
$2x + 3y = 6$
$3y = -2x + 6$
$y = -\dfrac{2}{3}x + 2$

The line containing $(1, -1)$ with slope $= -\dfrac{2}{3}$
$y + 1 = -\dfrac{2}{3}(x - 1)$
$y + 1 = -\dfrac{2}{3}x + \dfrac{2}{3}$
$y = -\dfrac{2}{3}x - \dfrac{1}{3}$

**45.** A line perpendicular to $y = 3x - 1$ will have a slope of $-\dfrac{1}{3}$. If it contains $(4, 7)$, its equation is

$y - 7 = -\dfrac{1}{3}(x - 4)$
$y - 7 = -\dfrac{1}{3}x + \dfrac{4}{3}$
$y = -\dfrac{1}{3}x + \dfrac{25}{3}$

**47.** A line perpendicular to $y - 2x = 5$ will have a slope of $-\dfrac{1}{2}$. If it contains $(-5, 1)$, its equation is

$y - 1 = -\dfrac{1}{2}(x + 5)$
$y - 1 = -\dfrac{1}{2}x - \dfrac{5}{2}$
$y = -\dfrac{1}{2}x - \dfrac{3}{2}$

**49.**

$AB$ has slope $= \dfrac{4}{2} = 2$ and equation
$y + 1 = 2(x + 1)$
$y = 2x + 1$

$BC$ has slope $= \dfrac{1}{-3} = -\dfrac{1}{3}$ and equation
$y - 2 = -\dfrac{1}{3}(x - 4)$
$y - 2 = -\dfrac{1}{3}x + \dfrac{4}{3}$
$y = -\dfrac{1}{3}x + \dfrac{10}{3}$

$CA$ has slope $= \dfrac{3}{5}$ and equation
$y + 1 = \dfrac{3}{5}(x + 1)$
$y = \dfrac{3}{5}x - \dfrac{2}{5}$

The altitude to $\overline{AC}$ will have slope
$= -\dfrac{5}{3}$ and contain $(1, 3)$. Its equation is
$y - 3 = -\dfrac{5}{3}(x - 1)$

$$y - 3 = -\frac{5}{3}x + \frac{5}{3}$$

$$y = -\frac{5}{3}x + \frac{14}{3}$$

The altitude to $\overline{CB}$ will have slope $= 3$ and contain $(-1, -1)$. Its equation is

$$y + 1 = 3(x + 1)$$

$$y = 3x + 2$$

The altitude to $\overline{BA}$ will have slope $= -\frac{1}{2}$ and contain $(4, 2)$. Its equation is

$$y - 2 = -\frac{1}{2}(x - 4)$$

$$y - 2 = -\frac{1}{2}x + 2$$

$$y = -\frac{1}{2}x + 4$$

**51.** The slope of the line formed by $(2, 2)$, $(5, -1)$ is $\frac{3}{-3} = -1$.

The slope of the line formed by $(5, 2)$, $(2, -1)$ is $\frac{3}{3} = 1$.

The slopes are negative reciprocals of each other, and the square's diagonals are perpendicular.

**53.** $y = f(x) = x + 3$

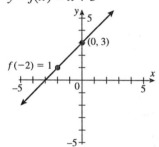

**55.** $y = f(x) = \frac{1}{2}x - 3$

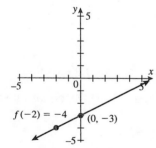

**57. (a)** If $c = 0$, then the origin $(0, 0)$ would satisfy the equation $ax + by = c$, and the line would contain the origin.

**(b)** $ax + by = c$ has an $x$-intercept $\left(\frac{c}{a}, 0\right)$ and a $y$-intercept of $\left(0, \frac{c}{b}\right)$.

**(c)** $ax + by = c$

$$\frac{ax}{c} + \frac{by}{c} = 1$$

$$\frac{a}{c}x + \frac{b}{c}y = 1$$

$$\frac{x}{\frac{c}{a}} + \frac{y}{\frac{c}{b}} = 1$$

$$\frac{x}{q} + \frac{y}{p} = 1$$

**(d)** $\left(\frac{3}{2}, 0\right)$, $(0, -5)$ are on the line described by the equation

$$\frac{x}{\frac{3}{2}} + \frac{y}{-5} = 1$$

$$\frac{2}{3}x - \frac{1}{5}y = 1$$

$$10x - 3y = 15$$

**(e)** Slope $= \frac{5}{\frac{3}{2}} = \frac{2}{3}(5) = \frac{10}{3}$

$$y = \frac{10}{3}x - 5$$

## Critical Thinking

**1.** $f(f(x)) = f(2x + 3) = 2(2x + 3) + 3$
$= 4x + 6 + 3 = 4x + 9$

**3.** In order for the line to pass through quadrants I, III, and IV, the line must be increasing, and have a $y$-intercept below the $x$-axis. Thus, $m > 0$ and $b < 0$.

## Exercises 2.4

**1.** $[99.1] = $ the greatest integer less than or equal to $99.1 = 99$.

**3.** $[-99.1] = $ the greatest integer less than or equal to $-99.1 = -100$.

**5.** 
$$\left[-\frac{7}{2}\right] = -4$$

**7.** $[\sqrt{2}] = [1.414...] = 1$

**9.** $y = f(x) = |x - 1|$
Domain:   All reals
Range:      $y \geq 0$

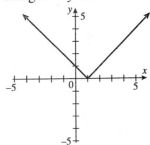

**11.** $y = f(x) = |2x|$
Domain:   All reals
Range:      $y \geq 0$

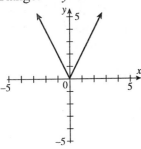

**13.** $y = f(x) = |3 - 2x|$
Domain:   All reals
Range:      $y \geq 0$

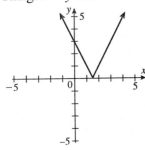

**15.** $y = \begin{cases} 3x \text{ if } -1 \leq x \leq 1 \\ -x \text{ if } 1 < x \end{cases}$
Domain:   $x \geq -1$
Range:      $y \leq 3$

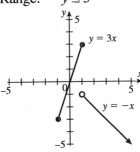

**17.** $y = \begin{cases} x \text{ if } -2 < x \leq 0 \\ 2x \text{ if } 0 < x \leq 2 \\ -x + 3 \text{ if } 2 < x \leq 3 \end{cases}$
Domain:   $-2 < x \leq 3$
Range:      $-2 < y \leq 4$

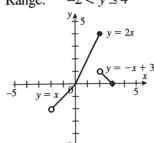

**19.** $y = \dfrac{x}{|x|}$ for all $x \neq 0$

$$y = \begin{cases} \dfrac{x}{-x} = -1 \text{ for all } x < 0 \\ \dfrac{x}{x} = 1 \text{ for all } x > 0 \end{cases}$$

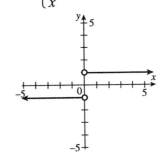

**21.** $y = \dfrac{x+3}{|x+3|}$ for all $x \neq -3$

$$y = \begin{cases} \dfrac{x+3}{-(x+3)} = -1 \text{ for all } x < -3 \\ \dfrac{x+3}{x+3} = 1 \text{ for all } x > -3 \end{cases}$$

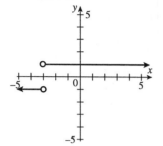

**23.** $y = [x]; -3 \leq x \leq 3$

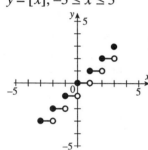

**25.** $y = 2[x]; 0 \leq x \leq 2$

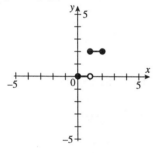

**27.** $y = [x-1]; -2 \leq x \leq 3$

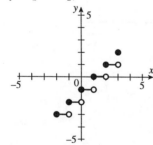

**29.** $y = 3x - 7$ for all $x \geq 0$

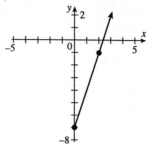

**31.** $y = 3x - 7$ for $x = -1, 0, 1, 2, 3$

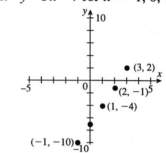

**33.** $x = -3$. This equation is not a function because there are infinitely many values $-4 \leq y \leq 4$ paired with $x = -3$.

**35.** $(-2, 0), (1, 3)$ establishes a line with slope $\dfrac{3-0}{1-(-2)} = \dfrac{3}{3} = 1$.

$y - 0 = 1(x+2) \Rightarrow y = x + 2 = f(x)$
Domain: $x \geq -2$
Range: $y \geq 0$

**37.** $y = \begin{cases} -1 \text{ for } -1 \leq x < 0 \\ 3 \text{ for } 0 \leq x < 3 \\ 1 \text{ for } 3 \leq x \leq 4 \end{cases}$

Domain: $-1 \leq x \leq 4$
Range: $\{-1, 1, 3\}$

**39.**

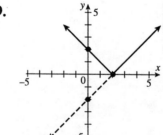

**41.** The least postage charged is 80 cents, which is charged for packages of weight up to but not including 1 pound. $10[x]$ is the amount in cents added to the 80-cent base postage.

Let $P(x)$ represent the cost in cents of postage for a package of weight $x$.

$P(x) = 80 + 10[x]$

$P(x) = 10(8 + [x])$ for $0 < x < 10$

If $P(x)$ represents the cost in dollars of postage for a package of weight $x$,

$P(x) = 0.80 + 0.10[x]$

$P(x) = \dfrac{1}{10}(8 + [x])$ for $0 < x < 10$

## Challenge Problems

**1.** $y = |[x]|$

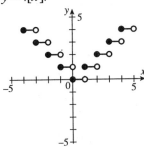

$y = [|x|]$

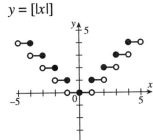

The two functions are equal when $x \geq 0$ and also when $x$ is a negative integer.

## Exercises 2.5

**1.** $4x - y = 6 \quad \Rightarrow \quad y = 4x - 6$

$2x + 3y = 10$

$2x + 3(4x - 6) = 10$

$14x - 18 = 10$

$14x = 28 \quad \Rightarrow \quad x = 2$

If $x = 2$ then

$4(2) - y = 6$

$8 - y = 6$

$2 = y$

Solution: $(2, 2)$

**3.** $s + 2t = 5 \quad \Rightarrow \quad s = 5 - 2t$

$-3(5 - 2t) + 10t = -7$

So $-15 + 6t + 10t = -7$

$16t = 8$

$t = \dfrac{1}{2}$

If $t = \dfrac{1}{2}$, then $s + 2\left(\dfrac{1}{2}\right) = 5$

$s = 4$

Solution: $\left(4, \dfrac{1}{2}\right)$

**5.** $4x - 5y = 3 \quad \Rightarrow \quad -16x + 20y = -12$

$16x + 2y = 3 \qquad \underline{\quad 16x + 2y = \phantom{-}3 \quad}$

$\phantom{16x + }22y = -9$

$y = -\dfrac{9}{22}$

If $y = -\dfrac{9}{22}$, $\quad 16x + 2\left(-\dfrac{9}{22}\right) = 3$

$16x - \dfrac{9}{11} = 3$

$16x = \dfrac{42}{11}$

$x = \left(\dfrac{42}{11}\right)\left(\dfrac{1}{16}\right)$

$x = \dfrac{21}{88}$

Solution: $\left(\dfrac{21}{88}, -\dfrac{9}{22}\right)$

**7.** $x - 2y = 3 \quad \Rightarrow \quad x = 2y + 3$

$y - 3x = -14 \qquad y - 3(2y + 3) = -14$

$y - 6y - 9 = -14$

$-5y = -5$

$y = 1$

If $y = 1, \quad x - 2(1) = 3$

$x - 2 = 3$

$x = 5$

Solution: $(5, 1)$

**9.** $-3x + 8y = 16 \implies -15x + 40y = 80$

$16x - 5y = 103$ $\quad\quad \underline{+128x - 40y = 824}$

$\quad\quad\quad\quad\quad\quad\quad\quad 113x \quad\quad = 904$

$\quad\quad\quad\quad\quad\quad\quad\quad\quad\; x \quad\quad = 8$

If $x = 8$, then $3(8) + 8y = 16$

$\quad\quad\quad\quad\quad 24 + 8y = 16$

$\quad\quad\quad\quad\quad\quad\; 8y = 40$

$\quad\quad\quad\quad\quad\quad\;\; y = 5$

Solution: $(8, 5)$

**11.** $16x - 5y = 103 \implies 112x - 35y = 721$

$7x + 19y = -188 \quad\quad \underline{112x + 304y = -3008}$

$\quad\quad\quad\quad\quad\quad\quad\quad\quad -339y = 3729$

$\quad\quad\quad\quad\quad\quad\quad\quad\quad\quad\;\; y = -11$

If $y = -11$, then $7x + 19(-11) = -188$

$\quad\quad\quad\quad\quad\quad 7x - 209 = -188$

$\quad\quad\quad\quad\quad\quad\quad\;\; 7x = 21$

$\quad\quad\quad\quad\quad\quad\quad\;\;\; x = 3$

Solution: $(3, -11)$

**13.** $3s + t - 3 = 0 \implies 9s + 3t - 9 = 0$

$2s - 3t - 2 = 0 \quad\quad \underline{+2s - 3t - 2 = 0}$

$\quad\quad\quad\quad\quad\quad\quad 11s \quad\quad - 11 = 0$

$\quad\quad\quad\quad\quad\quad\quad\quad\quad\quad 11s = 11$

$\quad\quad\quad\quad\quad\quad\quad\quad\quad\quad\;\; s = 1$

If $s = 1$, then $3(1) + t - 3 = 0$

$\quad\quad\quad\quad\quad\quad 3 + t - 3 = 0$

$\quad\quad\quad\quad\quad\quad\quad\quad\quad\; t = 0$

Solution: $(1, 0)$

**15.** $\dfrac{1}{4}x + \dfrac{1}{3}y = \dfrac{5}{12} \implies \dfrac{3}{4}x + y = \dfrac{15}{12}$

$\dfrac{1}{2}x + y = 1 \quad\quad\quad\quad -\left(\dfrac{1}{2}x + y = 1\right)$

$\quad\quad\quad\quad\quad\quad\quad\quad\quad\;\; \dfrac{1}{4}x \quad\quad = \dfrac{3}{12}$

$\quad\quad\quad\quad\quad\quad\quad\quad\quad\quad\;\; x \quad\quad = \dfrac{12}{12}$

$\quad\quad\quad\quad\quad\quad\quad\quad\quad\quad\;\; x \quad\quad = 1$

If $x = 1$, then $\dfrac{1}{2}(1) + y = 1$

$\quad\quad\quad\quad\quad\quad \dfrac{1}{2} + y = 1$

$\quad\quad\quad\quad\quad\quad\quad\quad\; y = \dfrac{1}{2}$

Solution: $\left(1, \dfrac{1}{2}\right)$

**17.** $\dfrac{x}{2} + \dfrac{y}{6} = \dfrac{1}{2} \implies 3x + y = 3 \implies 9x + 3y = 9$

$0.2x - 0.3y = 0.2 \quad\quad 2x - 3y = 2 \quad\quad \underline{+2x - 3y = 2}$

$\quad\quad\quad\quad\quad\quad\quad\quad\quad\quad\quad\quad\quad\quad\quad 11x \quad\quad = 11$

$\quad\quad\quad\quad\quad\quad\quad\quad\quad\quad\quad\quad\quad\quad\quad\;\; x \quad\quad = 1$

If $x = 1$, then $\dfrac{1}{2} + \dfrac{y}{6} = \dfrac{1}{2}$

$\quad\quad\quad\quad\quad\quad\quad\quad\; y = 0$

Solution: $(1, 0)$

**19.** $2(x - y - 1) = 1 - 2x$

$6(x - y) = 4 - 3(3y - x)$

| $2(x - y - 1) = 1 - 2x$ | $6(x - y) = 4 - 3(3y - x)$ |
|---|---|
| $2x - 2y - 2 = 1 - 2x$ | $6x - 6y = 4 - 9y + 3x$ |
| $4x - 2y = 3$ | $3x + 3y = 4$ |

If $x = \dfrac{17}{18}$, then $3\left(\dfrac{17}{18}\right) + 3y = 4$

$\quad\quad\quad\quad\quad\quad\;\; \dfrac{17}{6} + 3y = 4$

$\quad\quad\quad\quad\quad\quad\quad\quad\; 3y = \dfrac{7}{6}$

$\quad\quad\quad\quad\quad\quad\quad\quad\;\; y = \dfrac{7}{18}$

$$4x - 2y = 3 \implies 12x - 6y = 9$$
$$3x + 3y = 4 \qquad +6x + 6y = 8$$
$$\overline{\phantom{+6x}18x \qquad = 17} \implies x = \frac{17}{18}$$

Solution: $\left(\dfrac{17}{18}, \dfrac{7}{18}\right)$

**21.** $4x - 12y = 3 \implies 4x - 12y = 3$    Because a unique $x$-value exists, this system is *consistent*.
$$x + \frac{1}{3}y = 3 \qquad \frac{36x + 12y = 108}{40x \qquad = 111}$$
$$x \qquad = \frac{111}{40}$$

**23.** $2x + 5y = -20 \implies 2x + 5y = -20$    Because the equations in this system describe
$$x + \frac{5}{2}y = -10 \qquad \frac{-(2x + 5y = -20)}{0 = 0}$$
the same sets of values, this system is *dependent*.

**25.** $x - 5y = 15 \qquad \implies \qquad x - 5y = 15$    Because the equations in this system will never
$$0.01x - 0.05y = 0.5 \qquad \frac{-(x - 5y = 50)}{0 = -35}$$
share sets of values, this system is *inconsistent*.

**27.** $-x + y = 6c$      If $y = \dfrac{9}{2}c$
$$\frac{+x + y = 3c}{2y = 9c} \qquad \text{then } x + \frac{9}{2}c = 3c$$
$$y = \frac{9}{2}c \qquad\qquad x = -\frac{3}{2}c$$

Solution: $\left(-\dfrac{3}{2}c, \ \dfrac{9}{2}c\right)$

**29.**

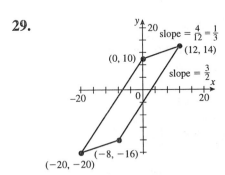

$-16 - y = 4$ and $-8 - x = 12$
$$-y = 20 \ \Big| \ -x = 20$$
$$y = -20 \ \Big| \ x = -20$$

By simultaneous equations:

$$y - 10 = \frac{3}{2}x \text{ and } y + 16 = \frac{1}{3}(x + 8)$$
$$y = \frac{3}{2}x + 10 \text{ and } y + 16 = \frac{1}{3}x + \frac{8}{3}$$

$$\left(\frac{3}{2}x + 10\right) + 16 = \frac{1}{3}x + \frac{8}{3}$$

If $x = -20$, then $y - 10 = \dfrac{3}{2}(-20)$

$$\frac{3}{2}x + 26 = \frac{1}{3}x + \frac{8}{3}$$
$$9x + 156 = 2x + 16$$

$$y - 10 = -30$$
$$y = -20$$

$$7x = -140$$
$$x = -20$$
Solution: $(-20, -20)$

**31.** $(6, 3)$ and $(-1, -1)$ lie on $ax + by = 3 \Rightarrow 6a + 3b = 3$ and $-a - b = 3$

$6a + 3b = 3 \Rightarrow \quad 6a + 3b = 3$     If $a = 4$, then $-4 - b = 3$

$-a - b = 3 \quad\quad \underline{+(-3a - 3b = 9)} \quad -b = 7$

$\qquad\qquad\qquad 3a \quad\quad\; = 12 \quad\quad b = -7$

$\qquad\qquad\qquad\; a \quad\quad = 4 \quad\quad$ Solution: $a = 4$, $b = -7$

**33.** Let $g$ = number of field goals made

$t$ = number of free throws made

$\left.\begin{array}{l} 2g + t = 96 \\ g = \dfrac{5}{2}t \end{array}\right\} \Rightarrow 2\left(\dfrac{5}{2}t\right) + t = 96 \qquad$ If $t = 16$, then $g = \dfrac{5}{2}(16)$

$\qquad\qquad\qquad\quad 5t + t = 96 \qquad\qquad\qquad g = 40$

$\qquad\qquad\qquad\quad\; 6t = 96 \qquad$ Solution: 40 field goals; 16 free throws.

$\qquad\qquad\qquad\quad\;\; t = 16$

**35.** Let $l$ = rectangle's length in centimeters, $w$ = rectangle's width in centimeters.

$2l + 2w = 60 \Rightarrow l + w = 30 \qquad$ If $w = 9$, then $l + 9 = 30$

$l = 2w + 3 \quad \Rightarrow \underline{-(l - 2w = 3)} \qquad\qquad l = 21$

$\qquad\qquad\qquad\qquad 3w = 27 \qquad$ Solution: length is 21 centimeters; width is 9 centimeters.

$\qquad\qquad\qquad\qquad\; w = 9$

**37.** Let $p$ = amount in pounds of potatoes, $s$ = amount in pounds of string beans.

$p + s = \dfrac{19}{2} \qquad \Rightarrow \quad s = \dfrac{19}{2} - p \qquad\qquad$ If $p = 8$, $8 + s = \dfrac{19}{2}$

$35p + 68s = 382 \qquad 35p + 68\left(\dfrac{19}{2} - p\right) = 382 \qquad\qquad s = \dfrac{3}{2}$

$\qquad\qquad\qquad\quad 35p + 646 - 68p = 382 \qquad$ Solution: 8 pounds of potatoes

$\qquad\qquad\qquad\qquad\quad -33p = -264 \qquad\qquad\qquad$ and $1\dfrac{1}{2}$ pounds of string

$\qquad\qquad\qquad\qquad\qquad\quad p = 8 \qquad\qquad\qquad\qquad$ beans purchased.

**39.** Let $t$ = cost of tuition, $r$ = cost of room and board.

$t + r = 8400 \quad \Rightarrow r + t = 8400 \qquad$ If $t = 5200$, then $r + 5200 = 8400$

$r = \dfrac{1}{2}t + 600 \qquad r - \dfrac{1}{2}t = 600 \qquad\qquad\qquad r = 3200$

$\qquad\qquad\qquad \underline{\qquad\qquad\qquad} \qquad$ Solution: \$5200 tuition; \$3200 for room and board

$\qquad\qquad\qquad\quad \dfrac{3}{2}t = 7800$

$\qquad\qquad\qquad\qquad t = 5200$

**41.** Let $n$ = number of nickels, $q$ = number of quarters.

$n + q = 22 \qquad \Rightarrow 5n + 5q = 110$

$5n + 25q = 390 \quad \underline{-(5n + 25q = 390)} \qquad$ If $q = 14$, then $n + 14 = 22$

$\qquad\qquad\qquad\qquad -20q = -280 \qquad\qquad\qquad n = 8$

$\qquad\qquad\qquad\qquad\quad\; q = 14 \qquad$ Solution: 14 quarters, 8 nickels

**43.** Let $n$ = number of attending non students, $s$ = number of attending students.

$$n + s = 560 \qquad \Rightarrow \quad n = 560 - s$$

$$225n + 125s = 91600 \qquad 225(560 - s) + 125s = 91600$$

$$126000 - 225s + 125s = 91600$$

$$126000 - 100s = 91600$$

$$-100s = -34400$$

$$s = 344 \text{ students attending}$$

**45.** Let $f$ = time traveled at 5 kph, $t$ = time traveled at 3 kph.

$$5f + 3t = 19 \Rightarrow 25f + 15t = 95 \qquad \text{If } f = 2\frac{3}{4}, \text{ then } 5\left(\frac{11}{4}\right) + 3t = 19$$

$$3f + 5t = 17 \qquad \underline{\quad 9f + 15t = 51 \quad}$$

$$16f \qquad = 44 \qquad \frac{55}{4} + 3t = 19$$

$$f \qquad = 2\frac{3}{4} \qquad 3t = \frac{21}{4}$$

$$t = \frac{7}{4} = 1\frac{3}{4}$$

Solution: total time traveled $= f + t = 2\frac{3}{4} + 1\frac{3}{4} = 4\frac{1}{2}$ hours.

**47.** Let $x$ = number of 25¢ letters, $y$ = number of 45¢ letters.

$$x + y = 910 \qquad \Rightarrow \quad 25x + 25y = 22750 \qquad \text{If } y = 360, \; x + 360 = 910$$

$$25x + 45y = 29950 \qquad \underline{-(25x + 45y = 29950)} \qquad x = 550$$

$$-20y = -7200 \qquad \text{Solution: 550 25¢ letters,}$$

$$y = \quad 360 \qquad\qquad\qquad 360 \text{ 45¢ letters}$$

**49.** Let $x$ = amount invested at 8%, $y$ = amount invested at $7\frac{1}{2}$%.

$$x + y = 6000 \qquad \Rightarrow \quad 0.08x + 0.080y = \quad 480 \qquad \text{If } y = 3200, \text{ then } x + 3200 = 6000$$

$$0.08x + 0.075y = 464 \qquad \underline{-(0.08x + 0.075y = \quad 464)} \qquad\qquad x = 2800$$

$$0.005y = \quad 16 \qquad \text{Solution: \$2800 at 8\%, \$3200 at } 7\frac{1}{2}\%.$$

$$y = 3200$$

**51.** Let $a$ = amount of Solution A, $b$ = amount of Solution B.

$$a + b = 32 \qquad\qquad \Rightarrow \quad 0.42a + 0.42b = 13.44 \qquad \text{If } b = 16, \text{ then } a + 16 = 32$$

$$0.42a + 0.18b = 0.30(32) \qquad \underline{-(0.42a + 0.18b = \quad 9.6)} \qquad\qquad a = 16$$

$$0.42a + 0.18b = 9.6 \qquad\qquad 0.24b = \quad 3.84 \qquad \text{Solution: 16 milliliters each}$$

$$b = 16 \qquad\qquad\qquad\qquad \text{of A and B.}$$

**53.** Let $c$ = time spent in the car, $t$ = time spent in the train.

$$c + t = 82 \text{ minutes}$$

$$\text{Distance traveled by car} = \left(\frac{36}{60} \text{ miles per minute}\right)(c) = \frac{3}{5}c \text{ miles}$$

Distance traveled by train $= \left(\dfrac{60}{60} \text{ miles per minute}\right)(t) = 1t \text{ miles}$

$c + t = 82 \qquad\qquad \Rightarrow \quad 9c + 9t = 738 \qquad\qquad$ If $t = 72$, then $c + 72 = 82$

$15\left(\dfrac{3}{5}c\right) + 6t = 522 \quad \underline{-(9c + 6t = 522)} \qquad\qquad\qquad c = 10$

$$3t = 216$$
$$t = 72$$

Solution: Distance traveled by car $= \dfrac{3}{5}(10) = 6$ miles, distance traveled by train $= 72$ miles.

**55.** Let $x =$ the larger number, $y =$ the smaller number.
Since these 2 equations are describing
the same sets of values, there are
infinitely many solutions; the system
is dependent.

**57.** If the clerk were right, then the system $\quad 6x + 12y = 234$
$\qquad\qquad\qquad\qquad\qquad\qquad\qquad\qquad\qquad\quad 2x + 4y = 77$
would be a dependent system. The system is not dependent because the second equation is
equivalent to $6x + 12y = 231$. This indicates that the smaller bag is the better buy.

**59.** From example 5, it is seen that the break-even point is at the production of 25 skateboards. The
profit will be based on the cost and revenue of $40 - 25 = 15$ skateboards.
Profit = Revenue − Cost $= 16(15) - 8(15) = 8(15) = \$120$.

**61.** If profit is to equal $\$520$, then $\quad 16(x - 25) - 8(x - 25) = 520$
$$8(x - 25) = 520$$
$$8x - 200 = 520$$
$$8x = 720$$
$$x = 90$$
Solution: 90 skateboards must be made and sold per day for a $\$520$ profit.

**63.** Let $x =$ number of units to sell in order to break even,
$y =$ total revenue needed in order to break even.
At what point will the Company's revenue cover its costs?
$12x + 720 = 20x = y \qquad\qquad$ If $x = 90$, then $y = 20(90)$
$720 = 8x \qquad\qquad\qquad\qquad\qquad\qquad y = 1800$
$90 = x \qquad\qquad\qquad\qquad$ The break-even point is $(90, 1800)$.

**65.** After 90 units are produced, the fixed costs are covered.
Profit = Revenue − Cost
$200 = 20(x - 90) - 12(x - 90)$, where $x =$ number of units sold
$200 = 8(x - 90)$
$200 = 8x - 720$
$920 = 8x$
$115 = x$
Solution: 115 units must be made and sold to net a profit of $\$200$.

**Challenge Problem**

$$-d(ax+by=c) \Rightarrow -adx-bdy=-cd$$
$$a(dx+ey=f) \Rightarrow \underline{adx+aey= af}$$
$$(ae-bd)y=af-cd$$
$$y=\frac{af-cd}{ae-bd}$$

The solution is: $x=\dfrac{bf-ec}{bd-ae}$, $y=\dfrac{af-cd}{ae-bd}$.

$$-e(ax+by=c) \Rightarrow -aex-eby=-ec$$
$$b(dx+ey=f) \Rightarrow \underline{bdx+eby= bf}$$
$$(bd-ae)x=bf-ec$$
$$x=\frac{bf-ec}{bd-ae}$$

## Exercises 2.6

**1.**

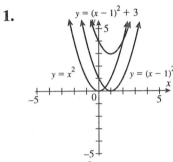

(a) $y=x^2$

(b) $y=(x-1)^2$
Shift $y=x^2$ one unit to the right.

(c) $y=(x-1)^2+3$
Shift $y=(x-1)^2$ three units upward.

**3.**

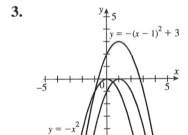

(a) $y=-x^2$

(b) $y=-(x-1)^2$
Shift $y=-x^2$ one unit to the right.

(c) $y=-(x-1)^2+3$
Shift $y=-(x-1)^2$ 3 units upward.

**5.**

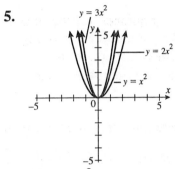

(a) $y=x^2$

(b) $y=2x^2$
Each $y$-value is two times what it would be for $y=x^2$.

(c) $y=3x^2$. Each $y$-value is three times what it would be for $y=x^2$.

**7.**

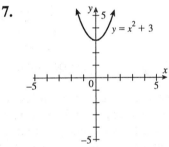

Shift $y=x^2$ up 3 units

**9.**

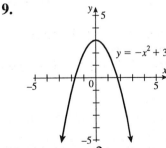

Invert $y=x^2$; then shift up 3 units.

**11.**

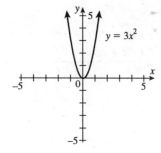

$y = 3x^2$

**13.**

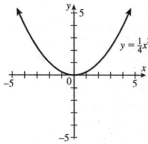

$y = \frac{1}{4}x^2$

Each $y$-value is $\frac{1}{4}$ what it would be for $y = x^2$.

**15.**

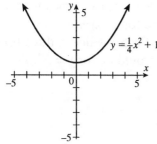

$y = \frac{1}{4}x^2 + 1$

Shift $y = \frac{1}{4}x^2$ (see #13) upward one unit.

**17.**

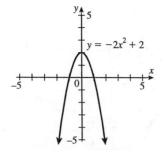

$y = -2x^2 + 2$

**19.** $f(x) = (x - 1)^2 + 2$
(1) Shift $f(x) = x^2$ to the right one unit to obtain $f(x) = (x - 1)^2$.
(2) Shift $f(x) = (x - 1)^2$ upward two units to obtain $f(x) = (x - 1)^2 + 2$.

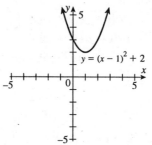

$y = (x - 1)^2 + 2$

The function is decreasing ($y$-values are getting smaller) on $(-\infty, 1]$.
The function is increasing ($y$-values are getting larger) on $[1, +\infty)$.
The function is concave up.

**21.** $f(x) = -(x + 1)^2 + 2$
The graph of $f(x) - x^2$ is shifted to the left one unit and then upward two units to obtain $f(x) = -(x + 1)^2 + 2$

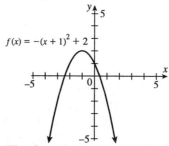

$f(x) = -(x + 1)^2 + 2$

The function increases from $(-\infty, -1]$ and decreases from $[-1, +\infty)$.
The function is concave down.

**23.** $f(x) = 2(x - 3)^2 - 1$

The graph of $f(x) = 2x^2$ is shifted to the right three units and down one unit.

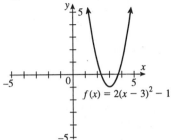

The function decreases on $(-\infty, 3]$ and increases on $[3, +\infty)$.
The function "bends" concave upward.

**25.** $y = f(x) = (x - 3)^2 + 5$

**(a)** The vertex may be found by shifting the vertex of $y = x^2$ $(0, 0)$ three units to the right then five units upward.

$(0, 0) \Rightarrow (0 + 3, 0 + 5) \Rightarrow (3, 5)$

**(b)** The axis of symmetry is the vertical line containing the vertex, and its equation is $x = 3$.

**(c)** Since $f(x)$ is a "shift" of $y = x^2$, its domain is all real numbers, just like $y = x^2$.

**(d)** The range is the set of all possible values obtained when $f(x)$ is evaluated for all possible $x$. Since $(x - 3)^2$ will never fall below 0, $(x - 3)^2 + 5$ will never fall below 5. The range of $f(x)$ is $f(x) = y \geq 5$.

**27.** $y = f(x) = -(x - 3)^2 + 5$

**(a)** The vertex of $y = -x^2$ is $(0, 0)$. Shift this vertex to the right three units and upward five units, and the vertex of $f(x) = -(x + 3)^2 + 5$ is obtained.

$(0 + 3, 0 + 5) \Rightarrow (3, 5)$

**(b)** The axis of symmetry is a vertical line containing the vertex: $x = 3$.

**(c)** The domain is all real numbers, since $f(x)$ may be found for any real number.

**(d)** The range of values will include no values for $f(x)$ greater than 5, regardless of $x$. So the range is $f(x) = y \leq 5$.

**29.** $y = f(x) = 2(x + 1)^2 - 3$

**(a)** The vertex of $y = 2x^2$ is $(0, 0)$. Shift this vertex one unit to the left and three units downward to obtain the vertex of $f(x) = 2(x + 1)^2 - 3$.

$(0 - 1, 0 - 3) \Rightarrow (-1, -3)$

**(b)** The axis of symmetry is $x = -1$.

**(c)** The domain is all real numbers.

**(d)** The range, the set of all values obtained when $f(x)$ is evaluated for all $x$, will never include values less than $-3$. So the range is $f(x) = y \geq -3$.

**31.** $y = f(x) = -2(x - 1)^2 + 2$

**(a)** $(0, 0)$ shifted one unit right, two units up yields a vertex for $f(x)$ of $(1, 2)$.

**(b)** The axis of symmetry is $x = 1$.

**(c)** The domain is all real numbers.

**(d)** The range will include no values greater than 2, so the range is $f(x) = y \leq 2$.

**33.** $y = f(x) = \frac{1}{4}(x + 2)^2 - 4$

**(a)** $(0, 0)$ shifted two units left, four units down yields a vertex of $(-2, -4)$.

**(b)** The axis of symmetry is $x = -2$.

**(c)** The domain is all real numbers.

**(d)** The range is $f(x) = y \geq -4$.

**35.** $f(x) = \begin{cases} x^2 \text{ if } -2 \leq x \leq 1 \\ x \text{ if } 1 < x \leq 3 \end{cases}$

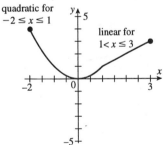

**37.** $x = y^2$. If $y$ were a function of $x$, each value $x$ would be paired with only one value $y$. The domain of $x = y^2$ is all $x \geq 0$, but for any $x > 0$ (for example, 25) two possible values for $y$ could be paired with $x$ (5 and $-5$). Thus, $y$ is not a function of $x$.

**39.**

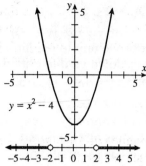

$$x^2 - 4 > 0 \Rightarrow (x+2)(x-2) > 0$$
$$x + 2 > 0 \text{ and } x - 2 > 0 \text{ or }$$
$$x + 2 < 0 \text{ and } x - 2 < 0$$

$$\underbrace{x > -2 \text{ and } x > 2}_{x > 2} \;\Big|\; \underbrace{x < -2 \text{ and } x < 2}_{x < -2}$$

$f(x)$ will take on positive values $[f(x) > 0]$
when $x > 2$ or when $x < -2$.

**41.** If $y = ax^2$ contains $(1, -2)$, then $-2 = a(1)^2$
$\Rightarrow -2 = a$.

**43.** If $y = (x-2)^2 + k$ contains $(5, 12)$, then
$$12 = (5-2)^2 + k$$
$$12 = 3^2 + k$$
$$12 = 9 + k$$
$$3 = k$$

**45.** $x = y^2 - 4$

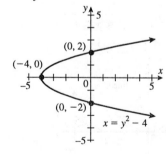

**47.** $x = (y-3)^2$
Shift the graph of $x = y^2$ up three units.

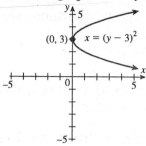

**49.** $x = 2y^2$
Each value of $x$ is twice that of $x = y^2$.

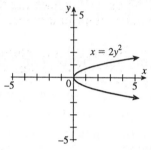

**51.**

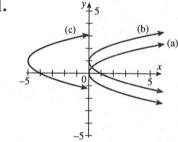

**(a)** $x = y^2$
**(b)** $x = (y-1)^2$ is a one unit shift upward
of $x = y^2$.
**(c)** $x = (y-1)^2 - 5$ is a five unit shift to
the left of $x = (y-1)^2$.

**53.** $y = x^2$ shifted to the right four units results
in $y = (x-4)^2$, then shifted up two units
results in $y = (x-4)^2 + 2$.

**55.** $x = y^2$ shifted up two units results in
$x = (y-2)^2$, then shifted to the left five
units results in $x = (y-2)^2 - 5$.

**57.** $y = \frac{1}{2}x^2$ shifted to the right three units

results in $y = \frac{1}{2}(x-3)^2$, then shifted down

two units results in $y = \frac{1}{2}(x-3)^2 - 2$.

**59.** Yes, because no vertical line will cross the graph more than once.

**61.** Yes, because no vertical line will cross the graph more than once.

**63.** No, because any vertical line (excluding tangents to the circle) intersecting the graph once must intersect it twice.

**65.** No, because any vertical line intersecting one side of the triangle will intersect the triangle in a second side.

**67.**

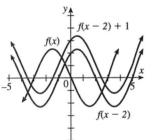

$f(x-2)$ shifts the entire graph of $f(x)$ two units to the right, and $f(x-2)+1$ shifts $f(x-2)$ upward one unit.

**69.**

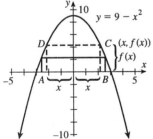

Infinitely many rectangles may be drawn inside the parabola $y = 9 - x^2$; the area of any of those rectangles will depend on the distance $x$ from the origin selected for the placement of points $A$ and $B$.
Area of rectangle $ABCD = (AB)(BC)$
$= (2x)(9-x^2) = 18x - 2x^3 = A(x)$

Domain of $A(x)$: $0 < x < 3$. (The domain excludes $-3 < x < 0$ from the original function because in the function $A(x)$, $x$ represents a segment length, which can't be considered to be negative.)

**Challenge Problem**

For any positive number $a$, $3 + a$ and $3 - a$ are equidistant from the line $x = 3$ and lie on opposite sides of it.
$f(3 + a) = (3 + a - 3)^2 = (a)^2 = a^2$
$f(3 - a) = (3 - a - 3)^2 = (-a)^2 = a^2$
Since $f(3 + a) = f(3 - a)$, the function $f(x) = (x - 3)^2$ is symmetric about the axis $x = 3$.

**Exercises 2.7**

**1.** $y = x^2 + 2x - 5$
$y = (x^2 + 2x + 1) - 1 - 5$
$y = (x + 1)^2 - 6$

**3.** $y = -x^2 - 6x + 2$
$y = -1(x^2 + 6x) + 2$
$y = -(x^2 + 6x + 9) + 9 + 2$
$y = -1(x + 3)^2 + 11$

**5.** $y = -x^2 + 3x - 4$
$y = -1(x^2 - 3x) - 4$
$y = -1\left(x^2 - 3x + \frac{9}{4}\right) + \frac{9}{4} - 4$
$y = -1\left(x - \frac{3}{2}\right)^2 - \frac{7}{4}$

**7.** $y = x^2 + 5x - 2$
$y = \left(x^2 + 5x + \frac{25}{4}\right) - \frac{25}{4} - 2$
$y = \left(x + \frac{5}{2}\right)^2 - \frac{33}{4}$

**9.** $y = 5 - 6x + 3x^2$
$y = 3(x^2 - 2x) + 5$
$y = 3(x^2 - 2x + 1) - 3 + 5$
$y = 3(x - 1)^2 + 2$

**11.** $y = -3x^2 - 6x + 5$
$y = -3(x^2 + 2x) + 5$
$y = -3(x^2 + 2x + 1) + 3 + 5$
$y = -3(x + 1)^2 + 8$

**13.** $y = x^2 + 2x - 1$
$y = (x^2 + 2x + 1) - 1 - 1$
$y = (x + 1)^2 - 2$
This parabola has vertex $(-1, -2)$ with axis of symmetry $x = -1$. The $y$-intercept occurs at $(0, -1)$.

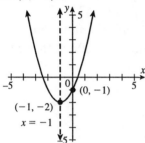

**15.** $y = -x^2 + 4x - 1$
$y = -1(x^2 - 4x + 4) + 4 - 1$
$y = -(x - 2)^2 + 3$
This parabola has vertex $(2, 3)$ and axis of symmetry $x = 2$. The $y$-intercept occurs at $(0, -1)$.

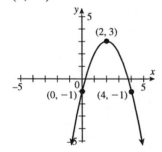

**17.** $y = 3x^2 + 6x - 3$
$y = 3(x^2 + 2x + 1) - 3 - 3$
$y = 3(x + 1)^2 - 6$
This parabola has vertex $(-1, -6)$ and axis of symmetry $x = -1$.
The $y$-intercept is $(0, -3)$.

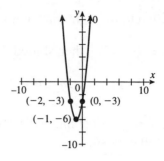

**19.** $x^2 - 5x + 6 = 0$
$(x - 2)(x - 3) = 0$
$x - 2 = 0$ or $x - 3 = 0$
$x = 2$ or $x = 3$
$x = 2, x = 3$

**21.** $x^2 - 10x = 0$
$x(x - 10) = 0$
$x = 0$ or $x - 10 = 0$
$x = 0$ or $x = 10$
$x = 0, x = 10$

**23.** $10x^2 - 13x - 3 = 0$
$(5x + 1)(2x - 3) = 0$
$5x + 1 = 0$ or $2x - 3 = 0$
$5x = -1$ or $2x = 3$
$x = -\dfrac{1}{5}$ or $x = \dfrac{3}{2}$
$x = -\dfrac{1}{5}, x = \dfrac{3}{2}$

**25.** $4x^2 = 32x - 64$
$4x^2 - 32x + 64 = 0$
$4(x^2 - 8x + 16) = 0$
$4(x - 4)^2 = 0$
$x - 4 = 0$
$x = 4$
$x = 4$ is the only solution.

**27.** $2x(x + 6) = 22x$
$2x(x + 6) - 22x = 0$
$2[x(x + 6) - 11x] = 0$
$2[x^2 + 6x - 11x] = 0$
$2(x^2 - 5x) = 0$
$2x(x - 5) = 0$
$2x = 0$ or $x - 5 = 0$
$x = 0$ or $x = 5$
$x = 0, x = 5$

**29.** $x^2 - 4x + 1 = 0$
$(x^2 - 4x + 4) = -1 + 4$
$(x - 2)^2 = 3$
$x - 2 = \pm\sqrt{3}$
$x = 2 \pm \sqrt{3}$

**31.** $x^2 - 3x - 10 = 0 \qquad a = 1, b = -3, c = -10$
$$x = \frac{3 \pm \sqrt{(-3)^2 - 4(-10)}}{2}$$
$$x = \frac{3 \pm \sqrt{9 + 40}}{2} = \frac{3 \pm \sqrt{49}}{2} = \frac{3 \pm 7}{2}$$
$$x = \frac{10}{2}, \ -\frac{4}{2}$$
$$x = 5, \ x = -2$$

**33.** $x^2 - 2x - 4 = 0 \qquad a = 1, b = -2, c = -4$
$$x = \frac{2 \pm \sqrt{(-2)^2 - 4(-4)}}{2}$$
$$x = \frac{2 \pm \sqrt{4 + 16}}{2} = \frac{2 \pm \sqrt{20}}{2} = \frac{2 \pm 2\sqrt{5}}{2}$$
$$x = 1 \pm \sqrt{5}$$
$$x = 1 + \sqrt{5} \doteq 3.24, \ x = 1 - \sqrt{5} \doteq -1.24$$

**35.** $3x^2 + 7x + 2 = 0 \qquad a = 3, b = 7, c = 2$
$$x = \frac{-7 \pm \sqrt{(7)^2 - (4)(3)(2)}}{2(3)}$$
$$= \frac{-7 \pm \sqrt{49 - 24}}{6}$$
$$= \frac{-7 \pm \sqrt{25}}{6} = \frac{-7 \pm 5}{6}$$
$$x = -\frac{12}{6} = -2, \ x = -\frac{2}{6} = -\frac{1}{3}$$

**37.** $x^2 - 6x + 6 = 0 \qquad a = 1, b = -6, c = 6$
$$x = \frac{-(-6) \pm \sqrt{(-6)^2 - 4(6)}}{2}$$
$$= \frac{6 \pm \sqrt{36 - 24}}{2} = \frac{6 \pm \sqrt{12}}{2}$$
$$= \frac{6 \pm 2\sqrt{3}}{2}$$
$$x = 3 - \sqrt{3} \doteq 1.27, \ x = 3 + \sqrt{3} \doteq 4.73$$

**39.** $-x^2 + 6x - 14 = 0 \qquad a = -1, b = 6, c = -14$
$$x = \frac{-6 \pm \sqrt{(6)^2 - (4)(-1)(-14)}}{2(-1)}$$
$$= \frac{-6 \pm \sqrt{36 - 56}}{-2} = \frac{-6 \pm \sqrt{-20}}{-2}$$
$$= \frac{-6 \pm 2i\sqrt{5}}{-2} = 3 \pm i\sqrt{5}$$
$$x = 3 \pm i\sqrt{5}$$

**41.** $3x + 1 = 2x^2$
$2x^2 - 3x - 1 = 0 \qquad a = 2, b = -3, c = -1$
$$x = \frac{-(-3) \pm \sqrt{(-3)^2 - (4)(2)(-1)}}{2(2)}$$
$$= \frac{3 \pm \sqrt{9 + 8}}{4} = \frac{3 \pm \sqrt{17}}{4}$$
$$x = \frac{3 + \sqrt{17}}{4} \doteq 1.78, \ x = \frac{3 - \sqrt{17}}{4} \doteq -0.28$$

**43.** $y = 2x^2 - 5x - 3$ will intercept the $x$-axis when $y = 0$.
$0 = 2x^2 - 5x - 3 \qquad a = 2, b = -5, c = -3$
$$x = \frac{-(-5) \pm \sqrt{(-5)^2 - (4)(2)(-3)}}{2(2)}$$
$$= \frac{5 \pm \sqrt{25 + 24}}{4} = \frac{5 \pm \sqrt{49}}{4} = \frac{5 \pm 7}{4}$$
The $x$-intercepts are at $(3, 0)$ and
$$\left(-\frac{1}{2}, \ 0\right).$$

**45.** $y = x^2 - x + 3$ will intercept the $x$-axis when $y = 0$.
$0 = x^2 - x + 3 \qquad a = 1, b = -1, c = 3$
$$x = \frac{-(-1) \pm \sqrt{(-1)^2 - (4)(1)(3)}}{2}$$
$$= \frac{1 \pm \sqrt{1 - 12}}{2} = \frac{1 \pm \sqrt{-11}}{2}$$
Because the discriminant is negative, $y = x^2 - x + 3$ will never intercept the $x$-axis.

**47.** $y = 3x^2 + x - 1$

$0 = 3x^2 + x - 1$ $\qquad a = 3, b = 1, c = -1$

$x = \dfrac{-1 \pm \sqrt{(1)^2 - 4(3)(-1)}}{2(3)}$

$= \dfrac{-1 \pm \sqrt{1 + 12}}{6} = \dfrac{-1 \pm \sqrt{13}}{6}$

The $x$-intercepts are at $\left(\dfrac{-1 + \sqrt{13}}{6}, 0\right)$ and

$\left(\dfrac{-1 - \sqrt{13}}{6}, 0\right)$.

**49.** $x^2 - 8x + 16 = 0$

$a = 1, b = -8, c = 16 \Rightarrow$ The discriminant
$= (-8)^2 - 4(16) = 64 - 64 = 0$
Because the discriminant is 0, the solution
to $x^2 - 8x + 16 = 0$ is a single real number
(a).

**51.** $-x^2 + 2x + 15 = 0$

$a = -1, b = 2, c = 15 \Rightarrow$ The discriminant
$= (2)^2 - 4(-1)(15) = 4 + 60 = 64$
Because the discriminant is a positive real
number and a perfect square other than 0,
the solution is two rational numbers (b).

**53.** $x^2 + 3x - 1 = 0$

$a = 1, b = 3, c = -1 \Rightarrow$ The discriminant
$= (3)^2 - 4(-1) = 9 + 4 = 13$
Because the discriminant is a positive real
number which is not a perfect square, the
solution has two irrational numbers (c).

**55.** $y = x^2 - 4x + 4$ is factorable as a trinomial
square $\Rightarrow y = (x - 2)^2$ so it can be
predicted without the discriminant that the
graph of $y = x^2 - 4x + 4$ will cross the
$x$-axis only once. To confirm this, a
discriminant of 0 is looked for:

$a = 1, b = -4, c = 4 \Rightarrow$ The discriminant
$= (-4)^2 - 4(1)(4) = 16 - 16 = 0$ $\checkmark$

Since the discriminant is 0, the *vertex* and
*x-intercept* are the same point:

$\left(\dfrac{-b}{2a}, 0\right) = \left(\dfrac{4}{2}, 0\right) = (2, 0)$

The $y$-intercept is $(0, 4)$.

**57.** $y = 9x^2 - 6x + 1$ is factorable as a trinomial
square $\Rightarrow y = (3x - 1)^2$ so it can be
predicted that the graph of $y = 9x^2 - 6x + 1$
will cross the $x$-axis only once. A
discriminant of 0 will confirm this
prediction:

$a = 9, b = -6, c = 1 \Rightarrow$ The discriminant
$= (-6)^2 - 4(9)(1) = 36 - 36 = 0$ $\checkmark$
Since the discriminant is 0, the *vertex* and
*x-intercept* are the same point:

$\left(\dfrac{-b}{2a}, 0\right) = \left(\dfrac{6}{18}, 0\right) = \left(\dfrac{1}{3}, 0\right)$

The $y$-intercept is $(0, 1)$.

**59.** $y = 2x^2 - 4x + 3$ $\qquad a = 2, b = -4, c = 3$

The discriminant $= (-4)^2 - 4(2)(3)$
$= 16 - 24 = -8$.
Since the discriminant is negative, there
are *no x-intercepts*. The *vertex* and
*y-intercept* can be identified by placing
$y = 2x^2 - 4x + 3$ in standard form:
$y = 2(x^2 - 2x + 1) - 2 + 3$
$y = 2(x - 1)^2 + 1$
The *vertex* is $(1, 1)$ and the $y$-intercept is
$(0, 3)$.

**61.** $f(x) = x^2 + bx + 9$ will have the $x$-axis as a
tangent when $b^2 - 4ac = 0$
$b^2 - 4(1)(9) = 0$
$b^2 - 36 = 0$
$b^2 = 36$
$b = \pm 6$
If $b = 6$ or $-6$, the $x$-axis will be tangent to
the parabola.

**63.** $f(x) = x^2 - bx + 7$ will have the $x$-axis as a
tangent when $b^2 - 4ac = 0$.
$b^2 - 4(1)(7) = 0$
$b^2 - 28 = 0$
$b^2 = 28$
$b = \pm\sqrt{28} = \pm 2\sqrt{7}$

If $b = 2\sqrt{7}$ or $-2\sqrt{7}$, the $x$-axis will be tangent to the parabola.

**65.** $f(x) = -x^2 + 4x + k$ will cross the $x$-axis twice when $b^2 - 4ac > 0$.
$(4)^2 - 4(-1)(k) > 0$
$16 + 4k > 0$
$4k > -16$
$k > -4$

**67.** $f(x) = kx^2 - x - 1$ will cross the $x$-axis twice when $b^2 - 4ac > 0$.
$(-1)^2 - 4(k)(-1) > 0$
$1 + 4k > 0$
$4k > -1$
$k > -\dfrac{1}{4}$

**69.** $f(x) = x^2 - 6x + t$ will not cross the $x$-axis when $b^2 - 4ac < 0$.
$(-6)^2 - 4(1)(t) < 0$
$36 - 4t < 0$
$36 < 4t$
$9 < t$ or $t > 9$

**71.** $f(x) = tx^2 - x - 1$ will not cross the $x$-axis when $b^2 - 4ac < 0$.
$(-1)^2 - 4(t)(-1) < 0$
$1 + 4t < 0$
$4t < -1$
$t < -\dfrac{1}{4}$

**73.** The 2 roots of $ax^2 + bx + c = 0$ are
$\dfrac{-b + \sqrt{b^2 - 4ac}}{2a}$ and $\dfrac{-b - \sqrt{b^2 - 4ac}}{2a}$.
Their sum is
$$\dfrac{-b + \sqrt{b^2 - 4ac} + \left(-b - \sqrt{b^2 - 4ac}\right)}{2a}$$
$$= \dfrac{-2b}{2a} = \dfrac{-b}{a}.$$
Their product is
$$\dfrac{\left(-b + \sqrt{b^2 - 4ac}\right)\left(-b - \sqrt{b^2 - 4ac}\right)}{(2a)^2}$$

$$= \dfrac{(-b)^2 - \left(\sqrt{b^2 - 4ac}\right)^2}{4a^2}$$
$$= \dfrac{b^2 - (b^2 - 4ac)}{4a^2} = \dfrac{4ac}{4a^2} = \dfrac{c}{a}$$

**75.** $x^2 - 2x - 3 < 0$
$(x - 3)(x + 1) < 0 \Rightarrow y = x^2 - 2x - 3$ has $x$-intercepts at $(3, 0)$ and $(-1, 0)$. Further, $y = x^2 - 2x - 3 = (x^2 - 2x + 1) - 1 - 3$
$= (x - 1)^2 - 4$, so this parabola opens upward, or is concave up. $x^2 - 2x - 3 < 0$ when $-1 < x < 3$.

**77.** $x^2 + 3x - 10 \geq 0$
$(x + 5)(x - 2) \geq 0 \Rightarrow y = x^2 + 3x - 10$ has $x$-intercepts at $(-5, 0)$ and $(2, 0)$. Further,
$$y = x^2 + 3x - 10 = \left(x^2 + 3x + \dfrac{9}{4}\right) - \dfrac{9}{4} - 10$$
$$= \left(x + \dfrac{3}{2}\right)^2 - \dfrac{49}{4},$$ so the parabola opens
upward, or is concave up. $x^2 + 3x - 10 \geq 0$ when $x \leq -5$ or $x \geq 2$.

**79.** $x^2 + 6x + 9 < 0$
$(x + 3)^2 < 0 \Rightarrow y = x^2 + 6x + 9$ has a single $x$-intercept at $(-3, 0)$. Further,
$y = x^2 + 6x + 9 = (x + 3)^2$ is concave up, so $y = x^2 + 6x + 9$ will never produce negative $y$-values. There are *no solutions* for $x^2 + 6x + 9 < 0$.

**81.** $f(x) = (x - 3)(x + 1)$ has *x-intercepts* at $(3, 0)$ and $(-1, 0)$. $f(x) = x^2 - 2x - 3$
$= (x^2 - 2x + 1) - 1 - 3 = (x - 1)^2 - 4$ has *vertex* $(1, -4)$ and $y$-intercept at $(0, -3)$. The parabola is *concave up*.

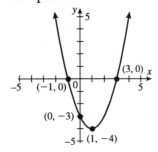

**83.** $f(x) = 4 - 4x - x^2 = -x^2 - 4x + 4$

$a = -1, b = -4, c = 4$

$f(x) = -(x^2 + 4x) + 4$

$f(x) = -(x^2 + 4x + 4) + 4 + 4$

$f(x) = -(x + 2)^2 + 8 \Rightarrow f(x)$ has *vertex*
$(-2, 8)$ and *y-intercept* $(0, 4)$, and *the
parabola is concave down.*

The *x-intercepts* are:

$$x = \frac{4 \pm \sqrt{(-4)^2 - 4(-1)(4)}}{-2} = \frac{4 \pm \sqrt{16 + 16}}{-2}$$

$$= \frac{4 \pm \sqrt{32}}{-2} = \frac{4 \pm 4\sqrt{2}}{-2}$$

The *x-intercepts* are:

$\left(-2 + 2\sqrt{2},\ 0\right)$ and $\left(-2 - 2\sqrt{2},\ 0\right)$.

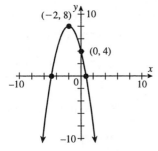

**85.** $f(x) = 3x^2 - 4x + 1$

$f(x) = (3x - 1)(x - 1) \Rightarrow f(x)$ has

*x-intercepts* at $(1, 0)$ and $\left(\frac{1}{3},\ 0\right)$.

$$f(x) = 3\left(x^2 - \frac{4}{3}x\right) + 1$$

$$f(x) = 3\left(x^2 - \frac{4}{3}x + \frac{4}{9}\right) - \frac{4}{3} + 1$$

$$f(x) = 3\left(x - \frac{2}{3}\right)^2 - \frac{1}{3} \Rightarrow f(x) \text{ has } vertex$$

$\left(\frac{2}{3},\ -\frac{1}{3}\right)$, *y-intercept* $(0, 1)$, and is
*concave up.*

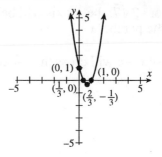

**87.** $2x^4 - 13x^2 - 7 = 0$. Let $u = x^2$.

$2(x^2)^2 - 13(x^2) - 7 = 0$

$2u^2 - 13u - 7 = 0,\ a = 2,\ b = -13,\ c = -7$

$$u = \frac{13 \pm \sqrt{(-13)^2 - 4(2)(-7)}}{4}$$

$$= \frac{13 \pm \sqrt{169 + 56}}{4} = \frac{13 \pm \sqrt{225}}{4}$$

$$= \frac{13 \pm 15}{4}$$

$u = \dfrac{28}{4} = 7$ or $u = -\dfrac{2}{4} = -\dfrac{1}{2}$

Therefore, $x^2 = 7$ or $x^2 = -\dfrac{1}{2}$

$$x = \pm\sqrt{7} \text{ or } x = \pm\sqrt{-\frac{1}{2}}$$

$$\text{or } x = \pm i\sqrt{\frac{1}{2}} = \pm i\frac{1}{\sqrt{2}} = \pm i\frac{\sqrt{2}}{2}$$

Solutions: $x = \pm\sqrt{7},\ x = \pm\dfrac{\sqrt{2}}{2}i$

**89.** $x^3 + 3x^2 - 4x - 12 = 0$

$x^2(x + 3) - 4(x + 3) = 0$

$(x^2 - 4)(x + 3) = 0$

$(x + 2)(x - 2)(x + 3) = 0$

Solutions: $x = -2,\ x = 2,\ x = -3$

**91.** $\dfrac{4x}{2x - 1} + \dfrac{3}{2x + 1} = 0$

$$\frac{4x(2x + 1) + 3(2x - 1)}{(2x + 1)(2x - 1)} = 0$$

$$\frac{8x^2 + 4x + 6x - 3}{(2x + 1)(2x - 1)} = 0$$

$$\frac{8x^2 + 10x - 3}{(2x + 1)(2x - 1)} = 0$$

When $8x^2 + 10x - 3 = 0$,
$a = 8$, $b = 10$, $c = -3$

$$x = \frac{-10 \pm \sqrt{(10)^2 - (4)(8)(-3)}}{16}$$

$$= \frac{-10 \pm \sqrt{100 + 96}}{16} = \frac{-10 \pm \sqrt{196}}{16}$$

$$= \frac{-10 \pm 14}{16}$$

Solutions: $x = \frac{4}{16} = \frac{1}{4}$ or $x = -\frac{24}{16} = -\frac{3}{2}$.

**93.** $f(x) = |9 - x^2|$

Start by graphing $f(x) = 9 - x^2$, then reflect across the $x$-axis those pieces of the parabola where $f(x) < 0$ to obtain a complete graph of $f(x) = |9 - x^2|$.
From $(-\infty, -3]$ and $[0, 3]$, $f(x)$ is *decreasing*.
From $[-3, 0]$ and $[3, \infty)$, $f(x)$ is *increasing*.
The curve is *concave up* on $(-\infty, -3)$ and $(3, \infty)$; *concave down* on $(-3, 3)$.
*Range*: $f(x) = y \geq 0$.

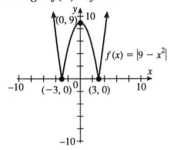

**95.** $f(x) = |x^2 - x - 6|$
$f(x)$ has $x$-intercepts at $(3, 0)$ and $(-2, 0)$.
$f(x) = |(x - 3)(x + 2)|$
Consider $f(x) = x^2 - x - 6$ first, then reflect across the $x$-axis those pieces of the parabola where $f(x) < 0$.
$f(x) = x^2 - x - 6$

$$f(x) = \left( x^2 - x + \frac{1}{4} \right) - \frac{1}{4} - 6$$

$$f(x) = \left( x - \frac{1}{2} \right)^2 - \frac{25}{4} \text{ has } vertex$$

$\left( \dfrac{1}{2}, -\dfrac{25}{4} \right)$ and *y-intercept* $(0, -6)$, and is *concave up*.

$f(x)$ is *decreasing* on $(-\infty, -2]$ and $\left[ \dfrac{1}{2}, 3 \right]$.

$f(x)$ is *increasing* on $\left[ -2, \dfrac{1}{2} \right]$ and $[3, \infty)$.

$f(x)$ is concave up on $(-\infty, -2)$ and $(3, \infty)$; concave down on $(-2, 3)$.
Range: $f(x) = y \geq 0$.

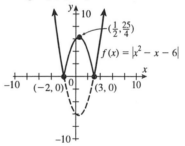

**97.** $f(x) = \begin{cases} 1 \text{ if } -3 \leq x < 0 \Rightarrow \text{ a horizontal line at } y = 1 \text{ for } -3 \leq x < 0 \\ x^2 - 4x + 1 \text{ if } 0 \leq x < 5 \Rightarrow \text{ a concave up parabola for } 0 \leq x < 5 \\ -2x + 16 \text{ if } 5 \leq x < 9 \Rightarrow \text{ a line of slope} = -2 \text{ for } 5 \leq x < 9 \end{cases}$

$f(x) = x^2 - 4x + 1 = (x^2 - 4x + 4) - 4 + 1$
$= (x - 2)^2 - 3$ has *vertex* $(2, -3)$ and
*y-intercept* $(0, 1)$.
Is there an *x-intercept* on the given interval?
The *x*-intercepts would be

$$x = \frac{4 \pm \sqrt{16 - 4}}{2} = \frac{4 \pm \sqrt{12}}{2} = \frac{4 \pm 2\sqrt{3}}{2}$$

$x = 2 + \sqrt{3}, 2 - \sqrt{3}$
$x \doteq 3.73, 0.27$
Both *x-intercepts* of the parabola fall in the given interval.

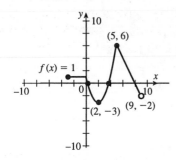

**99.** $y = a(x - h)^2 + k$ will be concave up with vertex $(h, 0)$ when $a > 0$ and $k = 0$.

**101.** $y = a(x - h)^2 + k$ will be concave up with range $y \geq 2$ when $a > 0$ and $k = 2$. If $k = 2$, the vertex will be $(h, 2)$, and since the parabola will be concave up, the lowest possible value for $y$ will occur at the vertex, where $y = 2$. Thus, the range for $y$ will be $y \geq 2$.

## Challenge Problems

1. First, consider the case where $b$ is even, and say $b = 2n$ for some integer $n$.
   Then $b^2 - 4ac = (2n)^2 - 4ac = 4n^2 - 4ac$
   $= 4(n^2 - ac)$, which is a multiple of 4, and so can't equal 23. Next, consider the case where $b$ is odd, and say $b = 2n + 1$ for some integer $n$. Then $b^2 - 4ac$
   $= (2n + 1)^2 - 4ac = 4n^2 + 4n + 1 - 4ac$
   $= 4(n^2 + n - ac) + 1$, which is one more than a multiple of 4. Since 23 is not one more than a multiple of 4, $b^2 - 4ac$ can't equal 23.

## Critical Thinking

1. The maximum or minimum value of the function $f(x) = ax^2 + bx + c$ will occur at the vertex of the parabola.

3. The graph of $x = (y + 2)^2 - 1$ is identical to the graph of $y = (x + 2)^2 - 1$ except that it has been rotated clockwise by $90°$.

5.

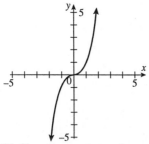

If $f$ is a one-to-one function and $f(x_1) = f(x_2)$, then $x_1 = x_2$.

## Exercises 2.8

1. Let $x$ represent the first integer. $x + 1$ represents the next consecutive integer. $x(x + 1)$ represents their product.
   $x(x + 1) = 210$
   $x^2 + x = 210$
   $x^2 + x - 210 = 0$
   $(x - 14)(x + 15) = 0$
   $x - 14 = 0$ or $x + 15 = 0$
   $x = 14$ or $x = -15$
   Reject this solution because the integers must be positive.
   If $x = 14$, then $x + 1 = 15$. The two consecutive positive integers with product 210 are 14 and 15.

3. Let $x$ represent a positive integer. $x + 3$ represents the other positive integer.
   $x^2 + (x + 3)^2$ represents the sum of their squares.
   $x^2 + (x + 3)^2 = 89$
   $x^2 + x^2 + 6x + 9 = 89$

$2x^2 + 6x + 9 = 89$
$2x^2 + 6x = 80$
$x^2 + 3x = 40$
$x^2 + 3x - 40 = 0$
$(x + 8)(x - 5) = 0$
$x + 8 = 0$ or $x - 5 = 0$
$x = -8$ or $x = 5$
$x = -8$ is rejected because it isn't positive.
Therefore, the two integers are 5 and 8.

5. A sketch will help to solve this problem. If a border of width $x$ is added on all sides, the expanded rectangle has length $(12 + 2x)$ and width $(8 + 2x)$. The original area is $8 \times 12 = 96$.

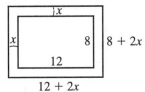

The new area is $(12 + 2x)(8 + 2x)$ and is twice the original area.
$(12 + 2x)(8 + 2x) = 2(96)$
$96 + 40x + 4x^2 = 192$
$4x^2 + 40x - 96 = 0$
$x^2 + 10x - 24 = 0$
$(x + 12)(x - 2) = 0$
$x + 12 = 0$ or $x - 2 = 0$
$x = -12$ or $x = 2$
Reject the $-12$ because the border must have a positive width. If the border is 2 feet, then the new rectangle will be $16 \times 12$ with an area of 192, which is twice 96.
Answer: The border should be 2 feet wide.

7. Let $x$ represent the rectangle's width, in centimeters. If the perimeter is 16 centimeters, then "length + width" = 8 centimeters. This indicates that $8 - x$ could represent the length.
$(8 - x)(x) = 15$
$8x - x^2 = 15$
$x^2 - 8x + 15 = 0$
$(x - 5)(x - 3) = 0$
$x - 5 = 0$ or $x - 3 = 0$
$x = 5$ or $x = 3$

Answer: The dimensions of the rectangle are 5 centimeters by 3 centimeters, with length and width interchangeable.

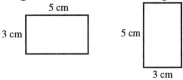

9. The area of the walk added to the area of the small rectangle equals the area of the large rectangle. $10 \times 18 = 180$ is the area of the small rectangle. $(18 + 2x)(10 + 2x)$ is the area of the large rectangle.
$(18 + 2x)(10 + 2x) = 52 + 180$
$180 + 56x + 4x^2 = 232$
$4x^2 + 56x - 52 = 0$
$x^2 + 14x - 13 = 0$
$$x = \frac{-(14) \pm \sqrt{(14)^2 - 4(1)(-13)}}{2(1)}$$
$$= \frac{-(14) \pm \sqrt{248}}{2} = \frac{-(14) \pm 2\sqrt{62}}{2}$$
$$= -7 \pm \sqrt{62}$$
$-7 + \sqrt{62} \approx 0.9$; $-7 - \sqrt{62} \approx -14.9$
Reject the negative solution. The width is $-7 + \sqrt{62}$ m, or about 0.9 m.

11. If $S = \frac{1}{2}n(n + 1)$ and $S = 120$,
$120 = \frac{1}{2}n^2 + \frac{1}{2}n$
$240 = n^2 + n$
$240 + \frac{1}{4} = n^2 + n + \frac{1}{4}$
$$\frac{961}{4} = \left(n + \frac{1}{2}\right)^2$$
$$\pm\sqrt{\frac{961}{4}} = n + \frac{1}{2}$$
$$\pm\frac{31}{2} = n + \frac{1}{2}$$
$$-\frac{1}{2} \pm \frac{31}{2} = n$$
$$n = \frac{-32}{2} \text{ or } n = \frac{30}{2}, \text{ so } n = -16 \text{ or } 15$$

Reject the negative solution, because $n$ represents a number of consecutive integers.
Answer: The sum of the first 15 integers is 120. Check by addition:
$$1 + 2 + 3 + 4 + 5 + 6 + 7 + 8 + 9 + 10 + 11$$
$$+ 12 + 13 + 14 + 15$$
$$= (1 + 15) + (2 + 14) + (3 + 13) + (4 + 12)$$
$$+ (5 + 11) + (6 + 10) + (7 + 9) + 8$$
$$= 7(16) + 8$$
$$= 112 + 8$$
$$= 120$$

13. Let $x$ = rate of the boat in still water. Then $x + 4$ = rate of the boat going downstream, and $x - 4$ = rate of the boat going upstream. Since time = distance ÷ rate,

time going downstream = $\dfrac{36}{x+4}$ and time

going upstream = $\dfrac{36}{x-4}$. The trip

downstream took $\dfrac{3}{4}$ of an hour less than the

trip upstream, or $\dfrac{36}{x+4} = \dfrac{36}{x-4} - \dfrac{3}{4}$.

Multiply the equation by $4(x+4)(x-4)$ on both sides:

$$4(x+4)(x-4) \cdot \dfrac{36}{x+4}$$
$$= 4(x+4)(x-4) \cdot \dfrac{36}{x-4}$$
$$\quad - 4(x+4)(x-4) \cdot \dfrac{3}{4}$$
$$4(36)(x-4)$$
$$= 4(x+4)(36) - 3(x+4)(x-4)$$
$$144x - 576$$
$$= 144x + 576 - 3x^2 + 48$$
$$3x^2 = 1200$$
$$x^2 = 400$$
$$x = \pm 20$$
Reject the negative solution.
The rate is 20 mph.

Time downstream $= \dfrac{36}{x+4} = \dfrac{36}{24} = 1\dfrac{1}{2}$ h.

Time upstream $= \dfrac{36}{x-4} = \dfrac{36}{16} = 2\dfrac{1}{4}$ h.

15. $f(x) = -x^2 + 10x - 18$
$f(x) = -(x^2 - 10x) - 18$
$f(x) = -(x^2 - 10x + 25) + 25 - 18$
$f(x) = -(x - 5)^2 + 7$
This parabola is concave down with vertex $(5, 7)$. Thus, the *maximum* value for $f(x) = 7$, which occurs at $x = 5$.

17. $f(x) = 16x^2 - 64x + 100$
$f(x) = 16(x^2 - 4x) + 100$
$f(x) = 16(x^2 - 4x + 4) + 100 - 16(4)$
$f(x) = 16(x - 2)^2 + 36$
Since $16 > 0$, the parabola opens upward, so 36 is the *minimum* and it occurs when $x = 2$.

19. $f(x) = 49 - 28x + 4x^2$
$f(x) = 4x^2 - 28x + 49$
$f(x) = 4(x^2 - 7x) + 49$
$f(x) = 4\left(x^2 - 7x + \dfrac{49}{4}\right) - 49 + 49$
$f(x) = 4\left(x - \dfrac{7}{2}\right)^2$
This parabola is concave up with a *minimum* value $f(x) = 0$, which occurs at $x = \dfrac{7}{2}$.

21. $f(x) = -x\left(\dfrac{2}{3} + x\right) = -x^2 - \dfrac{2}{3}x$
$f(x) = -\left(x^2 + \dfrac{2}{3}x + \dfrac{1}{9}\right) + \dfrac{1}{9}$
$f(x) = -\left(x + \dfrac{1}{3}\right)^2 + \dfrac{1}{9}$
Since $-1 < 0$, the parabola opens downward, so $\dfrac{1}{9}$ is the maximum and it occurs when $x = -\dfrac{1}{3}$. $\left(x + \dfrac{1}{3} = x - \left(-\dfrac{1}{3}\right)\right)$.

23. $C = n^2 - 120n + 4200$
At what point in production will the cost ($C$) be minimized?
$C = (n^2 - 120n + 3600) - 3600 + 4200$
$C = (n - 60)^2 + 600$ is a concave up parabola with *vertex* $(60, 600)$.

Answer: 60 statues must be produced so as to achieve the minimum cost of $600.

25. Let $x$ = one number. Then $12 - x$ = the other number. Their product is $x(12 - x)$.
$$y = x(12 - x) = -x^2 + 12x$$
$$y = -(x^2 - 12x)$$
$$y = -(x^2 - 12x + 36) + 36$$
$$y = -(x - 6)^2 + 36$$
The vertex of this function is at (6, 36). Since $-1 < 0$, the parabola opens downward. When $x = 6$ and $12 - x = 6$, the product is at its maximum, 36. The numbers are 6 and 6.

27. The first number is represented by $x$, and the second by $x - 22$. Their product:
$$P(x) = x(x - 22) = x^2 - 22$$
$$P(x) = (x^2 - 22x + 121) - 121$$
$$P(x) = (x - 11)^2 - 121$$
The minimum product, $-121$, will occur when $x = 11$ and $x - 22 = -11$.

29. Start with a sketch:

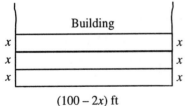

$$(100 - 2x) \text{ ft}$$
$$A(x) = (3x)(100 - 2x)$$
$$= -6(x - 25)^2 + 3750$$
Three rectangles: 25' by 50'; total area = 3750 sq. ft

31. $h = 128t - 16t^2$ describes the height $h$ above the ground of an object after $t$ seconds.
$$h = f(t) = -16t^2 + 128t$$
$$f(t) = -16(t^2 - 8t)$$
$$f(t) = -16(t^2 - 8t + 16) + 256$$
$$f(t) = -16(t - 4)^2 + 256$$
Answer: The maximum height of 256 feet will be reached in 4 seconds.

33. At what time will the object be at 192 feet?
$$192 = -16t^2 + 128t$$
$$-16t^2 + 128t - 192 = 0$$
$$-16(t^2 - 8t + 12) = 0$$
$$t^2 - 8t + 12 = 0$$
$$(t - 6)(t - 2) = 0 \Rightarrow t - 6 = 0 \text{ or } t - 2 = 0$$
$$t = 6 \text{ or } t = 2$$
Answer: After *2 seconds*, while the object's height is increasing, it will reach a height of 192 feet. 4 seconds later, or *6 seconds* after it left the ground, when the object's height is decreasing, it will again be at 192 feet.

35. Attendance is based on the number of quarters in the price. For example, $A(28) = 14,000$ because at a $7.00 price (28 quarters), 14,000 in attendance is expected.
$A(28 + x) = 14,000 - 280x$, because for each quarter added to the price, attendance decreases by 280.
Gate receipts = (Price)(Attendance), so
$$G(x) = (28 + x)(14000 - 280x)$$
$$G(x) = 392000 - 7840x + 14000x - 280x^2$$
$$G(x) = -280x^2 + 6160x + 392000$$
$$G(x) = -280(x^2 - 22x) + 392000$$
$$G(x) = -280(x^2 - 22x + 121) + 33,880 + 392000$$
$$G(x) = -280(x - 11)^2 + 425880$$
Remembering that $x$ = quarters added to $7.00, when $x = 11$, gate receipts of 425880 quarters ($106,470) will be achieved. The ticket price will be $28 + 11 = 39$ quarters = $9.75.

37. The area of the rectangle is given by $A = xy$. To express this area as a function of one variable we may begin by writing $y$ in terms of $x$. Since the bottom right triangle with height $y$ and base $2 - x$ is similar to the large right triangle with height 4 and base 2, we obtain this proportion $\dfrac{y}{2 - x} = \dfrac{4}{2}$.

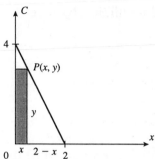

Solve this equation for $y$ to obtain
$y = 4 - 2x$ (note that $0 < x < 2$) and
substitute into $A = xy$ to get
$A(x) = x(4 - 2x)$. Now complete the square.
$A(x) = x(4 - 2x) = -2x^2 + 4x = -2(x^2 - 2x)$
$= -2(x^2 - 2x + 1) + 2 = -2(x - 1)^2 + 2$
The maximum area is obtained when $x = 1$
and $y = 4 - 2x = 4 - 2(1) = 2$, and the
coordinates of $P$ which would maximize
area is $P(1, 2)$.

**39.**

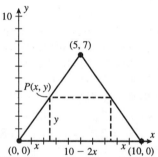

The isosceles triangle has a height of
seven units. The height $y$ of the triangle
formed in the lower left corner can be
expressed in terms of $x$ because
$\dfrac{7}{y} = \dfrac{5}{x} \Rightarrow y = \dfrac{7}{5}x \Rightarrow A(x)$ of the rectangle
$= \dfrac{7}{5}x(10 - 2x)$.

[See #38.] When will $A(x)$ be maximized?
$A(x) = \dfrac{7}{5}x(10 - 2x) = 14x - \dfrac{14}{5}x^2$

$A(x) = -\dfrac{14}{5}(x^2 - 5x)$

$A(x) = -\dfrac{14}{5}\left(x^2 - 5x + \dfrac{25}{4}\right) + \dfrac{35}{2}$

$A(x) = -\dfrac{14}{5}\left(x - \dfrac{5}{2}\right)^2 + \dfrac{35}{2}$

Answer: At $P\left(\dfrac{5}{2}, \dfrac{7}{5}\left(\dfrac{5}{2}\right)\right)$, a maximum
area for the rectangle of $\dfrac{35}{2}$ square units
will be achieved. $P\left(\dfrac{5}{2}, \dfrac{7}{2}\right)$ or $P(2.5, 3.5)$
will maximize area.

**41.** $y = 2x^2 - 6x + 9 \qquad a = 2,\ b = -6,\ c = 9$
has its vertex at
$\left(-\dfrac{-6}{2(2)},\ \dfrac{4(2)(9) - (-6)^2}{4(2)}\right)$
$= \left(\dfrac{6}{4},\ \dfrac{72 - 36}{8}\right) = \left(\dfrac{3}{2},\ \dfrac{36}{8}\right)$
$= \left(\dfrac{3}{2},\ \dfrac{9}{2}\right)$

Since this parabola is concave up, $\dfrac{9}{2}$ is the
minimum value for $y$.

**43.** $y = -\dfrac{1}{2}x^2 - \dfrac{1}{3}x + 1 \qquad a = -\dfrac{1}{2},\ b = -\dfrac{1}{3},\ c = 1$
has its vertex at
$\left(-\dfrac{-\frac{1}{3}}{2\left(-\frac{1}{2}\right)},\ \dfrac{(4)\left(-\frac{1}{2}\right)(1) - \left(-\frac{1}{3}\right)^2}{4\left(-\frac{1}{2}\right)}\right)$
$= \left(-\dfrac{1}{3},\ \dfrac{-2 - \frac{1}{9}}{-2}\right) = \left(-\dfrac{1}{3},\ \dfrac{-\frac{19}{9}}{-\frac{2}{1}}\right)$
$= \left(-\dfrac{1}{3},\ \dfrac{19}{18}\right)$

Since this parabola is concave down, $\dfrac{19}{18}$ is
the maximum value for $y$.

**45.** $y + \dfrac{2}{3}x^2 = 9$
$y = -\dfrac{2}{3}x^2 + 9 \qquad a = -\dfrac{2}{3},\ b = 0,\ c = 9$
has its vertex at
$\left(0,\ \dfrac{4\left(-\frac{2}{3}\right)(9)}{4\left(-\frac{2}{3}\right)}\right) = (0,\ 9)$

Since the parabola is concave down, 9 is
the maximum value for $y$.

**47.** $d = 0.045\,r^2 + 1.1\,r$

  **(a)** $d = 0.045(40)^2 + 1.1(40)$
  $d = 116$
  116 feet

  **(b)** $d = 0.045(55)^2 + 1.1(55)$
  $d = 196.625$
  197 feet

  **(c)** $d = 0.045(65)^2 + 1.1(65)$
  $d = 261.625$
  262 feet

**49.** Let $x =$ the number of dollars the price was decreased.
  $R(20 - x) = (100 + 10x)(20 - x)$
  $R = 2000 - 100x + 200x - 10x^2$
  $R = -10x^2 + 100x + 2000$
  $R = -10(x^2 - 10x) + 2000$
  $R = -10(x^2 - 10x + 25) + 250 + 2000$
  $R = -10(x - 5)^2 + 2250$
  Answer: At $x = \$5$ decrease, a maximum revenue of \$2250 will be realized. Therefore, the price should be set at $20 - 5 = \$15$.

## Chapter 2 Review Exercises

**1.** A *function* is a correspondence between two sets, the domain and the range, such that for each value of the domain there corresponds exactly one value in the range.

**3.** $y = \dfrac{1}{x^2 + 1}$ is a function of $x$ because each $x$-value will produce exactly one $y$-value. The domain is all real numbers.

**5.** $y^2 = x + 1$
  $y = \pm\sqrt{x + 1} \Rightarrow y$ is not a function of $x$.

**7.** $y = \pm x \Rightarrow y$ is not a function of $x$.

**9.** If $f(x) = x^2 + 3x - 2$,
  $2f(3) = 2[3^2 + 3(3) - 2] = 2[9 + 9 - 2]$
  $= 2[16] = 32$
  $f(6) = 6^2 + 3(6) - 2 = 36 + 18 - 2 = 52$
  $2f(3) \neq f(6)$

**11.** If $g(x) = x^2 + 1$, then
  $$\frac{g(x) - g(2)}{x - 2} = \frac{x^2 + 1 - 5}{x - 2} = \frac{x^2 - 4}{x - 2}$$
  $$= \frac{(x + 2)(x - 2)}{x - 2} = x + 2$$

**13.** If $f(x) = \dfrac{2}{x - 1}$

  **(a)** $f(3x) = \dfrac{2}{3x - 1}$

  **(b)** $3f(x) = (3)\left(\dfrac{2}{x - 1}\right) = \dfrac{6}{x - 1}$

**15.** The slope $m$ of a line between two points $(x_1, y_1)$ and $(x_2, y_2)$ is the difference of the $y$-values compared in a *ratio* to the difference of the $x$-values:
  $$m = \frac{y_2 - y_1}{x_2 - x_1},\; x_2 \neq x_1.$$

**17.** Two perpendicular lines have slopes which are the negative reciprocals of one another (the product of the slopes $= -1$).

**19.** If $x - 3y = 6$, the *x-intercept* is where $y = 0$:
  $x - 3(0) = 6$
  $x = 6 \Rightarrow (6, 0)$
  The *y-intercept* is where $x = 0$:
  $0 - 3y = 6$
  $y = -2 \Rightarrow (0, -2)$

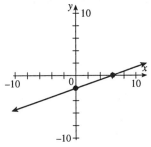

**21.**

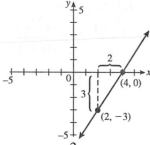

A slope of $\frac{3}{2}$ indicates that from $(2, -3)$,

the $y$-value will increase three units while the $x$-value will increase two units. $(2 + 2, -3 + 3)$ or $(4, 0)$.

**23.** The "equilibrium point" is the point at which supply will meet demand. Graphically, it's where $y = 2000 - 50x$ (demand) will intersect $y = 20x + 600$ (supply). When will supply "meet" demand?

$2000 - 50x = 20x + 600$
$2000 = 70x + 600$
$1400 = 70x$
$20 = x$

Answer: Supply and demand "meet" at $(20, 1000)$.

**25.** For any point $(x, y)$ and a specific point $(x_1, y_1)$ which both fall on the line of slope $m$, the general equation for the line is given by $y - y_1 = m(x - x_1)$ [called the point-slope form of a line].

**27.** $y = f(x) = 2x + 1$ has slope $m = 2$ and $y$-intercept $(0, b) \Rightarrow (0, 1)$.

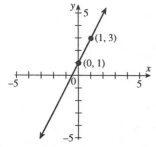

**29. (a)** Parallel to the $x$-axis containing $(-3, 2) \Rightarrow y = 2$ ($y$-value is "fixed").

**(b)** Parallel to the $y$-axis containing $(-3, 2) \Rightarrow x = -3$ ($x$-value is "fixed").

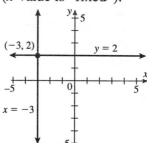

**31. (a)** If the line contains $(-3, 2)$ and $(-2, 5)$, its slope is $\frac{5 - 2}{-2 - (-3)} = \frac{3}{1} = 3$. Its equation in point-slope form is $y - 2 = 3(x + 3)$ or $y - 5 = 3(x + 2)$

**(b)** If $y - 2 = 3(x + 3)$
$y - 2 = 3x + 9$
$y = 3x + 11 \Rightarrow$ slope of 3, $y$-intercept of $(0, 11)$.

**(c)** In standard form: $3x - y = -11$

**33.** A line parallel to $2x + 3y = 4$ will have a slope equal to this line's slope, while a line perpendicular to $2x + 3y = 4$ will have a slope which is the negative reciprocal of this line's slope.

$2x + 3y = 4 \Rightarrow 3y = -2x + 4$

$y = -\frac{2}{3}x + 4 \Rightarrow 2x + 3y = 4$ has slope $-\frac{2}{3}$

A line containing $(-3, 1)$ with slope $-\frac{2}{3}$:

$$y - 1 = -\frac{2}{3}(x + 3)$$
$$y - 1 = -\frac{2}{3}x - 2$$
$$y = -\frac{2}{3}x - 1$$

A line containing $(-3, 1)$ with slope $\frac{3}{2}$:

$$y - 1 = \frac{3}{2}(x + 3)$$
$$y - 1 = \frac{3}{2}x + \frac{9}{2}$$
$$y = \frac{3}{2}x + \frac{11}{2}$$

**35.** Shift $y = |x|$ three units to the left.

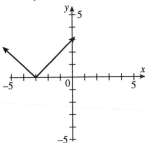

**37.** $y = f(x) = \begin{cases} x \text{ if } -2 \le x < 1 \\ x + 2 \text{ if } x \ge 1 \end{cases}$

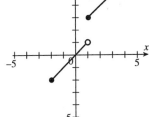

**39.** $y = f(x) = [x]$ for $-3 < x \le 1$
The domain of $f$: $-3 < x \le 1$
The range of $f$: $\{-3, -2, -1, 0, 1\}$

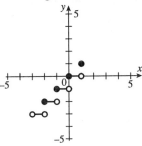

**41.** $3x + y = 9$
$x + 2y = 8 \Rightarrow x = -2y + 8$
If $x = -2y + 8$, then
$3(-2y + 8) + y = 9$
$-6y + 24 + y = 9$
$-5y = -15$
$y = 3$
If $y = 3$, then $x = -2(3) + 8$
$\qquad x = -6 + 8 = 2$
Answer: $(2, 3)$, or $x = 2$, $y = 3$

**43.** $2s - 5t = 15$
$-3s + 2t = -17 \Rightarrow \qquad -3s = -2t - 17$
$$s = \frac{2}{3}t + \frac{17}{3}$$
If $s = \frac{2}{3}t + \frac{17}{3}$, then
$$2\left(\frac{2}{3}t + \frac{17}{3}\right) - 5t = 15$$
$$\frac{4}{3}t + \frac{34}{3} - \frac{15}{3}t = 15$$
$$-\frac{11}{3}t = \frac{45 - 34}{3}$$
$$-\frac{11}{3}t = \frac{11}{3}$$
$$t = -1$$
If $t = -1$, then $s = \frac{2}{3}(-1) + \frac{17}{3}$
$$s = -\frac{2}{3} + \frac{17}{3} = \frac{15}{3} = 5$$
Answer: $s = 5$, $t = -1$

**45.** $2x + 4y = -2 \qquad 2x + 4y = -2$
$3x - 2y = -5 \Rightarrow \dfrac{6x - 4y = -10}{8x \qquad = -12}$
$$x = -\frac{12}{8} = -\frac{3}{2}$$
If $x = -\frac{3}{2}$,
$$2\left(-\frac{3}{2}\right) + 4y = -2$$
$$-3 + 4y = -2$$
$$4y = 1$$
$$y = \frac{1}{4}$$
Answer: $x = -\frac{3}{2}$, $y = \frac{1}{4}$

**47.** $3x + 2y = 6$
$y = -\frac{3}{2}x + 3 \Rightarrow \qquad 2y = -3x + 6$
$$\qquad\qquad 2y + 3x = 6,$$
which is equivalent to $3x + 2y = 6$.
Answer: Since the two given equations represent the same set of coordinates, the system is *dependent*.

**49.** $3x - y = 8 \Rightarrow 9x - 3y = 24$
$\quad\; 2x + 3y = 4 \qquad \underline{2x + 3y = \;\;4}$
$$11x \qquad = 28$$
$$x \qquad = \frac{28}{11} \Rightarrow$$
One unique solution exists for the system.
Answer: The system is *consistent*.

**51. (a)** If $x$ is the number of calculators produced in a day, the total cost to produce them is $C(x) = 6x + 240$.

**(b)** If $x$ calculators are sold at \$10 each, total revenues $R(x) = 10x$.

**(c)** The break-even point is where daily revenue will cover daily costs.
$R(x) = C(x)$,
$10x = 6x + 240$
$4x = 240$
$x = 60$
If $x = 60$, then $R(x) = 10x = 600$
Answer: The break-even point
$[(x, R(x)) = (60, 600) = (x, C(x))]$

**53.** A *quadratic function* is defined by a polynomial expression of degree 2.

**55. (a)** $y = x^2$

**(b)** $y = (x + 2)^2$ is a two unit shift to the left of $y = x^2$.

**(c)** $y = (x + 2)^2 - 3$ is a three unit shift down from $y = (x + 2)^2$.

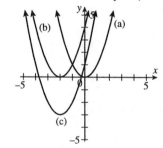

**57.** $y = -x^2 - 2$: Invert $y = x^2$, then shift the resulting graph two units down.

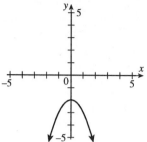

**59.** $y = 2x^2 - 1$: "Steepen" the graph of $y = x^2$ by doubling each $y$-value associated with $x$, and then shift that "steepened" graph down one unit.

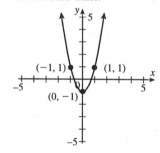

**61.** $y = 2(x - 1)^2$: Shift $y = x^2$ one unit to the right, then "steepen" the graph by doubling each $y$-value associated with $x$.

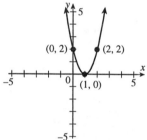

**63.** $f(x) = -3(x + 2)^2 + 3$ has *vertex* $(-2, 3)$ and *axis of symmetry* $x = -2$. Because $-3 < 0$, the parabola is *concave down*. From $(-\infty, -2]$, the function is *increasing*, and from $[-2, \infty)$ the function is *decreasing*.

**65.** If $y = a(x - h)^2 + k$: **(a)** the vertex is $(h, k)$; and **(b)** the axis of symmetry is $x = h$.

**67.** $x = (y-2)^2 + 3$ has vertex $(3, 2)$ and axis of symmetry $y = 2$.

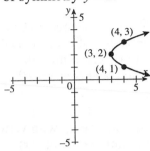

**69.** $y = -2x^2 - 12x + 11$
$y = -2(x^2 + 6x) + 11$
$y = -2(x^2 + 6x + 9) + 18 + 11$
$y = -2(x+3)^2 + 29$

**71.** If $ax^2 + bx + c = 0$, then
$$x = \frac{-b \pm \sqrt{b^2 - 4ac}}{2a}$$

**73.** $x^2 + 7x + 10 = 0 \qquad a = 1, \, b = 7, \, c = 10$
$$x = \frac{-7 \pm \sqrt{7^2 - 4(1)(10)}}{2(1)}$$
$$= \frac{-7 \pm \sqrt{49 - 40}}{2} = \frac{-7 \pm \sqrt{9}}{2}$$
$$= \frac{-7 \pm 3}{2}$$
$$x = -\frac{10}{2} = -5 \text{ or } x = -\frac{4}{2} = -2$$

**75.** $x^2 + 4x - 7 = 0 \qquad a = 1, \, b = 4, \, c = -7$
$$x = \frac{-4 \pm \sqrt{4^2 - 4(1)(-7)}}{2(1)}$$
$$= \frac{-4 \pm \sqrt{16 + 28}}{2} = \frac{-4 \pm \sqrt{44}}{2}$$
$$= \frac{-4 \pm 2\sqrt{11}}{2}$$
$$x = -2 + \sqrt{11} \text{ or } x = -2 - \sqrt{11}$$

**77.** $2x^2 - x - 3 = 0, \qquad a = 2, \, b = -1, \, c = -3$
$$x = \frac{1 \pm \sqrt{(-1)^2 - 4(2)(-3)}}{2(2)}$$
$$= \frac{1 \pm \sqrt{1 + 24}}{4} = \frac{1 \pm 5}{4}$$
$$x = \frac{6}{4} = \frac{3}{2} \text{ or } x = -\frac{4}{4} = -1$$

**79.** If $y = a(x - h)^2 + k$ opens downward, then $a < 0$. If it intersects the $x$-axis in two points, then $k > 0$, since $k$ represents the $y$-coordinate of the parabola's vertex. The domain is all real numbers, while the range is all real numbers $y \leq k$.

**81.** $f(x) = 2x^2 + x - 6$
**(a)** By factoring:
$f(x) = (2x - 3)(x + 2)$
When $f(x) = 0$, the parabola will cross the $x$-axis.
$(2x - 3)(x + 2) \Rightarrow$
$2x - 3 = 0 \text{ or } x + 2 = 0$
$2x = 3 \,\big|\, x = -2$
$x = \dfrac{3}{2}$

Answer: The $x$-intercepts are $\left(\dfrac{3}{2}, \, 0\right)$ and $(-2, 0)$.

**(b)** By completing the square:
$f(x) = 2x^2 + x - 6 = 0$
$$f(x) = 2\left(x^2 + \frac{1}{2}x\right) = 6$$
$$x^2 + \frac{1}{2}x = 3$$
$$x^2 + \frac{1}{2}x + \frac{1}{16} = 3 + \frac{1}{16}$$
$$\left(x + \frac{1}{4}\right)^2 = \frac{49}{16} \Rightarrow$$
$$x + \frac{1}{4} = \pm\frac{7}{4}$$
$$x = -\frac{1}{4} \pm \frac{7}{4}$$

Answer: The $x$-intercepts are

$\left(-\dfrac{8}{4},\ 0\right) = (-2,\ 0)$ and

$\left(\dfrac{6}{4},\ 0\right) = \left(\dfrac{3}{2},\ 0\right)$.

83. $4x^2 - 4x = -1 \Rightarrow 4x^2 - 4x + 1 = 0$
with $a = 4$, $b = -4$, $c = 1$.
The discriminant $= b^2 - 4ac$
$= (-4)^2 - 4(4)(1) = 16 - 16 = 0$.
Because the discriminant is 0, there is one real solution.

85. $y = |x^2 + 3x - 4|$
To graph this function, start with the graph of $y = x^2 + 3x - 4$, and then reflect about the $x$-axis any pieces of the graph falling below the $x$-axis, where $y < 0$.
$y = x^2 + 3x - 4$
$y = (x + 4)(x - 1)$ has $x$-intercepts at $(-4, 0)$ and $(1, 0)$.
Now complete the square to determine more:
$y = x^2 + 3x - 4$

$y = \left(x^2 + 3x + \dfrac{9}{4}\right) - \dfrac{9}{4} - 4$

$y = \left(x + \dfrac{3}{2}\right)^2 - \dfrac{25}{4}$

has *vertex* $\left(-\dfrac{3}{2},\ -\dfrac{25}{4}\right)$ and is *concave up*.
Remembering that the requested graph is $|x^2 + 3x - 4|$, the section of the graph related to the range $-\dfrac{25}{4} \le y < 0$ must be reflected across the $x$-axis:

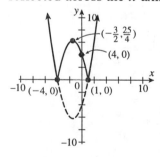

87. $f(x) = 2(x - 1)^2$ is concave up because $2 > 0$. Thus, the vertex of $(1, 0)$ is the

"minimum" of the graph. The minimum *value* of $f(x)$ is 0, which occurs at $x = 1$.

89. $f(x) = 3(x + 2)^2 - 1$ is concave up because $3 > 0$. Thus, the vertex of $(-2, -1)$ is the graph's "minimum". The minimum *value* of $f(x)$ is $-1$, which occurs at $x = -2$.

91. $f(x) = x^2 + 4x + 4$
$f(x) = (x + 2)^2$ is concave up with vertex $(-2, 0)$. The minimum *value* of $f(x) = 0$, which occurs at $x = -2$.

93. Let $x$ represent one number, with $40 - x$ representing the other. Their product is
$f(x) = 40x - x^2$
$f(x) = -(x^2 - 40x + 400) + 400$
$f(x) = -(x - 20)^2 + 400$
$(x, f(x)) = (20, 400)$ is the graph's vertex, and the graph is concave down. Thus, when each number is 20, a maximum product of 400 results.

95. $h = -16t^2 + 32t + 80$
(a) To maximize height, identify the vertex of the parabola formed when height is compared to time:
$h(t) = -16t^2 + 32t + 80$
$h(t) = -16(t^2 - 2t + 1) + 16 + 80$
$h(t) = -16(t - 1)^2 + 96$
The vertex $(t, h(t)) = (1, 96)$ is the "maximum" of this concave down parabola, so the maximum height reached by the object is 96 feet.
(b) The object reached 96 feet after 1 second.
(c) When will $h = 0$?
$0 = -16t^2 + 32t + 80$
$0 = -16(t^2 - 2t - 5)$
$0 = t^2 - 2t - 5$
$t^2 - 2t = 5$
$t^2 - 2t + 1 = 5 + 1$
$(t - 1)^2 = 6$
$t - 1 = \pm\sqrt{6}$
$t = 1 \pm \sqrt{6}$
Reject $1 - \sqrt{6}$, which is negative.
Answer: $1 + \sqrt{6} = 3.45$ seconds

# Chapter 2 Test: Standard Answer

**1.** $y = \dfrac{1}{\sqrt[3]{x^3 + 8}}$

The domain is restricted because $\sqrt[3]{x^3 + 8}$ would equal 0 if $x = -2$. So the domain is all real numbers $x \neq -2$.

**2.** $g(x) = \dfrac{3}{x}$

(a) $g(2 + x) = \dfrac{3}{2 + x}$

(b) $g(2) + g(x) = \dfrac{3}{2} + \dfrac{3}{x} = \dfrac{3x + 6}{2x}$

**3.** $g(x) = x^2 - x \Rightarrow \dfrac{g(x) - g(9)}{x - 9}$

$= \dfrac{x^2 - x - (9^2 - 9)}{x - 9} = \dfrac{x^2 - x - 72}{x - 9}$

$= \dfrac{(x - 9)(x + 8)}{x - 9} = x + 8$

**4.** If $3x - 2y = 6$, the $x$-intercept is
$3x - 2(0) = 6$
$3x = 6$
$x = 2$
The $y$-intercept is
$3(0) - 2y = 6$
$-2y = 6$
$y = -3$

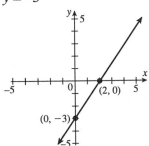

**5.** (a) $(2, -3)$ and $(-1, 4)$ form a line of slope
$m = \dfrac{4 - (-3)}{-1 - 2} = \dfrac{7}{-3} = -\dfrac{7}{3}.$

(b) $(-3, 2)$ and $(4, 2)$ form a horizontal line [no change in $y$]. Its slope is 0.

**6.** $y = 2x + 1$ has a domain and range of all real numbers.

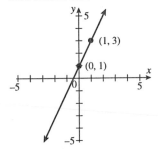

**7.** (a) Parallel to the $y$-axis and containing $(3, -2)$ is $x = 3$.

(b) Parallel to the $x$-axis and containing $(3, -2)$ is $y = -2$.

**8.** $m = \dfrac{1}{2}, \ b = -3 \Rightarrow y = \dfrac{1}{2}x - 3$

**9.** $2x - 3y = 5 \Rightarrow \quad -3y = -2x + 5$
$$y = \dfrac{2}{3}x - \dfrac{5}{3}$$
Slope $m = \dfrac{2}{3}$; $y$-intercept
$(0, b) = \left(0, \ -\dfrac{5}{3}\right).$

**10.** $(3, -5), (-2, 4)$ forms a line of slope
$\dfrac{-5 - 4}{3 - (-2)} = \dfrac{-9}{5}$
$y - 4 = -\dfrac{9}{5}(x + 2)$
$y - 4 = -\dfrac{9}{5}x - \dfrac{18}{5}$
$y = -\dfrac{9}{5}x + \dfrac{2}{5}$

**11.** Any line perpendicular to $y = -\dfrac{2}{5}x + 3$ will have slope $m = \dfrac{5}{2}$. A line containing $(2, 8)$ with slope $m = \dfrac{5}{2}$: $y - 8 = \dfrac{5}{2}(x - 2)$

**12.** $y = |x + 2|$

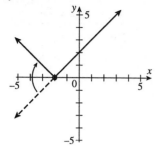

**13.** $y = \dfrac{|x+1|}{x+1}$

If $x = -1$, the function is undefined.

If $x > -1$, $y = \dfrac{|x+1|}{x+1} = \dfrac{x+1}{x+1} = 1$.

If $x < -1$, $y = \dfrac{|x+1|}{x+1} = \dfrac{-(x+1)}{x+1} = -1$.

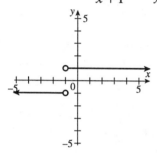

**14.** $y = 2 - x$ for $-1 \le x \le 2$.

$y = -x + 2$

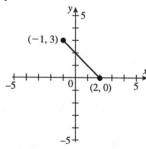

**15.** $y = (x-2)^2 + 3$ is a shift of $y = x^2$ two units right and three units up.

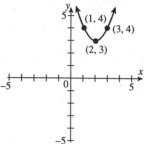

**16.** $y = f(x) = x^2 - 9$ is a shift of $y = x^2$ nine units down.

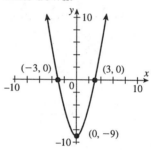

**17.** $y = -5x^2 + 20x - 1$

(a) $y = -5(x^2 - 4x) - 1$
$y = -5(x^2 - 4x + 4) + 20 - 1$
$y = -5(x-2)^2 + 19$

(b) The vertex is $(2, 19)$.

(c) The axis of symmetry is $x = 2$.

(d) The domain is all real numbers (no restrictions on $x$). The range is all real numbers $y \le 19$; because the parabola is concave down, the vertex holds the maximum value for $y$.

**18.** $3x^2 - 8x - 3 = 0 \qquad a = 3,\ b = -8,\ c = -3$

$x = \dfrac{-(-8) \pm \sqrt{(-8)^2 - 4(3)(-3)}}{2(3)}$

$= \dfrac{8 \pm \sqrt{64 + 36}}{6} = \dfrac{8 \pm \sqrt{100}}{6}$

$= \dfrac{8 \pm 10}{6}$

Answer: $x = \dfrac{18}{6} = 3$ or

$x = \dfrac{-2}{6} = -\dfrac{1}{3}$.

**19.** $y = -x^2 + 4x + 7$ will cross the $x$-axis when $y = 0$.

$0 = -(x^2 - 4x - 7)$

$0 = x^2 - 4x - 7 \qquad a = 1, b = -4, c = -7$

$x = \dfrac{-(-4) \pm \sqrt{(-4)^2 - 4(1)(-7)}}{2(1)}$

$= \dfrac{4 \pm \sqrt{16 + 28}}{2} = \dfrac{4 \pm \sqrt{44}}{2}$

$= \dfrac{4 \pm 2\sqrt{11}}{2}$

Answer: $x = 2 + \sqrt{11}$ or $x = 2 - \sqrt{11}$.

**20.** Let $x$ represent the first odd positive integer. $x + 2$ represents the next consecutive odd integer.

$x^2 + (x + 2)^2 = 202$

$x^2 + x^2 + 4x + 4 = 202$

$2x^2 + 4x + 4 = 202$

$x^2 + 2x + 2 = 101$

$x^2 + 2x = 99$

$x^2 + 2x + 1 = 100$

$(x + 1)^2 = 100$

$x + 1 = \pm\sqrt{100}$

$x + 1 = \pm 10$

$x = -1 \pm 10 = 9$ or $-11$.

Disregard $-11$, as the first odd integer must be positive.

Answer: The two odd, positive, consecutive integers are 9 and 11.

**21. (a)** $y = x^2 + 3x + 1 \qquad a = 1, b = 3, c = 1$

has discriminant

$(3)^2 - 4(1)(1) = 9 - 4 = 5$

Since the discriminant is a real number $> 0$ which is not a perfect square, the parabola intercepts the $x$-axis at two different irrational values.

**(b)** $y = 6x^2 + 5x - 6 \qquad a = 6, b = 5, c = -6$

has discriminant

$5^2 - 4(6)(-6) = 25 + 144 = 169$

Since the discriminant is a real number $> 0$ which *is* a perfect square, the parabola intercepts the $x$-axis at two different rational values.

**22.** $3x + 4y = 7 \quad \Rightarrow \quad 9x + 12y = 21$

$2x - 3y = 16 \qquad \underline{8x - 12y = 64}$

$\qquad\qquad\qquad\qquad 17x \qquad = 85$

$\qquad\qquad\qquad\qquad\quad x \qquad = 5$

If $x = 5$,

$3(5) + 4y = 7$

$15 + 4y = 7$

$4y = -8$

$y = -2$

Answer: $x = 5$, $y = -2$

**23.** $f(x) = -\dfrac{1}{2}x^2 - 6x + 2$

$f(x) = -\dfrac{1}{2}(x^2 + 12x) + 2$

$f(x) = -\dfrac{1}{2}(x^2 + 12x + 36) + 18 + 2$

$f(x) = -\dfrac{1}{2}(x + 6)^2 + 20$

$f(x)$ is concave down because $-\dfrac{1}{2} < 0$; the vertex $(-6, 20)$ shows that the maximum value for $f(x) = 20$, which occurs at $x = -6$.

**24.** Let $PQ = x$ since $PQRC$ is a rectangle, $RC = x$ also. If $PQ + QR + RC + CB = 28$ and $CB = 4$:

$x + QR + x + 4 = 28$

$QR = 28 - 4 - 2x = 24 - 2x$

$A(x) = x(24 - 2x)$

$A(x) = -2x^2 + 24x$

$A(x) = -2(x^2 - 12x)$

$A(x) = -2(x^2 - 12x + 36) + 72$

$A(x) = -2(x - 6)^2 + 72$

Answer: The maximum area for rectangle $PQRC$, 72 square feet, is achieved at $x = 6$ feet.

**25.** $h(t) = 64t - 16t^2$

$h(t) = -16(t^2 - 4t)$

$h(t) = -16(t^2 - 4t + 4) + 64$

$h(t) = -16(t - 2)^2 + 64$

Answer: The maximum height achieved by the object is 64 feet, which occurs at $t = 2$ seconds.

## Chapter 2 Test: Multiple Choice

1. $2x - 3y = 5$
   $-3y = -2x + 5$
   $y = \frac{2}{3}x - \frac{5}{3}$ has slope $m = \frac{2}{3}$
   A line parallel to $2x - 3y = 5$ will also have
   slope $= \frac{2}{3}$. If it also contains $(-8, 4)$, its
   equation will be $y - 4 = \frac{2}{3}(x + 8)$.
   The answer is (b).

2. If $f(x) = x - 2$, then
   $f(x - 2) = (x - 2) - 2 = x - 4$.
   The answer is (d).

3. If $g(x) = (x - 2)^2$, then
   $$\frac{g(x) - g(7)}{x - 7} = \frac{(x - 2)^2 - (7 - 2)^2}{x - 7}$$
   $$= \frac{x^2 - 4x + 4 - 25}{x - 7} = \frac{x^2 - 4x - 21}{x - 7}$$
   $$= \frac{(x - 7)(x + 3)}{x - 7} = x + 3$$
   The answer is (d).

4. A line containing $(2, -3)$ parallel to the
   $y$-axis will have a "fixed" $x$-value at $x = 2$
   (a vertical line).
   The answer is (a).

5. I is false—a horizontal line has slope 0.
   Thus, II is also false. III is true, because
   the product of slopes of perpendicular
   lines equals $-1$.
   The answer is (c).

6. $2x - 3y = 6$
   $-3y = -2x + 6$
   $y = \frac{2}{3}x - 2$ has slope $m = \frac{2}{3}$. A line
   perpendicular to $2x - 3y = 6$ will have
   slope $= -\frac{3}{2}$.
   The answer is (b).

7. Through $(2, -3)$ and $(-1, 6)$, the line will
   have slope $m = \frac{-3 - 6}{2 - (-1)} = \frac{-9}{3} = -3$.
   $y - (-3) = -3(x - 2)$
   $y + 3 = -3x + 6$
   $y = -3x + 3$
   The answer is (b).

8. A sketch is helpful:

   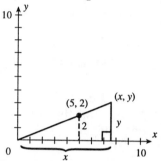

   Two similar triangles are formed, which
   will help define $y$ in terms of $x$:
   $$\frac{2}{y} = \frac{5}{x} \implies 2x = 5y$$
   $$\frac{2}{5}x = y$$
   The triangle's area can be determined by
   $A(x) = \frac{1}{2}(x)(y)$
   $$A(x) = \frac{1}{2}(x)\left(\frac{2}{5}x\right)$$
   $$A(x) = \frac{1}{5}x^2$$
   The answer is (d).

9. $(0, -4)$ and $(2, 0)$ creates a line with slope
   $\frac{-4 - 0}{0 - 2} = \frac{-4}{-2} = 2$
   $y - 0 = 2(x - 2)$
   $y = 2x - 4 \implies 2x - y = 4$
   The answer is (e).

10. $f(x) = \begin{cases} x - 2 \text{ if } -3 \le x \le 1 \\ -x + 1 \text{ if } 1 < x \end{cases}$
    $f(-3) = -3 - 2 = -5$ and $f(1) = 1 - 2 = -1$
    $\implies f(-3) + f(1) = -5 - 1 = -6$
    The answer is (a).

**11.** $y = f(x) = \dfrac{|x-1|}{x-1}$ is undefined at $x = 1$.

For $x > 1$, $\dfrac{|x-1|}{x-1} = \dfrac{x-1}{x-1} = 1$.

For $x < 1$, $\dfrac{|x-1|}{x-1} = \dfrac{-(x-1)}{x-1} = -1$.

Domain: All real numbers $x \neq 1$.
Range: $\{1, -1\}$
The answer is (b).

**12.** $y = [|x|]$. Select $x = -1.5$, which is such that $-2 < x < -1$. $|-1.5| = 1.5$, and $[1.5] = 1$, the greatest integer less than or equal to 1.5.
The answer is (d).

**13.** $y = (x-2)^2 - 5$ has vertex $(2, -5)$ and axis of symmetry $x = 2$.
The answer is (c).

**14.** $y = -2(x+1)^2 - 3$ is concave down because $-2 < 0$; thus, the function achieves its maximum value at the vertex $(-1, -3)$. All $y \leq -3$ make up the range of this function.
The answer is (e).

**15.** The answer is (b).

**16.** $f(x) = x^2 - 8x + 10$
$f(x) = (x^2 - 8x + 16) - 16 + 10$
$f(x) = (x-4)^2 - 6$
achieves its *minimum* value of $-6$ at $x = 4$.
The answer is (a).

**17.**

$A(x) = (10 + 2x)(6 + 2x) = 60 + 80 = 140$
$A(x) = 60 + 32x + 4x^2 = 140$
$A(x) = 4x^2 + 32x = 80$
$A(x) = x^2 + 8x = 20$
$A(x) = x^2 + 8x - 20 = 0$
The answer is (d)

**18.** $y = -4x^2 + 20x - 25$
$y = -(4x^2 - 20x) - 25$
$y = -4(x^2 - 5x) - 25$
$y = -4\left(x^2 - 5x + \dfrac{25}{4}\right) + 25 - 25$
$y = -4\left(x - \dfrac{5}{2}\right)^2$ is concave down with a single $x$-intercept (the vertex).
The answer is (c).

**19.** $3x^2 + 6x + 2 = 0 \qquad a = 3, \, b = 6, \, c = 2$

$x = \dfrac{-6 \pm \sqrt{6^2 - 4(3)(2)}}{2(3)}$

$= \dfrac{-6 \pm \sqrt{36 - 24}}{6} = \dfrac{-6 \pm \sqrt{12}}{6}$

$= \dfrac{-6 \pm 2\sqrt{3}}{6} = \dfrac{-3 \pm \sqrt{3}}{3}$

The answer is (a).

**20.** $2x - y = 4 \Rightarrow 6x - 3y = 12$
$\phantom{2}x + 3y = 7 \qquad\quad x + 3y = \phantom{1}7$
$\phantom{2x+3y=7 \Rightarrow} \overline{\phantom{6}7x \phantom{-3y} = 19}$
$\phantom{2x+3y=7 \Rightarrow 6} x \phantom{-3y} = \dfrac{19}{7}$

If $x = \dfrac{19}{7}$, $\dfrac{19}{7} + 3y = 7 = \dfrac{49}{7}$

$3y = \dfrac{30}{7}$

$y = \dfrac{30}{21} = \dfrac{10}{7}$

The system has a unique solution at $\left(\dfrac{19}{7}, \dfrac{10}{7}\right)$.
The answer is (c).

# Chapter 3: Polynomial and Rational Functions

## Exercises 3.1

**1.**

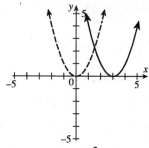

For $y = (x - 3)^2$:
Domain: all reals
Range: all $y \geq 0$
Decreasing on: $(-\infty, 3]$
Increasing on: $[3, \infty)$
Concave up on: $(-\infty, \infty)$

**3.**

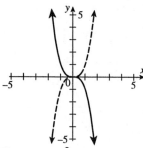

For $y = -x^3$:
Domain: all reals
Range: all reals
Decreasing on: $(-\infty, \infty)$
Concave up on: $(-\infty, 0)$
Concave down on: $(0, \infty)$

**5.**

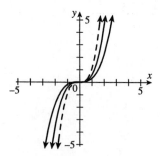

For $h(x) = \frac{1}{4}x^3$:
Domain: all reals
Range: all reals
Increasing on: $(-\infty, \infty)$
Concave down on: $(-\infty, 0)$
Concave up on: $(0, \infty)$

**7.**

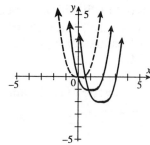

For $h(x) = (x - 2)^4 - 2$:
Domain: all reals
Range: all $y \geq -2$
Decreasing on: $(-\infty, 2]$
Increasing on: $[2, \infty)$
Concave up on: $(-\infty, \infty)$

**9.**

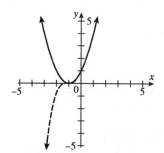

**11.**

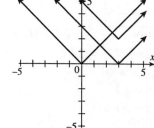

**13.**

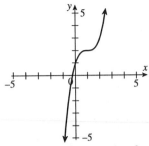

Translate the graph of $y = x^3$ one unit to the right and two units up.

**15.**

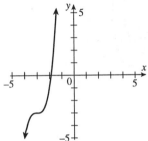

First sketch $y = 2x^3$ by multiplying the ordinates of $y = x^3$ by 2. Then translate $y = 2x^3$ three units left, and three units down.

**17.**

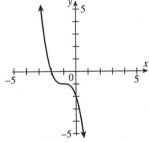

Reflect the graph of $y = x^3$ in the $x$-axis. Then translate one unit to the left and one unit down.

**19.** $y = (x - 3)^4 + 2$

**21.** $y = \left| x + \dfrac{3}{4} \right|$

**23.** $y = \left| x^3 - 1 \right|$

**25.**

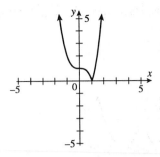

**27.** $y = x^3 + 3x^2 + 3x + 1 = (x + 1)^3$

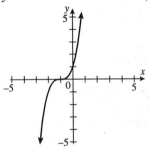

**29.** $y = -x^3 + 3x^2 - 3x + 1$
$= -(x^3 - 3x^2 + 3x - 1) = -(x - 1)^3$

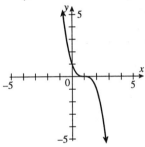

**31.** $f(x) = x^3$ and $f(3) = 27$

$$\frac{f(x) - f(3)}{x - 3} = \frac{x^3 - 27}{x - 3}$$

$$= \frac{(x - 3)(x^2 + 3x + 9)}{(x - 3)} = x^2 + 3x + 9$$

**33.** $f(x) = x^4, f(1) = 1,$
$f(1 + h) = (1 + h)^4$
$= 1 + 4h + 6h^2 + 4h^3 + h^4$

$$\frac{f(1 + h) - f(1)}{h}$$

$$= \frac{1 + 4h + 6h^2 + 4h^3 + h^4 - 1}{h}$$

$$= \frac{4h + 6h^2 + 4h^3 + h^4}{h}$$

$$= \frac{h(4 + 6h + 4h^2 + h^3)}{h}$$

$$= 4 + 6h + 4h^2 + h^3$$

## Challenge Problem

**1.** Translating $y = mx + b$ to the right $h$ units gives the equation $y' = m(x - h) + b$ or $y = mx + b - mh$. This is equivalent to translating $y = mx + b$ down by $mh$ units.

## Exercises 3.2

**1.**

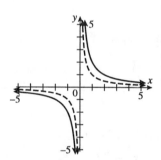

**3.**

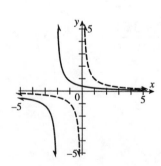

**5.**

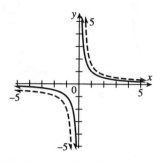

**7.**

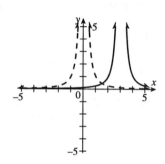

**9.**

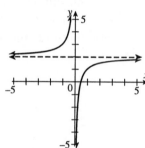

Asymptotes: $x = 0$, $y = 2$
Domain: all $x \neq 0$
Range: all $y \neq 2$
Increasing and concave up on: $(-\infty, 0)$
Increasing and concave down on: $(0, \infty)$

**11.**

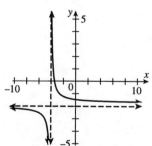

Asymptotes: $x = -4$, $y = -2$
Domain: all $x \neq -4$
Range: all $y \neq -2$

Decreasing and concave down on:
$(-\infty, -4)$
Decreasing and concave up on: $(-4, \infty)$

**13.**

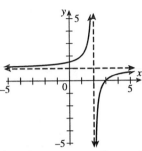

Asymptotes: $x = 2$, $y = 1$
Domain: all $x \neq 2$
Range: all $y \neq 1$
Increasing and concave up on: $(-\infty, 2)$
Increasing and concave down on: $(2, \infty)$

**15.**

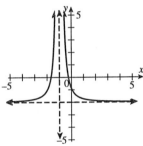

Asymptotes: $x = -1$, $y = -2$
Domain: all $x \neq -1$
Range: all $y > -2$
Increasing and concave up on: $(-\infty, -1)$
Decreasing and concave up on: $(-1, \infty)$

**17.**

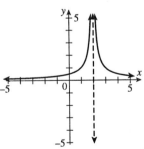

Asymptotes: $x = 2$, $y = 0$
Domain: all $x \neq 2$
Range: all $y > 0$
Increasing and concave up on: $(-\infty, 2)$
Decreasing and concave up on: $(2, \infty)$

**19.** $xy = 3$

$y = \dfrac{3}{x}$

Asymptotes: $x = 0$, $y = 0$

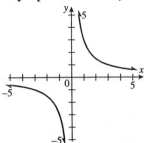

**21.** $xy - y = 1$
$y(x - 1) = 1$

$y = \dfrac{1}{x - 1}$

Asymptotes: $x = 1$, $y = 0$

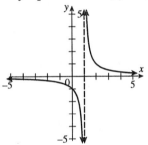

**23.** $f(x) = \dfrac{x^2 - 9}{x - 3} = \dfrac{(x - 3)(x + 3)}{x - 3} = (x + 3),$

$x \neq 3$
$f(x) = x + 3$, $x \neq 3$
No asymptotes.

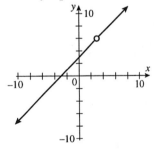

**25.**

$$\frac{x^2 - x - 6}{x - 3} = \frac{(x-3)(x+2)}{x-3} = x+2$$

$y = f(x) = x + 2,\ x \neq 3$

No asymptotes.

**27.**

$$\frac{x+1}{x^2 - 1} = \frac{x+1}{(x-1)(x+1)} = \frac{1}{x-1}$$

$$y = f(x) = \frac{1}{x-1},\ x \neq -1$$

Asymptotes: $x = 1,\ y = 0$

**29.**

$$\frac{x^3 - 8}{x - 2} = \frac{(x-2)(x^2 + 2x + 4)}{(x-2)}$$

$$= x^2 + 2x + 4$$

$y = f(x) = x^2 + 2x + 4,\ x \neq 2$

No asymptotes.

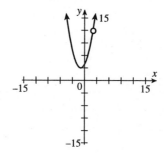

**31.**

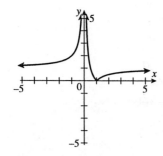

**33.** $f(x) = \dfrac{1}{x},\ f(3) = \dfrac{1}{3},$

$$\frac{f(x) - f(3)}{x - 3} = \frac{\frac{1}{x} - \frac{1}{3}}{x - 3} \cdot \frac{3x}{3x} = \frac{3 - x}{3x(x-3)} = \frac{-(x-3)}{3x(x-3)} = -\frac{1}{3x}$$

## Exercises 3.3

1. $f(x) = (x - 1)(x - 2)(x - 3)$. Roots of $f(x)$ are 1, 2, 3.

| Interval | $(-\infty, 1)$ | $(1, 2)$ | $(2, 3)$ | $(3, \infty)$ |
|---|---|---|---|---|
| Sign of $(x - 1)$ | − | + | + | + |
| Sign of $(x - 2)$ | − | − | + | + |
| Sign of $(x - 3)$ | − | − | − | + |
| Sign of $f(x)$ | − | + | − | + |

$f(x) < 0$ on $(-\infty, 1)$ and $(2, 3)$; $f(x) > 0$ on $(1, 2)$ and $(3, \infty)$.

3. $f(x) = \dfrac{(3x - 1)(x + 4)}{x^2(x - 2)}$

Numerator or denominator equals 0 when $x = \dfrac{1}{3}, -4, 0, 2$.

| Interval | $(-\infty, -4)$ | $(-4, 0)$ | $\left(0, \dfrac{1}{3}\right)$ | $\left(\dfrac{1}{3}, 2\right)$ | $(2, \infty)$ |
|---|---|---|---|---|---|
| Sign of $(3x - 1)$ | − | − | − | + | + |
| Sign of $(x + 4)$ | − | + | + | + | + |
| Sign of $x^2$ | + | + | + | + | + |
| Sign of $(x - 2)$ | − | − | − | − | + |
| Sign of $f(x)$ | − | + | + | − | + |

$f(x) < 0$ on $(-\infty, -4)$ and $\left(\dfrac{1}{3}, 2\right)$; $f(x) > 0$ on $(-4, 0)$, $\left(0, \dfrac{1}{3}\right)$, and $(2, \infty)$.

5. $f(x) = (x^2 + 2)(x - 4)(x + 1)$. Roots of $f(x)$ are 4 and −1.

| Interval | $(-\infty, -1)$ | $(-1, 4)$ | $(4, \infty)$ |
|---|---|---|---|
| Sign of $(x^2 + 2)$ | + | + | + |
| Sign of $(x - 4)$ | − | − | + |
| Sign of $(x + 1)$ | − | + | + |
| Sign of $f(x)$ | + | − | + |

$f(x) < 0$ on $(-1, 4)$; $f(x) > 0$ on $(-\infty, -1)$ and $(4, \infty)$.

7. $f(x) = \dfrac{x - 10}{3(x + 1)(5x - 1)}$

Numerator or denominator equals 0 when $x = 10, -1, \dfrac{1}{5}$.

| Interval | $(-\infty, -1)$ | $\left(-1, \dfrac{1}{5}\right)$ | $\left(\dfrac{1}{5}, 10\right)$ | $(10, \infty)$ |
|---|---|---|---|---|
| Sign of $(x - 10)$ | − | − | − | + |
| Sign of $(x + 1)$ | − | + | + | + |
| Sign of $(5x - 1)$ | − | − | + | + |
| Sign of $f(x)$ | − | + | − | + |

$f(x) < 0$ on $(-\infty, -1)$ and $\left(\dfrac{1}{5}, 10\right)$; $f(x) > 0$ on $\left(-1, \dfrac{1}{5}\right)$ and $(10, \infty)$.

**9.**

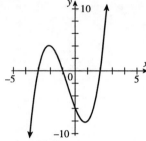

x-intercepts:
$$0 = (x + 3)(x + 1)(x - 2)$$
$$x = -3, -1, 2$$
y-intercept:
$$x = 0$$
$$y = (0 + 3)(0 + 1)(0 - 2) = -6$$

**11.**

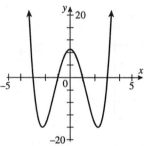

x-intercepts:
$$0 = (x + 3)(x + 1)(x - 1)(x - 3)$$
$$x = -3, -1, 1, 3$$
y-intercept:
$$x = 0$$
$$y = (0 + 3)(0 + 1)(0 - 1)(0 - 3) = 9$$

**13.**

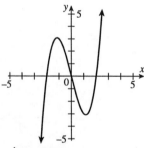

x-intercepts:
$$0 = x^3 - 4x$$
$$0 = x(x^2 - 4)$$
$$0 = x(x - 2)(x + 2)$$
$$x = 0, 2, -2$$
y-intercept:
$$x = 0$$
$$y = 0^3 - 4 \cdot 0 = 0$$

**15.**

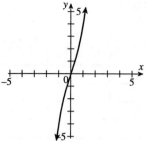

x-intercept:
$$0 = x^3 + 3x$$
$$0 = x(x^2 + 3)$$
$$x = 0$$
y-intercept:
$$x = 0$$
$$y = 0^3 + 3 \cdot 0 = 0$$

**17.**

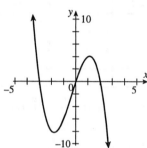

x-intercepts:
$$0 = -x^3 - x^2 + 6x$$
$$0 = -x(x^2 + x - 6)$$
$$0 = -x(x + 3)(x - 2)$$
$$x = 0, -3, 2$$
y-intercept:
$$x = 0$$
$$y = -0^3 - 0^2 + 6 \cdot 0 = 0$$

**19.**

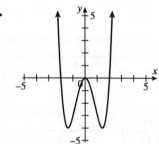

x-intercepts:
$$0 = x^4 - 4x^2$$
$$0 = x^2(x^2 - 4)$$

$$0 = x^2(x-2)(x+2)$$
$$x = 0, 2, -2$$
$y$-intercept:
$$x = 0$$
$$y = 0^4 - 4 \cdot 0^2 = 0$$

**21.**

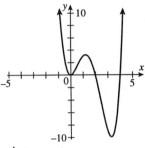

$x$-intercepts:
$$0 = x^4 - 6x^3 + 8x^2$$
$$0 = x^2(x^2 - 6x + 8)$$
$$0 = x^2(x-4)(x-2)$$
$$x = 0, 2, 4$$
$y$-intercept:
$$x = 0$$
$$y = 0^4 - 6 \cdot 0^3 + 8 \cdot 0^2 = 0$$

**23.**

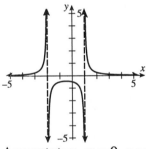

Asymptotes: $y = 0$, $x = -2$, $x = 1$
$x$-intercepts: none
$y$-intercept: $x = 0$,
$$y = \frac{1}{(0-1)(0+2)} = -\frac{1}{2}$$

**25.**

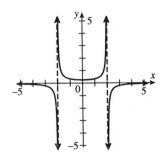

Asymptotes: $x = -2$, $x = 2$, $y = 0$
$x$-intercepts: none
$y$-intercept: $x = 0$, $y = \dfrac{1}{4 - 0^2} = \dfrac{1}{4}$

**27.**

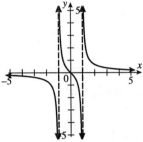

Asymptotes: $x = -1$, $x = 1$, $y = 0$
$x$-intercept: $x = 0$
$y$-intercept: $x = 0$, $.y = \dfrac{0}{0^2 - 1} = 0$

**29.**

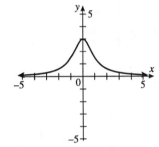

Asymptote: $y = 0$
$x$-intercepts: none
$y$-intercept: $x = 0$, $y = \dfrac{3}{0^2 + 1} = 3$

**31.**

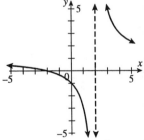

Asymptotes: $x = 2$, $y = 1$
$x$-intercept: $x = -2$
$y$-intercept: $x = 0$, $y = \dfrac{0+2}{0-2} = -1$

**33.**

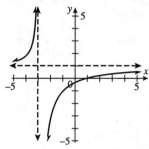

Asymptotes: $x = -3$, $y = 1$

$x$-intercept: $x = 1$

$y$-intercept: $x = 0$, $\quad y = \dfrac{0 - 1}{0 + 3} = -\dfrac{1}{3}$

**35.**

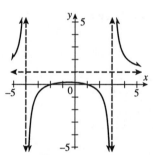

$y = \dfrac{x^2 + x - 2}{x^2 + x - 12} = \dfrac{(x+2)(x-1)}{(x+4)(x-3)}$

Asymptotes: $x = -4$, $x = 3$, $y = 1$

$x$-intercepts: $x = -2$, $1$

$y$-intercept: $x = 0$, $\quad y = \dfrac{0^2 + 0 - 2}{0^2 + 0 - 12} = \dfrac{1}{6}$

**37.**

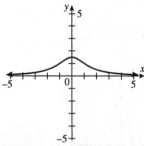

Horizontal asymptote: $y = 0$

$x$-intercepts: none

$y$-intercept: $x = 0$, $\quad y = \dfrac{9}{3 \cdot 0^2 + 6} = \dfrac{9}{6} = \dfrac{3}{2}$

**39.**

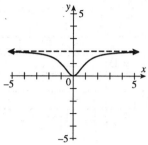

Horizontal asymptote: $y = 2$

$x$-intercept: $x = 0$

$y$-intercept: $x = 0$, $\quad y = \dfrac{2 \cdot 0^2}{0^2 + 1} = 0$

**41.** $\quad f(x) = \dfrac{x^2 + 3}{x} = x + \dfrac{3}{x}$

Oblique asymptote: $y = x$

**43.**

$$\require{enclose}\begin{array}{r}x - 2 \phantom{)} \\ x - 1 \enclose{longdiv}{x^2 - 3x + 1} \\ \underline{x^2 - x \phantom{+1}} \\ -2x + 1 \\ \underline{-2x + 2} \\ -1 \end{array}$$

$f(x) = \dfrac{x^2 - 3x + 1}{x - 1} = x - 2 - \dfrac{1}{x - 1}$

Oblique asymptote: $y = x - 2$

**45.**

$$\begin{array}{r}x - 1 \phantom{)} \\ x^2 - 1 \enclose{longdiv}{x^3 - x^2 - x - 2} \\ \underline{x^3 \phantom{-x^2} - x \phantom{-2}} \\ -x^2 \phantom{+} - 2 \\ \underline{-x^2 \phantom{-2} + 1} \\ -3 \end{array}$$

$f(x) = \dfrac{x^3 - x^2 - x - 2}{x^2 - 1} = x - 1 - \dfrac{3}{x^2 - 1}$

Oblique asymptote: $y = x - 1$

**47.** $g(x) = 2(2x-1)(x-2) + 2(x-2)^2 = (x-2)[2(2x-1) + 2(x-2)] = (x-2)(4x-2+2x-4)$
$= (x-2)(6x-6) = 6(x-2)(x-1)$
The roots of $g(x)$ are $x = 2, 1$.

| Interval | $(-\infty, 1)$ | $(1, 2)$ | $(2, \infty)$ |
|---|---|---|---|
| Sign of $(x-2)$ | $-$ | $-$ | $+$ |
| Sign of $(x-1)$ | $-$ | $+$ | $+$ |
| Sign of $g(x)$ | $+$ | $-$ | $+$ |

$g(x) < 0$ on $(1, 2)$; $g(x) > 0$ on $(-\infty, 1)$ and $(2, \infty)$.

**49.** $g(x) = \dfrac{x^2 - 2x(x+2)}{x^4} = \dfrac{x[x - 2(x+2)]}{x^4} = \dfrac{x - 2x - 4}{x^3} = -\dfrac{x+4}{x^3}$

Numerator or denominator are equal to 0 when $x = 0, -4$.

| Interval | $(-\infty, -4)$ | $(-4, 0)$ | $(0, \infty)$ |
|---|---|---|---|
| Sign of $-(x+4)$ | $+$ | $-$ | $-$ |
| Sign of $x^3$ | $-$ | $-$ | $+$ |
| Sign of $g(x)$ | $-$ | $+$ | $-$ |

$g(x) > 0$ on $(-4, 0)$; $g(x) < 0$ on $(-\infty, -4)$ and $(0, \infty)$.

**51.** $g(x) = 2\left(\dfrac{x-5}{x+2}\right)\dfrac{(x+2)-(x-5)}{(x+2)^2} = \dfrac{2(x-5)[x+2-x+5]}{(x+2)^3} = \dfrac{14(x-5)}{(x+2)^3}$

Numerator or denominator are equal to 0 when $x = -2, 5$.

| Interval | $(-\infty, -2)$ | $(-2, 5)$ | $(5, \infty)$ |
|---|---|---|---|
| Sign of $(x-5)$ | $-$ | $-$ | $+$ |
| Sign of $(x+2)^3$ | $-$ | $+$ | $+$ |
| Sign of $g(x)$ | $+$ | $-$ | $+$ |

$g(x) > 0$ on $(-\infty, -2)$ and $(5, \infty)$; $g(x) < 0$ on $(-2, 5)$.

**53.** $f(x) = \dfrac{2x^2}{3 - 5x^2} \cdot \dfrac{\frac{1}{x^2}}{\frac{1}{x^2}} = \dfrac{\frac{2x^2}{x^2}}{\frac{3}{x^2} - \frac{5x^2}{x^2}} = \dfrac{2}{\frac{3}{x^2} - 5}$

As $x \to \pm\infty$, $\dfrac{3}{x^2} \to 0$ and $g(x) \to -\dfrac{2}{5}$. Thus $y = -\dfrac{2}{5}$ is a horizontal asymptote.

**55.** (a)

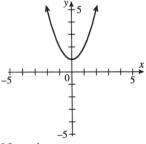

No $x$-intercepts
One turning point

(b)

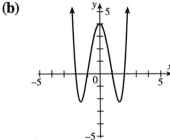

Four $x$-intercepts
Three turning points

**(c)**

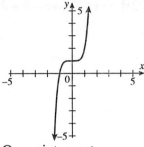

One $x$-intercept
No turning points

**(d)**

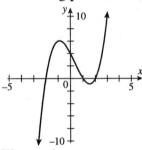

Three $x$-intercepts
Two turning points

## Challenge Problem

To show that $(x')^2 = (a-y)y$, use the fact that $(x', y)$ lies on the circle of radius $\dfrac{a}{2}$, centered at $\left(0, \dfrac{a}{2}\right)$.

So, $(x')^2 + \left(y - \dfrac{a}{2}\right)^2 = \left(\dfrac{a}{2}\right)^2$,

$$(x')^2 + y^2 - ay + \frac{a^2}{4} = \frac{a^2}{4}$$

$$(x')^2 = ay - y^2$$

$$(x')^2 = (a-y)y.$$

Plugging in $(x')^2 = (a-y)y$ into $\dfrac{a^2}{y^2} = \dfrac{x^2}{(x')^2}$

gives:

$$\frac{a^2}{y^2} = \frac{x^2}{(a-y)y}$$

$$a^2(a-y)y = x^2 y^2$$

$$a^3 y - a^2 y^2 = x^2 y^2$$

$$a^3 y = x^2 y^2 + a^2 y^2$$

$$a^3 y = y^2(a^2 + x^2)$$

$$\frac{a^3 y}{a^2 + x^2} = y^2$$

$$\frac{a^3}{a^2 + x^2} = y \text{ or } y = \frac{a^3}{a^2 + x^2}.$$

## Critical Thinking

**1.**

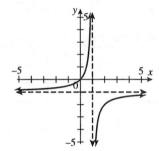

One such function is $f(x) = -\dfrac{x}{x-1}$.

**3.**

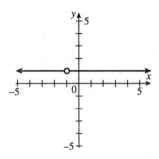

A counter example to this statement is the function $f(x) = \dfrac{x+1}{x+1}$.

## Exercises 3.4

**1.** $\dfrac{x}{2} - \dfrac{x}{5} = 6$

$10\left(\dfrac{x}{2} - \dfrac{x}{5}\right) = 10 \cdot 6$

$5x - 2x = 60$

$3x = 60$

$x = 20$

**3.** $\dfrac{x-1}{2} = \dfrac{x+2}{4}$

$4\left(\dfrac{x-1}{2}\right) = 4\left(\dfrac{x+2}{4}\right)$

$2(x-1) = x+2$

$2x-2 = x+2$

$x = 4$

**5.** $\dfrac{5}{x} - \dfrac{3}{4x} = 1$

$4x\left(\dfrac{5}{x} - \dfrac{3}{4x}\right) = 4x \cdot 1$

$20 - 3 = 4x$

$17 = 4x$

$\dfrac{17}{4} = x \text{ or } x = \dfrac{17}{4}$

**7.** $\dfrac{2x+1}{5} - \dfrac{x-2}{3} = 1$

$15\left(\dfrac{2x+1}{5}\right) - 15\left(\dfrac{x-2}{3}\right) = 15 \cdot 1$

$3(2x+1) - 5(x-2) = 15$

$6x+3-5x+10 = 15$

$x+13 = 15$

$x = 2$

**9.** $\dfrac{x+4}{2x-10} = \dfrac{8}{7}$

$7(2x-10)\left(\dfrac{x+4}{2x-10}\right) = 7(2x-10)\left(\dfrac{8}{7}\right)$

$7(x+4) = (2x-10)8$

$7x+28 = 16x-80$

$-9x = -108$

$x = 12$

**11.** $\dfrac{2}{x+6} - \dfrac{2}{x-6} = 0$

$(x+6)(x-6)\left(\dfrac{2}{x+6}\right)$

$\qquad - (x+6)(x-6)\left(\dfrac{2}{x-6}\right)$

$\qquad = (x+6)(x-6) \cdot 0$

$2(x-6) - 2(x+6) = 0$

$2x-12-2x-12 = 0$

$-24 = 0$

No solutions

**13.** $\dfrac{x+1}{x+10} = \dfrac{1}{2x}$

$2x(x+10) \cdot \dfrac{x+1}{x+10} = 2x(x+10) \cdot \dfrac{1}{2x}$

$2x(x+1) = (x+10) \cdot 1$

$2x^2 + 2x = x + 10$

$2x^2 + x - 10 = 0$

$(2x+5)(x-2) = 0$

$2x+5 = 0 \text{ or } x-2 = 0$

$x = -\dfrac{5}{2} \text{ or } x = 2$

**15.** $\dfrac{x^2}{2} - \dfrac{3x}{2} + 1 = 0$

$2 \cdot \dfrac{x^2}{2} - 2 \cdot \dfrac{3x}{2} + 2 \cdot 1 = 2 \cdot 0$

$x^2 - 3x + 2 = 0$

$(x-1)(x-2) = 0$

$x-1 = 0 \text{ or } x-2 = 0$

$x = 1 \text{ or } x = 2$

**17.** $\dfrac{5}{x^2-9} = \dfrac{3}{x+3} - \dfrac{2}{x-3}$

$(x+3)(x-3) \cdot \dfrac{5}{x^2-9}$

$\qquad = (x+3)(x-3) \cdot \dfrac{3}{x+3}$

$\qquad - (x+3)(x-3) \cdot \dfrac{2}{x-3}$

$5 = 3(x-3) - 2(x+3)$

$5 = 3x-9-2x-6$

$5 = x-15$

$20 = x \text{ or } x = 20$

**19.** $\dfrac{1}{x^2+4}+\dfrac{1}{x^2-4}=\dfrac{18}{x^4-16}$

$(x^2+4)(x^2-4)\cdot\dfrac{1}{x^2+4}$

$\qquad +(x^2+4)(x^2-4)\cdot\dfrac{1}{x^2-4}$

$=(x^2+4)(x^2-4)\cdot\dfrac{18}{x^4-16}$

$x^2-4+x^2+4=18$

$2x^2=18$

$x^2=9$

$x^2-9=0$

$(x+3)(x-3)=0$

$x+3=0 \text{ or } x-3=0$

$x=-3 \text{ or } x=3$

**21.** $x$-intercepts:

$0=\dfrac{2x-5}{x+1}-\dfrac{3}{x^2+x}$

$x(x+1)\cdot 0$

$=x(x+1)\cdot\dfrac{2x-5}{x+1}-x(x+1)\cdot\dfrac{3}{x^2+x}$

$0=x(2x-5)-3$

$0=2x^2-5x-3$

$0=(2x+1)(x-3)$

$2x+1=0 \text{ or } x-3=0$

$x=-\dfrac{1}{2} \text{ or } x=3$

Domain: all $x\neq -1,\,0$

**23.** $x$-intercepts:

$0=\dfrac{10-5x}{3x}-\dfrac{2}{x+5}-\dfrac{8-4x}{x+5}$

$3x(x+5)\cdot 0$

$=3x(x+5)\cdot\dfrac{10-5x}{3x}-3x(x+5)$

$\qquad \cdot\dfrac{2}{x+5}-3x(x+5)\cdot\dfrac{8-4x}{x+5}$

$0=(x+5)(10-5x)-3x\cdot 2-3x(8-4x)$

$0=-5x^2-15x+50-6x-24x+12x^2$

$0=7x^2-45x+50$

$0=(7x-10)(x-5)$

$7x-10=0 \text{ or } x-5=0$

$x=\dfrac{10}{7} \text{ or } x=5$

Domain: all $x\neq 0,\,-5$

**25.** $x$-intercepts:

$0=\dfrac{3}{2x^2-3x-2}-\dfrac{x+2}{2x+1}-\dfrac{2x}{10-5x}$

$5(2x+1)(x-2)\cdot 0$

$=5(2x+1)(x-2)\cdot\dfrac{3}{2x^2-3x-2}$

$\qquad -5(2x+1)(x-2)\cdot\dfrac{x+2}{2x+1}$

$\qquad -5(2x+1)(x-2)\cdot\dfrac{2x}{10-5x}$

$0=15-5(x-2)(x+2)+2x(2x+1)$

$0=15-5x^2+20+4x^2+2x$

$0=-x^2+2x+35$

$0=(-x-5)(x-7)$

$-x-5=0 \text{ or } x-7=0$

$x=-5 \text{ or } x=7$

Domain: all $x\neq -\dfrac{1}{2},\,2$

**27.** $\dfrac{v^2}{K}=\dfrac{2g}{m}$

$\dfrac{mv^2}{K}=2g$

$mv^2=2gK$

$m=\dfrac{2gK}{v^2}$

**29.** $S=\pi(r_1+r_2)s$

$S=\pi r_1 s+\pi r_2 s$

$S-\pi r_2 s=\pi r_1 s$

$\dfrac{S-\pi r_2 s}{\pi s}=r_1$

$\dfrac{S}{\pi s}-r_2=r_1 \text{ or } r_1=\dfrac{S}{\pi s}-r_2$

**31.** $d = \dfrac{s-a}{n-1}$

$d(n-1) = s - a$

$d(n-1) + a = s$ or $s = a + (n-1)d$

**33.** $\dfrac{1}{f} = \dfrac{1}{m} + \dfrac{1}{p}$

$fmp \cdot \dfrac{1}{f} = fmp \cdot \dfrac{1}{m} + fmp \cdot \dfrac{1}{p}$

$mp = fp + fm$

$mp - fm = fp$

$m(p - f) = fp$

$m = \dfrac{fp}{p-f}$

**35.** $\dfrac{x}{2} - \dfrac{x}{3} \le 5$

$6 \cdot \dfrac{x}{2} - 6 \cdot \dfrac{x}{3} \le 6 \cdot 5$

$3x - 2x \le 30$

$x \le 30$

**37.** $\dfrac{x+3}{4} - \dfrac{x}{2} > 1$

$4 \cdot \dfrac{x+3}{4} - 4 \cdot \dfrac{x}{2} > 4 \cdot 1$

$x + 3 - 2x > 4$

$-x > 1$

$x < -1$

**39.** $\dfrac{1}{2}(x+1) - \dfrac{2}{3}(x-2) < \dfrac{1}{6}$

$6 \cdot \dfrac{1}{2}(x+1) - 6 \cdot \dfrac{2}{3}(x-2) < 6 \cdot \dfrac{1}{6}$

$3(x+1) - 4(x-2) < 1$

$3x + 3 - 4x + 8 < 1$

$-x + 11 < 1$

$-x < -10$

$x > 10$

**41.** $\dfrac{x-2}{x+3} \ge 0$

$x - 2 = 0 \qquad x + 3 = 0$

$x = 2 \qquad\qquad x = -3$

There are three intervals:

$(-\infty, -3), (-3, 2], [2, \infty)$

Test $x = -4$: $\dfrac{-4-2}{-4+3} = \dfrac{-6}{-1} = 6 \ge 0$

Test $x = 0$: $\dfrac{0-2}{0+3} = -\dfrac{2}{3} \not\ge 0$

Test $x = 3$: $\dfrac{3-2}{3+3} = \dfrac{1}{6} \ge 0$

So $x < -3$ or $x \ge 2$

**43.** $\dfrac{(6-x)(3+x)}{x+1} \le 0$

$(6-x)(3+x) = 0 \qquad\qquad x + 1 = 0$

$6 - x = 0$ or $3 + x = 0 \qquad x = -1$

$x = 6$ or $x = -3$

There are four intervals:

$(-\infty, -3], [-3, -1), (-1, 6], [6, \infty)$

Test $x = -4$:

$\dfrac{(6-(-4))(3-4)}{-4+1} = \dfrac{-10}{-3} = \dfrac{10}{3} \not\le 0$

Test $x = -2$:

$\dfrac{(6-(-2))(3-2)}{-2+1} = \dfrac{8}{-1} = -8 \le 0$

Test $x = 0$: $\dfrac{(6-0)(3+0)}{0+1} = \dfrac{18}{1} = 18 \not\le 0$

Test $x = 7$: $\dfrac{(6-7)(3+7)}{7+1} = \dfrac{-10}{8} \le 0$

So, $-3 \le x < -1$ or $x \ge 6$

**45.** $\dfrac{x}{x-1} > 2$

$\dfrac{x}{x-1} - 2 \cdot \dfrac{x-1}{x-1} > 0$

$\dfrac{x - 2(x-1)}{x-1} > 0$

$\dfrac{x - 2x + 2}{x-1} > 0$

$\dfrac{-x+2}{x-1} > 0$

$-x + 2 = 0 \qquad\qquad x - 1 = 0$

$x = 2 \qquad\qquad\quad x = 1$

There are three intervals:
$(-\infty, 1), (1, 2), (2, \infty)$

Test $x = 0$: $\dfrac{-0+2}{0-1} = -2 \not> 0$

Test $x = \dfrac{3}{2}$: $\dfrac{-\frac{3}{2}+2}{\frac{3}{2}-1} = \dfrac{\frac{1}{2}}{\frac{1}{2}} = 1 > 0$

Test $x = 3$: $\dfrac{-3+2}{3-1} = \dfrac{-1}{2} \not> 0$

So, $1 < x < 2$

47. $\dfrac{11-x}{23-x} = \dfrac{2}{5}$

$5(23-x) \cdot \dfrac{11-x}{23-x} = 5(23-x) \cdot \dfrac{2}{5}$

$5(11-x) = (23-x) \cdot 2$

$55 - 5x = 46 - 2x$

$-3x = -9$

$x = 3$

49. Let $x$ = time (in hours) to do the job together. Then $\dfrac{1}{x}$ = portion of job done in 1 hour. Also, $\dfrac{1}{3}$ = portion of job done by first pipe in 1 hour, $\dfrac{1}{4}$ = portion of job done by second pipe in 1 hour.

$\dfrac{1}{3} + \dfrac{1}{4} = \dfrac{1}{x}$

$12x \cdot \dfrac{1}{3} + 12x \cdot \dfrac{1}{4} = 12x \cdot \dfrac{1}{x}$

$4x + 3x = 12$

$7x = 12$

$x = \dfrac{12}{7}$ or $x = 1\dfrac{5}{7}$ hours

51. Let $x$ be the larger fraction. Then the smaller is $\dfrac{1}{2}x$.

$x + \dfrac{1}{2}x = \dfrac{2}{3}$

$\dfrac{3}{2}x = \dfrac{2}{3}$

$x = \dfrac{2}{3} \cdot \dfrac{2}{3} = \dfrac{4}{9}$

The larger fraction is $\dfrac{4}{9}$ and the smaller is $\dfrac{2}{9}$.

53. Let $h$ be the height of the tree.

$\dfrac{h}{20} = \dfrac{1}{\frac{1}{3}}$

$h = 20(3) = 60$

The tree is 60 feet tall.

55. Using the hint with $N = 561$, $Q = 29$, and $R = 10$,

$\dfrac{561}{D} = 29 + \dfrac{10}{D}$

$D \cdot \dfrac{561}{D} = D \cdot 29 + D \cdot \dfrac{10}{D}$

$561 = 29D + 10$

$551 = 29D$

$19 = D$

The number is 19.

57. Let $x$ be the numerator of the fraction. The denominator is then $x + 1$.

$\dfrac{x+2.5}{x+1} = \dfrac{x+1}{x}$

$x(x+1) \cdot \dfrac{x+2.5}{x+1} = x(x+1) \cdot \dfrac{x+1}{x}$

$x(x+2.5) = (x+1)(x+1)$

$x^2 + 2.5x = x^2 + 2x + 1$

$0.5x = 1$

$x = 2$

The fraction is $\dfrac{2}{3}$.

59. $\dfrac{a}{b} = \dfrac{c}{d}$ implies $ad = bc$. Then,

$ad + bd = bc + bd$; $d(a + b) = b(c + d)$.

Therefore, $\dfrac{a+b}{b} = \dfrac{c+d}{d}$.

**61.** Let $x$ be the number of hardcover copies the store should order. Then $2x$ is the number of paperback copies.

$$\frac{30+2x}{50+x} = \frac{4}{3}$$

$$3(50+x) \cdot \frac{30+2x}{50+x} = 3(50+x) \cdot \frac{4}{3}$$

$$3(30+2x) = (50+x) \cdot 4$$

$$90+6x = 200+4x$$

$$2x = 110$$

$$x = 55$$

The store should order 55 hardcover copies and 110 paperback copies.

**63.** Let $x$ be the number of fish in the lake.

$$\frac{x}{200} = \frac{160}{4}$$

$$x = 200 \cdot \frac{160}{4}$$

$$x = 8000$$

There are 8000 fish in the lake.

**65.** Let $x$ be the numerical value of the grade in the fifth course.

$$3.4 = \frac{4 \cdot 4 + 4 \cdot 3 + 3 \cdot 3 + 1 \cdot 2 + x \cdot 3}{15}$$

$$3.4 = \frac{16+12+9+2+3x}{15}$$

$$(3.4)(15) = 39+3x$$

$$51 = 39+3x$$

$$12 = 3x$$

$$4 = x$$

She needs an $A$ in the fifth course.

**67.** $\dfrac{1}{R} = \dfrac{1}{R_1} + \dfrac{1}{R_2}$

$$\frac{1}{4000} = \frac{1}{20,000} + \frac{1}{R_2}$$

$$20,000R_2 \cdot \frac{1}{4000}$$

$$= 20,000R_2 \cdot \frac{1}{20,000} + 20,000R_2 \cdot \frac{1}{R_2}$$

$$5R_2 = R_2 + 20,000$$

$$4R_2 = 20,000$$

$$R_2 = 5000$$

A resistance of 5000 ohms is needed.

**69.** $\dfrac{1}{f} = \dfrac{1}{p} + \dfrac{1}{q}$

$$\frac{1}{f} = \frac{1}{30} + \frac{1}{15}$$

$$30f \cdot \frac{1}{f} = 30f \cdot \frac{1}{30} + 30f \cdot \frac{1}{15}$$

$$30 = f + 2f$$

$$30 = 3f$$

$$10 = f$$

The focal length is 10 cm.

**71.** $2x+3y = 7$

$$y = \frac{1}{x}$$

$$2x+3\left(\frac{1}{x}\right) = 7$$

$$x \cdot 2x + x \cdot 3 \cdot \frac{1}{x} = x \cdot 7$$

$$2x^2 + 3 = 7x$$

$$2x^2 - 7x + 3 = 0$$

$$(2x-1)(x-3) = 0$$

$$2x - 1 = 0 \text{ or } x - 3 = 0$$

$$x = \frac{1}{2} \text{ or } x = 3$$

For $x = \dfrac{1}{2}$, $y = \dfrac{1}{\frac{1}{2}} = 2$

For $x = 3$, $y = \dfrac{1}{3}$

$$\left(\frac{1}{2}, 2\right), \left(3, \frac{1}{3}\right)$$

**73.**

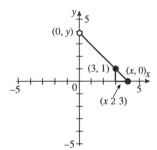

Using the fact that the two triangles in the diagram are similar,

$$\frac{y}{x} = \frac{1}{x-3}$$

$$y = \frac{x}{x-3}$$

The area of the triangle is:

$$A = \frac{1}{2}xy = \frac{1}{2}x\left(\frac{x}{x-3}\right)$$

$$A = \frac{x^2}{2(x-3)}$$

When $x = 3.1$,

$$A = \frac{(3.1)^2}{2(3.1-3)} = \frac{9.61}{0.2} = 48.05$$

## Challenge Problem

Using the hint, $x = 1 + \dfrac{1}{x}$,

$$x \cdot x = x \cdot 1 + x \cdot \frac{1}{x}$$

$$x^2 = x + 1$$

$$x^2 - x - 1 = 0$$

$$x = \frac{1 \pm \sqrt{1+4}}{2}$$

$$x = \frac{1 \pm \sqrt{5}}{2}$$

The golden ratio is $\dfrac{1+\sqrt{5}}{2} = \dfrac{\sqrt{5}+1}{2}$.

The reciprocal is

$$\frac{2}{1+\sqrt{5}} \cdot \frac{1-\sqrt{5}}{1-\sqrt{5}} = \frac{2\left(1-\sqrt{5}\right)}{1-5} = \frac{2\left(1-\sqrt{5}\right)}{-4}$$

$$= \frac{\sqrt{5}-1}{2}.$$

## Exercises 3.5

1. $P = 4s, k = 4$

3. $A = 5l, k = 5$

5. $z = kxy^3$

7. $z = \dfrac{kx}{y^3}$

9. $w = \dfrac{kx^2}{yz}$

11. $s = kt^2$

$$50 = k(10)^2$$

$$50 = 100k$$

$$\frac{1}{2} = k \text{ or } k = \frac{1}{2}$$

13. $u = kvw$

$$2 = k \cdot 15 \cdot \frac{2}{3}$$

$$2 = 10k$$

$$\frac{2}{10} = k \text{ or } k = \frac{1}{5}$$

15. $a = \dfrac{k}{b^2}$

$$10 = \frac{k}{5^2}$$

$$250 = k$$

$$a = \frac{250}{b^2} = \frac{250}{(25)^2} = \frac{250}{625} = \frac{2}{5}$$

17. $s = klw^2$

$$\frac{10}{3} = k \cdot 12 \cdot \left(\frac{5}{6}\right)^2$$

$$\frac{10}{3} = k \cdot 12 \cdot \frac{25}{36}$$

$$\frac{10}{3} = k \cdot \frac{25}{3}$$

$$\frac{3}{25} \cdot \frac{10}{3} = k \text{ or } k = \frac{2}{5}$$

$$s = \frac{2}{5}lw^2$$

$$s = \frac{2}{5} \cdot 15 \cdot \left(\frac{9}{4}\right)^2 = 6 \cdot \frac{81}{16} = \frac{486}{16}$$

$$s = \frac{243}{8}$$

**19.** $d = k \cdot t^2$
$12 = k \cdot 2^2$
$12 = 4k$
$3 = k$
$d = 3t^2$
$d = 3 \cdot 3^2 = 27$
The ball will roll 27 feet.

**21.** $F = kx$
$2.4 = k(1.8)$
$\dfrac{4}{3} = k$

$F = \dfrac{4}{3} \cdot 3 = 4$
A force of 4 pounds is required.

**23.** $P = \dfrac{k}{v}$

$50 = \dfrac{k}{150}$

$7500 = k$

$P = \dfrac{7500}{100} = 75$
The pressure exerted is 75 pounds per square inch.

**25.** $d = kt^2$
$256 = k \cdot 4^2$
$16 = k$
$d = 16 \cdot 7^2 = 784$
The object will fall 784 feet in 7 seconds.

**27.** $V = kr^3$
$288\pi = k \cdot 6^3$
$\dfrac{4}{3}\pi = k$

$V = \dfrac{4}{3}\pi \cdot 2^3 = \dfrac{32\pi}{3}$

The volume of the sphere is $\dfrac{32\pi}{3}$ cubic

inches.

**29.** $R = \dfrac{kl}{d^2}$

$64 = \dfrac{k \cdot 200}{(0.16)^2}$

$8.192 \times 10^{-3} = k$

$R = \dfrac{8.192 \times 10^{-3} \cdot 100}{(0.4)^2} = 5.12$ ohms

$R = \dfrac{8.192 \times 10^{-3} \cdot 100}{(0.04)^2} = 512$ ohms

The resistance is 5.12 ohms when the diameter is 0.4 inch, and 512 ohms when the diameter is 0.04 inch.

**31.** $x^2 + y^2 = (2.5)^2 = \left(\dfrac{5}{2}\right)^2$

(a) $s(y) = kyx^2 = ky\left(\dfrac{25}{4} - y^2\right)$

(b) $s(x) = kyx^2 = kx^2\sqrt{\dfrac{25}{4} - x^2}$

$= kx^2\sqrt{\dfrac{25 - 4x^2}{4}} = \dfrac{1}{2}kx^2\sqrt{25 - 4x^2}$

(c) If $x = y$, $s(x) = s(y)$

$ky\left(\dfrac{25}{4} - y^2\right) = \dfrac{1}{2}ky^2\sqrt{25 - 4y^2}$

$\dfrac{25}{4} - y^2 = \dfrac{1}{2}y\sqrt{25 - 4y^2}$

$\left[\dfrac{1}{4}(25 - 4y^2)\right]^2 = \left[\dfrac{1}{2}y\sqrt{25 - 4y^2}\right]^2$

$\dfrac{1}{16}(25 - 4y^2)^2 = \dfrac{1}{4}y^2(25 - 4y^2)$

$25 - 4y^2 = 4y^2$

$25 = 8y^2$

$\dfrac{25}{8} = y^2$

$\dfrac{5}{2\sqrt{2}} = y$

$y = \dfrac{5\sqrt{2}}{4}$

$x = y = \dfrac{5\sqrt{2}}{4}$

$$s = k \cdot \left(\frac{5\sqrt{2}}{4}\right)\left(\frac{5\sqrt{2}}{4}\right)^2$$

$$= k \cdot \frac{125 \cdot 2\sqrt{2}}{64} = 5.52k$$

**33.** $w = kd$

$150 = k \cdot 4000$

$0.0375 = k$

$w = (0.0375)(3900)$

$w = 146.25$

The object weighs 146 pounds.

**35.** $x = k_1z$, $y = k_2z$. So, $x + y = k_1z + k_2z$
$= (k_1 + k_2)z = kz$
Hence, $x + y$ varies directly as $z$.

**37.** $x = k_1z$, $y = k_2z$ and $xy = k_1zk_2z = k_1k_2z^2$
$= kz^2$. The product varies directly as the square of $z$.

**39.** $I = \dfrac{k}{d^2}$

$$20 = \frac{k}{4^2}$$

$320 = k$

$$I = \frac{320}{8^2} = 5$$

The illumination is 5 foot-candles.

**41.** The base of the triangle has length $x$ and the height has length $y = mx$. The area is given by $A = \dfrac{1}{2}x \cdot mx = \dfrac{1}{2}mx^2$. The area varies jointly as the square of $x$ and the slope of the line $y = mx$.

## Exercises 3.6

**1.**

$$\begin{array}{r|rrrr} +3 & 1 & -2 & -5 & +6 \\ & & +3 & +3 & -6 \\ \hline & 1 & +1 & -2 & +0 \end{array}$$

$x^2 + x - 2$; $r = 0$

**3.**

$$\begin{array}{r|rrrr} -1 & 2 & +1 & -3 & +7 \\ & & -2 & +1 & +2 \\ \hline & 2 & -1 & -2 & +9 \end{array}$$

$2x^2 - x - 2$; $r = 9$

**5.**

$$\begin{array}{r|rrrr} +2 & 1 & +5 & -7 & +8 \\ & & +2 & +14 & +14 \\ \hline & 1 & +7 & +7 & +22 \end{array}$$

$x^2 + 7x + 7$; $r = 22$

**7.**

$$\begin{array}{r|rrrrr} -2 & 1 & -3 & +7 & -2 & +1 \\ & & -2 & +10 & -34 & +72 \\ \hline & 1 & -5 & +17 & -36 & +73 \end{array}$$

$x^3 - 5x^2 + 17x - 36$; $r = 73$

**9.**

$$\begin{array}{r|rrrrr} +1 & 2 & +0 & -3 & +4 & -2 \\ & & +2 & +2 & -1 & +3 \\ \hline & 2 & +2 & -1 & +3 & +1 \end{array}$$

$2x^3 + 2x^2 - x + 3$; $r = 1$

**11.**

$$\begin{array}{r|rrrr} +3 & 1 & +0 & +0 & -27 \\ & & +3 & +9 & +27 \\ \hline & 1 & +3 & +9 & +0 \end{array}$$

$x^2 + 3x + 9$; $r = 0$

**13.**

$$\begin{array}{r|rrrr} -3 & 1 & +0 & +0 & +27 \\ & & -3 & +9 & -27 \\ \hline & 1 & -3 & +9 & +0 \end{array}$$

$x^2 - 3x + 9$; $r = 0$

**15.**

$$\begin{array}{r|rrrrr} +2 & 1 & +0 & +0 & +0 & -16 \\ & & +2 & +4 & +8 & +16 \\ \hline & 1 & +2 & +4 & +8 & +0 \end{array}$$

$x^3 + 2x^2 + 4x + 8$; $r = 0$

**17.**

$$\begin{array}{r|rrrrr} -2 & 1 & +0 & +0 & +0 & +16 \\ & & -2 & +4 & -8 & +16 \\ \hline & 1 & -2 & +4 & -8 & +32 \end{array}$$

$x^3 - 2x^2 + 4x - 8$; $r = 32$

**19.**

$$
\begin{array}{r|rrrrr}
+1 & 1 & -\dfrac{1}{2} & +\dfrac{1}{3} & -\dfrac{1}{4} & +\dfrac{1}{5} \\
   &   & +1 & +\dfrac{1}{2} & +\dfrac{5}{6} & +\dfrac{7}{12} \\
\hline
   & 1 & +\dfrac{1}{2} & +\dfrac{5}{6} & +\dfrac{7}{12} & \boxed{+\dfrac{47}{60}}
\end{array}
$$

$$x^3 + \frac{1}{2}x^2 + \frac{5}{6}x + \frac{7}{12};\; r = \frac{47}{60}$$

**21.**  $\dfrac{6x^3 - 5x^2 - 3x + 4}{2x - 1} = \dfrac{6x^3 - 5x^2 - 3x + 4}{2\left(x - \frac{1}{2}\right)}$

$$= \frac{1}{2}\left(\frac{6x^3 - 5x^2 - 3x + 4}{x - \frac{1}{2}}\right)$$

$$
\begin{array}{r|rrrr}
+\dfrac{1}{2} & 6 & -5 & -3 & +4 \\
   &   & +3 & -1 & -2 \\
\hline
   & 6 & -2 & -4 & \boxed{+2}
\end{array}
$$

$$\frac{6x^3 - 5x^2 - 3x + 4}{2x - 1}$$

$$= \frac{1}{2}\left(6x^2 - 2x - 4 + \frac{2}{x - \frac{1}{2}}\right)$$

$$= 3x^2 - x - 2 + \frac{2}{2x - 1}$$

$$3x^2 - x - 2;\; r = 2$$

**23.**  $\dfrac{6x^3 + 7x^2 + x + 8}{2x + 3}$

$$= \frac{6x^3 + 7x^2 + x + 8}{2\left(x + \frac{3}{2}\right)}$$

$$= \frac{1}{2}\left(\frac{6x^3 + 7x^2 + x + 8}{x + \frac{3}{2}}\right)$$

$$
\begin{array}{r|rrrr}
-\dfrac{3}{2} & 6 & +7 & +1 & +8 \\
   &   & -9 & +3 & -6 \\
\hline
   & 6 & -2 & +4 & \boxed{+2}
\end{array}
$$

$$\frac{6x^3 + 7x^2 + x + 8}{2x + 3}$$

$$= \frac{1}{2}\left(6x^2 - 2x + 4 + \frac{2}{x + \frac{3}{2}}\right)$$

$$= 3x^2 - x + 2 + \frac{2}{2x + 3}$$

$$3x^2 - x + 2;\; r = 2$$

**25.**

$$
\begin{array}{r|rrrr}
+2 & 1 & -1 & +3 & -2 \\
   &   & +2 & +2 & +10 \\
\hline
   & 1 & +1 & +5 & \boxed{+8}
\end{array}
$$

$$f(2) = 8$$

**27.**

$$
\begin{array}{r|rrrrr}
+3 & 1 & +0 & -3 & +1 & +2 \\
   &   & +3 & +9 & +18 & +57 \\
\hline
   & 1 & +3 & +6 & +19 & \boxed{+59}
\end{array}
$$

$$f(3) = 59$$

**29.**

$$
\begin{array}{r|rrrrrr}
+1 & 1 & +0 & -1 & +2 & +1 & -3 \\
   &   & +1 & +1 & +0 & +2 & +3 \\
\hline
   & 1 & +1 & +0 & +2 & +3 & \boxed{+0}
\end{array}
$$

$$f(1) = 0$$

**31.**  Let $f(x) = x^3 - 2x^2 + 3x - 5$
$f(2) = 2^3 - 2 \cdot 2^2 + 3 \cdot 2 - 5$
$= 8 - 8 + 6 - 5 = 1$
$r = 1$

**33.**  Let $f(x) = 2x^3 + 3x^2 - 5x + 1$
$f(3) = 2 \cdot 3^3 + 3 \cdot 3^2 - 5 \cdot 3 + 1$
$= 54 + 27 - 15 + 1 = 67$
$r = 67$

**35.**  Let $f(x) = 4x^5 - x^3 - 3x^2 + 2$
$f(-1) = 4 \cdot (-1)^5 - (-1)^3 - 3(-1)^2 + 2$
$= -4 + 1 - 3 + 2 = -4$
$r = -4$

**37.**

$$
\begin{array}{r|rrrr}
-1 & 1 & +6 & +11 & +6 \\
   &   & -1 & -5 & -6 \\
\hline
   & 1 & +5 & +6 & \boxed{+0}
\end{array}
$$

$x^3 + 6x^2 + 11x + 6 = (x^2 + 5x + 6)(x + 1)$
$= (x + 2)(x + 3)(x + 1)$
So, $x + 2 = 0$ or $x + 3 = 0$ or $x + 1 = 0$
$x = -2$ or $x = -3$ or $x = -1$

**39.**

$$
\begin{array}{r|rrrr}
+2 & 1 & +5 & -2 & -24 \\
  &   & +2 & +14 & +24 \\
\hline
  & 1 & +7 & +12 & +0 \\
\end{array}
$$

$x^3 + 5x^2 - 2x - 24 = (x^2 + 7x + 12)(x - 2)$
$= (x + 3)(x + 4)(x - 2)$
$x + 3 = 0$ or $x + 4 = 0$ or $x - 2 = 0$
$x = -3$ or $x = -4$ or $x = 2$

**41.**

$$
\begin{array}{r|rrrr}
-2 & -1 & +0 & +7 & +6 \\
   &    & +2 & -4 & -6 \\
\hline
   & -1 & +2 & +3 & +0 \\
\end{array}
$$

$-x^3 + 7x + 6 = (-x^2 + 2x + 3)(x + 2)$
$= -(x^2 - 2x - 3)(x + 2)$
$= -(x - 3)(x + 1)(x + 2)$
$x - 3 = 0$ or $x + 1 = 0$ or $x + 2 = 0$
$x = 3$ or $x = -1$ or $x = -2$

**43.**

$$
\begin{array}{r|rrrr}
+5 & 6 & -25 & -29 & +20 \\
   &   & +30 & +25 & -20 \\
\hline
   & 6 & +5 & -4 & +0 \\
\end{array}
$$

$6x^3 - 25x^2 - 29x + 20$
$= (6x^2 + 5x - 4)(x - 5)$
$= (3x + 4)(2x - 1)(x - 5)$
$3x + 4 = 0$ or $2x - 1 = 0$ or $x - 5 = 0$
$3x = -4 \qquad 2x = 1$
$x = -\dfrac{4}{3}$ or $x = \dfrac{1}{2}$ or $x = 5$

**45.**

$$
\begin{array}{r|rrrrr}
-2 & 1 & +4 & +3 & -4 & -4 \\
   &   & -2 & -4 & +2 & +4 \\
\hline
   & 1 & +2 & -1 & -2 & +0 \\
\end{array}
$$

$x^4 + 4x^3 + 3x^2 - 4x - 4$
$= (x^3 + 2x^2 - x - 2)(x + 2)$
$= (x^2(x + 2) - 1(x + 2))(x + 2)$
$= (x^2 - 1)(x + 2)(x + 2)$
$= (x + 1)(x - 1)(x + 2)^2$
$x + 1 = 0$ or $x - 1 = 0$ or $x + 2 = 0$
$x = -1$ or $x = 1$ or $x = -2$ (a double root)

**47.**

$$
\begin{array}{r|rrrrrrr}
-3 & 1 & +6 & +8 & -6 & -9 & +0 & +0 \\
   &   & -3 & -9 & +3 & +9 & +0 & +0 \\
\hline
   & 1 & +3 & -1 & -3 & +0 & +0 & +0 \\
\end{array}
$$

$x^6 + 6x^5 + 8x^4 - 6x^3 - 9x^2$
$= (x^5 + 3x^4 - x^3 - 3x^2)(x + 3)$
$= x^2(x^3 + 3x^2 - x - 3)(x + 3)$
$= x^2(x^2(x + 3) - (x + 3))(x + 3)$
$= x^2(x^2 - 1)(x + 3)(x + 3)$
$= x^2(x + 1)(x - 1)(x + 3)^2$
$x = 0$ or $x + 1 = 0$ or $x - 1 = 0$ or $x + 3 = 0$
$x = 0$ (a double root) or $x = -1$ or $x = 1$
or $x = -3$ (a double root)

**49.** $p(x) = a(x + 2)(x - 2)(x - 3)$
$p(1) = 18 = a(1 + 2)(1 - 2)(1 - 3) = 6a$
$a = 3$
$p(x) = 3(x + 2)(x - 2)(x - 3)$
$= 3(x^2 - 4)(x - 3)$
$= 3x^3 - 9x^2 - 12x + 36$

**51.**

$$
\begin{array}{r|ccc}
-n & 1 & +5 & -2 \\
   &   & -n & n^2 - 5n \\
\hline
   & 1 & +5 - n & +n^2 - 5n - 2 \\
\end{array}
$$

$n^2 - 5n - 2 = -8$
$n^2 - 5n + 6 = 0$
$(n - 2)(n - 3) = 0$
$n - 2 = 0$ or $n - 3 = 0$
$n = 2$ or $n = 3$

**53.**

$$
\begin{array}{r|cccc}
+2 & 1 & +b & -13 & +10 \\
   &   & +2 & +2b + 4 & +4b - 18 \\
\hline
   & 1 & +b + 2 & +2b - 9 & +4b - 8 \\
\end{array}
$$

$4b - 8 = 0$
$4b = 8$
$b = 2$

**55.**

$$
\begin{array}{r|rrrrrr}
-2 & 1 & +1 & +0 & +5 & -1 & -6 \\
   &   & -2 & +2 & -4 & -2 & +6 \\
\hline
   & 1 & -1 & +2 & +1 & -3 & +0 \\
\end{array}
$$

$$
\begin{array}{r|rrrrr}
-1 & 1 & -1 & +2 & +1 & -3 \\
   &   & -1 & +2 & -4 & +3 \\
\hline
   & 1 & -2 & +4 & -3 & +0 \\
\end{array}
$$

$$
\begin{array}{r|rrrr}
+1 & 1 & -2 & +4 & -3 \\
   &   & +1 & -1 & +3 \\
\hline
   & 1 & -1 & +3 & +0 \\
\end{array}
$$

$$x^5 + x^4 + 5x^2 - x - 6$$
$$= (x+2)(x+1)(x-1)(x^2 - x + 3)$$

**57.** If $p(x)$ has a factor of the form $x - c$, then $p(c) = 0$. $p(c) = 2c^4 + 5c^2 + 20 > 0$ since $c^4 \geq 0$ and $c^2 \geq 0$. So, $p(x)$ has no factor of the form $x - c$ where $c$ is a real number.

**59.** $x + 1$ is a factor of $x^n + 1$ with $n$ an odd integer if $-1$ is a root of $p(x) = x^n + 1$. This is equivalent to $p(-1) = 0$. Since $n$ is odd, $(-1)^n = -1$, so $p(-1) = (-1)^n + 1 = -1 + 1 = 0$. Therefore, $x + 1$ is a factor of $x^n + 1$.

## Challenge Problem

If $x - r$ is a factor of $x^3 + bx - 54$, then

$$
\begin{array}{r|rrrr}
+r & 1 & +0 & +b & -54 \\
   &   & +r & +r^2 & +br + r^3 \\
\hline
   & 1 & +r & +b + r^2 & +r^3 + br - 54 \\
\end{array}
$$

$r^3 + br - 54 = 0$
Since $x - r$ is a double factor, then

$$
\begin{array}{r|rrr}
+r & 1 & +r & +b + r^2 \\
   &   & +r & +2r^2 \\
\hline
   & 1 & +2r & +b + 3r^2 \\
\end{array}
$$

$b + 3r^2 = 0$
$b = -3r^2$
$b = -3r^2$ and $r^3 + br - 54 = 0$
$\qquad r^3 - 3r^3 = 54$
$\qquad -2r^3 = 54$
$\qquad r^3 = -27$
$\qquad r = -3$
$b = -3(-3)^2 = -27$
$b = -27$

## Exercises 3.7

**1.** $x^3 + x^2 - 21x - 45 = 0$
The possible rational roots are: $\pm 1, \pm 3, \pm 5, \pm 9, \pm 15, \pm 45$

$$
\begin{array}{r|rrrr}
-3 & 1 & +1 & -21 & -45 \\
   &   & -3 & +6 & +45 \\
\hline
   & 1 & -2 & -15 & +0 \\
\end{array}
$$

$x^3 + x^2 - 21x - 45 = (x + 3)(x^2 - 2x - 15)$
$= (x + 3)(x + 3)(x - 5)$
The roots are $-3$ (a double root) and $5$.

**3.** $3x^3 + 2x^2 - 75x - 50 = 0$
The possible rational roots are: $\pm 1, \pm \frac{1}{3}, \pm 2,$
$\pm \frac{2}{3}, \pm 5, \pm \frac{5}{3}, \pm 10, \pm \frac{10}{3}, \pm \frac{25}{3}, \pm 25, \pm 50, \pm \frac{50}{3}.$

$$
\begin{array}{r|rrrr}
+5 & 3 & +2 & -75 & -50 \\
   &   & +15 & +85 & +50 \\
\hline
   & 3 & +17 & +10 & +0 \\
\end{array}
$$

$3x^3 + 2x^2 - 75x - 50$
$= (x - 5)(3x^2 + 17x + 10)$
$= (x - 5)(3x + 2)(x + 5)$
The roots are $-\frac{2}{3}, -5, 5.$

**5.** $x^4 + 3x^3 + 3x^2 + x = 0$
$x^4 + 3x^3 + 3x^2 + x = x(x^3 + 3x^2 + 3x + 1)$
The possible rational roots of
$x^3 + 3x^2 + 3x + 1$ are $\pm 1.$

$$
\begin{array}{r|rrrr}
-1 & 1 & +3 & +3 & +1 \\
   &   & -1 & -2 & -1 \\
\hline
   & 1 & +2 & +1 & +0 \\
\end{array}
$$

$x^4 + 3x^3 + 3x^2 + x = x(x + 1)(x^2 + 2x + 1)$
$= x(x + 1)(x + 1)(x + 1)$
The roots are $0$ and $-1$ (a triple root).

**7.** $x^4 + 6x^3 + 2x^2 - 18x - 15 = 0$
The possible rational roots are: $\pm 1, \pm 3, \pm 5, \pm 15.$

$$
\begin{array}{r|rrrrr}
-5 & 1 & +6 & +2 & -18 & -15 \\
   &   & -5 & -5 & +15 & +15 \\
\hline
   & 1 & +1 & -3 & -3 & +0 \\
\end{array}
$$

$$
\begin{array}{r|rrrr}
-1 & 1 & +1 & -3 & -3 \\
   &   & -1 & +0 & +3 \\
\hline
   & 1 & +0 & -3 & +0 \\
\end{array}
$$

$x^4 + 6x^3 + 2x^2 - 18x - 15$
$= (x + 5)(x + 1)(x^2 - 3)$
$= (x + 5)(x + 1)(x + \sqrt{3})(x - \sqrt{3})$
The roots are $-5, -1, -\sqrt{3}, \sqrt{3}.$

**9.** $x^4 + 2x^3 - 7x^2 - 18x - 18 = 0$
The possible rational roots are: $\pm 1, \pm 2, \pm 3, \pm 6, \pm 9, \pm 18.$

$$
\begin{array}{r|rrrrr}
+3 & 1 & +2 & -7 & -18 & -18 \\
   &   & +3 & +15 & +24 & +18 \\
\hline
   & 1 & +5 & +8 & +6 & +0 \\
\end{array}
$$

$$\begin{array}{r|rrrr}
-3 & 1 & +5 & +8 & +6 \\
& & -3 & -6 & -6 \\
\hline
& 1 & +2 & +2 & \enclose{box}{+0}
\end{array}$$

$x^4 + 2x^3 - 7x^2 - 18x - 18$
$= (x-3)(x+3)(x^2+2x+2)$
$x^2 + 2x + 2$ has no real roots, so the real roots are $x = 3$ and $x = -3$.

**11.** $-x^5 + 5x^4 - 3x^3 - 15x^2 + 18x = 0$
$-x(x^4 - 5x^3 + 3x^2 + 15x - 18) = 0$
The possible rational roots are: $\pm 1, \pm 2, \pm 3,$
$\pm 6, \pm 9, \pm 18.$

$$\begin{array}{r|rrrrr}
+2 & 1 & -5 & +3 & +15 & -18 \\
& & +2 & -6 & -6 & +18 \\
\hline
& 1 & -3 & -3 & +9 & \enclose{box}{+0}
\end{array}$$

$$\begin{array}{r|rrrr}
+3 & 1 & -3 & -3 & +9 \\
& & +3 & +0 & -9 \\
\hline
& 1 & +0 & -3 & \enclose{box}{+0}
\end{array}$$

$-x^5 + 5x^4 - 3x^3 - 15x^2 + 18x$
$= -x(x-2)(x-3)(x^2-3)$
The real roots are: $0, 2, 3, -\sqrt{3}, \sqrt{3}.$

**13.** $2x^3 - 5x - 3 = 0$

The possible rational roots are: $\pm 1, \pm\frac{1}{2}, \pm 3,$
$\pm\frac{3}{2}.$

$$\begin{array}{r|rrrr}
-1 & 2 & +0 & -5 & -3 \\
& & -2 & +2 & +3 \\
\hline
& 2 & -2 & -3 & \enclose{box}{+0}
\end{array}$$

$2x^3 - 5x - 3 = (x+1)(2x^2 - 2x - 3)$

The roots of $2x^2 - 2x - 3$ are $\dfrac{1\pm\sqrt{7}}{2}$, so

the roots of $2x^3 - 5x - 3$ are $-1, \dfrac{1+\sqrt{7}}{2},$

$\dfrac{1-\sqrt{7}}{2}.$

**15.** $3x^4 - 11x^3 - 3x^2 - 6x + 8 = 0$

The possible rational roots are: $\pm 1, \pm\frac{1}{3}, \pm 2,$

$\pm\frac{2}{3}, \pm 4, \pm\frac{4}{3}, \pm 8, \pm\frac{8}{3}.$

$$\begin{array}{r|rrrrr}
+4 & 3 & -11 & -3 & -6 & +8 \\
& & +12 & +4 & +4 & -8 \\
\hline
& 3 & +1 & +1 & -2 & \enclose{box}{+0}
\end{array}$$

$$\begin{array}{r|rrrr}
+\frac{2}{3} & 3 & +1 & +1 & -2 \\
& & +2 & +2 & +2 \\
\hline
& 3 & +3 & +3 & \enclose{box}{+0}
\end{array}$$

$3x^4 - 11x^3 - 3x^2 - 6x + 8$
$= (x-4)\left(x - \frac{2}{3}\right)(3x^2 + 3x + 3)$

$= 3(x-4)\left(x - \frac{2}{3}\right)(x^2 + x + 1)$

$= (x-4)(3x-2)(x^2 + x + 1)$
$x^2 + x + 1$ has no real roots.

The real roots are: $4, \dfrac{2}{3}.$

**17.** The possible rational roots are: $\pm 1, \pm 2, \pm 4,$
$\pm 5, \pm 8, \pm 10, \pm 16, \pm 20, \pm 40, \pm 80.$

$$\begin{array}{r|rrrr}
+5 & -1 & -3 & +24 & +80 \\
& & -5 & -40 & -80 \\
\hline
& -1 & -8 & -16 & \enclose{box}{+0}
\end{array}$$

$-x^3 - 3x^2 + 24x + 80$
$= (x-5)(-x^2 - 8x - 16)$
$= -(x-5)(x^2 + 8x + 16)$
$= -(x-5)(x+4)(x+4)$
$= -(x+4)^2(x-5)$

**19.** $6x^4 + 9x^3 + 9x - 6 = 3(2x^4 + 3x^3 + 3x - 2)$. The possible rational roots are: $\pm 1, \pm\frac{1}{2}, \pm 2.$

$$\begin{array}{r|rrrrr}
-2 & 2 & +3 & +0 & +3 & -2 \\
& & -4 & +2 & -4 & +2 \\
\hline
& 2 & -1 & +2 & -1 & \enclose{box}{+0}
\end{array}$$

$6x^4 + 9x^3 + 9x - 6 = 3(x+2)(2x^3 - x^2 + 2x - 1) = 3(x+2)(x^2(2x-1) + 2x - 1)$
$= 3(x+2)(2x-1)(x^2+1)$

**21.** The only possible rational roots are $\pm 1, \pm \frac{1}{2}, \pm 2, \pm 4, \pm 8$. Using synthetic division, we see that none of these are roots.

**23.** $p(x) = x^4 - 3x^3 + 5x^2 - x - 10 = 0$. The only possible rational roots are $\pm 1, \pm 2, \pm 5, \pm 10$.

$$
\begin{array}{r|rrrrr}
-1 & 1 & -3 & +5 & -1 & -10 \\
   &   & -1 & +4 & -9 & +10 \\
\hline
   & 1 & -4 & +9 & -10 & +0 \\
\end{array}
$$

$$
\begin{array}{r|rrrr}
2 & 1 & -4 & +9 & -10 \\
  &   & +2 & -4 & +10 \\
\hline
  & 1 & -2 & +5 & +0 \\
\end{array}
$$

$x^4 - 3x^3 + 5x^2 - x - 10 = (x + 1)(x - 2)(x^2 - 2x + 5)$
The roots of $x^2 - 2x + 5$ are $1 \pm 2i$. The roots of $p(x)$ are $-1, 2, 1 + 2i, 1 - 2i$.

**25.** $p(x) = 3x^3 - 5x^2 + 2x - 8 = 0$. The only possible rational roots are $\pm 1, \pm \frac{1}{3}, \pm 2, \pm \frac{2}{3}, \pm 4, \pm \frac{4}{3}, \pm 8, \pm \frac{8}{3}$.

$$
\begin{array}{r|rrrr}
+2 & 3 & -5 & +2 & -8 \\
   &   & +6 & +2 & +8 \\
\hline
   & 3 & +1 & +4 & +0 \\
\end{array}
$$

$3x^3 - 5x^2 + 2x - 8 = (x - 2)(3x^2 + x + 4)$. The roots of $3x^2 + x + 4$ are $\dfrac{-1 \pm \sqrt{47}i}{6}$. The roots of

$p(x)$ are $2, -\dfrac{1}{6} + \dfrac{\sqrt{47}}{6}i, -\dfrac{1}{6} - \dfrac{\sqrt{47}}{6}i$

**27.** $p(x) = x^5 - 9x^4 + 31x^3 - 49x^2 + 36x - 10 = 0$. The only possible rational roots are $\pm 1, \pm 2, \pm 5, \pm 10$.

$$
\begin{array}{r|rrrrrr}
+1 & 1 & -9 & +31 & -49 & +36 & -10 \\
   &   & +1 & -8 & +23 & -26 & +10 \\
\hline
   & 1 & -8 & +23 & -26 & +10 & +0 \\
\end{array}
$$

$$
\begin{array}{r|rrrrr}
+1 & 1 & -8 & +23 & -26 & +10 \\
   &   & +1 & -7 & +16 & -10 \\
\hline
   & 1 & -7 & +16 & -10 & +0 \\
\end{array}
$$

$$
\begin{array}{r|rrrr}
+1 & 1 & -7 & +16 & -10 \\
   &   & +1 & -6 & +10 \\
\hline
   & 1 & -6 & +10 & +0 \\
\end{array}
$$

$x^5 - 9x^4 + 31x^3 - 49x^2 + 36x - 10 = (x - 1)^3(x^2 - 6x + 10)$. The roots of $x^2 - 6x + 10$ are $3 \pm i$, so the roots of $p(x)$ are 1 (a triple root), $3 + i, 3 - i$.

**29.** $p(x)$ varies in sign twice, so there are either 2 or 0 positive zeros.
$p(-x) = 4(-x)^3 - 3(-x)^2 - 7(-x) + 9 = -4x^3 - 3x^2 + 7x + 9$ varies in sign once, so there is only one negative zero.

| Number of Positive Zeros | Number of Negative Zeros | Number of Imaginary Zeros | Total |
|:---:|:---:|:---:|:---:|
| 2 | 1 | 0 | 3 |
| 0 | 1 | 2 | 3 |

**31.** $p(x)$ varies in sign three times, so there are either 3 or 1 positive zeros.

$p(-x) = (-x)^5 + 4(-x)^4 - 3(-x)^3 - (-x)^2 + (-x) - 1 = -x^5 + 4x^4 + 3x^3 - x^2 - x - 1$ varies in sign twice, so there are either 2 or 0 negative zeros.

| Number of Positive Zeros | Number of Negative Zeros | Number of Imaginary Zeros | Total |
|:---:|:---:|:---:|:---:|
| 3 | 2 | 0 | 5 |
| 3 | 0 | 2 | 5 |
| 1 | 2 | 2 | 5 |
| 1 | 0 | 4 | 5 |

**33.** $y = x^3 - 3x^2 + 3x - 1$

$y = 7x - 13$

$x^3 - 3x^2 + 3x - 1 = 7x - 13$

$x^3 - 3x^2 - 4x + 12 = 0$

The only possible rational roots are $\pm1, \pm2,$ $\pm3, \pm4, \pm6, \pm12$

$$\begin{array}{r|rrrr} +3 & 1 & -3 & -4 & +12 \\ & & +3 & +0 & -12 \\ \hline & 1 & +0 & -4 & +0 \end{array}$$

$x^3 - 3x^2 - 4x + 12 = (x-3)(x^2 - 4)$

$= (x-3)(x-2)(x+2) = 0$

$x - 3 = 0$ or $x - 2 = 0$ or $x + 2 = 0$

$x = 3$ or $x = 2$ or $x = -2$

For $x = 3$, $y = 7 \cdot 3 - 13 = 8$

For $x = 2$, $y = 7 \cdot 2 - 13 = 1$

For $x = -2$, $y = 7 \cdot (-2) - 13 = -27$

The solutions are $(-2, -27)$, $(2, 1)$, $(3, 8)$.

**35.** $y = 4x^3 - 7x^2 + 10$

$y = x^3 + 43x - 5$

$4x^3 - 7x^2 + 10 = x^3 + 43x - 5$

$3x^3 - 7x^2 - 43x + 15 = 0$

The only possible rational solutions are

$\pm1, \pm\frac{1}{3}, \pm3, \pm5, \pm\frac{5}{3}, \pm15$

$$\begin{array}{r|rrrr} -3 & 3 & -7 & -43 & +15 \\ & & -9 & +48 & -15 \\ \hline & 3 & -16 & +5 & +0 \end{array}$$

$3x^3 - 7x^2 - 43x + 15$

$= (x+3)(3x^2 - 16x + 5) = 0$

$(x+3)(3x-1)(x-5) = 0$

$x + 3 = 0$ or $3x - 1 = 0$ or $x - 5 = 0$

$x = -3$ or $x = \frac{1}{3}$ or $x = 5$

For $x = -3$, $y = (-3)^3 + 43(-3) - 5 = -161$

For $x = \frac{1}{3}$, $y = \left(\frac{1}{3}\right)^3 + 43\left(\frac{1}{3}\right) - 5 = \frac{253}{27}$

For $x = 5$, $y = 5^3 + 43(5) - 5 = 335$

The solutions are $(-3, -161)$, $\left(\frac{1}{3}, \frac{253}{27}\right)$, $(5, 335)$.

**37.** $\overline{z + w} = \overline{(-6 + 8i) + \left(\frac{1}{2} + \frac{1}{2}i\right)} = \overline{-\frac{11}{2} + \frac{17}{2}i}$

$= -\frac{11}{2} - \frac{17}{2}i$

$\overline{z} + \overline{w} = \overline{(-6 + 8i)} + \overline{\left(\frac{1}{2} + \frac{1}{2}i\right)}$

$= (-6 - 8i) + \left(\frac{1}{2} - \frac{1}{2}i\right) = -\frac{11}{2} - \frac{17}{2}i$

So, $\overline{z + w} = \overline{z} + \overline{w}$.

**39.** $\overline{zw} = \overline{(-6+8i)\left(\dfrac{1}{2}+\dfrac{1}{2}i\right)} = \overline{(-7+i)} = -7-i$

$$\overline{z} \cdot \overline{w} = \overline{(-6+8i)} \cdot \overline{\left(\dfrac{1}{2}+\dfrac{1}{2}i\right)}$$

$$= (-6-8i)\left(\dfrac{1}{2}-\dfrac{1}{2}i\right) = -7-i$$

So, $\overline{zw} = \overline{z} \cdot \overline{w}$.

**41.** $\overline{z+w} = \overline{(a+bi)+(c+di)}$

$= \overline{(a+c)+(b+d)i} = (a+c)-(b+d)i$

$\overline{z}+\overline{w} = \overline{(a+bi)}+\overline{(c+di)}$

$= (a-bi)+(c-di) = (a+c)-(b+d)i$

So, $\overline{z+w} = \overline{z}+\overline{w}$.

**43.** $\overline{\left(\dfrac{z}{w}\right)} = \overline{\left(\dfrac{a+bi}{c+di}\right) \cdot \dfrac{c-di}{c-di}}$

$$= \overline{\dfrac{ac+bd}{c^2+d^2} + \dfrac{bc-ad}{c^2+d^2}i}$$

$$= \dfrac{ac+bd}{c^2+d^2} - \dfrac{bc-ad}{c^2+d^2}i$$

$$\dfrac{\overline{z}}{\overline{w}} = \dfrac{\overline{a+bi}}{\overline{c+di}} = \dfrac{a-bi}{c-di} \cdot \dfrac{c+di}{c+di}$$

$$= \dfrac{ac+bd}{c^2+d^2} - \dfrac{bc-ad}{c^2+d^2}i$$

So, $\overline{\left(\dfrac{z}{w}\right)} = \dfrac{\overline{z}}{\overline{w}}$.

**45.** Since $0 = p(z)$, $\overline{0} = \overline{p(z)}$. Then, since $\overline{0} = 0$, we have

$0 = \overline{p(z)} = \overline{a_n z^n + a_{n-1}z^{n-1}+\ldots+a_1 z + a_0}$

$= \overline{a_n z^n} + \overline{a_{n-1}z^{n-1}} +\ldots+ \overline{a_1 z} + \overline{a_0}$   (Ex. 41)

$= \overline{a_n}\ \overline{z^n} + \overline{a_{n-1}}\ \overline{z^{n-1}} +\ldots+ \overline{a_1}\ \overline{z} + \overline{a_0}$ (Ex. 44)

$= a_n\ \overline{z^n} + a_{n-1}\ \overline{z^{n-1}} +\ldots+ a_1\ \overline{z} + a_0$

(The $a_i$ are real numbers)

$= a_n\left(\overline{z}\right)^n + a_{n-1}\left(\overline{z}\right)^{n-1} +\ldots+ a_1\left(\overline{z}\right) + a_0$

(Ex. 44)

$= p\left(\overline{z}\right)$

Thus $p\left(\overline{z}\right) = 0$

**47.** $f(1) = 1^5 - 3 \cdot 1^2 + 2 \cdot 1 - 3 = -3$ and $f(2) = 2^5 - 3 \cdot 2^2 + 2 \cdot 2 - 3 = 21$; thus there exists a $c$ in $(1, 2)$ such that $f(c) = 0$.

**49.** By the rational root theorem, the only possible rational roots of $x^2 - 5 = 0$ are $\pm 1$ and $\pm 5$. Since none of these are roots, and $\sqrt{5}$ is a root, it follows that $\sqrt{5}$ is an irrational number.

## Challenge Problem

**1.** $p(x) = 2x^3 + 5x^2 - 6x - 2$. Using synthetic division,

| 1 | 2 | +5 | −6 | −2 |
|---|---|----|----|----|
|   |   | +2 | +7 | +1 |
|   | 2 | +7 | +1 | −1 |

and

| 2 | 2 | +5 | −6 | −2 |
|---|---|----|----|----|
|   |   | +4 | +18 | +24 |
|   | 2 | +9 | +12 | +22 |

Hence, 1 is not an upper bound but 2 is and therefore, 2 is the smallest integer which is an upper bound. Also,

| −3 | 2 | +5 | −6 | −2 |
|----|---|----|----|----|
|    |   | −6 | +3 | +9 |
|    | 2 | −1 | −3 | +7 |

and

| −4 | 2 | +5 | −6 | −2 |
|----|---|----|----|----|
|    |   | −8 | +12 | −24 |
|    | 2 | −3 | +6 | −26 |

Since −3 is not a lower bound and −4 is, −4 is the largest integer which is a lower bound.

## Critical Thinking

**1.** It is possible to solve an equation by multiplying by the common denominator. However, the solutions obtained must be checked since they may not actually satisfy the equation.

**3.** Let $p(x) = x^n - y^n$. Then $x + y$ is a factor of $x^n - y^n$ if $p(-y) = 0$.

$p(-y) = (-y)^n - y^n = y^n - y^n$

since $n$ is even

$= 0$

Hence, $x + y$ is a factor of $x^n - y^n$.

**5.**

$$\begin{array}{r|rrrrr} 5 & 2 & +3 & +4 & +2 & +3 \\ & & +10 & +65 & +345 & +1735 \\ \hline & 2 & +13 & +69 & +347 & \boxed{+1738} \end{array}$$

So, $23423_{\text{five}} = 1738$.

## Exercises 3.8

**1.** $\dfrac{2x}{(x+1)(x-1)} = \dfrac{A}{x+1} + \dfrac{B}{x-1}$

$2x = A(x-1) + B(x+1)$

Let $x = 1$:   $2 = A \cdot 0 + B \cdot 2$

$\qquad\qquad 2 = 2B$

$\qquad\qquad B = 1$

Let $x = -1$:   $-2 = A(-2) + B \cdot 0$

$\qquad\qquad\quad -2 = -2A$

$\qquad\qquad\quad A = 1$

$\dfrac{2x}{(x+1)(x-1)} = \dfrac{1}{x+1} + \dfrac{1}{x-1}$

**3.** $\dfrac{x+7}{x^2 - x - 6} = \dfrac{x+7}{(x-3)(x+2)} = \dfrac{A}{x-3} + \dfrac{B}{x+2}$

$x + 7 = A(x+2) + B(x-3)$

Let $x = -2$:   $-2 + 7 = A(0) + B(-5)$

$\qquad\qquad\quad 5 = -5B$

$\qquad\qquad\quad B = -1$

Let $x = 3$:   $3 + 7 = A(5) + B(0)$

$\qquad\qquad\quad 10 = 5A$

$\qquad\qquad\quad A = 2$

$\dfrac{x+7}{x^2 - x - 6} = \dfrac{2}{x-3} - \dfrac{1}{x+2}$

**5.** $\dfrac{5x^2 + 9x - 56}{(x-4)(x-2)(x+1)}$

$= \dfrac{A}{x-4} + \dfrac{B}{x-2} + \dfrac{C}{x+1}$

$5x^2 + 9x - 56 = A(x-2)(x+1)$
$\qquad\qquad\qquad + B(x-4)(x+1) + C(x-4)(x-2)$

Let $x = 2$:   $5 \cdot 2^2 + 9 \cdot 2 - 56$

$\qquad\qquad = A \cdot 0 + B(2-4)(2+1)$
$\qquad\qquad\quad + C \cdot 0$

$\qquad\qquad -18 = -6B$

$\qquad\qquad\quad B = 3$

Let $x = 4$:   $5 \cdot 4^2 + 9 \cdot 4 - 56$

$\qquad\qquad = A(4-2)(4+1) + B \cdot 0$
$\qquad\qquad\quad + C \cdot 0$

$\qquad\qquad 60 = 10A$

$\qquad\qquad 6 = A$

Let $x = -1$:   $5(-1)^2 + 9(-1) - 56$

$\qquad\qquad = A \cdot 0 + B \cdot 0$
$\qquad\qquad\quad + C(-1-4)(-1-2)$

$\qquad\qquad -60 = 15C$

$\qquad\qquad C = -4$

$\dfrac{5x^2 + 9x - 56}{(x-4)(x-2)(x+1)}$

$= \dfrac{6}{x-4} + \dfrac{3}{x-2} - \dfrac{4}{x+1}$

**7.** $\dfrac{3x-3}{(x-2)^2} = \dfrac{A}{x-2} + \dfrac{B}{(x-2)^2}$

$3x - 3 = A(x-2) + B$

Let $x = 2$:   $3 \cdot 2 - 3 = A \cdot 0 + B$

$\qquad\qquad\quad 3 = B$

Let $x = 3$:   $3 \cdot 3 - 3 = A(3-2) + 3$

$\qquad\qquad\quad 6 = A + 3$

$\qquad\qquad\quad 3 = A$

$\dfrac{3x-3}{(x-2)^2} = \dfrac{3}{x-2} + \dfrac{3}{(x-2)^2}$

**9.** $\dfrac{3x-30}{15x^2 - 14x - 8} = \dfrac{3x-30}{(5x+2)(3x-4)}$

$= \dfrac{A}{5x+2} + \dfrac{B}{3x-4}$

$3x - 30 = A(3x-4) + B(5x+2)$

Let $x = \dfrac{4}{3}$:   $3 \cdot \dfrac{4}{3} - 30$

$\qquad\qquad\qquad = A(0) + B\left(5 \cdot \dfrac{4}{3} + 2\right)$

$\qquad\qquad -26 = \dfrac{26}{3}B$

$\qquad\qquad\quad B = -3$

Let $x = -\dfrac{2}{5}$:    $3\left(-\dfrac{2}{5}\right) - 30$

$$= A\left(3\left(-\dfrac{2}{5}\right) - 4\right) + B \cdot 0$$

$$-\dfrac{156}{6} = -\dfrac{26}{5}A$$

$$A = 6$$

$$\dfrac{3x - 30}{15x^2 - 14x - 8} = \dfrac{6}{5x + 2} - \dfrac{3}{3x - 4}$$

**11.** $\dfrac{x^2 - x - 4}{x(x+2)^2} = \dfrac{A}{x} + \dfrac{B}{x+2} + \dfrac{C}{(x+2)^2}$

$x^2 - x - 4 = A(x+2)^2 + Bx(x+2) + Cx$

Let $x = 0$:   $0^2 - 0 - 4$
$\qquad = A(0+2)^2 + B \cdot 0 + C \cdot 0$
$\qquad -4 = 4A$
$\qquad A = -1$

Let $x = -2$:   $(-2)^2 - (-2) - 4$
$\qquad = A \cdot 0 + B \cdot 0 + C(-2)$
$\qquad 2 = -2C$
$\qquad C = -1$

Let $x = -1$:   $(-1)^2 - (-1) - 4$
$\qquad = -1(-1+2)^2 + B(-1)(-1+2)$
$\qquad\qquad + (-1)(-1)$
$\qquad -2 = -1 - B + 1$
$\qquad -2 = -B$
$\qquad B = 2$

$$\dfrac{x^2 - x - 4}{x(x+2)^2} = -\dfrac{1}{x} + \dfrac{2}{x+2} - \dfrac{1}{(x+2)^2}$$

**13.**

$$x^2 - 1\overline{\big)\,x^3 - x + 2}$$

quotient $x$

$$\underline{x^3 - x}$$
$$0x + 2$$

$$\dfrac{x^3 - x + 2}{x^2 - 1} = x + \dfrac{2}{x^2 - 1}$$

$$\dfrac{2}{x^2 - 1} = \dfrac{2}{(x+1)(x-1)} = \dfrac{A}{x+1} + \dfrac{B}{x-1}$$

$2 = A(x - 1) + B(x + 1)$

Let $x = 1$:   $2 = A \cdot 0 + B(1 + 1)$
$\qquad 2 = 2B$
$\qquad B = 1$

Let $x = -1$:   $2 = A(-1 - 1) + B \cdot 0$
$\qquad 2 = -2A$
$\qquad A = -1$

$$\dfrac{x^3 - x + 2}{x^2 - 1} = x - \dfrac{1}{x+1} + \dfrac{1}{x-1}$$

**15.**

$$4x^2 - 4x + 1\overline{\big)\,12x^4 - 12x^3 + 7x^2 - 2x - 3}$$

quotient $3x^2 \quad\quad + 1$

$$\underline{12x^4 - 12x^3 + 3x^2}$$
$$4x^2 - 2x - 3$$
$$\underline{4x^2 - 4x + 1}$$
$$2x - 4$$

$$\dfrac{12x^4 - 12x^3 + 7x^2 - 2x - 3}{4x^2 - 4x + 1}$$

$$= 3x^2 + 1 + \dfrac{2x - 4}{4x^2 - 4x + 1}$$

$$= 3x^2 + 1 + \dfrac{2x - 4}{(2x - 1)^2}$$

$$\dfrac{2x - 4}{(2x - 1)^2} = \dfrac{A}{2x - 1} + \dfrac{B}{(2x - 1)^2}$$

$2x - 4 = A(2x - 1) + B$

Let $x = \dfrac{1}{2}$:   $2\left(\dfrac{1}{2}\right) - 4 = A \cdot 0 + B$
$\qquad -3 = B$

Let $x = 0$:   $2 \cdot 0 - 4 = A(2 \cdot 0 - 1) + (-3)$
$\qquad -4 = -A - 3$
$\qquad -1 = -A$
$\qquad A = 1$

$$\dfrac{12x^4 - 12x^3 + 7x^2 - 2x - 3}{4x^2 - 4x + 1}$$

$$= 3x^2 + 1 + \dfrac{1}{2x - 1} - \dfrac{3}{(2x - 1)^2}$$

**17.**

$$\begin{array}{r} 10x-5 \\ x^2-x-6\overline{\smash{\big)}\,10x^3-15x^2-35x} \\ \underline{10x^3-10x^2-60x} \\ -5x^2+25x \\ \underline{-5x^2+5x+30} \\ 20x-30 \end{array}$$

$$\frac{10x^3-15x^2-35x}{x^2-x-6}=10x-5+\frac{20x-30}{x^2-x-6}$$

$$=10x-5+\frac{20x-30}{(x-3)(x+2)}$$

$$\frac{20x-30}{(x-3)(x+2)}=\frac{A}{x-3}+\frac{B}{x+2}$$

$$20x-30=A(x+2)+B(x-3)$$

Let $x=3$:   $20\cdot 3-30=A(3+2)+B\cdot 0$

$$30=5A$$
$$A=6$$

Let $x=-2$:   $20(-2)-30$
$$=A\cdot 0+B(-2-3)$$
$$-70=-5B$$
$$B=14$$

$$\frac{10x^3-15x^2-35x}{x^2-x-6}$$

$$=10x-5+\frac{6}{x-3}+\frac{14}{x+2}$$

**19.** The possible rational roots of
$x^3+4x^2-31x-70$ are $\pm1, \pm2, \pm5, \pm7,$
$\pm10, \pm14, \pm35, \pm70$

$$\begin{array}{r|rrrr} +5 & 1 & +4 & -31 & -70 \\ & & +5 & +45 & +70 \\ \hline & 1 & +9 & +14 & \underline{+0} \end{array}$$

$x^3+4x^2-31x-70=(x-5)(x^2+9x+14)$
$=(x-5)(x+2)(x+7)$

$$\frac{5x^2-24x-173}{x^3+4x^2-31x-70}$$

$$=\frac{A}{x+7}+\frac{B}{x-5}+\frac{C}{x+2}$$

$5x^2-24x-173$
$=A(x-5)(x+2)+B(x+7)(x+2)$
$\quad+C(x+7)(x-5)$

Let $x=5$:   $5\cdot 5^2-24\cdot 5-173$
$$=A\cdot 0+B(5+7)(5+2)$$
$$+C\cdot 0$$
$$-168=84B$$
$$B=-2$$

Let $x=-2$:   $5(-2)^2-24(-2)-173$
$$=A\cdot 0+B\cdot 0$$
$$+C(-2+7)(-2-5)$$
$$-105=-35C$$
$$C=3$$

Let $x=-7$:   $5(-7)^2-24(-7)-173$
$$=A(-7-5)(-7+2)+B\cdot 0$$
$$+C\cdot 0$$
$$240=60A$$
$$A=4$$

$$\frac{5x^2-24x-173}{x^3+4x^2-31x-70}$$

$$=\frac{4}{x+7}-\frac{2}{x-5}+\frac{3}{x+2}$$

**21.**  $\dfrac{-4x}{(x^2+3)(x-1)}=\dfrac{Ax+B}{x^2+3}+\dfrac{C}{x-1}$

$-4x=(Ax+B)(x-1)+C(x^2+3)$

Let $x=1$:   $-4=(A+B)(0)+C(1^2+3)$
$$-4=4C$$
$$C=-1$$

Let $x=0$:   $0=(A\cdot 0+B)(0-1)$
Use $C=-1$,   $+(-1)(0^2+3)$
$$0=-B-3$$
$$B=-3$$

Let $x=-1$:   $-4(-1)=(A(-1)-3)(-1-1)$
Use $C=-1$,   $+(-1)((-1)^2+3)$

$B=-3$        $4=2A+6-4$
$$2=2A$$
$$A=1$$

$$\frac{-4x}{(x^2+3)(x-1)}=\frac{x-3}{x^2+3}-\frac{1}{x-1}$$

**23.**  $\dfrac{4x^2+5}{(x-1)(x^2+x+1)}=\dfrac{A}{x-1}+\dfrac{Bx+C}{x^2+x+1}$

$4x^2+5=A(x^2+x+1)+(Bx+C)(x-1)$

Let $x=1$:   $4\cdot 1^2+5$
$$=A(1^2+1+1)+(B+C)0$$
$$9=3A$$
$$A=3$$

Let $x = 0$:   $4 \cdot 0^2 + 5$
Use $A = 3$   $= 3(0^2 + 0 + 1)$
                $+ (B \cdot 0 + C)(0 - 1)$
            $5 = 3 - C$
            $C = -2$
Let $x = -1$:   $4(-1)^2 + 5$
Use $A = 3$,   $= 3((-1)^2 - 1 + 1)$
$C = -2$         $+ (B(-1) + (-2))(-1 - 1)$
            $9 = 3 + 2B + 4$
            $2 = 2B$
            $B = 1$

$$\frac{4x^2 + 5}{(x-1)(x^2 + x + 1)} = \frac{3}{x-1} + \frac{x-2}{x^2 + x + 1}$$

## Chapter 3 Review Exercises

**1.**

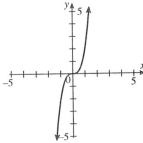

Domain: all $x$
Range: all $y$

**3.**

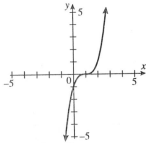

Domain: all $x$
Range: all $y$

**5.**

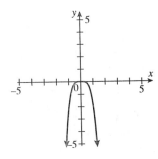

Domain: all $x$
Range: $y \le 0$

**7.** The graph of a function $y = f(x)$ is symmetric through the origin if $f(-x) = -f(x)$.

**9.** The graph of $y = f(x + h)$ can be obtained by shifting $y = f(x)$, $h$ units to the left if $h > 0$, or $h$ units to the right if $h < 0$.

**11.** A rational expression is a ratio of polynomials.

**13.**

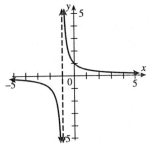

Asymptotes: $x = -1$, $y = 0$
Domain: $x \ne -1$
Range: $y \ne 0$
Concave down: $(-\infty, -1)$
Concave up: $(-1, \infty)$

**15.**

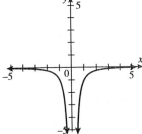

Asymptotes: $x = 0$, $y = 0$
Domain: $x \ne 0$
Range: $y < 0$
Concave down: $(-\infty, 0)$, $(0, \infty)$

**17.**

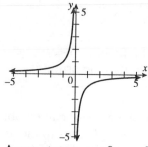

Asymptotes: $x = 0$, $y = 0$
Domain: $x \neq 0$
Range: $y \neq 0$
Concave up: $(-\infty, 0)$
Concave down: $(0, \infty)$

**19.** The line $x = c$ is a vertical asymptote for the graph of a function $f$ if

$$f(x) \to \infty \text{ or } f(x) \to -\infty$$

as $x$ approaches $c$ from either the left or the right.

**21.** $f(x) = x^2 + 3x - 10 = (x + 5)(x - 2)$
The zeros of $f(x)$ are $-5, 2$.

| Interval | $(-\infty, -5)$ | $(-5, 2)$ | $(2, \infty)$ |
|---|---|---|---|
| Sign of $(x + 5)$ | $-$ | $+$ | $+$ |
| Sign of $(x - 2)$ | $-$ | $-$ | $+$ |
| Sign of $f(x)$ | $+$ | $-$ | $+$ |

$f(x) < 0$ on $(-5, 2)$;
$f(x) > 0$ on $(-\infty, -5)$ and $(2, \infty)$.

**23.** $g(x) = \dfrac{2x^2}{3x^2 - 1}$

Horizontal asymptote: $y = \dfrac{2}{3}$

Vertical asymptotes: $\quad 3x^2 - 1 = 0$
$$3x^2 = 1$$
$$x^2 = \frac{1}{3}$$
$$x = \pm\sqrt{\frac{1}{3}} = \pm\frac{\sqrt{3}}{3}$$

**25.** $r(x) = \dfrac{x}{2x - x^2} = \dfrac{x}{x(2 - x)} = \dfrac{1}{2 - x}$, $x \neq 0$

Horizontal asymptote: $y = 0$
Vertical asymptote: $x = 2$

**27.**

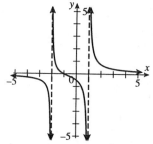

Asymptotes: $x = 1$, $x = -2$, $y = 0$

**29.** $f(x) = \dfrac{x^2 - 2x + 3}{x} = x - 2 + \dfrac{3}{x}$

Oblique asymptote: $y = x - 2$

**31.** $\dfrac{x}{x - 1} - \dfrac{2}{x(x - 1)} = \dfrac{4}{x} - \dfrac{1}{x - 1}$

$$x(x - 1) \cdot \frac{x}{x - 1} - x(x - 1) \cdot \frac{2}{x(x - 1)}$$

$$= x(x - 1) \cdot \frac{4}{x} - x(x - 1) \cdot \frac{1}{x - 1}$$

$$x^2 - 2 = 4x - 4 - x$$

$$x^2 - 2 = 3x - 4$$

$$x^2 - 3x + 2 = 0$$

$(x-1)(x-2)=0$
$x-1=0$ or $x-2=0$
$x=1$ or $x=2$, but $x \neq 1$, so $x=2$

**33.** If $\dfrac{a}{b}=\dfrac{c}{d}$, then $ad=bc$.

**35.** $f(x)=\dfrac{8}{x}-2x-6$

Domain: $x \neq 0$

$x$-intercepts: $\quad 0=\dfrac{8}{x}-2x-6$

$$x \cdot 0 = x \cdot \dfrac{8}{x} - x \cdot 2x - x \cdot 6$$
$$0 = 8 - 2x^2 - 6x$$
$$2x^2 + 6x - 8 = 0$$
$$2(x^2 + 3x - 4) = 0$$
$$2(x+4)(x-1) = 0$$
$$x+4 = 0 \text{ or } x-1 = 0$$
$$x = -4 \text{ or } x = 1$$

$x$-intercepts: $\quad x=-4,\ x=1$

**37.** $r=\dfrac{ab}{a+b+c}$

$r(a+b+c)=ab$
$ra+rb+rc=ab$
$rc=ab-ra-rb$
$c=\dfrac{ab-ra-rb}{r}$

**39.** Numerator or denominator are 0 when

$x=\dfrac{2}{3}$ and $x=2$. There are three intervals:

$$\left(-\infty, \dfrac{2}{3}\right], \left[\dfrac{2}{3}, 2\right), (2, \infty)$$

Test $x=0$: $\quad \dfrac{3 \cdot 0 - 2}{2 - 0} = \dfrac{-2}{2} = -1 \leq 0$

Test $x=1$: $\quad \dfrac{3 \cdot 0 - 2}{2 - 1} = \dfrac{1}{1} = 1 \nleq 0$

Test $x=3$: $\quad \dfrac{3 \cdot 3 - 2}{2 - 3} = \dfrac{7}{-1} = -7 \leq 0$

So, $x \leq \dfrac{2}{3}$ or $x>2$.

**41.** Let $x$ be the height of the tree. Then,

$$\dfrac{x}{15}=\dfrac{3}{\frac{1}{2}}$$
$$\dfrac{1}{2}x=45$$
$$x=90$$

The tree is 90 feet high.

**43. (a)** $y$ varies directly as $x$ if $y=kx$ for some constant of variation $k$.

**(b)** $y$ varies inversely as $x$ if $y=\dfrac{k}{x}$ for some constant of variation $k$.

**45.** $A=k \cdot bh;\ k=\dfrac{1}{2}$

**47.** $y=\dfrac{k}{x}$

$4=\dfrac{k}{3}$

$12=k$

$y=\dfrac{12}{x}=\dfrac{12}{4}$

$y=3$

**49.** $P=\dfrac{k}{V}$

$40=\dfrac{k}{300}$

$12,000=k$

$P=\dfrac{k}{V}=\dfrac{12,000}{V}=\dfrac{12,000}{200}$

$P=60$ pounds per square inch.

**51.**

$$
\begin{array}{r|rrr}
+1 & 1 & +1 & -7 & +5 \\
 &  & +1 & +2 & -5 \\
\hline
 & 1 & +2 & -5 & +0 \\
\end{array}
$$

$x^2+2x-5;\ r=0$

**53.**

$$
\begin{array}{r|rrrr}
-2 & 2 & -1 & +5 & +7 \\
 &  & -4 & +10 & -30 \\
\hline
 & 2 & -5 & +15 & -23 \\
\end{array}
$$

$2x^2-5x+15;\ r=-23$

**55.**

$$
\begin{array}{r|rrrrr}
+3 & 1 & -3 & +0 & +2 & -8 \\
& & +3 & +0 & +0 & +6 \\
\hline
& 1 & +0 & +0 & +2 & \!\!\!\!\boxed{-2}
\end{array}
$$

$x^3 + 2;\ r = -2$

**57.** If a polynomial $p(x)$ is divided by $x - c$, the remainder is equal to $p(c)$.

**59.** $p(1) = 2 \cdot 1^3 - 5 \cdot 1^2 + 9 \cdot 1 - 3$
$= 2 - 5 + 9 - 3 = 3$

**61.**

$$
\begin{array}{r|rrrr}
-5 & 1 & +2 & -13 & +10 \\
& & -5 & +15 & -10 \\
\hline
& 1 & -3 & +2 & \!\!\!\!\boxed{+0}
\end{array}
$$

$p(x) = (x + 5)(x^2 - 3x + 2)$
$= (x + 5)(x - 1)(x - 2)$

**63.** The possible rational roots are $\pm 1, \pm\dfrac{1}{2}, \pm\dfrac{1}{3},$
$\pm\dfrac{1}{6}, \pm 2, \pm\dfrac{2}{3}, \pm 4, \pm\dfrac{4}{3}, \pm 8, \pm\dfrac{8}{3}.$

**65.** The only possible roots are $\pm 1, \pm 2$.

$$
\begin{array}{r|rrrrr}
-1 & 1 & +2 & -1 & -4 & -2 \\
& & -1 & -1 & +2 & +2 \\
\hline
& 1 & +1 & -2 & -2 & \!\!\!\!\boxed{+0}
\end{array}
$$

$p(x) = (x + 1)(x^3 + x^2 - 2x - 2)$
$= (x + 1)(x^2(x + 1) - 2(x + 1))$
$= (x + 1)(x + 1)(x^2 - 2)$
$= (x + 1)^2(x + \sqrt{2})(x - \sqrt{2})$
The roots are $-1$ (double root), $\pm\sqrt{2}$.

**67.** If $p(x)$ is a polynomial of degree $n \geq 1$, then $p(x)$ has at least one complex zero.

**69.** $\overline{zw} = \overline{(2 - 3i)(-3 + 4i)} = \overline{6 + 17i} = 6 - 17i$
$\bar{z} \cdot \bar{w} = \overline{(2 - 3i)}\ \overline{(-3 + 4i)}$
$= (2 + 3i)(-3 - 4i) = 6 - 17i$

**71.** $\dfrac{5x + 7}{x^2 + 2x - 3} = \dfrac{5x + 7}{(x + 3)(x - 1)}$

$= \dfrac{A}{x + 3} + \dfrac{B}{x - 1}$

$5x + 7 = A(x - 1) + B(x + 3)$

Let $x = 1$:   $5 + 7 = A \cdot 0 + B \cdot 4$
$12 = 4B$
$B = 3$

Let $x = -3$:   $-15 + 7 = A(-4) + B \cdot 0$
$-8 = -4A$
$A = 2$

$\dfrac{5x + 7}{x^2 + 2x - 3} = \dfrac{2}{x + 3} + \dfrac{3}{x - 1}$

**73.** $\dfrac{3x^2 + 15x + 8}{(x + 1)(x + 2)(x - 3)}$

$= \dfrac{A}{x + 1} + \dfrac{B}{x + 2} + \dfrac{C}{x - 3}$

$3x^2 + 15x + 8$
$= A(x + 2)(x - 3) + B(x + 1)(x - 3)$
$\quad + C(x + 1)(x + 2)$

Let $x = -1$:   $3(-1)^2 + 15(-1) + 8$
$= A(-1 + 2)(-1 - 3) + B \cdot 0$
$\quad + C \cdot 0$
$-4 = -4A$
$A = 1$

Let $x = -2$:   $3(-2)^2 + 15(-2) + 8$
$= A \cdot 0 + B(-2 + 1)(-2 - 3)$
$\quad + C \cdot 0$
$-10 = 5B$
$B = -2$

Let $x = 3$:   $3 \cdot 3^2 + 15 \cdot 3 + 8$
$= A \cdot 0 + B \cdot 0$
$\quad + C(3 + 1)(3 + 2)$
$80 = 20C$
$C = 4$

$\dfrac{3x^2 - 15x + 8}{(x + 1)(x + 2)(x - 3)}$

$= \dfrac{1}{x + 1} - \dfrac{2}{x + 2} + \dfrac{4}{x - 3}$

**75.**

$$\begin{array}{r} 2x+5 \\ x^2-2x-3\overline{)2x^3+\phantom{0}x^2-\phantom{0}3x+5} \\ \underline{2x^3-4x^2-\phantom{0}6x} \\ 5x^2+\phantom{0}3x+5 \\ \underline{5x^2-10x-15} \\ 13x+20 \end{array}$$

$$\frac{2x^3+x^2-3x+5}{x^2-2x-3}=2x+5+\frac{13x+20}{x^2-2x-3}$$

$$=2x+5+\frac{13x+20}{(x-3)(x+1)}$$

$$\frac{13x+20}{(x-3)(x+1)}=\frac{A}{x-3}+\frac{B}{x+1}$$

$$13x+20=A(x+1)+B(x-3)$$

Let $x=-1$:  $13(-1)+20$
$$=A\cdot 0+B(-1-3)$$
$$7=-4B$$
$$B=-\frac{7}{4}$$

Let $x=3$:  $13\cdot 3+20=A(3+1)+B\cdot 0$
$$59=4A$$
$$A=\frac{59}{4}$$

$$\frac{2x^3+x^2-3x+5}{x^2-2x-3}$$

$$=2x+5+\frac{59}{4(x-3)}-\frac{7}{4(x+1)}$$

**1.**

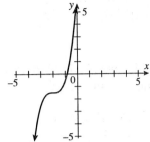

No asymptotes

**2.**

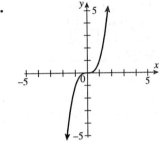

No asymptotes

**3.**

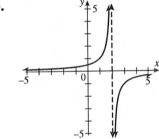

Asymptotes: $x=2$, $y=0$

**4.**

$$f(x)=\frac{x-1}{x^2-x-2}=\frac{x-1}{(x-2)(x+1)}$$

Asymptotes: $x=-1$, $x=2$, $y=0$

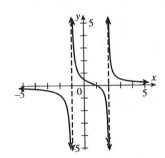

**5.**

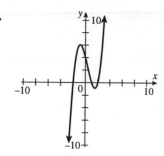

**6.** $f(x) = \dfrac{x^2 - 2x}{x+3} = \dfrac{x(x-2)}{x+3}$. Numerator or denominator are 0 when $x = -3, 0, 2$, so there are four intervals:

| Interval | $(-\infty, -3)$ | $(-3, 0)$ | $(0, 2)$ | $(2, \infty)$ |
|---|---|---|---|---|
| Sign of $x$ | $-$ | $-$ | $+$ | $+$ |
| Sign of $(x-2)$ | $-$ | $-$ | $-$ | $+$ |
| Sign of $(x+3)$ | $-$ | $+$ | $+$ | $+$ |
| Sign of $f(x)$ | $-$ | $+$ | $-$ | $+$ |

$f(x) < 0$ on both $(-\infty, -3)$ and $(0, 2)$; $f(x) > 0$ on both $(-3, 0)$ and $(2, \infty)$.

**7.**

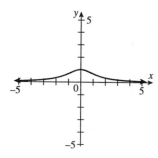

**8.** $\dfrac{x+2}{x-1} \le 0$

| Interval | $(-\infty, -2]$ | $[-2, 1)$ | $(1, \infty)$ |
|---|---|---|---|
| Sign of $\dfrac{x+2}{x-1}$ | $+$ | $-$ | $+$ |

**9.** $S = \dfrac{n}{2}[2a + (n-1)d]$

$S = an + \dfrac{n}{2}(n-1)d$

$S - \dfrac{n}{2}(n-1)d = an$

$\dfrac{S}{n} - \dfrac{1}{2}(n-1)d = a$ or $a = \dfrac{S}{n} - \dfrac{1}{2}(n-1)d$

**10.** $\dfrac{6}{x} = 2 + \dfrac{3}{x+1}$

$x(x+1) \cdot \dfrac{6}{x} = x(x+1) \cdot 2 + x(x+1) \cdot \dfrac{3}{x+1}$

$6(x+1) = 2x(x+1) + 3x$

$6x + 6 = 2x^2 + 2x + 3x$

$0 = 2x^2 - x - 6$

$0 = (2x+3)(x-2)$

$2x + 3 = 0$ or $x - 2 = 0$

$x = -\dfrac{3}{2}$ or $x = 2$

**11.** $\dfrac{x}{2} - \dfrac{3x+1}{3} > 2$

$6 \cdot \dfrac{x}{2} - 6 \cdot \dfrac{3x+1}{3} > 6 \cdot 2$

$3x - 2(3x+1) > 12$

$3x - 6x - 2 > 12$

$-3x > 14$

$x < -\dfrac{14}{3}$

**12.** $\dfrac{6}{x^2-9} - \dfrac{2}{x-3} = \dfrac{1}{x+3}$

$(x+3)(x-3) \cdot \dfrac{6}{x^2-9}$

$\quad - (x+3)(x-3) \cdot \dfrac{2}{x-3}$

$\quad = (x+3)(x-3) \cdot \dfrac{1}{x+3}$

$6 - 2(x+3) = x - 3$

$6 - 2x - 6 = x - 3$

$-3x = -3$

$x = 1$

**13.** Let $x$ be the length of the smaller piece. Then the length of the larger piece is $10 - x$.

$\dfrac{x}{10-x} = \dfrac{3}{4}$

$4x = 3(10 - x)$

$4x = 30 - 3x$

$7x = 30$

$x = \dfrac{30}{7}$

The length of the larger piece is $10 - \dfrac{30}{7}$ or $5\dfrac{5}{7}$ feet.

**14.** Let $x$ be the number of minutes in which Ellen can do the job. Then she can do $\dfrac{1}{x}$ of the job in one minute, Dave can do $\dfrac{1}{45}$ of the job in one minute, and together they can do $\dfrac{1}{30}$ the job in one minute. Hence,

$\dfrac{1}{x} + \dfrac{1}{45} = \dfrac{1}{30}$

$90x \cdot \dfrac{1}{x} + 90x \cdot \dfrac{1}{45} = 90x \cdot \dfrac{1}{30}$

$90 + 2x = 3x$

$90 = x$

Ellen can wash the car in 90 minutes.

**15.** $z = \dfrac{kx}{y}$

$\dfrac{2}{3} = \dfrac{k \cdot 2}{15}$

$30 = 6k$

$5 = k$

$z = \dfrac{5x}{y}$

$z = \dfrac{5 \cdot 4}{10} = 2$

**16.** $V = kr^3$

$36\pi = k \cdot 3^3$

$k = \dfrac{36\pi}{27} = \dfrac{4}{3}\pi$

$V = \dfrac{4}{3}\pi \cdot 6^3$

$V = 288\pi$

**17.**

$$
\begin{array}{r|rrrrrr}
-3 & 2 & +5 & +0 & -1 & -21 & +7 \\
   &   & -6 & +3 & -9 & +30 & -27 \\
\hline
   & 2 & -1 & +3 & -10 & +9 & -20 \\
\end{array}
$$

Quotient: $2x^4 - x^3 + 3x^2 - 10x + 9$
Remainder: $-20$

**18.**

$$
\begin{array}{r|rrrrr}
\frac{1}{3} & 27 & -36 & +18 & -4 & +1 \\
   &   & +9 & -9 & +3 & -\frac{1}{3} \\
\hline
   & 27 & -27 & +9 & -1 & +\frac{2}{3} \\
\end{array}
$$

$p\left(\dfrac{1}{3}\right) = \dfrac{2}{3}$

**19.** Since $p\left(\dfrac{1}{3}\right) \neq 0$, the factor theorem says that $x - \dfrac{1}{3}$ is not a factor of $p(x)$.

**20.**

$$\begin{array}{r|rrrrr} +2 & 1 & -4 & +7 & -12 & +12 \\ & & +2 & -4 & +6 & -12 \\ \hline & 1 & -2 & +3 & -6 & +0 \end{array}$$

$$\begin{aligned} p(x) &= x^4 - 4x^3 + 7x^2 - 12x + 12 \\ &= (x-2)(x^3 - 2x^2 + 3x - 6) \\ &= (x-2)[x^2(x-2) + 3(x-2)] \\ &= (x-2)(x-2)(x^2+3) \\ &= (x-2)^2(x^2+3) \end{aligned}$$

**21.** The possible rational roots are $\pm 1, \pm 3, \pm 9$

$$\begin{array}{r|rrrrr} -3 & 1 & +5 & +4 & -3 & +9 \\ & & -3 & -6 & +6 & -9 \\ \hline & 1 & +2 & -2 & +3 & +0 \end{array}$$

$$\begin{array}{r|rrrr} -3 & 1 & +2 & -2 & +3 \\ & & -3 & +3 & -3 \\ \hline & 1 & -1 & +1 & +0 \end{array}$$

$$\begin{aligned} f(x) &= x^4 + 5x^3 + 4x^2 - 3x + 9 \\ &= (x+3)^2(x^2 - x + 1) \end{aligned}$$

**22.** The possible rational roots are $\pm 1, \pm 2, \pm 3, \pm 6$.

$$\begin{array}{r|rrrrr} -1 & 1 & +3 & -3 & -11 & -6 \\ & & -1 & -2 & +5 & +6 \\ \hline -1 & 1 & +2 & -5 & -6 & +0 \\ & & -1 & -1 & +6 & \\ \hline & 1 & +1 & -6 & +0 & \end{array}$$

$$\begin{aligned} p(x) &= x^4 + 3x^3 - 3x^2 - 11x - 6 \\ &= (x+1)^2(x^2 + x - 6) \\ &= (x+1)^2(x+3)(x-2) \end{aligned}$$

The roots are $-1$ (a double root), $-3, 2$.

**23.** The possible rational roots are $\pm 1, \pm 3$

$$\begin{array}{r|rrrrr} -1 & 1 & -2 & -2 & -2 & -3 \\ & & -1 & +3 & -1 & +3 \\ \hline & 1 & -3 & +1 & -3 & +0 \end{array}$$

$$\begin{aligned} p(x) &= (x+1)(x^3 - 3x^2 + x - 3) \\ &= (x+1)[x^2(x-3) + 1(x-3)] \\ &= (x+1)(x-3)(x^2+1) \end{aligned}$$

The roots of $x^2 + 1$ are $\pm i$, so the roots of $p(x)$ are $-1, 3, \pm i$.

**24.**
$$\frac{x-15}{x^2-25} = \frac{x-15}{(x+5)(x-5)} = \frac{A}{x+5} + \frac{B}{x-5}$$
$$x - 15 = A(x-5) + B(x+5)$$

Let $x = 5$:  $\quad 5 - 15 = A \cdot 0 + B(5+5)$
$$-10 = 10B$$
$$B = -1$$

Let $x = -5$:  $\quad -5 - 15 = A(-5-5) + B \cdot 0$
$$-20 = -10A$$
$$A = 2$$

$$\frac{x-15}{x^2-25} = \frac{2}{x+5} - \frac{1}{x-5}$$

**25.** The possible roots of $x^3 - 2x^2 - 5x + 6$ are $\pm 1, \pm 2, \pm 3, \pm 6$

$$\begin{array}{r|rrrr} 1 & 1 & -2 & -5 & +6 \\ & & +1 & -1 & -6 \\ \hline & 1 & -1 & -6 & +0 \end{array}$$

$$\begin{aligned} x^3 - 2x^2 - 5x + 6 &= (x-1)(x^2 - x - 6) \\ &= (x-1)(x-3)(x+2) \end{aligned}$$

$$\frac{6x^2 - 2x + 2}{x^3 - 2x^2 - 5x + 6}$$
$$= \frac{A}{x-1} + \frac{B}{x-3} + \frac{C}{x+2}$$

$$6x^2 - 2x + 2$$
$$= A(x-3)(x+2) + B(x-1)(x+2)$$
$$\quad + C(x-1)(x-3)$$

Let $x = 1$:  $\quad 6 \cdot 1^2 - 2 \cdot 1 + 2$
$$= A(1-3)(1+2) + B \cdot 0$$
$$\quad + C \cdot 0$$
$$6 = -6A$$
$$A = -1$$

Let $x = 3$:  $\quad 6 \cdot 3^2 - 2 \cdot 3 + 2$
$$= A \cdot 0 + B(3-1)(3+2)$$
$$\quad + C \cdot 0$$
$$50 = 10B$$
$$B = 5$$

Let $x = -2$:  $\quad 6(-2)^2 - 2(-2) + 2$
$$= A \cdot 0 + B \cdot 0$$
$$\quad + C(-2-1)(-2-3)$$
$$30 = 15C$$
$$C = 2$$

$$\frac{6x^2 - 2x + 2}{x^3 - 2x^2 - 5x + 6}$$
$$= -\frac{1}{x-1} + \frac{5}{x-3} + \frac{2}{x+2}$$

## Chapter 3 Test: Multiple Choice

1. I is true, II is true, but III is false. The answer is (d).

2. The horizontal asymptote is $y = 0$, so the answer is (b).

3. The answer is (a).

4. I is false, and so is II since the horizontal asymptote is $y = 0$. III is true, so the answer is (c).

5.

| Interval | Sign of $f(x)$ |
|---|---|
| $(-\infty, -3)$ | $-$ |
| $(-3, -1)$ | $+$ |
| $(-1, 2)$ | $-$ |
| $(2, \infty)$ | $+$ |

The answer is (b).

6. There are no vertical asymptotes, so the answer is (e).

7. Let $x$ be the amount of time it takes for them to complete the job together. They can complete $\dfrac{1}{x}$ of the job in one hour. Amy can complete $\dfrac{1}{5}$ of the job in one hour and Julie can complete $\dfrac{1}{4}$ of the job in one hour. $\dfrac{1}{4} + \dfrac{1}{5} = \dfrac{1}{x}$.
The answer is (d).

8. $c = \dfrac{a + 2b}{ab}$
$cab = a + 2b$
$cab - 2b = a$
$b(ca - 2) = a$
$b = \dfrac{a}{ca - 2} = \dfrac{a}{ac - 2}$
The answer is (a).

9. $\dfrac{9x + 14}{2x^2 - x - 6} - \dfrac{1}{x - 2} = \dfrac{2}{2x + 3}$

$(x - 2)(2x + 3) \cdot \dfrac{9x + 14}{2x^2 - x - 6}$

$- (x - 2)(2x + 3) \cdot \dfrac{1}{x - 2}$

$= (x - 2)(2x + 3) \cdot \dfrac{2}{2x + 3}$

$9x + 14 - (2x + 3) = (x - 2) \cdot 2$
$9x + 14 - 2x - 3 = 2x - 4$
$5x = -15$
$x = -3$
There is only one solution, so the answer is (b).

10. $d = kt^2$
$64 = k \cdot 2^2$
$16 = k$
$d = 16 \cdot t^2$
$d = 16 \cdot 6^2 = 16 \cdot 36 = 576$ ft.
The answer is (a).

11. $z = \dfrac{kx}{y^2}$

$60 = \dfrac{k \cdot 3}{\left(\frac{1}{2}\right)^2}$

$15 = 3k$
$k = 5$

$z = \dfrac{5x}{y^2} = \dfrac{5 \cdot 6}{4^2} = \dfrac{30}{16} = \dfrac{15}{8}$

The answer is (d).

12.

```
-2 | 1   +3   -5    +7
   |     -2   -2   +14
   ----------------------
     1   +1   -7  | +21
```

The answer is (c).

13. The answer is (b).

**14.** $\dfrac{6x^2 + x - 37}{(x-3)(x+2)(x-1)}$

$= \dfrac{A}{x-3} + \dfrac{B}{x+2} + \dfrac{C}{x-1}$

$6x^2 + x - 37$
$= A(x+2)(x-1) + B(x-3)(x-1)$
$\qquad + C(x-3)(x+2)$

Let $x = 3$:  $6 \cdot 3^2 + 3 - 37$
$\qquad\qquad = A(3+2)(3-1) + B \cdot 0$
$\qquad\qquad\quad + C \cdot 0$
$\qquad\quad 20 = 10A$
$\qquad\quad A = 2$

The answer is (c).

**15.**

$$
\begin{array}{r|rrrrr}
2 & 1 & -6 & +14 & -16 & +8 \\
  &   & +2 & -8  & +12 & -8 \\
\hline
2 & 1 & -4 & +6  & -4  & \boxed{+0} \\
  &   & +2 & -4  & +4  & \\
\hline
  & 1 & -2 & +2  & \boxed{+0} & \\
\end{array}
$$

$x^4 - 6x^3 + 14x^2 - 16x + 8$
$= (x-2)^2(x^2 - 2x + 2)$

The roots of $x^2 - 2x + 2$ are $1 \pm i$.
The answer is (a).

**16.** The answer is (a).

**17.** The answer is (c).

**18.**

$$
\begin{array}{r|rrr}
+2 & 1 & +3 & -1 \\
   &   & +2 & +10 \\
\hline
   & 1 & +5 & \boxed{+9} \\
\end{array}
$$

$f(x) = \dfrac{x^2 + 3x - 1}{x - 2} = x + 5 + \dfrac{9}{x-2}$

The oblique asymptote is $y = x + 5$, so the answer is (d).

**19.** Since $p(x)$ varies in sign twice, there are either 2 or 0 positive zeros.
The answer is (b).

**20.** $f(x)$ doesn't vary in sign, so there are no positive intercepts.
$f(-x) = (-x)^3 + 2(-x) + 1 = -x^3 - 2x + 1$
has one variation in sign, so $f(x)$ has one negative intercept. Together, $f(x)$ has only one intercept, so the answer is (b).

# Chapter 4: Circles, Additional Curves, and the Algebra Functions

**Exercises 4.1**

**1.**

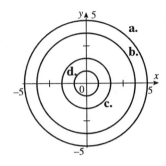

**3.** $x^2 - 4x + y^2 = 21$
$x^2 - 4x + 4 + y^2 = 21 + 4$
$(x-2)^2 + y^2 = 25$
$C(2, 0)$, $r = 5$

**5.** $x^2 - 2x + y^2 - 6y = -9$
$x^2 - 2x + 1 + y^2 - 6y + 9 = -9 + 1 + 9$
$(x-1)^2 + (y-3)^2 = 1$
$C(1, 3)$, $r = 1$

**7.** $x^2 - 4x + y^2 - 10y = -28$
$x^2 - 4x + 4 + y^2 - 10y + 25$
$\quad = -28 + 4 + 25$
$(x-2)^2 + (y-5)^2 = 1$
$C(2, 5)$, $r = 1$

**9.** $x^2 - 8x + y^2 = -14$
$x^2 - 8x + 16 + y^2 = -14 + 16$
$(x-4)^2 + y^2 = 2$
$C(4, 0)$, $r = \sqrt{2}$

**11.** $x^2 - 20x + y^2 + 20y = -100$
$x^2 - 20x + 100 + y^2 + 20y + 100$
$\quad = -100 + 100 + 100$
$(x-10)^2 + (y+10)^2 = 100$
$C(10, -10)$, $r = 10$

**13.** $16x^2 + 24x + 16y^2 - 32y = 119$
$x^2 + \frac{3}{2}x + y^2 - 2y = \frac{119}{16}$

$x^2 + \frac{3}{2}x + \frac{9}{16} + y^2 - 2y + 1$
$= \frac{119}{16} + \frac{9}{16} + 1$

$\left(x + \frac{3}{4}\right)^2 + (y-1)^2 = 9$

$C\left(-\frac{3}{4}, 1\right)$, $r = 3$

**15.** $(x-2)^2 + (y-0)^2 = 2^2$
$(x-2)^2 + y^2 = 4$

**17.** $[x - (-3)]^2 + (y-3)^2 = (\sqrt{7})^2$
$(x+3)^2 + (y-3)^2 = 7$

**19.**

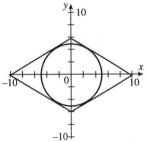

The center of the circle is $(0, 0)$
For $(3, 4)$: slope of radius is
$\frac{4-0}{3-0} = \frac{4}{3}$
Slope of tangent line is $-\frac{3}{4}$

Tangent line: $y - 4 = -\frac{3}{4}(x-3)$
For $(-3, 4)$: slope of radius is
$\frac{4-0}{-3-0} = -\frac{4}{3}$
Slope of tangent line is $\frac{3}{4}$

Tangent line: $y - 4 = \frac{3}{4}(x+3)$
For $(3, -4)$: slope of radius is
$\frac{-4-0}{3-0} = -\frac{4}{3}$
Slope of tangent line is $\frac{3}{4}$

Tangent line: $y + 4 = \frac{3}{4}(x-3)$

For $(-3, -4)$: slope of radius is
$$\frac{-4 - 0}{-3 - 0} = \frac{4}{3}$$

Slope of tangent line is $-\dfrac{3}{4}$

Tangent line: $y + 4 = -\dfrac{3}{4}(x + 3)$

**21.** $x = 2,\ x = -2$

**23.**

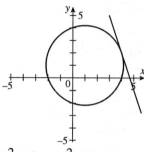

$x^2 - 2x + y^2 - 2y = 8$
$x^2 - 2x + 1 + y^2 - 2y + 1 = 8 + 1 + 1$
$(x - 1)^2 + (y - 1)^2 = 10$
Center: $(1, 1)$
Slope of radius is $\dfrac{2 - 1}{4 - 1} = \dfrac{1}{3}$
Slope of tangent line is $-3$
Tangent line: $y - 2 = -3(x - 4)$
$y - 2 = -3x + 12$
$y = -3x + 14$

**25.**

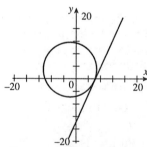

$x^2 + 4x + y^2 - 6y = 60$
$x^2 + 4x + 4 + y^2 - 6y + 9 = 60 + 4 + 9$
$(x + 2)^2 + (y - 3)^2 = 73$
Center: $(-2, 3)$
Slope of radius is $\dfrac{0 - 3}{6 - (-2)} = -\dfrac{3}{8}$
Slope of tangent line is $\dfrac{8}{3}$

Tangent line: $y - 0 = \dfrac{8}{3}(x - 6)$
$$y = \frac{8}{3}(x - 6)$$

**27.**

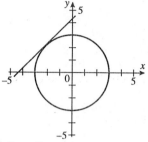

$x^2 + y^2 = 9$
Center: $(0, 0)$

Slope of radius is $\dfrac{\sqrt{5} - 0}{-2 - 0} = -\dfrac{\sqrt{5}}{2}$

Slope of tangent line is $\dfrac{2}{\sqrt{5}} = \dfrac{2\sqrt{5}}{5}$

Tangent line: $y - \sqrt{5} = \dfrac{2\sqrt{5}}{5}\left[x - (-2)\right]$

$$y - \sqrt{5} = \frac{2\sqrt{5}}{5}x + \frac{4\sqrt{5}}{5}$$

$$y = \frac{2\sqrt{5}}{5}x + \frac{9\sqrt{5}}{5}$$

**29.** $\left(\dfrac{-2 + 3}{2},\ \dfrac{4 - 8}{2}\right) = \left(\dfrac{1}{2},\ -2\right)$

**31.** $\left(\dfrac{-8 + 3}{2},\ \dfrac{7 - 6}{2}\right) = \left(-\dfrac{5}{2},\ \dfrac{1}{2}\right)$

**33.** If $x = 4,\ 4^2 + y^2 = 80$
$y^2 = 64$
$y = 8$ since $(x, y)$ is in the first quadrant.
Slope of radius is $\dfrac{8 - 0}{4 - 0} = 2$

Slope of tangent line is $-\dfrac{1}{2}$

Tangent line: $y - 8 = -\dfrac{1}{2}(x - 4)$

$$y - 8 = -\frac{1}{2}x + 2$$

$$y = -\frac{1}{2}x + 10$$

**35.** $x^2 + 14x + y^2 + 18y = 39$
$x^2 + 14x + 49 + y^2 + 18y + 81$
$= 39 + 49 + 81$
$(x + 7)^2 + (y + 9)^2 = 169$
Center: $(-7, -9)$
If $x = -2$, $(-2 + 7)^2 + (y + 9)^2 = 169$
$(y + 9)^2 = 144$
$y + 9 = \pm 12$
$y = -9 \pm 12 = 3$ since $(x, y)$ is in the second quadrant.
Slope of radius is $\dfrac{3 + 9}{-2 + 7} = \dfrac{12}{5}$
Slope of tangent line is $-\dfrac{5}{12}$
Tangent line:  $y - 3 = -\dfrac{5}{12}(x + 2)$
$12y - 36 = -5x - 10$
$5x + 12y = 26$

**37. (a)** Original height is
$\sqrt{13^2 - 5^2} = \sqrt{169 - 25}$
$= \sqrt{144} = 12$
$(y + 5)^2 + (12 - x)^2 = 13^2$
$(y + 5)^2 = 13^2 - (12 - x)^2$
$y + 5 = \sqrt{169 - 144 + 24x - x^2}$
$y(x) = \sqrt{25 + 24x - x^2} - 5$
**(b)** $y(7) = \sqrt{25 + 24 \cdot 7 - 7^2} - 5$
$= \sqrt{144} - 5 = 12 - 5 = 7$

**39.** $s = (x - 6)^2 + (y - 6)^2$
$= (x - 6)^2 + (-2x + 4 - 6)^2$
$= x^2 - 12x + 36 + 4x^2 + 8x + 4$
$= 5x^2 - 4x + 40$
$x = -\dfrac{-4}{2 \cdot 5} = \dfrac{2}{5}$
$y = -2\left(\dfrac{2}{5}\right) + 4 = \dfrac{16}{5}$
$\left(\dfrac{2}{5}, \dfrac{16}{5}\right)$

**41.** $A = 2xy$
$x^2 + y^2 = 12^2$
$y^2 = 144 - x^2$ or $y = \sqrt{144 - x^2}$
$A(x) = 2x\sqrt{144 - x^2}$

**43.** The base of the small triangle is
$\sqrt{5^2 - h^2} = \sqrt{25 - h^2}$
$A(h) = \dfrac{1}{2}h\sqrt{25 - h^2}$
$+ \dfrac{1}{2}h\sqrt{25 - h^2} + 10h$
$= 10h + h\sqrt{25 - h^2}$
$= h\left(10 + \sqrt{25 - h^2}\right)$

**45.** By the Pythagorean Theorem, $AB = 5$. The radii on $BO$, $OA$ and $AB$ are $2, \dfrac{3}{2}$, and $\dfrac{5}{2}$, respectively. The sum of the areas of the semicircles on the legs are
$\dfrac{1}{2}\pi(2)^2 + \dfrac{1}{2}\pi\left(\dfrac{3}{2}\right)^2 = 2\pi + \dfrac{9}{8}\pi = \dfrac{25}{8}\pi$
The area of the semicircle on $AB$ is
$\dfrac{1}{2}\pi\left(\dfrac{5}{2}\right)^2 = \dfrac{25}{8}\pi$

**47.** The diameter of the circular stained glass window is $24 - 8 = 16$ inches, the radius is 8 inches, and the center is at $(0, 12)$. The equation is
$(x - 0)^2 + (y - 12)^2 = 8^2$
$x^2 + (y - 12)^2 = 64$  (in inches)
$x^2 + (y - 1)^2 = \dfrac{4}{9}$  ( in feet)

**49.**

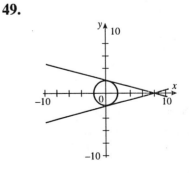

The points of tangency are the intersections of the two circles.

$x^2 + y^2 = 4$  or  $y^2 = 4 - x^2$

$(x-4)^2 + y^2 = 16$  or  $y^2 = 16 - (x-4)^2$

$4 - x^2 = 16 - (x-4)^2$

$4 - x^2 = 16 - x^2 + 8x - 16$

$4 = 8x$

$x = \dfrac{1}{2}$

$y^2 = 4 - \left(\dfrac{1}{2}\right)^2 = \dfrac{15}{4}$

$y = \pm\sqrt{\dfrac{15}{4}} = \pm\dfrac{\sqrt{15}}{2}$

$\left(\dfrac{1}{2},\ \pm\dfrac{\sqrt{15}}{2}\right)$

## Challenge Problem

The center of the circle is located at the intersection of the perpendicular bisectors of any two chords.

Chord $BC$:

Midpoint: $\left(\dfrac{5-1}{2},\ \dfrac{13+11}{2}\right) = (2,\ 12)$

Slope of $BC$: $\dfrac{13-11}{5-(-1)} = \dfrac{2}{6} = \dfrac{1}{3}$

Slope of bisector is $-3$

Bisector: $y - 12 = -3(x - 2)$

$y - 12 = -3x + 6$

$y = -3x + 18$

Chord $AB$:

Midpoint: $\left(\dfrac{-1-3}{2},\ \dfrac{11-3}{2}\right) = (-2,\ 4)$

Slope of $AB$: $\dfrac{-3-11}{-3+1} = \dfrac{-14}{-2} = 7$

Slope of bisector is $-\dfrac{1}{7}$

Bisector: $y - 4 = -\dfrac{1}{7}(x + 2)$

$y - 4 = -\dfrac{1}{7}x - \dfrac{2}{7}$

$y = -\dfrac{1}{7}x + \dfrac{26}{7}$

$-3x + 18 = -\dfrac{1}{7}x + \dfrac{26}{7}$

$-21x + 126 = -x + 26$

$-20x = -100$

$x = 5$

$y = -3(5) + 18 = -15 + 18 = 3$

Center: $(5, 3)$

Radius: $\sqrt{(5+3)^2 + (3+3)^2} = \sqrt{(8)^2 + (6)^2}$

$= \sqrt{64 + 36} = \sqrt{100} = 10$

The length of the radius is 10 inches.

## Exercises 4.2

1.

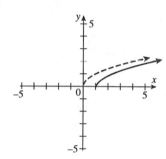

3.

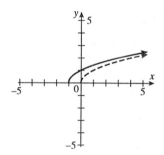

5.

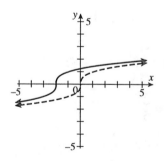

**7.**

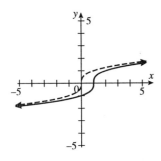

**9.**

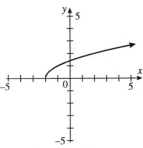

Domain: $x \geq -2$
Increasing for $x \geq -2$
Concave down for $x > -2$

**11.**

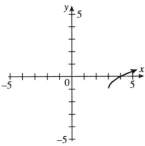

Domain: $x \geq 3$
Increasing for $x \geq 3$
Concave down for $x > 3$

**13.**

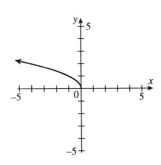

Domain: $x \leq 0$
Decreasing for $x \leq 0$
Concave down for $x < 0$

**15.**

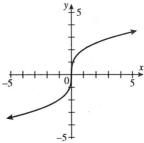

Domain: all real $x$
Increasing for all $x$
Concave up for $x < 0$
Concave down for $x > 0$

**17.**

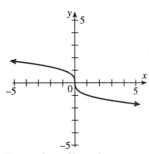

Domain: all real $x$
Decreasing for all $x$
Concave down for $x < 0$
Concave up for $x > 0$

**19.**

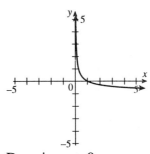

Domain: $x > 0$
Asymptotes: $x = 0$, $y = -1$
Decreasing and concave up for $x > 0$.

**21.** **(a)** $f(-x) = \dfrac{1}{\sqrt[3]{-x}} = \dfrac{1}{-\sqrt[3]{x}}$

$= -\dfrac{1}{\sqrt[3]{x}} = -f(x)$

Since $f(-x) = -f(x)$, $f(x)$ is symmetric through the origin.

**(b)** all $x \neq 0$

**(c)**

| $x$ | $\dfrac{1}{27}$ | $\dfrac{1}{8}$ | 1 | 8 |
|-----|-----------------|----------------|---|---|
| $y$ | 3 | 2 | 1 | $\dfrac{1}{2}$ |

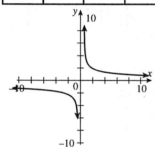

**(d)** $x = 0;\ y = 0$

**23.** $y = \sqrt[4]{x}$ is the equivalent to $y^4 = x$ for $x \geq 0$.

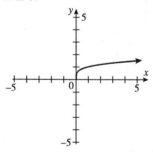

**25.**

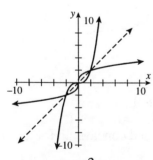

$x = y^3$ or $y = \sqrt[3]{x}$

**27.**

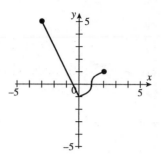

**29.** $\dfrac{f(4+h) - f(4)}{h} = \dfrac{\sqrt{4+h} - 2}{h}$

$= \dfrac{(\sqrt{4+h} - 2)(\sqrt{4+h} + 2)}{h(\sqrt{4+h} + 2)}$

$= \dfrac{4 + h - 4}{h(\sqrt{4+h} + 2)}$

$= \dfrac{h}{h(\sqrt{4+h} + 2)}$

$= \dfrac{1}{\sqrt{4+h} + 2}$

**31.** **(a)** $d(x) = \sqrt{5^2 + x^2}$

$+ \sqrt{10^2 + (20 - x)^2}$

$= \sqrt{25 + x^2} + \sqrt{100 + (20 - x)^2}$

**(b)** $t(x) = \dfrac{\sqrt{25 + x^2}}{12}$

$+ \dfrac{\sqrt{100 + (20 - x)^2}}{10}$

**(c)** $t(5) = \dfrac{\sqrt{25 + 5^2}}{12} + \dfrac{\sqrt{100 + (15)^2}}{10}$

$= \dfrac{\sqrt{50}}{12} + \dfrac{\sqrt{325}}{10} = 2.4$ hours

**33.** $d(x) = \sqrt{(x - 0)^2 + (y - 0)^2}$

$= \sqrt{x^2 + y^2} = \sqrt{x^2 + \left(\dfrac{1}{\sqrt{x}}\right)^2}$

$= \sqrt{x^2 + \dfrac{1}{x}}$

**Challenge Problem**

From Exercise 34, $d(x) = \sqrt{10x^2 - 10x + 5}$, so that $d^2 = 10x^2 - 10x + 5$. The distance will be the shortest when $10x^2 - 10x + 5$ is the smallest.

$10x^2 - 10x + 5 = 10(x^2 - x) + 5$

$= 10\left(x^2 - x + \dfrac{1}{4}\right) + 5 - \dfrac{10}{4}$

$= 10\left(x^2 - x + \dfrac{1}{4}\right) + \dfrac{5}{2}$

$= 10\left(x - \dfrac{1}{2}\right)^2 + \dfrac{5}{2} \geq \dfrac{5}{2}$

The smallest $d^2$ is $\dfrac{5}{2}$, so the smallest distance

is $\sqrt{\dfrac{5}{2}} = \dfrac{\sqrt{10}}{2}$.

**Exercises 4.3**

**1.** $x$-intercepts: $0 = \sqrt{x-1} - 4$

$4 = \sqrt{x-1}$

$16 = x - 1$

$x = 17$

Domain: $x \geq 1$

**3.** $x$-intercepts: $0 = \sqrt{x^2 + 2x}$

$0 = x^2 + 2x$

$0 = x(x + 2)$

$x = 0$ or $x = -2$

Domain: $x^2 + 2x \geq 0$

$x(x + 2) \geq 0$

$x \geq 0$ and $x + 2 \geq 0$ or $x \leq 0$

and $x + 2 \leq 0$

$x \geq 0$ or $x \leq -2$

**5.** $x$-intercepts: $0 = \sqrt{x^2 - 5x - 6}$

$0 = x^2 - 5x - 6$

$0 = (x - 6)(x + 1)$

$x - 6 = 0$ or $x + 1 = 0$

$x = 6$ or $x = -1$

Domain: $x^2 - 5x - 6 \geq 0$

$(x - 6)(x + 1) \geq 0$

$x - 6 \geq 0$ and $x + 1 \geq 0$ or

$x - 6 \leq 0$ and $x + 1 \leq 0$

$x \geq 6$ or $x \leq -1$

**7.** $\sqrt{4x + 9} - 7 = 0$

$\sqrt{4x + 9} = 7$

$\left(\sqrt{4x + 9}\right)^2 = 7^2$

$4x + 9 = 49$

$4x = 40$

$x = 10$

**9.** $(3x + 1)^{1/2} = (2x + 6)^{1/2}$

$\left[(3x + 1)^{1/2}\right]^2 = \left[(2x + 6)^{1/2}\right]^2$

$3x + 1 = 2x + 6$

$x = 5$

**11.** $\sqrt{x^2 - 36} = 8$

$\left(\sqrt{x^2 - 36}\right)^2 = 8^2$

$x^2 - 36 = 64$

$x^2 = 100$

$x = \pm 10$

**13.** $\sqrt{x^2 + \dfrac{1}{2}} = \dfrac{1}{\sqrt{3}}$

$\left(\sqrt{x^2 + \dfrac{1}{2}}\right)^2 = \left(\dfrac{1}{\sqrt{3}}\right)^2$

$x^2 + \dfrac{1}{2} = \dfrac{1}{3}$

$x^2 = -\dfrac{1}{6}$

No solutions.

**15.** $\dfrac{8}{\sqrt{x + 2}} = 4$

$8 = 4\sqrt{x + 2}$

$8^2 = (4\sqrt{x + 2})^2$

$64 = 16(x + 2)$

$4 = x + 2$

$x = 2$

**17.** $\dfrac{1}{\sqrt{2x-1}} = \dfrac{3}{\sqrt{5-3x}}$

$\sqrt{5-3x} = 3\sqrt{2x-1}$

$\left(\sqrt{5-3x}\right)^2 = \left(3\sqrt{2x-1}\right)^2$

$5 - 3x = 9(2x-1)$

$5 - 3x = 18x - 9$

$-21x = -14$

$x = \dfrac{2}{3}$

**19.** $\sqrt[4]{1-3x} = \dfrac{1}{2}$

$\left(\sqrt[4]{1-3x}\right)^4 = \left(\dfrac{1}{2}\right)^4$

$1 - 3x = \dfrac{1}{16}$

$-3x = -\dfrac{15}{16}$

$x = \dfrac{5}{16}$

**21.** $\sqrt{x} + \sqrt{x-5} = 5$

$\sqrt{x-5} = 5 - \sqrt{x}$

$\left(\sqrt{x-5}\right)^2 = \left(5 - \sqrt{x}\right)^2$

$x - 5 = 25 - 10\sqrt{x} + x$

$10\sqrt{x} = 30$

$\sqrt{x} = 3$

$\left(\sqrt{x}\right)^2 = 3^2$

$x = 9$

**23.** $\sqrt{x-1} + \sqrt{3x-2} = 3$

$\sqrt{3x-2} = 3 - \sqrt{x-1}$

$\left(\sqrt{3x-2}\right)^2 = \left(3 - \sqrt{x-1}\right)^2$

$3x - 2 = 9 - 6\sqrt{x-1} + x - 1$

$2x - 10 = -6\sqrt{x-1}$

$(2x-10)^2 = \left(-6\sqrt{x-1}\right)^2$

$4x^2 - 40x + 100 = 36(x-1)$

$4x^2 - 40x + 100 = 36x - 36$

$4x^2 - 76x + 136 = 0$

$4(x^2 - 19x + 34) = 0$

$4(x-2)(x-17) = 0$

$x - 2 = 0 \quad \text{or} \quad x - 17 = 0$

$x = 2 \quad \text{or} \quad x = 17, \text{ but } x = 17 \text{ isn't a solution.}$

$x = 2.$

**25.** $\sqrt{4x+1} + \sqrt{x+7} = 6$

$\sqrt{x+7} = 6 - \sqrt{4x+1}$

$\left(\sqrt{x+7}\right)^2 = \left(6 - \sqrt{4x+1}\right)^2$

$x + 7 = 36 - 12\sqrt{4x+1} + 4x + 1$

$-3x - 30 = -12\sqrt{4x+1}$

$x + 10 = 4\sqrt{4x+1}$

$(x+10)^2 = \left(4\sqrt{4x+1}\right)^2$

$x^2 + 20x + 100 = 16(4x+1)$

$x^2 + 20x + 100 = 64x + 16$

$x^2 - 44x + 84 = 0$

$(x-42)(x-2) = 0$

$x - 42 = 0 \text{ or } x - 2 = 0$

$x = 42 \text{ or } x = 2, \text{ but } x = 42 \text{ isn't a solution.}$

$x = 2$

**27.** $x\sqrt{4-x} - \sqrt{9x-36} = 0$

$x\sqrt{4-x} = \sqrt{9x-36}$

$\left(x\sqrt{4-x}\right)^2 = \left(\sqrt{9x-36}\right)^2$

$x^2(4-x) = 9x - 36$

$4x^2 - x^3 = 9x - 36$

$-x^3 + 4x^2 - 9x + 36 = 0$

$-x^2(x-4) - 9(x-4) = 0$

$-(x^2 + 9)(x-4) = 0$

$x^2 + 9 = 0 \text{ or } x - 4 = 0$

$x^2 + 9 = 0 \text{ not possible, so } x = 4$

**29.** $x = 8 - 2\sqrt{x}$

$x - 8 = -2\sqrt{x}$

$(x-8)^2 = \left(-2\sqrt{x}\right)^2$

$x^2 - 16x + 64 = 4x$

$x^2 - 20x + 64 = 0$

$(x-16)(x-4) = 0$

$x - 16 = 0 \text{ or } x - 4 = 0$

$x = 16$ or $x = 4$, but $x = 16$ is not a solution.
$x = 4$

**31.** $\sqrt{x^2 - 6x} = x - \sqrt{2x}$

$\left(\sqrt{x^2 - 6x}\right)^2 = \left(x - \sqrt{2x}\right)^2$

$x^2 - 6x = x^2 - 2x\sqrt{2x} + 2x$

$-8x = -2x\sqrt{2x}$

$2x\sqrt{2x} - 8x = 0$

$x(2\sqrt{2x} - 8) = 0$

$x = 0$ or $2\sqrt{2x} = 8$

$\sqrt{2x} = 4$

$2x = 16$

$x = 8$

$x = 0$ or $x = 8$

**33.** $4x^{2/3} - 12x^{1/3} + 9 = 0$

Let $u = x^{1/3}$

$4\left(x^{1/3}\right)^2 - 12x^{1/3} + 9 = 0$

$4u^2 - 12u + 9 = 0$

$(2u - 3)(2u - 3) = 0$

$2u - 3 = 0$

$2u = 3$

$u = \dfrac{3}{2}$

$x^{1/3} = \dfrac{3}{2}$

$x = \left(\dfrac{3}{2}\right)^3 = \dfrac{27}{8}$

**35.** $(5x - 6)^{1/5} + \dfrac{x}{(5x - 6)^{4/5}} = 0$

$(5x - 6)^{4/5} \cdot (5x - 6)^{1/5}$

$\quad + (5x - 6)^{4/5} \cdot \dfrac{x}{(5x - 6)^{4/5}} = 0$

$5x - 6 + x = 0$

$6x = 6$

$x = 1$

**37.** $x^{-3/2} - \dfrac{1}{9}x^{-1/2} = 0$

$x^{3/2} \cdot x^{-3/2} - x^{3/2} \cdot \dfrac{1}{9}x^{-1/2} = x^{3/2} \cdot 0$

$1 - \dfrac{1}{9}x = 0$

$-\dfrac{1}{9}x = -1$

$x = 9$

**39.** $x^{2/3} + \dfrac{2}{3}x^{-1/3}(x - 10) = 0$

$x^{1/3} \cdot x^{2/3} + x^{1/3} \cdot \dfrac{2}{3}x^{-1/3}(x - 10) = x^{1/3} \cdot 0$

$x + \dfrac{2}{3}(x - 10) = 0$

$x + \dfrac{2}{3}x - \dfrac{20}{3} = 0$

$\dfrac{5}{3}x = \dfrac{20}{3}$

$x = 4$

**41.** **(a)** All $x \geq 2$

**(b)** $f(x) > 0$ for all $x > 2$

**(c)** $0 = (x + 4)\sqrt{x - 2}$

$x + 4 = 0$ or $\sqrt{x - 2} = 0$

$x = -4$ or $x = 2$

But $x = -4$ is not a solution, so $x = 2$.

**43.** **(a)** All real numbers.

**(b)** $f(x) = \left[(x - 2)^{1/3}\right]^2$, so $f(x) > 0$ for all $x \neq 2$.

**(c)** $x = 2$

**45.** **(a)** All $x > 0$

**(b)** $f(x) > 0$ on $(0, \infty)$

**(c)** $0 = \dfrac{9 + x}{9\sqrt{x}}$

$0 = 9 + x$

$x = -9$, which isn't in the domain, so there are no roots of $f(x) = 0$.

**47.** **(a)** All $x \neq 0$

**(b)** There are four intervals: $(-\infty, -4)$, $(-4, 0)$, $(0, 2)$, $(2, \infty)$

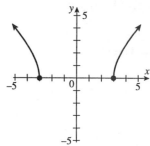

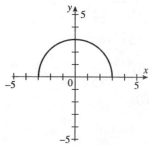

| Interval | $(-\infty, -4)$ | $(-4, 0)$ | $(0, 2)$ | $(2, \infty)$ |
|---|---|---|---|---|
| Sign of $x + 4$ | – | + | + | + |
| Sign of $\sqrt[3]{x-2}$ | – | – | – | + |
| Sign of $\sqrt[3]{x}$ | – | – | + | + |
| Sign of $f(x)$ | – | + | – | + |

$f(x) < 0$ on both $(-\infty, -4)$, $(0, 2)$;
$f(x) > 0$ on both $(-4, 0)$, $(2, \infty)$.
**(c)** $x = -4, 2$

**49.** $x^{1/2} + (x-4)\dfrac{1}{2}x^{-1/2}$

$= \dfrac{x^{1/2}2x^{1/2}}{2x^{1/2}} + \dfrac{x-4}{2x^{1/2}}$

$= \dfrac{2x + x - 4}{2\sqrt{x}}$

$= \dfrac{3x - 4}{2\sqrt{x}}$

**51.** $\dfrac{1}{2}(4 - x^2)^{-1/2}(-2x) = \dfrac{1}{2}(-2x)(4 - x^2)^{-1/2}$

$= \dfrac{-2x}{2\sqrt{4 - x^2}}$

$= -\dfrac{x}{\sqrt{4 - x^2}}$

**53.** $\dfrac{x}{3(x-1)^{2/3}} + (x-1)^{1/3}$

$= \dfrac{x}{3(x-1)^{2/3}} + \dfrac{3 \cdot (x-1)^{2/3}(x-1)^{1/3}}{3(x-1)^{2/3}}$

$= \dfrac{x + 3x - 3}{3(\sqrt[3]{x-1})^2} = \dfrac{4x - 3}{3(\sqrt[3]{x-1})^2}$

**55.** $f(x) = \sqrt{x} - \dfrac{1}{\sqrt{x}} = \dfrac{\sqrt{x} \cdot \sqrt{x}}{\sqrt{x}} - \dfrac{1}{\sqrt{x}} = \dfrac{x - 1}{\sqrt{x}}$

There are only two intervals: $(0, 1)$ and $(1, \infty)$.

| Interval | $(0, 1)$ | $(1, \infty)$ |
|---|---|---|
| Sign of $x - 1$ | – | + |
| Sign of $\sqrt{x}$ | + | + |
| Sign of $f(x)$ | – | + |

$f(x) < 0$ on $(0, 1)$, $f(x) > 0$ on $(1, \infty)$

**57.** $f(x) = \dfrac{x^2}{2}(x-2)^{-1/2} + 2x(x-2)^{1/2}$

$= \dfrac{x^2}{2\sqrt{x-2}} + \dfrac{2x(x-2)^{1/2}2(x-2)^{1/2}}{2\sqrt{x-2}}$

$= \dfrac{x^2 + 4x(x-2)}{2\sqrt{x-2}}$

$= \dfrac{5x^2 - 8x}{2\sqrt{x-2}} = \dfrac{x(5x-8)}{2\sqrt{x-2}}$

For $x > 2$, $f(x) > 0$, so
$f(x) > 0$ on $(2, \infty)$

**59.**

Domain: $x^2 - 9 \geq 0$
$x^2 \geq 9$
$x \geq 3$ or $x \leq -3$

**61.**

Domain: $9 - x^2 \geq 0$

$$9 \geq x^2$$
$$-3 \leq x \leq 3$$

**63.**

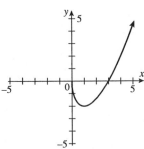

Domain: $x \geq 0$

**65.** $r = \dfrac{1}{2}\sqrt{\dfrac{A}{\pi}}$

$2r = \sqrt{\dfrac{A}{\pi}}$

$(2r)^2 = \left(\sqrt{\dfrac{A}{\pi}}\right)^2$

$4r^2 = \dfrac{A}{\pi}$

$A = 4\pi r^2$

$A = 4\pi \cdot 2^2$

$A = 16\pi$

**67.** $s = \sqrt{r^2 + h^2}$

$s^2 = \left(\sqrt{r^2 + h^2}\right)^2$

$s^2 = r^2 + h^2$

$s^2 - r^2 = h^2$

$\sqrt{s^2 - r^2} = h$

$h = \sqrt{(17.23)^2 - (8.96)^2}$

$= \sqrt{216.59} = 14.72 \text{ cm}$

**69.** $\pi r^2 h = 5$

$r^2 = \dfrac{5}{\pi h}$

$r = \sqrt{\dfrac{5}{\pi h}} = \dfrac{\sqrt{5\pi h}}{\pi h}$

$s = 2\pi r^2 + 2\pi r h$

$= 2\pi \cdot \dfrac{5}{\pi h} + 2\pi h \cdot \dfrac{\sqrt{5\pi h}}{\pi h}$

$= \dfrac{10}{h} + 2\sqrt{5\pi h}$

**71.** $3\sqrt{5} = \sqrt{(x-3)^2 + (\sqrt{x} - 0)^2}$

$3\sqrt{5} = \sqrt{(x-3)^2 + x}$

$\left(3\sqrt{5}\right)^2 = \left(\sqrt{x^2 - 5x + 9}\right)^2$

$45 = x^2 - 5x + 9$

$0 = x^2 - 5x - 36$

$0 = (x-9)(x+4)$

$x - 9 = 0 \text{ or } x + 4 = 0$

$x = 9 \text{ or } x = -4$

$x = -4$ is not possible, so $x = 9$,

$y = \sqrt{9} = 3$

$(9, 3)$

**73.**

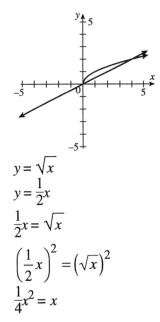

$y = \sqrt{x}$

$y = \dfrac{1}{2}x$

$\dfrac{1}{2}x = \sqrt{x}$

$\left(\dfrac{1}{2}x\right)^2 = \left(\sqrt{x}\right)^2$

$\dfrac{1}{4}x^2 = x$

$$\frac{1}{4}x^2 - x = 0$$

$$x\left(\frac{1}{4}x - 1\right) = 0$$

$$x = 0 \text{ or } \frac{1}{4}x - 1 = 0$$

$$\frac{1}{4}x = 1$$

$$x = 0 \text{ or } x = 4$$

If $x = 0$, $y = \frac{1}{2} \cdot 0 = 0$

If $x = 4$, $y = \frac{1}{2} \cdot 4 = 2$

$(0, 0)$ and $(4, 2)$

**75.** $x = \frac{3}{5}\sqrt{25 - y^2}$

$$x^2 = \left(\frac{3}{5}\sqrt{25 - y^2}\right)^2$$

$$x^2 = \frac{9}{25}\left(25 - y^2\right)$$

$$x^2 = 9 - \frac{9}{25}y^2$$

$$\frac{9}{25}y^2 = 9 - x^2$$

$$y^2 = \frac{25}{9}\left(9 - x^2\right)$$

$$y = \sqrt{\frac{25}{9}}\sqrt{9 - x^2}$$

$$y = \frac{5}{3}\sqrt{9 - x^2}$$

**77.** $\frac{1}{\sqrt{xy}}(xf + y) = f$

$$(xf + y) = f\sqrt{xy}$$

$$xf - f\sqrt{xy} = -y$$

$$f(x - \sqrt{xy}) = -y$$

$$f = \frac{-y}{x - \sqrt{xy}}$$

$$f = \frac{y}{\sqrt{xy} - x}$$

**Challenge Problem**

$$y_1 - y_2 = x - \sqrt{x^2 - 4}$$

$$= \frac{\left(x - \sqrt{x^2 - 4}\right)\left(x + \sqrt{x^2 - 4}\right)}{x + \sqrt{x^2 - 4}}$$

$$= \frac{x^2 - \left(x^2 - 4\right)}{x + \sqrt{x^2 - 4}}$$

$$= \frac{x^2 - x^2 + 4}{x + \sqrt{x^2 - 4}}$$

$$= \frac{4}{x + \sqrt{x^2 - 4}}$$

As $x \to \infty$, $y_1 - y_2 \to 0$, so $y = x$ is an oblique asymptote.

**Exercises 4.4**

**1. (a)** $f(1) = 2 \cdot 1 - 3 = -1$
$g(1) = 3 \cdot 1 + 2 = 5$
$f(1) + g(1) = -1 + 5 = 4$

**(b)** $(f + g)(x) = f(x) + g(x)$
$= 2x - 3 + 3x + 2 = 5x - 1$
Domain: all reals

**(c)** $(f + g)(1) = 5 \cdot 1 - 1 = 4$

**3. (a)** $f\left(\frac{1}{2}\right) = 2 \cdot \frac{1}{2} - 3 = -2$

$g\left(\frac{1}{2}\right) = 3 \cdot \frac{1}{2} + 2 = \frac{7}{2}$

$f\left(\frac{1}{2}\right) \cdot g\left(\frac{1}{2}\right) = -2 \cdot \frac{7}{2} = -7$

**(b)** $(f \cdot g)(x) = f(x) \cdot g(x)$
$= (2x - 3)(3x + 2)$
$= 6x^2 - 5x - 6$
Domain: all reals

**(c)** $(f \cdot g)\left(\frac{1}{2}\right) = 6 \cdot \left(\frac{1}{2}\right)^2 - 5 \cdot \frac{1}{2} - 6$

$= \frac{3}{2} - \frac{5}{2} - \frac{12}{2}$

$= -7$

**5. (a)** $g(0) = 3 \cdot 0 + 2 = 2$
$f(g(0)) = f(2) = 2 \cdot 2 - 3 = 1$

**(b)** $(f \circ g)(x) = f(g(x))$
$= f(3x + 2)$
$= 2(3x + 2) - 3$
$= 6x + 1$
Domain: all reals
**(c)** $(f \circ g)(0) = 6 \cdot 0 + 1 = 1$

**7. (a)** $(f + g)(x) = f(x) + g(x)$
$= x^2 + \sqrt{x}$
Domain: $x \geq 0$

**(b)** $\left(\dfrac{f}{g}\right)(x) = \dfrac{f(x)}{g(x)} = \dfrac{x^2}{\sqrt{x}} = x^{3/2}$
Domain: $x > 0$

**(c)** $(f \circ g)(x) = f(g(x)) = f\left(\sqrt{x}\right)$
$= \left(\sqrt{x}\right)^2 = x$
Domain: $x \geq 0$

**9. (a)** $(f + g)(x) = f(x) + g(x) = x^3 - 1 + \dfrac{1}{x}$
Domain: all $x \neq 0$

**(b)** $\left(\dfrac{f}{g}\right)(x) = \dfrac{f(x)}{g(x)} = \dfrac{x^3 - 1}{\frac{1}{x}} = x^4 - x$
Domain: all $x \neq 0$

**(c)** $(f \circ g)(x) = f(g(x)) = f\left(\dfrac{1}{x}\right) = \dfrac{1}{x^3} - 1$
Domain: all $x \neq 0$

**11. (a)** $(f + g)(x) = f(x) + g(x)$
$= x^2 + 6x + 8 + \sqrt{x - 2}$
Domain: $x \geq 2$

**(b)** $\left(\dfrac{f}{g}\right)(x) = \dfrac{f(x)}{g(x)} = \dfrac{x^2 + 6x + 8}{\sqrt{x - 2}}$
Domain: $x > 2$

**(c)** $(f \circ g)(x) = f(g(x)) = f\left(\sqrt{x - 2}\right)$
$= \left(\sqrt{x - 2}\right)^2 + 6\sqrt{x - 2} + 8$
$= x - 2 + 6\sqrt{x - 2} + 8$
$= x + 6 + 6\sqrt{x - 2}$
Domain: $x \geq 2$

**13. (a)** $(g - f)(x) = g(x) - f(x)$
$= 4x - 1 - (-2x + 5) = 6x - 6$
Domain: all reals

**(b)** $(g \cdot f)(x) = g(x) \cdot f(x)$
$= (4x - 1)(-2x + 5)$
$= -8x^2 + 22x - 5$
Domain: all reals.

**(c)** $(g \circ f)(x) = g(f(x)) = g(-2x + 5)$
$= 4(-2x + 5) - 1 = -8x + 19$
Domain: all reals.

**15. (a)** $(g - f)(x) = g(x) - f(x)$
$= \dfrac{1}{2x} - (2x^2 - 1)$
$= \dfrac{1}{2x} - 2x^2 + 1$
Domain: all $x \neq 0$

**(b)** $(g \cdot f)(x) = g(x) \cdot f(x)$
$= \left(\dfrac{1}{2x}\right)(2x^2 - 1)$
$= x - \dfrac{1}{2x}$
Domain: all $x \neq 0$

**(c)** $(g \circ f)(x) = g(f(x)) = g(2x^2 - 1)$
$= \dfrac{1}{2(2x^2 - 1)} = \dfrac{1}{4x^2 - 2}$
Domain: All $x \neq \pm\dfrac{1}{\sqrt{2}}$

**17.** $(f \circ g)(x) = f(g(x)) = f(x - 1) = (x - 1)^2$
$(g \circ f)(x) = g(f(x)) = g(x^2) = x^2 - 1$

**19.** $(f \circ g)(x) = f(g(x)) = f\left(\dfrac{x + 3}{x}\right)$
$= \dfrac{\frac{x+3}{x}}{\frac{x+3}{x} - 2} \cdot \dfrac{x}{x} = \dfrac{x + 3}{x + 3 - 2x} = \dfrac{x + 3}{3 - x}$
$(g \circ f)(x) = g(f(x)) = g\left(\dfrac{x}{x - 2}\right)$
$= \dfrac{\frac{x}{x-2} + 3}{\frac{x}{x-2}} \cdot \dfrac{x - 2}{x - 2}$
$= \dfrac{x + 3(x - 2)}{x} = \dfrac{4x - 6}{x} = 4 - \dfrac{6}{x}$

**21.** $(f \circ g)(x) = f(g(x)) = f(x^4 - 1)$
$= \sqrt{x^4 - 1 + 1} = \sqrt{x^4} = x^2$

$$(g \circ f)(x) = g(f(x)) = g(\sqrt{x+1})$$
$$= \left(\sqrt{x+1}\right)^4 - 1 = (x+1)^2 - 1$$
$$= x^2 + 2x + 1 - 1 = x^2 + 2x$$

**23.** $(f \circ g)(x) = f(g(x)) = f(4) = \sqrt{4} = 2$
$(g \circ f)(x) = g(f(x)) = g(\sqrt{x}) = 4$

**25. (a)** $(f \circ g \circ h)(x) = f(g(h(x)))$
$$= f(g(x^{1/3})) = f(2x^{1/3} - 1)$$
$$= \frac{1}{2x^{1/3} - 1} = \frac{1}{2\sqrt[3]{x} - 1}$$
**(b)** $(g \circ f \circ h) = g(f(h(x)))$
$$= g(f(x^{1/3})) = g\left(\frac{1}{\sqrt[3]{x}}\right) = \frac{2}{\sqrt[3]{x}} - 1$$
**(c)** $(h \circ f \circ g)(x) = h(f(g(x)))$
$$= h(f(2x-1)) = h\left(\frac{1}{2x-1}\right)$$
$$= \sqrt[3]{\frac{1}{2x-1}} = \frac{1}{\sqrt[3]{2x-1}}$$

**27.** $(f \circ f)(x) = f(f(x)) = f\left(\frac{1}{x}\right) = \frac{1}{\frac{1}{x}} = x$

$(f \circ f \circ f)(x) = f(f \circ f(x)) = f(x) = \frac{1}{x}$

**29.** $g(x) = 3x + 1; \ f(x) = x^2$

**31.** $g(x) = 1 - 4x; \ f(x) = \sqrt{x}$

**33.** $g(x) = \frac{x+1}{x-1}; \ f(x) = x^2$

**35.** $g(x) = 3x^2 - 1; \ f(x) = x^{-3}$

**37.** $g(x) = \frac{x}{x-1}; \ f(x) = \sqrt{x}$

**39.** $g(x) = (x^2 - x - 1)^3; \ f(x) = \sqrt{x}$

**41.** $g(x) = 4 - x^2; \ f(x) = \frac{2}{\sqrt{x}}$

**43.** $f(x) = 2x + 1; \ g(x) = x^{1/2}; \ h(x) = x^3$

**45.** $f(x) = \frac{x}{x+1}; \ g(x) = x^5; \ h(x) = x^{1/2}$

**47.** $f(x) = x^2 - 9; \ g(x) = x^2; \ h(x) = x^{1/3}$

**49.** $f(x) = x^2 - 4x + 7; \ g(x) = x^3; \ h(x) = -\sqrt{x}$

**51.** $f(x) = 2x - 11; \ g(x) = 1 + \sqrt{x}; \ h(x) = x^2$

**53.** Let $f(x) = (x+1)^2$. Then
$$(f \circ f)(x) = f(f(x)) = f((x+1)^2)$$
$$= \left((x+1)^2 + 1\right)^2 \text{ So, } f(x) = (x+1)^2$$

**55.** $(f \circ g)(x) = f(g(x)) = [g(x)]^3 = x$
$g(x) = \sqrt[3]{x}$
Notice that
$$(g \circ f)(x) = g(f(x)) = g(x^3) = \sqrt[3]{x^3} = x$$

**57.** $(f \circ g)(x) = f(g(x)) = 2[g(x)] + 1$
$= 2x^2 - 4x + 1$
$2g(x) = 2x^2 - 4x$
$g(x) = x^2 - 2x$

**59.** $V = \frac{4}{3}\pi r^3$

$\frac{3V}{4\pi} = r^3$

$r(V) = \sqrt[3]{\frac{3V}{4\pi}}$

$V(t) = 50t$
$(r \circ V)(t) = r(V(t)) = r(50t)$

$$= \sqrt[3]{\frac{3 \cdot 50t}{4\pi}} = \sqrt[3]{\frac{150t}{4\pi}}$$

$(r \circ V)(t)$ is the length of the radius in feet after $t$ seconds.

$$(r \circ V)(10) = \sqrt[3]{\frac{150 \cdot 10}{4\pi}} = \sqrt[3]{\frac{375}{\pi}} \text{ which is}$$
approximately 4.9 feet.

## Challenge Problem

$$f(x) = \frac{1}{x^2 - 4x + 4} = \frac{1}{(x-2)^2}.$$

Let $g(x) = \frac{1}{x} + 2$.

Then $(f \circ g)(x) = f(g(x)) = f\left(\frac{1}{x} + 2\right)$

$$= \frac{1}{\left(\frac{1}{x} + 2 - 2\right)^2} = \frac{1}{\left(\frac{1}{x}\right)^2} = \frac{1}{\frac{1}{x^2}} = x^2$$

so $g(x) = \frac{1}{x} + 2$ is one such function.

## Critical Thinking

1. One example of a situation in which an operation on an equation can produce extraneous roots is the equation $\sqrt{x} = -1$ which obviously has no solutions. But, if both sides of this equation are squared, the extraneous root $x = 1$ is obtained. One example of a situation in which a root is lost is the equation $x(x + 1) = 2x$. If both sides of this equation are divided by $x$, the result is $x + 1 = 2$ which has one solution, $x = 1$. However, the original equation has two solutions, $x = 0, 1$. In this case, a solution is lost.

3. $f(g(x)) = f\left(\frac{x+2}{3}\right) = 3\left(\frac{x+2}{3}\right) - 2$

   $= x + 2 - 2 = x$

   $g(f(x)) = g(3x - 2) = \frac{(3x - 2) + 2}{3}$

   $= \frac{3x}{3} = x$

   The fact that $f(g(x)) = g(f(x)) = x$ means that the graph of $f(x)$ is a reflection of the graph of $g(x)$ about the line $y = x$ and vice-versa.

5. Yes, it is true that $\sqrt[3]{|x|} = \left|\sqrt[3]{x}\right|$.

   If $x \geq 0$, $\sqrt[3]{x} \geq 0$ and $\sqrt[3]{|x|} = \sqrt[3]{x} = \left|\sqrt[3]{x}\right|$

   If $x < 0$, then $\sqrt[3]{x} < 0$ and $\sqrt[3]{|x|} = \sqrt[3]{-x}$

   $= -\sqrt[3]{x} = \left|\sqrt[3]{x}\right|$

## Exercises 4.5

1. One-to-one

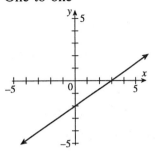

3. Not one-to-one

5. One-to-one

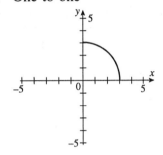

7. Not one-to-one

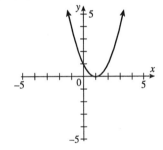

9. One-to-one

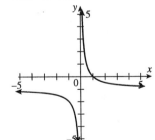

**11.** One-to-one

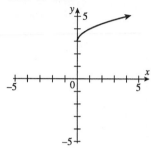

**13.** Using condition 2, assume that $f(x_1) = f(x_2)$, then

$$\frac{3}{5}(x_1 - 4) + 6 = \frac{3}{5}(x_2 - 4) + 6$$

$$\frac{3}{5}(x_1 - 4) = \frac{3}{5}(x_2 - 4)$$

$$\frac{5}{3} \cdot \frac{3}{5}(x - 4) = \frac{5}{3} \cdot \frac{3}{5}(x_2 - 4)$$

$$x_1 - 4 = x_2 - 4$$

$$x_1 = x_2$$

Therefore, $f(x)$ is one-to-one.

**15.** $(f \circ g)(x) = f(g(x)) = f(3x + 9)$

$$= \frac{1}{3}(3x + 9) - 3 = x + 3 - 3 = x$$

$$(g \circ f)(x) = g(f(x)) = g\left(\frac{1}{3}x - 3\right)$$

$$= 3\left(\frac{1}{3}x - 3\right) + 9 = x - 9 + 9 = x$$

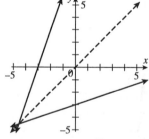

**17.** $(f \circ g)(x) = f(g(x)) = f\left(\sqrt[3]{x} - 1\right)$

$$= \left(\sqrt[3]{x} - 1 + 1\right)^3 = \left(\sqrt[3]{x}\right)^3 = x$$

$$(g \circ f)(x) = g(f(x)) = g\left((x + 1)^3\right)$$

$$= \sqrt[3]{(x + 1)^3} - 1 = x + 1 - 1 = x$$

**19.** $(f \circ g)(x) = f(g(x)) = f\left(\frac{1}{x} + 1\right)$

$$= \frac{1}{\frac{1}{x} + 1 - 1} = \frac{1}{\frac{1}{x}} = x$$

$$(g \circ f)(x) = g(f(x)) = g\left(\frac{1}{x - 1}\right)$$

$$= \frac{1}{\frac{1}{x-1}} + 1 = x - 1 + 1 = x$$

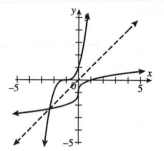

**21.** $x = (y - 5)^3$

$$\sqrt[3]{x} = y - 5$$

$$y = \sqrt[3]{x} + 5$$

$$g(x) = \sqrt[3]{x} + 5$$

**23.** $x = \frac{2}{3}y - 1$

$$x + 1 = \frac{2}{3}y$$

$$y = \frac{3}{2}(x + 1)$$

$$y = \frac{3}{2}x + \frac{3}{2}$$

$$g(x) = \frac{3}{2}x + \frac{3}{2}$$

**25.** $x = (y-1)^5$
$\sqrt[5]{x} = y - 1$
$y = \sqrt[5]{x} + 1$
$g(x) = \sqrt[5]{x} + 1$

**27.** $x = y^{3/5}$
$x^{5/3} = (y^{3/5})^{5/3}$
$y = x^{5/3}$
$g(x) = x^{5/3}$

**29.** $x = \dfrac{2}{y-2}$
$x(y-2) = 2$
$xy - 2x = 2$
$xy = 2 + 2x$
$y = \dfrac{2+2x}{x}$
$f^{-1}(x) = 2 + \dfrac{2}{x}$

$f(f^{-1}(x)) = f\left(2 + \dfrac{2}{x}\right) = \dfrac{2}{2 + \frac{2}{x} - 2}$

$= \dfrac{2}{\frac{2}{x}} = 2 \cdot \dfrac{x}{2} = x$

$f^{-1}(f(x)) = f^{-1}\left(\dfrac{2}{x-2}\right) = 2 + \dfrac{2}{\frac{2}{x-2}}$

$= 2 + 2 \cdot \dfrac{x-2}{2} = 2 + x - 2 = x$

**31.** $x = \dfrac{3}{y+2}$
$x(y+2) = 3$
$xy + 2x = 3$
$xy = 3 - 2x$
$y = \dfrac{3-2x}{x}$
$f^{-1}(x) = \dfrac{3}{x} - 2$

$f(f^{-1}(x)) = f\left(\dfrac{3}{x} - 2\right) = \dfrac{3}{\frac{3}{x} - 2 + 2} = \dfrac{3}{\frac{3}{x}}$

$= 3 \cdot \dfrac{x}{3} = x$

$f^{-1}(f(x)) = f^{-1}\left(\dfrac{3}{x+2}\right) = \dfrac{3}{\frac{3}{x+2}} - 2$

$= 3 \cdot \dfrac{x+2}{3} - 2 = x + 2 - 2 = x$

**33.** $x = y^{-5}$
$x^{-1/5} = \left(y^{-5}\right)^{-1/5}$
$y = x^{-1/5}$
$f^{-1}(x) = x^{-1/5}$

$f(f^{-1}(x)) = f\left(x^{-1/5}\right) = \left(x^{-1/5}\right)^{-5} = x$

$f^{-1}(f(x)) = f^{-1}\left(x^{-5}\right) = \left(x^{-5}\right)^{-1/5} = x$

**35.** $(f \circ f)(x) = f(f(x)) = f\left(\dfrac{1}{x}\right) = \dfrac{1}{\frac{1}{x}} = x$

**37.** $(f \circ f)(x) = f(f(x)) = f\left(\dfrac{x}{x-1}\right)$

$= \dfrac{\frac{x}{x-1}}{\frac{x}{x-1} - 1} \cdot \dfrac{x-1}{x-1} = \dfrac{x}{x - (x-1)}$

$= \dfrac{x}{x - x + 1} = x$

**39.** $x = (y+1)^2$
$\sqrt{x} = y + 1$
$y = \sqrt{x} - 1$
$g(x) = \sqrt{x} - 1$
Domain of $g$: $x \geq 0$
Range of $g$: $y \geq -1$

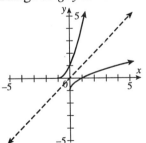

**41.**

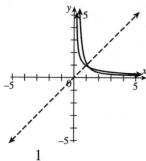

$$x = \frac{1}{\sqrt{y}}$$

$$x^2 = \frac{1}{y}$$

$$x^2 y = 1$$

$$y = \frac{1}{x^2}$$

$$g(x) = \frac{1}{x^2}$$

Domain of $g$: $x > 0$

Range of $g$: $y > 0$

**43.**

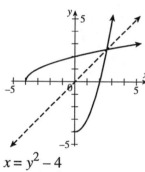

$$x = y^2 - 4$$

$$x + 4 = y^2$$

$$y = \sqrt{x+4}$$

Domain of $g$: $x \geq -4$

Range of $g$: $y \geq 0$

**45.**

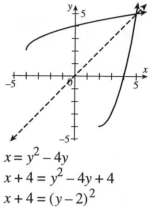

$$x = y^2 - 4y$$

$$x + 4 = y^2 - 4y + 4$$

$$x + 4 = (y - 2)^2$$

$$\sqrt{x+4} = y - 2$$

$$y = \sqrt{x+4} + 2$$

$$g(x) = \sqrt{x+4} + 2$$

Domain of $g$: $x \geq -4$

Range of $g$: $y \geq 2$

**47.** Let $y = f(x) = mx + k$ be a linear
function which is its own inverse. Then,
$f(f(x)) = x$
$f(mx + k) = x$
$m(mx + k) + k = x$
$m^2 x + mk + k = x$
Since this must hold for all values of $x$,
let $x = 0$. Then,
$mk + k = 0$
$k(m + 1) = 0$
$k = 0$ or $m = -1$
If $k = 0$, then $m^2 x = x$ for all values of $x$,
so if $x = 1$,
$m^2 = 1$
$m = \pm 1$
The only linear function which are
inverses of themselves are: $y = x$
$y = mx + k$, where $m = -1$

**Challenge Problem**

Using condition 1, let $x_1 \neq x_2$. Then either
$x_1 > x_2$ or $x_1 < x_2$. If $x_1 > x_2$, then
$f(x_1) > f(x_2)$ since $f$ is increasing. If $x_1 < x_2$,
then $f(x_1) < f(x_2)$. In either case,
$f(x_1) \neq f(x_2)$, and $f$ is one-to-one.

## Chapter 4  Review Exercises

1. $AB = \sqrt{(x_1 - x_2)^2 + (y_1 - y_2)^2}$

3. (a) $x^2 + y^2 = r^2$
   (b) $(x-h)^2 + (y-k)^2 = r^2$

5. $\left(\dfrac{-4+6}{2}, \dfrac{3-1}{2}\right) = (1,\ 1)$

7. Center: $(-3, 1)$
   Radius: 3

9. Center: $(3, 2)$

   Slope of radius: $\dfrac{1-2}{2-3} = \dfrac{-1}{-1} = 1$

   Slope of tangent: $-\dfrac{1}{1} = -1$

   Tangent line: $y - 1 = -1(x-2)$
   $\phantom{Tangent line: }y - 1 = -x + 2$
   $\phantom{Tangent line: }y = -x + 3$

11.

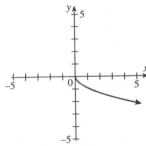

Domain: $x \geq 0$

13.

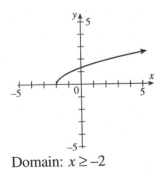

Domain: $x \geq -2$

15.

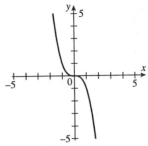

Domain: all $x$

17. $y = \dfrac{1}{\sqrt{x-1}}$

   Asymptotes: $x = 1,\ y = 0$

19. $\sqrt{x-4} + 3 = 5$
   $\sqrt{x-4} = 2$
   $\left(\sqrt{x-4}\right)^2 = 2^2$
   $x - 4 = 4$
   $x = 8$

21. $\sqrt{x-5} + \sqrt{x} = 5$
   $\sqrt{x-5} = 5 - \sqrt{x}$
   $\left(\sqrt{x-5}\right)^2 = \left(5 - \sqrt{x}\right)^2$
   $x - 5 = 25 - 10\sqrt{x} + x$
   $10\sqrt{x} = 30$
   $\sqrt{x} = 3$
   $\left(\sqrt{x}\right)^2 = 3^2$
   $x = 9$

23.

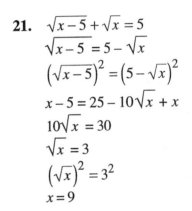

Domain: $x^2 - 9 \geq 0$
$\phantom{Domain: }x^2 \geq 9$
$\phantom{Domain: }x \leq -3 \text{ or } x \geq 3$

*Chapter 4: Circles, Additional Curves, and the Algebra Functions* **133**

**25.**

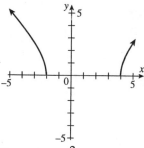

Domain: $x^2 - 2x - 8 \geq 0$
$(x-4)(x+2) \geq 0$
$x-4 \geq 0$ and $x+2 \geq 0$ or
$x-4 \leq 0$ and $x+2 \leq 0$
$x \geq 4$ or $x \leq -2$

**27.** Domain: $x > -1$
$f(x) = 0$

$$\frac{x^2-1}{\sqrt{x+1}} + \sqrt{x+1} = 0$$

$$\sqrt{x+1} \cdot \frac{x^2-1}{\sqrt{x+1}} + \sqrt{x+1} \cdot \sqrt{x+1}$$

$$= \sqrt{x+1} \cdot 0$$

$x^2 - 1 + x + 1 = 0$

$x^2 + x = 0$

$x(x+1) = 0$

$x = 0$ or $x - 1$, but $-1$ isn't in the domain of $f$. So, $x = 0$ is the only root of $f(x) = 0$.

| Interval | $(-1, 0)$ | $(0, \infty)$ |
|---|---|---|
| **Sign of $f(x)$** | $-$ | $+$ |

$f(x) < 0$ on $(-1, 0)$ and $f(x) > 0$ on $(0, \infty)$

**29.** $\sqrt[3]{x^2} + \sqrt[3]{x} - 6 = 0$

$\left(\sqrt[3]{x}\right)^2 + \sqrt[3]{x} - 6 = 0$

Let $u = \sqrt[3]{x}$. Then,

$u^2 + u - 6 = 0$

$(u+3)(u-2) = 0$

$u + 3 = 0$ or $u - 2 = 0$

$u = -3$ or $u = 2$

$\sqrt[3]{x} = -3$ or $\sqrt[3]{x} = 2$

$x = -27$ or $x = 8$

**31.** $\dfrac{x}{(5x-6)^{4/5}} + (5x-6)^{1/5}$

$$= \frac{x}{(5x-6)^{4/5}} + \frac{(5x-6)^{1/5} \cdot (5x-6)^{4/5}}{(5x-6)^{4/5}}$$

$$= \frac{x + (5x-6)}{(5x-6)^{4/5}} = \frac{6x-6}{(5x-6)^{4/5}}$$

$$= \frac{6(x-1)}{(5x-6)^{4/5}}$$

**33.** $(f+g)(x) = f(x) + g(x) = \sqrt{x+1} + \dfrac{1}{x+1}$
Domain: $x > -1$

**35.** $(f \cdot g)(x) = f(x) \cdot g(x) = \sqrt{x+1} \cdot \dfrac{1}{x+1}$

$$= \frac{\sqrt{x+1}}{x+1}$$

Domain: $x > -1$

**37.** $f(g(0)) = f\left(\dfrac{1}{0+1}\right) = f(1) = \sqrt{1+1} = \sqrt{2}$

**39.** $(f \circ g)(x) = f(g(x))$

$$= f\left(\frac{1}{x+1}\right) = \sqrt{\frac{1}{x+1} + 1}$$

$$= \sqrt{\frac{1}{x+1} + \frac{x+1}{x+1}} = \sqrt{\frac{x+2}{x+1}}$$

**41.** $(f \circ g \circ h)(x) = f(g(h(x))) = f\left(g(\sqrt{x})\right)$

$$= f\left(\sqrt{x} - 1\right) = \frac{1}{\left(\sqrt{x}-1\right)^2} = \frac{1}{x - 2\sqrt{x} + 1}$$

**43.** Consider $t(g(f(x)))$ where $f(x) = x^2 + x$,
$g(x) = x^3$ and $t(x) = \sqrt{x}$

**45.** $d(y) = \sqrt{\left(\dfrac{1}{4}\right)^2 + y^2} = \sqrt{\dfrac{1}{16} + \dfrac{16y^2}{16}}$

$$= \sqrt{\frac{1}{16}\left(16y^2 + 1\right)}$$

$$= \frac{1}{4}\sqrt{16y^2 + 1}$$

$y(t) = 45t$

$(d \circ y)(t) = d(y(t)) = d(45t)$

$= \dfrac{1}{4}\sqrt{16(45t)^2 + 1} = \dfrac{1}{4}\sqrt{32400t^2 + 1}$

$(d \circ y)(t)$ is the distance in miles that the car is from point $P$ after $t$ minutes.

At $t = 3$, $d = \dfrac{1}{4}\sqrt{32400 \cdot 3^2 + 1}$

$= \dfrac{1}{4}\sqrt{291,600 + 1} = \dfrac{1}{4}\sqrt{291,601}$

$d \approx 135$ miles

**47.** A function $f$ is one-to-one if and only if the horizontal lines through the range values intersect the graph of $f$ in exactly one point.

**49. (a)**

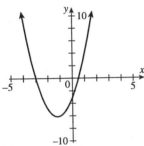

Not one-to-one

**(b)**

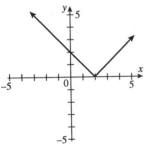

Not one-to-one

**(c)**

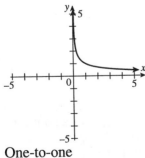

One-to-one

**51.**

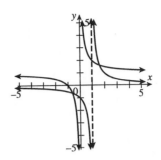

**53.** $f(g(x)) = f\left(\dfrac{1}{2} - \dfrac{x}{2}\right) = 1 - 2\left(\dfrac{1}{2} - \dfrac{x}{2}\right)$

$= 1 - 1 + x = x$

$g(f(x)) = g(1 - 2x) = \dfrac{1}{2} - \dfrac{(1 - 2x)}{2}$

$= \dfrac{1}{2} - \dfrac{1}{2} + \dfrac{2x}{2} = x$

**55.** $x = y^{2/3} + 1$

$x - 1 = y^{2/3}$

$(x - 1)^{3/2} = \left(y^{2/3}\right)^{3/2}$

$y = (x - 1)^{3/2}$

$g(x) = (x - 1)^{3/2}$

Domain: $x \geq 1$

Range: $y \geq 0$

**Chapter 4 Test: Standard Answer**

**1. (a)**

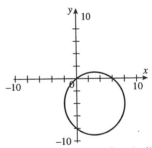

**(b)** Slope of radius: $\dfrac{0 - (-4)}{6 - 3} = \dfrac{4}{3}$

Slope of tangent: $-\dfrac{3}{4}$

Tangent line: $y - 0 = -\dfrac{3}{4}(x - 6)$

$$y = -\dfrac{3}{4}x + \dfrac{9}{2}$$

**2.** $4x^2 + 4x + 4y^2 - 56y = -97$

$$x^2 + x + y^2 - 14y = -\dfrac{97}{4}$$

$$x^2 + x + \dfrac{1}{4} + y^2 - 14y + 49$$

$$= -\dfrac{97}{4} + \dfrac{1}{4} + 49$$

$$\left(x + \dfrac{1}{2}\right)^2 + (y - 7)^2 = 25$$

$$C\left(-\dfrac{1}{2},\ 7\right);\ r = 5$$

**3.**

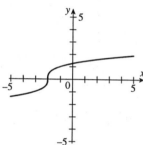

Domain: all reals

**4.** Domain: $x > 0$

Asymptotes: $y = 2,\ x = 0$

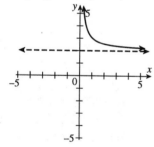

**5.**

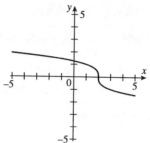

Domain: all reals

**6.** Domain: $x > 3$

Asymptotes: $x = 3,\ y = 0$

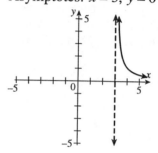

**7.**

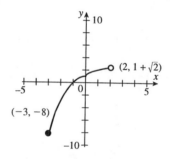

**8.** $\sqrt{18x + 5} - 9x = 1$

$$\sqrt{18x + 5} = 1 + 9x$$

$$\left(\sqrt{18x + 5}\right)^2 = (1 + 9x)^2$$

$$18x + 5 = 1 + 18x + 81x^2$$

$$4 = 81x^2$$

$$\dfrac{4}{81} = x^2$$

$$x = \pm\dfrac{2}{9}$$

Since $x = -\dfrac{2}{9}$ isn't a solution, the only

solution is $x = \dfrac{2}{9}$.

**9.** $6x^{2/3} + 5x^{1/3} - 4 = 0$

Let $u = x^{1/3}$. Then,

$6u^2 + 5u - 4 = 0$

$(3u + 4)(2u - 1) = 0$

$3u + 4 = 0$ or $2u - 1 = 0$

$u = -\dfrac{4}{3}$ or $u = \dfrac{1}{2}$

$x^{1/3} = -\dfrac{4}{3}$ or $x^{1/3} = \dfrac{1}{2}$

$x = -\dfrac{64}{27}$ or $x = \dfrac{1}{8}$

**10.** $\sqrt{x-7} + \sqrt{x+9} = 8$

$\sqrt{x-7} = 8 - \sqrt{x+9}$

$\left(\sqrt{x-7}\right)^2 = \left(8 - \sqrt{x+9}\right)^2$

$x - 7 = 64 - 16\sqrt{x+9} + (x+9)$

$-80 = -16\sqrt{x+9}$

$5 = \sqrt{x+9}$

$5^2 = \left(\sqrt{x+9}\right)^2$

$25 = x + 9$

$x = 16$

**11.** $(x+1)^{1/2}(2x) + (x+1)^{-1/2}\left(\dfrac{x^2}{2}\right)$

$= 2x(x+1)^{1/2} + \dfrac{x^2}{2(x+1)^{1/2}}$

$= \dfrac{2x\sqrt{x+1} \cdot 2\sqrt{x+1}}{2\sqrt{x+1}} + \dfrac{x^2}{2\sqrt{x+1}}$

$= \dfrac{4x(x+1) + x^2}{2\sqrt{x+1}}$

$= \dfrac{5x^2 + 4x}{2\sqrt{x+1}} = \dfrac{x(5x+4)}{2\sqrt{x+1}}$

**12.** $f(x) = \dfrac{\sqrt[3]{x-4}}{x-2}$

Numerator and denominator are 0 when $x = 2$ and $x = 4$, so there are three intervals.

| Interval | $(-\infty, 2)$ | $(2, 4)$ | $(4, \infty)$ |
|---|---|---|---|
| Sign of $\sqrt[3]{x-4}$ | $-$ | $-$ | $+$ |
| Sign of $x - 2$ | $-$ | $+$ | $+$ |
| Sign of $f(x)$ | $+$ | $-$ | $+$ |

$f(x) < 0$ on $(2, 4)$

$f(x) > 0$ on $(-\infty, 2)$ and $(4, \infty)$.

**13.** $f(x) = x^{1/2} - \dfrac{1}{2}x^{-1/2}(x+4)$

$= \dfrac{x^{1/2} \cdot 2x^{1/2}}{2x^{1/2}} - \dfrac{x+4}{2x^{1/2}}$

$= \dfrac{2x - x - 4}{2\sqrt{x}} = \dfrac{x-4}{2\sqrt{x}}$

The domain of $f(x)$ is $x > 0$, and the only root in the domain is $x = 4$.

| Interval | $(0, 4)$ | $(4, \infty)$ |
|---|---|---|
| Sign of $x - 4$ | $-$ | $+$ |
| Sign of $\sqrt{x}$ | $+$ | $+$ |
| Sign of $f(x)$ | $-$ | $+$ |

$f(x) < 0$ on $(0, 4)$; $f(x) > 0$ on $(4, \infty)$

**14.** $x$-intercepts: $0 = \sqrt{x^2 + x - 2}$

$0^2 = \left(\sqrt{x^2 + x - 2}\right)^2$

$0 = x^2 + x - 2$

$0 = (x+2)(x-1)$

$x + 2 = 0$ or $x - 1 = 0$

$x = -2$ or $x = 1$

Domain: $x^2 + x - 2 \geq 0$

$(x+2)(x-1) \geq 0$

$x + 2 \geq 0$ and $x - 1 \geq 0$

or $x + 2 \leq 0$ and $x - 1 \leq 0$

$x \geq 1$ or $x \leq -2$

**15. (a)** $(f + g)(x) = f(x) + g(x)$

$= \dfrac{1}{x^2 - 1} + \sqrt{x+2}$

Domain: all $x \geq -2$ and $x \neq \pm 1$

**(b)** $(f - g)(x) = f(x) - g(x)$

$$= \frac{1}{x^2 - 1} - \sqrt{x + 2}$$

Domain: all $x \geq -2$ and $x \neq \pm 1$

**16. (a)** $\dfrac{f}{g}(x) = \dfrac{f(x)}{g(x)} = \dfrac{\frac{1}{x^2 - 1}}{\sqrt{x + 2}}$

$$= \frac{1}{\left(x^2 - 1\right)\sqrt{x + 2}}$$

Domain: all $x > -2$ and $x \neq \pm 1$

**(b)** $(f \cdot g)(x) = f(x) \cdot g(x)$

$$= \frac{1}{x^2 - 1} \cdot \sqrt{x + 2}$$

$$= \frac{\sqrt{x + 2}}{x^2 - 1}$$

Domain: all $x \geq -2$ and $x = \pm 1$

**17.** The graph of $g(x)$ can be obtained from the graph of $f(x)$ by shifting it 4 units to the right and reflecting it in the $x$-axis.

**18. (a)** $f(4) = \dfrac{1}{4^2 + 1} = \dfrac{1}{17}$

$g(4) = 2\sqrt{4} = 4$

$(f \cdot g)(4) = f(4) \cdot g(4) = \dfrac{1}{17} \cdot 4 = \dfrac{4}{17}$

**(b)** $f(g(9)) = f(2\sqrt{9}) = f(6) = \dfrac{1}{6^2 + 1}$

$$= \frac{1}{37}$$

$g(f(9)) = g\left(\dfrac{1}{9^2 + 1}\right) = g\left(\dfrac{1}{82}\right)$

$$= 2\sqrt{\frac{1}{82}} = \frac{2}{\sqrt{82}}$$

**19.** $(f \circ g)(x) = f(g(x)) = f(\sqrt{x})$

$$= \frac{1}{1 - \left(\sqrt{x}\right)^2} = \frac{1}{1 - x}$$

Domain: all $x \geq 0$ and $x \neq 1$

$(g \circ f)(x) = g(f(x)) = g\left(\dfrac{1}{1 - x^2}\right)$

$$= \sqrt{\frac{1}{1 - x^2}}$$

Domain: $1 - x^2 > 0$

$-x^2 > -1$

$x^2 < 1$

$-1 < x < 1$

**20.** $g(x) = x - 2;\ f(x) = x^{2/3}$

**21.** $h(x) = 2x - 1;\ g(x) = x^{3/2};\ f(x) = \dfrac{1}{x}$

**22. (a)** $x = 3y - 2$

$x + 2 = 3y$

$y = \dfrac{1}{3}x + \dfrac{2}{3}$

$f^{-1}(x) = \dfrac{1}{3}x + \dfrac{2}{3} = \dfrac{x + 2}{3}$

**(b)**

**23.** $x = \sqrt[3]{y} - 1$

$x + 1 = \sqrt[3]{y}$

$(x + 1)^3 = \left(\sqrt[3]{y}\right)^3$

$y = (x + 1)^3$

$g(x) = (x + 1)^3$

$(f \circ g)(x) = f(g(x)) = f\left((x + 1)^3\right)$

$$= \sqrt[3]{(x + 1)^3} - 1 = x + 1 - 1 = x$$

**24.** Domain: all real numbers

x-intercepts: $0 = \sqrt[3]{3x+1} - x - 1$

$$x + 1 = \sqrt[3]{3x+1}$$
$$(x+1)^3 = \left(\sqrt[3]{3x+1}\right)^3$$
$$x^3 + 3x^2 + 3x + 1 = 3x + 1$$
$$x^3 + 3x^2 = 0$$
$$x^2(x+3) = 0$$
$$x = 0 \text{ or } x + 3 = 0$$
$$x = -3$$

**25.** $y = 2\sqrt{x-1}$

$y = \frac{1}{2}x + 1$

$$2\sqrt{x-1} = \frac{1}{2}x + 1$$
$$4\sqrt{x-1} = x + 2$$
$$\left(4\sqrt{x-1}\right)^2 = (x+2)^2$$
$$16(x-1) = x^2 + 4x + 4$$
$$16x - 16 = x^2 + 4x + 4$$
$$0 = x^2 - 12x + 20$$
$$0 = (x-2)(x-10)$$
$$x - 2 = 0 \text{ or } x - 10 = 0$$
$$x = 2 \text{ or } x = 10$$

If $x = 2$, $y = \frac{1}{2} \cdot 2 + 1 = 2$

If $x = 10$, $y = \frac{1}{2} \cdot 10 + 1 = 6$

$(2, 2)$ and $(10, 6)$

### Chapter 4 Test: Multiple Choice

**1.** $x^2 + y^2 + 3y = -\frac{1}{4}$

$$x^2 + y^2 + 3y + \frac{9}{4} = -\frac{1}{4} + \frac{9}{4}$$
$$x^2 + \left(y + \frac{3}{2}\right)^2 = 2$$

Center: $\left(0, -\frac{3}{2}\right)$

Radius: $\sqrt{2}$
The answer is (d).

**2.** The equation is $(x-2)^2 + (y+3)^2 = 4^2$, so the answer is (b).

**3.** $\sqrt{x^2 - 9x} + \sqrt{3x} = x$

$$\sqrt{x^2 - 9x} = x - \sqrt{3x}$$
$$\left(\sqrt{x^2 - 9x}\right)^2 = \left(x - \sqrt{3x}\right)^2$$
$$x^2 - 9x = x^2 - 2x\sqrt{3x} + 3x$$
$$-12x = -2x\sqrt{3x}$$
$$6x = x\sqrt{3x}$$
$$(6x)^2 = \left(x\sqrt{3x}\right)^2$$
$$36x^2 = 3x^3$$
$$0 = 3x^3 - 36x^2$$
$$0 = 3x^2(x - 12)$$
$$x = 0 \text{ or } x = 12$$

There are two solutions, so the answer is (c).

**4.** (d)

**5.** The answer is (a).

**6.** x-intercepts: $0 = \sqrt{2x^2 - 8x - 24}$

$$0^2 = \left(\sqrt{2x^2 - 8x - 24}\right)^2$$
$$0 = 2x^2 - 8x - 24$$
$$0 = 2\left(x^2 - 4x - 12\right)$$
$$0 = 2(x - 6)(x + 2)$$
$$x - 6 = 0 \text{ or } x + 2 = 0$$
$$x = 6 \text{ or } x = -2$$

The answer is (e).

**7.** $\sqrt[3]{x^2} - \sqrt[3]{x} - 6 = 0$

Let $u = \sqrt[3]{x}$. Then

$$\left(\sqrt[3]{x}\right)^2 - \sqrt[3]{x} - 6 = 0$$
$$u^2 - u - 6 = 0$$
$$(u - 3)(u + 2) = 0$$
$$u - 3 = 0 \text{ or } u + 2 = 0$$
$$u = 3 \text{ or } u = -2$$
$$\sqrt[3]{x} = 3 \text{ or } \sqrt[3]{x} = -2$$
$$x = 27 \text{ or } x = -8$$

The answer is (c).

**8.** $r = \sqrt{\dfrac{V}{\pi h}}$

$r^2 = \left(\sqrt{\dfrac{V}{\pi h}}\right)^2$

$r^2 = \dfrac{V}{\pi h}$

$hr^2 = \dfrac{V}{\pi}$

$h = \dfrac{V}{\pi r^2}$

The answer is (b).

**9.** The denominator is always positive. The numerator is equal to zero only when $x = 2$, so there are only two intervals, $(-\infty, 2), (2, \infty)$.

| Interval | $(-\infty, 2)$ | $(2, \infty)$ |
|---|---|---|
| **Sign of $x - 2$** | – | + |
| **Sign of $\sqrt{x^2 + 4}$** | + | + |
| **Sign of $f(x)$** | – | + |

$f(x) < 0$ in $(-\infty, 2)$ and $f(x) > 0$ on $(2, \infty)$, so the answer is (c).

**10.** Domain: $1 - x^2 > 0$

$x^2 < 1$

$-1 < x < 1$

The answer is (d).

**11.** I is not correct.
II is not correct.
III is not correct.
The answer is (e).

**12.** $\dfrac{f}{g}(4) = \dfrac{f(4)}{g(4)} = \dfrac{\frac{1}{4^2 - 1}}{\sqrt{4}} = \dfrac{\frac{1}{15}}{2} = \dfrac{1}{30}$

The answer is (c).

**13.** $x^{1/3} + (x - 8)\dfrac{1}{3}x^{-2/3}$

$= \dfrac{x^{1/3} \cdot 3x^{2/3}}{3x^{2/3}} + \dfrac{x - 8}{3x^{2/3}} = \dfrac{3x + x - 8}{3\sqrt[3]{x^2}}$

$= \dfrac{4x - 8}{3\sqrt[3]{x^2}} = \dfrac{4(x - 2)}{3\sqrt[3]{x^2}}$

The answer is (d).

**14.** $f\big(g(2)\big) = f(2 \cdot 2 - 3) = f(1) = \sqrt{1^3 + 1}$

$= \sqrt{2}$

The answer is (a).

**15.** $g(f(x)) = g\big(\sqrt[3]{x}\big) = \dfrac{1}{\left(\sqrt[3]{x}\right)^3 - 1} = \dfrac{1}{x - 1}$

The answer is (d).

**16.** $y = \dfrac{1}{x}$ is one-to-one, so the answer is (b).

**17.** $x = 3y + 2$

$x - 2 = 3y$

$y = \dfrac{1}{3}(x - 2) = \dfrac{1}{3}x - \dfrac{2}{3}$

The answer is (b).

**18.** If $f(x) = x^{3/2}$ and $g(x) = 4 - x^2$, then

$(f \circ g)(x) = f\big(g(x)\big) = f\big(4 - x^2\big)$

$= \sqrt{\big(4 - x^2\big)^3}$. The answer is (c).

**19.** $x = \dfrac{2y}{y - 3}$

$x(y - 3) = 2y$

$xy - 3x = 2y$

$xy - 2y = 3x$

$y(x - 2) = 3x$

$y = \dfrac{3x}{x - 2}$

$f^{-1}(x) = \dfrac{3x}{x - 2}$

The answer is (c).

**20.** Center: $(3, -4)$

Slope of radius: $\dfrac{0-(-4)}{6-3} = \dfrac{4}{3}$

Slope of tangent: $-\dfrac{3}{4}$

Tangent line: $y - 0 = -\dfrac{3}{4}(x - 6)$

$4y = -3x + 18$

$3x + 4y = 18$

The answer is (c).

# Chapter 5: Exponential and Logarithmic Functions

**Exercises 5.1**

**1.**

| $x$ | $-2$ | $-1$ | 0 | 1 | 2 |
|-----|------|------|---|---|---|
| $y = 2^x$ | $\frac{1}{4}$ | $\frac{1}{2}$ | 1 | 2 | 4 |

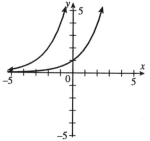

Horizontal asymptote: $y = 0$

**3.**

| $x$ | $-2$ | $-1$ | 0 | 1 | 2 |
|-----|------|------|---|---|---|
| $y = 4^x$ | $\frac{1}{16}$ | $\frac{1}{4}$ | 1 | 4 | 16 |

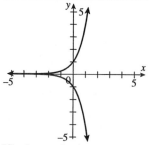

Horizontal asymptote: $y = 0$

**5.**

| $x$ | $-2$ | $-1$ | 0 | 1 | 2 |
|-----|------|------|---|---|---|
| $y = \left(\frac{3}{2}\right)^x$ | $\frac{4}{9}$ | $\frac{2}{3}$ | 1 | $\frac{3}{2}$ | $\frac{9}{4}$ |

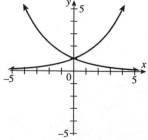

Horizontal asymptote: $y = 0$

**7.**

| $x$ | $-2$ | $-1$ | 0 | 1 | 2 |
|-----|------|------|---|---|---|
| $y = 3^x$ | $\frac{1}{9}$ | $\frac{1}{3}$ | 1 | 3 | 9 |

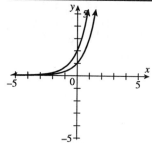

Horizontal asymptote: $y = 0$

**9.**

| $x$ | $-2$ | $-1$ | 0 | 1 | 2 |
|-----|------|------|---|---|---|
| $y = 2^{x/2}$ | $\frac{1}{2}$ | $\frac{1}{\sqrt{2}}$ | 1 | $\sqrt{2}$ | 2 |

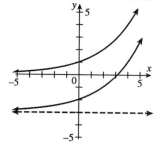

Horizontal asymptote: $y = 0$, $y = -3$

**11.**

| $x$ | $-2$ | $-1$ | 0 | 1 | 2 |
|-----|------|------|---|---|---|
| $y = 4^{-x}$ | 16 | 4 | 1 | $\frac{1}{4}$ | $\frac{1}{16}$ |

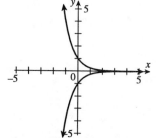

Horizontal asymptote: $y = 0$

**13.**

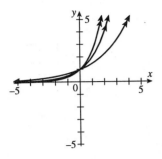

**15.**

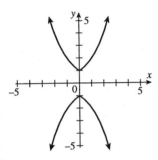

**17. (a)** $y = 2^{x-1}$
**(b)** $y = 2^{x+2} + 1$

**19.** $2^x = 64$
$2^x = 2^6$
$x = 6$

**21.** $2^{x^2} = 512$
$2^{x^2} = 2^9$
$x^2 = 9$
$x = \pm 3$

**23.** $5^{2x+1} = 125$
$5^{2x+1} = 5^3$
$2x + 1 = 3$
$2x = 2$
$x = 1$

**25.** $7^{x^2+x} = 49$
$7^{x^2+x} = 7^2$
$x^2 + x = 2$
$x^2 + x - 2 = 0$
$(x+2)(x-1) = 0$

$x + 2 = 0$  or  $x - 1 = 0$
$x = -2$  or  $x = 1$

**27.** $\dfrac{1}{2^x} = 32$
$2^{-x} = 2^5$
$-x = 5$
$x = -5$

**29.** $9^x = 3$
$\left(3^2\right)^x = 3$
$3^{2x} = 3^1$
$2x = 1$
$x = \dfrac{1}{2}$

**31.** $9^x = 27$
$\left(3^2\right)^x = 3^3$
$3^{2x} = 3^3$
$2x = 3$
$x = \dfrac{3}{2}$

**33.** $\left(\dfrac{1}{49}\right)^x = 7$
$\left(7^{-2}\right)^x = 7^1$
$7^{-2x} = 7^1$
$-2x = 1$
$x = -\dfrac{1}{2}$

**35.** $\left(\dfrac{27}{8}\right)^x = \dfrac{9}{4}$
$\left[\left(\dfrac{3}{2}\right)^3\right]^x = \left(\dfrac{3}{2}\right)^2$
$\left(\dfrac{3}{2}\right)^{3x} = \left(\dfrac{3}{2}\right)^2$
$3x = 2$
$x = \dfrac{2}{3}$

**37.** $3^{2x-1} = \dfrac{729}{9^{x+1}}$

$3^{2x-1} = \dfrac{3^6}{\left(3^2\right)^{x+1}}$

$3^{2x-1} = 3^{6-(2x+2)}$
$2x - 1 = 6 - 2x - 2$
$4x = 5$
$x = \dfrac{5}{4}$

**39.** $2^{x^2+x} = 4^{1+x}$

$2^{x^2+x} = \left(2^2\right)^{1+x}$

$2^{x^2+x} = 2^{2+2x}$
$x^2 + x = 2 + 2x$
$x^2 - x - 2 = 0$
$(x - 2)(x + 1) = 0$
$x - 2 = 0 \quad$ or $\quad x + 1 = 0$
$x = 2 \qquad$ or $\quad x = -1$

**41.**

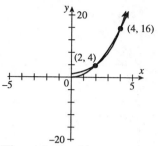

The points of intersection are (2, 4) and (4, 16)

**43.** $(5^{2x+1})(7^{2x}) = 175$
$5 \cdot 5^{2x} \cdot 7^{2x} = 175$
$5^{2x} \cdot 7^{2x} = 35$
$(35)^{2x} = 35^1$
$2x = 1$
$x = \dfrac{1}{2}$

**45.** **(a)**

| $x$ | 1.4 | 1.41 | 1.414 | 1.4142 | 1.41421 |
|---|---|---|---|---|---|
| $3^x$ | 4.6555 | 4.7070 | 4.7277 | 4.7287 | 4.7288 |

estimate: 4.729; calculator: 4.728804 . . .

**(b)**

| $x$ | 1.7 | 1.73 | 1.732 | 1.7320 | 1.73205 |
|---|---|---|---|---|---|
| $3^x$ | 6.4730 | 6.6899 | 6.7046 | 6.7046 | 6.7050 |

estimate: 6.705; calculator: 6.704991 . . .

**(c)**

| $x$ | 2.2 | 2.23 | 2.236 | 2.2360 | 2.23606 |
|---|---|---|---|---|---|
| $2^x$ | 4.5948 | 4.6913 | 4.7109 | 4.7109 | 4.7111 |

estimate: 4.711; calculator: 4.711113 . . .

**(d)**

| $x$ | 3.1 | 3.14 | 3.141 | 3.1415 | 3.14159 |
|---|---|---|---|---|---|
| $4^x$ | 73.5167 | 77.7085 | 77.8163 | 77.8702 | 77.8799 |

estimate: 77.880; calculator: 77.88023 . . .

**47.** Shifting $y = 2^x$ left by three units gives the function $y = 2^{x+3}$, so $y = 2^{x+3} = 2^x \cdot 2^3 = 8 \cdot 2^x$. Hence, $a = 8$.

# Exercises 5.2

**1.** $g(x) = \log_4 x$

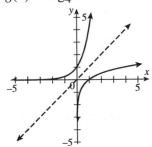

**3.** $g(x) = \log_{1/3} x$

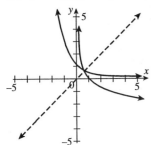

**5.** Shift $g(x)$ 2 units left.
Domain: $x > -2$
Vertical asymptote: $x = -2$

**7.** Shift $g(x)$ 2 units upward.
Domain: $x > 0$
Vertical asymptote: $x = 0$

**9.** Domain: all $x > 0$

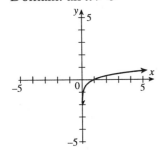

**11.** Domain: all $x > 0$

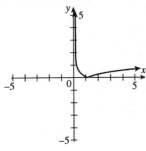

**13.** Domain: all $x \neq 0$

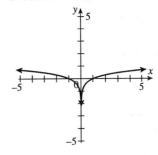

**15.** **(a)** $y = \log_2(x + 2)$
**(b)** $y = \log_2 x - 2$

**17.** $\log_2 256 = 8$

**19.** $\log_{1/3} 3 = -1$

**21.** $\log_{17} 1 = 0$

**23.** $10^{-4} = 0.0001$

**25.** $\left(\sqrt{2}\right)^2 = 2$

**27.** $12^{-3} = \dfrac{1}{1728}$

**29.** Let $y = \log_{10} 10{,}000$. Then
$$10^y = 10{,}000$$
$$10^y = 10^4$$
$$y = 4$$

**31.** Let $y = \log_5 625$. Then
$$5^y = 625$$
$$5^y = 5^4$$
$$y = 4$$

**33.** Let $y = \log_{2/3} \dfrac{8}{27}$. Then

$$\left(\frac{2}{3}\right)^y = \frac{8}{27}$$

$$\left(\frac{2}{3}\right)^y = \left(\frac{2}{3}\right)^3$$

$$y = 3$$

**35.** Let $y = \log_{0.3} \dfrac{1000}{27}$. Then

$$(0.3)^y = \frac{1000}{27}$$

$$\left(\frac{3}{10}\right)^y = \left(\frac{10}{3}\right)^3$$

$$\left(\frac{10}{3}\right)^{-y} = \left(\frac{10}{3}\right)^3$$

$$-y = 3$$

$$y = -3$$

**37.** $\log_2 16 = y$

$$2^y = 16$$

$$2^y = 2^4$$

$$y = 4$$

**39.** $\log_{1/3} 27 = y$

$$\left(\frac{1}{3}\right)^y = 27$$

$$3^{-y} = 3^3$$

$$-y = 3$$

$$y = -3$$

**41.** $\log_{1/6} x = 3$

$$\left(\frac{1}{6}\right)^3 = x$$

$$x = \frac{1}{216}$$

**43.** $\log_b 125 = 3$

$$b^3 = 125$$

$$b^3 = 5^3$$

$$b = 5$$

**45.** $\log_b \dfrac{1}{8} = -\dfrac{3}{2}$

$$b^{-3/2} = \frac{1}{8}$$

$$(b^{-3/2})^{-2/3} = \left(\frac{1}{8}\right)^{-2/3}$$

$$b = 4$$

**47.** $\log_{27} 3 = y$

$$(27)^y = 3$$

$$(3^3)^y = 3$$

$$3^{3y} = 3^1$$

$$3y = 1$$

$$y = \frac{1}{3}$$

**49.** $\log_b \dfrac{16}{81} = 4$

$$b^4 = \frac{16}{81}$$

$$b = \sqrt[4]{\frac{16}{81}}$$

$$b = \frac{2}{3}$$

**51.** $\log_b \dfrac{1}{27} = -\dfrac{3}{2}$

$$b^{-3/2} = \frac{1}{27}$$

$$(b^{-3/2})^{-2/3} = \left(\frac{1}{27}\right)^{-2/3}$$

$$b = 9$$

**53.** $\log_{\sqrt{8}} \left(\dfrac{1}{8}\right) = y$

$$(\sqrt{8})^y = \frac{1}{8}$$

$$(\sqrt{8})^y = 8^{-1}$$

$$\left(\sqrt{8}\right)^y = \left(\sqrt{8}\right)^{-2}$$

$$y = -2$$

**55.** $\log_{0.001} 10 = y$

$(0.001)^y = 10$

$\left(10^{-3}\right)^y = 10$

$10^{-3y} = 10^1$

$-3y = 1$

$y = -\dfrac{1}{3}$

**57.** $\log_9 x = 1$

$9^1 = x$

$x = 9$

**59.** $\log_{10} 1000 = \dfrac{y}{2}$

$10^{y/2} = 1000$

$10^{y/2} = 10^3$

$\dfrac{y}{2} = 3$

$y = 6$

**61.** Let $y = \log_{1/27} \dfrac{1}{81}$. Then

$\left(\dfrac{1}{27}\right)^y = \dfrac{1}{81}$

$\left(\dfrac{1}{3}\right)^{3y} = \left(\dfrac{1}{3}\right)^4$

$3y = 4$

$y = \dfrac{4}{3}$

So, $\log_{3/4}\left(\log_{1/27}\dfrac{1}{81}\right) = \log_{3/4}\left(\dfrac{4}{3}\right)$

Let $x = \log_{3/4}\left(\dfrac{4}{3}\right)$. Then

$\left(\dfrac{3}{4}\right)^x = \dfrac{4}{3}$

$\left(\dfrac{3}{4}\right)^x = \left(\dfrac{3}{4}\right)^{-1}$

$x = -1$

Therefore, $\log_{3/4}\left(\log_{1/27}\dfrac{1}{81}\right) = -1$

**63.** $x = \log_3(y+3)$

$3^x = y + 3$

$y = 3^x - 3$

$g(x) = 3^x - 3$

$(f \circ g)(x) = f(g(x)) = f(3^x - 3)$

$= \log_3(3^x - 3 + 3) = \log_3 3^x = x$

$(g \circ f)(x) = g(f(x)) = g(\log_3(x+3))$

$= 3^{\log_3(x+3)} - 3 = (x+3) - 3 = x$

**65.**

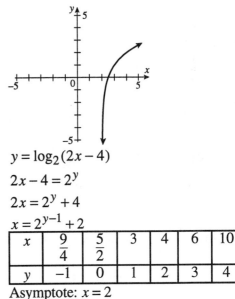

$y = \log_2(2x - 4)$

$2x - 4 = 2^y$

$2x = 2^y + 4$

$x = 2^{y-1} + 2$

| $x$ | $\dfrac{9}{4}$ | $\dfrac{5}{2}$ | 3 | 4 | 6 | 10 |
|---|---|---|---|---|---|---|
| $y$ | $-1$ | 0 | 1 | 2 | 3 | 4 |

Asymptote: $x = 2$

**67.** $f(g(x)) = f(2^{x-1} - 2)$

$= \log_2(2(2^{x-1} - 2) + 4)$

$= \log_2(2^x - 4 + 4)$

$= \log_2(2^x) = x$

$g(f(x)) = g(\log_2(2x + 4))$

$= 2^{\log_2(2x+4)-1} - 2$

$= 2^{-1} \cdot 2^{\log_2(2x+4)} - 2$

$= 2^{-1}(2x + 4) - 2$

$= x + 2 - 2 = x$

## Challenge

$$\log_4(2x+3) = -\frac{5}{2}$$

$$\log_4(2x+3) = \log_4\left(4^{-5/2}\right)$$

since $x = \log_b b^x$

$$\log_4(2x+3) = \log_4\left(\frac{1}{32}\right)$$

$2x+3 = \frac{1}{32}$ since log is one-to-one

$$2x = \frac{1}{32} - 3$$

$$2x = -\frac{95}{32}$$

$$x = -\frac{95}{64}$$

## Exercises 5.3

**1.** $\log_b \dfrac{3x}{x+1} = \log_b 3x - \log_b(x+1)$
$= \log_b 3 + \log_b x - \log_b(x+1)$

**3.** $\log_b \dfrac{\sqrt{x^2-1}}{x} = \log_b \sqrt{x^2-1} - \log_b x$

$= \dfrac{1}{2}\log_b\left(x^2-1\right) - \log_b x$

$= \dfrac{1}{2}\log_b(x-1)(x+1) - \log_b x$

$= \dfrac{1}{2}\log_b(x-1) + \dfrac{1}{2}\log_b(x+1) - \log_b x$

**5.** $\log_b \dfrac{1}{x^2} = \log_b x^{-2} = -2\log_b x$

**7.** $\log_b\left(\dfrac{2x-5}{x^3}\right) = \log_b(2x-5) - \log_b x^3$

$= \log_b(2x-5) - 3\log_b x$

**9.** $\log_b(x+1) - \log_b(x+2) = \log_b \dfrac{x+1}{x+2}$

**11.** $\dfrac{1}{2}\log_b\left(x^2-1\right) - \dfrac{1}{2}\log_b\left(x^2+1\right)$

$= \dfrac{1}{2}\log_b \dfrac{x^2-1}{x^2+1} = \log_b\sqrt{\dfrac{x^2-1}{x^2+1}}$

**13.** $3\log_b x - \log_b 2 - \log_b(x+5)$

$= \log_b x^3 - \log_b 2 - \log_b(x+5)$

$= \log_b \dfrac{x^3}{2(x+5)}$

**15.** $\log_b\left(x^2-x-6\right) - \log_b(x+2)$

$= \log_b\left(\dfrac{x^2-x-6}{x+2}\right)$

$= \log_b\left(\dfrac{(x-3)(x+2)}{x+2}\right)$

$= \log_b(x-3)$

**17.** $\dfrac{1}{3}\left[\log_b\left(x^3-8\right) - 2\log_b(x-2)\right]$

$= \dfrac{1}{3}[\log_b(x-2)\left(x^2+2x+4\right)$

$\quad - \log_b(x-2)^2]$

$= \dfrac{1}{3}\left[\log_b\left(\dfrac{(x-2)\left(x^2+2x+4\right)}{(x-2)^2}\right)\right]$

$= \log_b\left(\dfrac{x^2+2x+4}{x-2}\right)^{1/3}$

**19.** $\log_b 27 + \log_b 3 = \log_b 81$    (Law 1)
$\log_b 243 - \log_b 3 = \log_b 81$   (Law 2)

**21.** $-2\log_b \dfrac{4}{9} = \log_b\left(\dfrac{4}{9}\right)^{-2}$    (Law 3)

$= \log_b \dfrac{81}{16}$

**23. (a)** $\log 4 = \log 2^2 = 2\log 2$
$= 2(0.3010) = 0.6020$

**(b)** $\log 8 = \log 2^3 = 3 \log 2 = 3(0.3010)$
$= 0.9030$

**(c)** $\log \dfrac{1}{2} = \log 2^{-1} = -\log 2 = -(0.3010)$
$= -0.3010$

**25. (a)** $\log 48 = \log 16 \cdot 3 = \log 16 + \log 3$
$= \log 2^4 + \log 3 = 4 \log 2 + \log 3$
$= 4(0.3010) + 0.4771 = 1.6811$

**(b)** $\log \dfrac{2}{3} = \log 2 - \log 3$
$0.3010 - 0.4771 = -0.1761$

**(c)** $\log 125 = \log 5^3 = 3 \log 5$
$= 3(0.6990) = 2.0970$

**27. (a)** $\log \sqrt[3]{5} = \dfrac{1}{3} \log 5$
$= \dfrac{1}{3}(0.6990) = 0.2330$

**(b)** $\log \sqrt{20^3} = \dfrac{3}{2} \log 20 = \dfrac{3}{2} \log 4 \cdot 5$
$= \dfrac{3}{2} \cdot \log 2^2 + \dfrac{3}{2} \log 5$
$= 3 \log 2 + \dfrac{3}{2} \log 5$
$= 3(0.3010) + \dfrac{3}{2}(0.6990) = 1.9515$

**(c)** $\log \sqrt{900} = \log 30 = \log 2 \cdot 3 \cdot 5$
$= \log 2 + \log 3 + \log 5$
$= 0.3010 + 0.4771 + 0.6990$
$= 1.4771$

**29.** $\log_4 120 = \dfrac{\log_{10} 120}{\log_{10} 4} = 3.4534$

**31.** $\log_8 64 = \dfrac{\log_{10} 64}{\log_{10} 8} = 2$

**33.** $b^{3 \log_b 4} = b^{\log_b 4^3} = 4^3 = 64$

**35.** $\left(6^{\log_6 x}\right)^3 = x^3$

**37.** $\log_{10} x + \log_{10} 5 = 2$
$\log_{10} 5x = 2$
$5x = 10^2$
$5x = 100$
$x = 20$

**39.** $\log_{10} 5 - \log_{10} x = 2$
$\log_{10} \dfrac{5}{x} = 2$
$\dfrac{5}{x} = 10^2$
$5 = 100x$
$x = \dfrac{5}{100} = \dfrac{1}{20}$

**41.** $\log_{12}(x - 5) + \log_{12}(x - 5) = 2$
$2 \log_{12}(x - 5) = 2$
$\log_{12}(x - 5) = 1$
$x - 5 = 12^1$
$x = 17$

**43.** $\log_{16} x + \log_{16}(x - 4) = \dfrac{5}{4}$
$\log_{16} x(x - 4) = \dfrac{5}{4}$
$x(x - 4) = 16^{5/4}$
$x^2 - 4x = 32$
$x^2 - 4x - 32 = 0$
$(x - 8)(x + 4) = 0$
$x - 8 = 0$ or $x + 4 = 0$
$x = 8$ or $x = -4$, but $x = -4$ is not a
solution, so $x = 8$

**45.** $\log_{10}(3 - x) - \log_{10}(12 - x) = -1$
$\log_{10} \left( \dfrac{3 - x}{12 - x} \right) = -1$
$\dfrac{3 - x}{12 - x} = 10^{-1} = \dfrac{1}{10}$
$10(3 - x) = 12 - x$
$30 - 10x = 12 - x$
$-9x = -18$
$x = 2$

**47.** $\log_{1/7} x + \log_{1/7} (5x - 28) = -2$
$\log_{1/7} x(5x - 28) = -2$

$$x(5x - 28) = (1/7)^{-2}$$
$$5x^2 - 28x = 49$$
$$5x^2 - 28x - 49 = 0$$
$$(5x + 7)(x - 7) = 0$$
$$5x + 7 = 0 \quad \text{or} \quad x - 7 = 0$$
$$x = -\frac{7}{5} \quad \text{or} \quad x = 7$$

but $x = -\frac{7}{5}$ is not a solution,

so $x = 7$

**49.** $\log_{10} (x^3 - 1) - \log_{10} (x^2 + x + 1) = -2$

$$\log_{10} \left( \frac{x^3 - 1}{x^2 + x + 1} \right) = -2$$

$$\frac{(x - 1)\left(x^2 + x + 1\right)}{x^2 + x + 1} = 10^{-2}$$

$$x - 1 = 0.01$$

$x^2 + x + 1$ must be $> 0$, so $x = 1.01$

**51.** $2\log_{25} x - \log_{25} (25 - 4x) = \frac{1}{2}$

$$\log_{25} x^2 - \log_{25} (25 - 4x) = \frac{1}{2}$$

$$\log_{25} \frac{x^2}{25 - 4x} = \frac{1}{2}$$

$$\frac{x^2}{25 - 4x} = 25^{1/2}$$

$$x^2 = 5(25 - 4x)$$
$$x^2 = 125 - 20x$$
$$x^2 + 20x - 125 = 0$$
$$(x + 25)(x - 5) = 0$$
$$x + 25 = 0 \quad \text{or} \quad x - 5 = 0$$
$$x = -25 \quad \text{or} \quad x = 5$$

but $x = -25$ is not a solution, so $x = 5$

**53.** Let $\log_b M = r$ and $\log_b N = s$.

Then $M = b^r$ and $N = b^s$
Divide the two equations:

$$\frac{M}{N} = \frac{b^r}{b^s} = b^{r-s}$$

Convert to logarithmic form:

$$\log_b \frac{M}{N} = r - s = \log_b M - \log_b N$$

**55.** $(x + 2) \log_b b^x = x$
$$(x + 2)x = x$$
$$x^2 + 2x = x$$
$$x^2 + x = 0$$
$$x(x + 1) = 0$$
$$x = 0 \quad \text{or} \quad x + 1 = 0$$
$$x = -1$$

**57.** $\log_x (2x)^{3x} = 4x$
$$3x \log_x (2x) = 4x$$

$$\log_x (2x) = \frac{4}{3} \text{ since } x \text{ cannot equal } 0$$

$$2x = x^{4/3}$$
$$2 = x^{1/3}$$
$$x = 2^3 = 8$$

**59.** $100 = 10 \log \left( \dfrac{I}{10^{-16}} \right)$

$$10 = \log \left( \frac{I}{10^{-16}} \right)$$

$$10^{10} = \frac{I}{10^{-16}}$$

$$10^{-6} = I$$

The intensity of the train is $10^{-6}$ watts per sq cm.

**61.** $120 = 10 \log \left( \dfrac{I}{10^{-16}} \right)$

$$12 = \log \left( \frac{I}{10^{-16}} \right)$$

$$10^{12} = \frac{I}{10^{-16}}$$

$$10^{-4} = I$$

The intensity of a 120 decibel sound is $10^{-4}$ watts per sq cm, and the intensity of a 100 decibel sound is $10^{-6}$ watts per sq cm (Exercise 59). Therefore, the 120 decibel sound is 100 times greater in intensity than a 100 decibel sound.

**63.** Domain: $x > 0$
Asymptote: $x = 0$
$x$-intercept: $0 = \log_5 x^3$
$5^0 = x^3$
$x^3 = 1$
$x = 1$

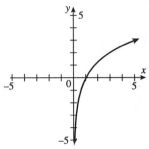

**65.** Domain: $x > -1$
Asymptote: $x = -1$
$y$-intercept: $y = \log_4 \sqrt{0 + 1}$
$= \log_4 \sqrt{1}$
$= \log_4 1 = 0$
$x$-intercept: $0 = \log_4 \sqrt{x + 1}$
$4^0 = \sqrt{x + 1}$
$1 = \sqrt{x + 1}$
$\left(\sqrt{x + 1}\right)^2 = 1^2$
$x + 1 = 1$
$x = 0$

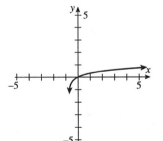

**67.** Domain: $x > 2$
Asymptote: $x = 2$
$x$-intercept: $0 = \log_5 \dfrac{1}{x - 2}$
$5^0 = \dfrac{1}{x - 2}$
$x - 2 = 1$
$x = 3$

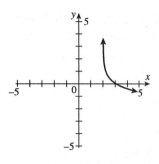

**69.** $y = \log_2 (x^2 + x) - \log_2 x$
$= \log_2 \left(\dfrac{x^2 + x}{x}\right)$
$= \log_2 (x + 1), \quad x > 0$
Domain: $x > 0$
No asymptotes or intercepts

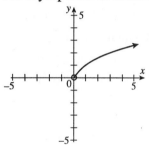

**Challenge**

$$\log_9 16 = \frac{\log_3 16}{\log_3 9} = \frac{\log_3 4^2}{\log_3 3^2}$$
$$= \frac{2 \log_3 4}{2}$$
$$= \log_3 4$$

**Critical Thinking**

**1.** If $x \neq 0$, then $2^x \neq 3^x$ since if $2^x = 3^x$,
then $(2^x)^{1/x} = (3^x)^{1/x}$ or $2 = 3$,
which is false. Also, since $2^0 = 3^0$, then
$2^x = 3^x$ only when $x = 0$. Consider
$f(x) = 2^x - 3^x$, and construct a sign chart:

| Interval | $(-\infty, 0)$ | $(0, \infty)$ |
|----------|----------------|---------------|
| Sign of $f(x)$ | $+$ | $-$ |

So, for $x < 0$, $2^x - 3^x > 0$ or $2^x > 3^x$;
for $x = 0$, $2^x = 3^x$; and for $x > 0$, $2^x < 3^x$.

**3.** $(f \circ f \circ f)(x) = f(f(f(x))) = f(f(\log_2 x))$
$= f(\log_2 (\log_2 x)) = \log_2(\log_2(\log_2 x))$
$f(f(f(x))) = 2$
$\log_2(\log_2(\log_2 x)) = 2$
$\log_2(\log_2 x) = 2^2 = 4$
$\log_2 x = 2^4 = 16$
$x = 2^{16} = 65,536$
For any base $b$,
$\log_b (\log_b (\log_b x)) = 2$
$\log_b (\log_b x) = b^2$
$\log_b x = b^{b^2}$
$x = b^{b^{b^2}}$

**5.** Since $\log_{10} x$ is increasing,
$\log_{10} x > 3$ when $x > 10^3 = 1000$
$\log_{10} x > 30$ when $x > 10^{30}$
and $\log_{10} x > 300$ when $x > 10^{300}$.
So, $\log_{10} x$ grows very slowly.

## Exercises 5.4

**1.**

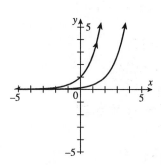

**3.**

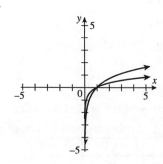

**5.**

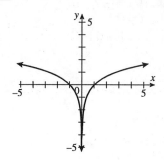

**7.**

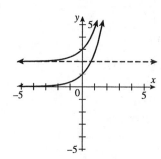

**9.**

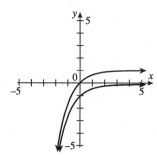

**11.**

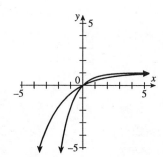

**13.** $f(x) = \ln ex = \ln e + \ln x = 1 + \ln x$
Shift $y = \ln x$ one unit upward.

**15.** $f(x) = \ln \sqrt{x} = \frac{1}{2} \ln x$

Multiply the ordinates of $y = \ln x$ by $\frac{1}{2}$.

**17.** $f(x) = \ln(x^2 - 1) - \ln(x + 1)$

$\quad = \ln\left(\dfrac{x^2 - 1}{x + 1}\right) = \ln\left(\dfrac{(x+1)(x-1)}{x+1}\right)$

$\quad = \ln(x - 1)$

Shift $y = \ln x$ one unit to the right.

**19.** Domain: all $x > -2$

$x$-intercept: $0 = \ln(x + 2)$

$x + 2 = e^0 = 1$

$x = -1$

**21.** Domain: all $x > \dfrac{1}{2}$

$x$-intercept: $0 = \ln(2x - 1)$

$\quad 2x - 1 = e^0 = 1$

$\quad\quad 2x = 2$

$\quad\quad\ x = 1$

**23.** Domain: all $x > 1$ except $x = 2$

$x$-intercept: $0 = \dfrac{\ln(x - 1)}{x - 2}$

$\quad 0 = \ln(x - 1)$

$\quad x - 1 = e^0 = 1$

$\quad\quad x = 2$

Since $x = 2$ is not in the domain of $f(x)$, there are no $x$-intercepts.

**25.** $\ln f(x) = \ln\left(\dfrac{5x}{x^2 - 4}\right) = \ln 5x - \ln(x^2 - 4)$

$\quad = \ln 5 + \ln x - \ln(x + 2)(x - 2)$

$\quad = \ln 5 + \ln x - \ln(x + 2) - \ln(x - 2)$

**27.** $\ln f(x) = \ln\left(\dfrac{(x-1)(x+3)^2}{\sqrt{x^2 + 2}}\right)$

$\quad = \ln(x - 1) + \ln(x + 3)^2 - \ln\sqrt{x^2 + 2}$

$\quad = \ln(x - 1) + 2\ln(x + 3) - \dfrac{1}{2}\ln(x^2 + 2)$

**29.** $\ln f(x) = \ln\sqrt{x^3(x + 1)} = \dfrac{1}{2}\ln(x^3(x + 1))$

$\quad = \dfrac{1}{2}\ln x^3 + \dfrac{1}{2}\ln(x + 1)$

$\quad = \dfrac{3}{2}\ln x + \dfrac{1}{2}\ln(x + 1)$

**31.** $\dfrac{1}{2}\ln x + \ln(x^2 + 5) = \ln\sqrt{x} + \ln(x^2 + 5)$

$\quad\quad\quad\quad\quad\quad\quad\quad = \ln\sqrt{x}(x^2 + 5)$

**33.** $3\ln(x + 1) + 3\ln(x - 1) = \ln(x + 1)^3$

$\quad + \ln(x - 1)^3 = \ln(x + 1)^3(x - 1)^3$

$\quad = \ln(x^2 - 1)^3$

**35.** $\dfrac{1}{2}\ln x - 2\ln(x - 1) - \dfrac{1}{3}\ln(x^2 + 1)$

$\quad = \ln\sqrt{x} - \ln(x - 1)^2 - \ln\sqrt[3]{x^2 + 1}$

$\quad = \ln\dfrac{\sqrt{x}}{(x - 1)^2\sqrt[3]{x^2 + 1}}$

**37.** $e^{\ln\sqrt{x}} = \sqrt{x}$

**39.** $e^{-2\ln x} = e^{\ln x^{-2}} = x^{-2} = \dfrac{1}{x^2}$

**41.** $\ln\left(\dfrac{e^x}{e^{x-1}}\right) = \ln e^x - \ln e^{x-1} = x - (x - 1)$

$\quad\quad\quad\quad\quad\quad = x - x + 1 = 1$

**43.** $\quad e^{-0.01x} = 27$

$\quad \ln e^{-0.01x} = \ln 27$

$\quad\quad -0.01x = \ln 27$

$\quad\quad\quad x = -\dfrac{\ln 27}{0.01} = -100\ln 27$

**45.** $\quad e^{\ln(1-x)} = 2x$

$\quad\quad 1 - x = 2x$

$\quad\quad\quad 1 = 3x$

$\quad\quad\quad x = \dfrac{1}{3}$

**47.** $\ln(x + 1) = 0$

$\quad\quad x + 1 = e^0 = 1$

$\quad\quad\quad x = 0$

**49.** $\quad \ln e^{\sqrt{x+1}} = 3$

$\quad\quad \sqrt{x + 1} = 3$

$$\left(\sqrt{x+1}\right)^2 = 3^2$$
$$x + 1 = 9$$
$$x = 8$$

**51.** $\ln(x^2 - 4) - \ln(x + 2) = 0$
$$\ln\left(\frac{x^2 - 4}{x + 2}\right) = 0$$
$$\ln\left(\frac{(x+2)(x-2)}{x+2}\right) = 0$$
$$\ln(x - 2) = 0$$
$$x - 2 = e^0 = 1$$
$$x = 3$$

**53.** $\ln x = \dfrac{1}{2}\ln 4 + \dfrac{2}{3}\ln 8$
$$\ln x = \ln\sqrt{4} + \ln 8^{2/3}$$
$$\ln x = \ln 2 + \ln 4$$
$$\ln x = \ln 8$$
$$x = 8$$

**55.** $\ln x = 2 + \ln(1 - x)$
$$\ln x - \ln(1 - x) = 2$$
$$\ln\frac{x}{1 - x} = 2$$
$$\frac{x}{1 - x} = e^2$$
$$x = e^2(1 - x)$$
$$x = e^2 - e^2 x$$
$$x + e^2 x = e^2$$
$$x(1 + e^2) = e^2$$
$$x = \frac{e^2}{1 + e^2}$$

**57.** Domain: all $x > 4$
Asymptote: $x = 4$
x-intercept: $0 = \ln(x - 4)$
$$x - 4 = e^0 = 1$$
$$x = 5$$

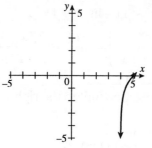

**59.** Domain: all $x < 4$
Asymptote: $x = 4$
y-intercept: $y = \ln(4 - 0)$
$$y = \ln 4$$
$$y = 1.4$$
x-intercept: $0 = \ln(4 - x)$
$$4 - x = e^0 = 1$$
$$-x = -3$$
$$x = 3$$

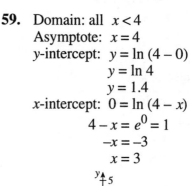

**61.** Domain: all $x > 0$
Asymptote: $x = 0$
x-intercept: $0 = \ln x^3$
$$x^3 = e^0 = 1$$
$$x = 1$$

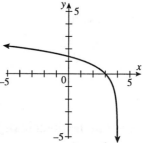

**63.** Let $g(x) = -x^2 + x$ and $f(x) = e^x$. Then
$$(f \circ g)(x) = f(g(x)) = f(-x^2 + x)$$
$$= e^{-x^2 + x} = h(x)$$

**65.** Let $g(x) = \dfrac{x}{x+1}$ and $f(x) = \ln x$. Then

$$(f \circ g)(x) = f(g(x)) = f\left(\dfrac{x}{x+1}\right)$$

$$= \ln\left(\dfrac{x}{x+1}\right) = h(x)$$

**67.** Let $g(x) = \ln x$ and $f(x) = \sqrt[3]{x}$.
Then $(f \circ g)(x) = f(g(x)) = f(\ln x)$

$$= \sqrt[3]{\ln x} = h(x)$$

**69.** Let $h(x) = 3x - 1$, $g(x) = x^2$, and $f(x) = e^x$.
Then $(f \circ g \circ h)(x) = f(g(h(x)))$

$$= f(g(3x-1)) = f((3x-1)^2)$$

$$= e^{(3x-1)^2} = F(x)$$

**71.** Let $h(x) = e^x + 1$, $g(x) = \sqrt{x}$, and
$f(x) = \ln x$. Then
$(f \circ g \circ h)(x) = f(g(h(x)))$

$$= f(g(e^x + 1)) = f\left(\sqrt{e^x + 1}\right)$$

$$= \ln\sqrt{e^x + 1} = F(x)$$

**73.** Domain: all real numbers
$0 = e^{2x} - 2x\, e^{2x}$
$0 = e^{2x}(1 - 2x)$
$1 - 2x = 0$ since $e^{2x} > 0$ for all $x$
$-2x = -1$
$x = \dfrac{1}{2}$

**75.** Domain: all $x > 0$
$0 = 1 + \ln x$
$-1 = \ln x$
$x = e^{-1} = \dfrac{1}{e}$

**77.** $\ln\left(\dfrac{x}{4} - \dfrac{\sqrt{x^2 - 4}}{4}\right) = \ln\left(\dfrac{x - \sqrt{x^2 - 4}}{4}\right)$

$$= \ln\left(\dfrac{x - \sqrt{x^2 - 4}}{4} \cdot \dfrac{x + \sqrt{x^2 - 4}}{x + \sqrt{x^2 - 4}}\right)$$

$$= \ln\left(\dfrac{x^2 - (x^2 - 4)}{4\left(x + \sqrt{x^2 - 4}\right)}\right)$$

$$= \ln\left(\dfrac{4}{4\left(x + \sqrt{x^2 - 4}\right)}\right)$$

$$= \ln\dfrac{1}{x + \sqrt{x^2 - 4}}$$

$$= -\ln\left(x + \sqrt{x^2 - 4}\right)$$

**79.** Let $u = e^x$. Then

$$y = \dfrac{u}{2} - \dfrac{1}{2u}$$

$$2uy = u^2 - 1$$

$$0 = u^2 - 2uy - 1$$

$$u = \dfrac{2y \pm \sqrt{4y^2 + 4}}{2} = y \pm \dfrac{\sqrt{4\left(y^2 + 1\right)}}{2}$$

$$= y \pm \dfrac{2\sqrt{y^2 + 1}}{2} = y \pm \sqrt{y^2 + 1}$$

$e^x = y + \sqrt{y^2 + 1}$ or $e^x = y - \sqrt{y^2 + 1}$
Since $y - \sqrt{y^2 + 1} < 0$, this is not
possible.

$$x = \ln\left(y + \sqrt{y^2 + 1}\right)$$

**81.** $\dfrac{g(2+h) - g(2)}{h} = \dfrac{\ln(2+h) - \ln 2}{h}$

$$= \dfrac{\ln\left(\dfrac{2+h}{2}\right)}{h} = \dfrac{1}{h}\ln\left(1 + \dfrac{h}{2}\right)$$

**83.** $\dfrac{g\left(e^{3x}\right)-g(e^{6})}{x-2}=\dfrac{\ln e^{3x}-\ln e^{6}}{x-2}$

$$=\dfrac{3x-6}{x-2}=\dfrac{3(x-2)}{x-2}=3$$

**85.** $e^{x}>500$

$x>\ln 500$ since $e^{x}$ is increasing

## Challenge

**1.** $\log_{b}(\log_{b} nx)=1$

$\log_{b} nx=b^{1}=b$

$nx=b^{b}$

$x=\dfrac{b^{b}}{n}$

## Exercises 5.5

**1.** $\dfrac{\ln 6}{\ln 2}=2.585$

**3.** $\dfrac{\ln 8}{\ln 0.2}=-1.292$

**5.** $\dfrac{\ln 15}{2\ln 3}=1.232$

**7.** $\dfrac{\ln 100}{-4\ln 10}=-0.5$

**9.** $2^{x}=45$

$\ln 2^{x}=\ln 45$

$x\ln 2=\ln 45$

$x=\dfrac{\ln 45}{\ln 2}$

$x=5.492$

**11.** $2^{-x}=125$

$\ln 2^{-x}=\ln 125$

$-x\ln 2=\ln 125$

$x=-\dfrac{\ln 125}{\ln 2}$

$x=-6.966$

**13.** $2^{3x}=80$

$\ln 2^{3x}=\ln 80$

$3x\ln 2=\ln 80$

$x=\dfrac{\ln 80}{3\ln 2}$

$x=2.107$

**15.** $80e^{x}=120$

$e^{x}=\dfrac{3}{2}$

$\ln e^{x}=\ln\dfrac{3}{2}$

$x=\ln\dfrac{3}{2}$

$x=0.405$

**17.** $y=100e^{(0.75)4}$

$=100e^{3}$

$2008.55$

**19.** $y=1000e^{(-1.8)2}$

$=1000e^{-3.6}$

$27.32$

**21.** $5000=50e^{2k}$

$100=e^{2k}$

$\ln 100=\ln e^{2k}$

$\ln 100=2k$

$k=\dfrac{1}{2}\ln 100$

**23.** $\dfrac{A}{3}=Ae^{4k}$

$\dfrac{1}{3}=e^{4k}$

$\ln\dfrac{1}{3}=\ln e^{4k}$

$\ln\dfrac{1}{3}=4k$

$k=\dfrac{1}{4}\ln\dfrac{1}{3}$

**25.** $y=10,000e^{(0.6)7}$

$=10,000e^{4.2}$

$=667,000$

**27.** $30,000 = 10,000e^{0.6x}$

$3 = e^{0.6x}$

$\ln 3 = \ln e^{0.6x}$

$\ln 3 = 0.6x$

$x = \dfrac{\ln 3}{0.6} = 1.83$ days

**29.** $25 = 50e^{-0.04t}$

$\dfrac{1}{2} = e^{-0.04t}$

$\ln \dfrac{1}{2} = \ln e^{-0.04t}$

$\ln \dfrac{1}{2} = -0.04t$

$t = -\dfrac{\ln \frac{1}{2}}{0.04} = 17.33$ years

**31.** **(a)** $8 = 10e^{k \cdot 5}$

$\dfrac{4}{5} = e^{5k}$

$\ln \dfrac{4}{5} = \ln e^{5k}$

$\ln \dfrac{4}{5} = 5k$

$k = \dfrac{1}{5} \ln \dfrac{4}{5}$

**(b)** $10e^{\left(\frac{1}{5} \ln \frac{4}{5}\right)10}$

$= 10e^{2\ln\left(\frac{4}{5}\right)}$

$= 10e^{\ln\left(\frac{4}{5}\right)^2}$

$10 \cdot \left(\dfrac{4}{5}\right)^2$

$\dfrac{160}{25} = 6.4$ grams

**(c)** $5 = 10e^{\left(\frac{1}{5} \ln \frac{4}{5}\right)x}$

$\dfrac{1}{2} = e^{\frac{x}{5} \ln \frac{4}{5}}$

$\ln \dfrac{1}{2} = \ln e^{\frac{x}{5} \ln \frac{4}{5}}$

$\ln \dfrac{1}{2} = \dfrac{x}{5} \ln \dfrac{4}{5}$

$x = \dfrac{5\ln \frac{1}{2}}{\ln \frac{4}{5}} = 15.5$ years

**33.** $4 = 5e^{k \cdot 30}$

$\dfrac{4}{5} = e^{30k}$

$\ln \dfrac{4}{5} = \ln e^{30k}$

$\ln \dfrac{4}{5} = 30k$

$k = \dfrac{1}{30} \ln \dfrac{4}{5}$

$\dfrac{5}{2} = 5e^{\left(\frac{1}{30} \ln \frac{4}{5}\right)x}$

$\dfrac{1}{2} = e^{\frac{x}{30} \ln \frac{4}{5}}$

$\ln \dfrac{1}{2} = \ln e^{\frac{x}{30} \ln \frac{4}{5}}$

$\ln \dfrac{1}{2} = \dfrac{x}{30} \ln \dfrac{4}{5}$

$x = \dfrac{30\ln \frac{1}{2}}{\ln \frac{4}{5}} = 93.2$ seconds

**35.** $44,000 = (22,000)\left(10^{0.0163t}\right)$

$2 = 10^{0.0163t}$

$\ln 2 = \ln 10^{0.0163t}$

$\ln 2 = (0.0163t)(\ln 10)$

$t = \dfrac{\ln 2}{(0.0163)\ln 10} = 18.47$ years

**37.** $\dfrac{3}{5}A = Ae^{\frac{\ln 0.5}{5750}x}$

$\dfrac{3}{5} = e^{\frac{\ln 0.5}{5750}x}$

$$\ln \frac{3}{5} = \ln e^{\frac{\ln 0.5}{5750}x}$$

$$\ln \frac{3}{5} = \frac{\ln 0.5}{5750}x$$

$$x = \frac{5750 \cdot \ln \frac{3}{5}}{\ln 0.5} = 4200 \text{ years}$$

**39.** $\quad \dfrac{1}{1,000,000} A = A e^{\frac{\ln 0.5}{5750}x}$

$$\frac{1}{1,000,000} = e^{\frac{\ln 0.5}{5750}x}$$

$$\ln \frac{1}{1,000,000} = \ln e^{\frac{\ln 0.5}{5750}x}$$

$$\ln \frac{1}{1,000,000} = \frac{\ln 0.5}{5750}x$$

$$x = \frac{5750 \cdot \ln \frac{1}{1,000,000}}{\ln 0.5} = 115,000 \text{ years}$$

**41.** $\quad 11,600 = 14,000 e^{k \cdot 1}$

$$\frac{29}{35} = e^k$$

$$k = \ln \frac{29}{35}$$

$$8000 = 14,000 e^{\left(\ln \frac{29}{35}\right)t}$$

$$\frac{4}{7} = e^{\left(\ln \frac{29}{35}\right)t}$$

$$\ln \frac{4}{7} = \ln e^{\left(\ln \frac{29}{35}\right)t}$$

$$\ln \frac{4}{7} = t \cdot \ln \frac{29}{35}$$

$$t = \frac{\ln \frac{4}{7}}{\ln \frac{29}{35}} = 3 \text{ years}$$

**43.** $\quad 3 = 5 e^{k \cdot 2}$

$$\frac{3}{5} = e^{2k}$$

$$\ln \frac{3}{5} = \ln e^{2k}$$

$$\ln \frac{3}{5} = 2k$$

$$k = \frac{1}{2} \ln \frac{3}{5}$$

$$\frac{1}{4}(5) = 5 e^{\left(\frac{1}{2} \ln \frac{3}{5}\right)t}$$

$$\frac{1}{4} = e^{\frac{t}{2} \ln \frac{3}{5}}$$

$$\ln \frac{1}{4} = \ln e^{\frac{t}{2} \ln \frac{3}{5}}$$

$$\ln \frac{1}{4} = \frac{t}{2} \ln \frac{3}{5}$$

$$t = \frac{2 \ln \frac{1}{4}}{\ln \frac{3}{5}} = 5.43 \text{ hours}$$

**45. (a)** $\quad A = 10,000 \left(1 + \dfrac{0.09}{4}\right)^{4 \cdot 5} = \$15,605$

**(b)** $\quad A = 10,000 \left(1 + \dfrac{0.09}{12}\right)^{12 \cdot 5} = \$15,657$

**(c)** $\quad A = 10,000 \left(1 + \dfrac{0.09}{52}\right)^{52 \cdot 5} = \$15,677$

**(d)** $\quad A = 10,000 \left(1 + \dfrac{0.09}{365}\right)^{365 \cdot 5}$
$\qquad = \$15,682$

**(e)** $\quad A = 10,000 e^{(0.09)5} = \$15,683$

**47. (a)** $\quad A = 10,000 \left(1 + \dfrac{0.09}{4}\right)^{4(3.5)}$
$\qquad = \$13,655$

**(b)** $\quad A = 10,000 \left(1 + \dfrac{0.09}{12}\right)^{12(3.5)}$
$\qquad = \$13,686$

**(c)** $\quad A = 10,000 \left(1 + \dfrac{0.09}{52}\right)^{52(3.5)}$
$\qquad = \$13,699$

**(d)** $\quad A = 10,000 \left(1 + \dfrac{0.09}{365}\right)^{365(3.5)}$
$\qquad = \$13,702$

**(e)** $\quad 10,000 e^{(0.09)(3.5)} = \$13,703$

**49.** $\quad 10,000 = 5000 e^{0.09t}$

$$2 = e^{0.09t}$$

$$\ln 2 = \ln e^{0.09t}$$
$$\ln 2 = 0.09t$$
$$t = \frac{\ln 2}{0.09} = 7.7 \text{ years}$$
$$10,000 = 5000e^{0.12t}$$
$$2 = e^{0.12t}$$
$$\ln 2 = \ln e^{0.12t}$$
$$\ln 2 = 0.12t$$
$$t = \frac{\ln 2}{0.12} = 5.8 \text{ years}$$

**51.** $2000 = 1000e^{\frac{r}{100} \cdot 5}$

$$2 = e^{\frac{r}{20}}$$
$$\ln 2 = \ln e^{\frac{r}{20}}$$
$$\ln 2 = \frac{r}{20}$$
$$r = 20 \ln 2 = 13.86\%$$

**53.** $8000 = 4000\left(1 + \frac{0.08}{12}\right)^{12t}$

$$2 = \left(1 + \frac{0.08}{12}\right)^{12t}$$
$$\ln 2 = \ln\left(1 + \frac{0.08}{12}\right)^{12t}$$
$$\ln 2 = 12t \ln\left(1 + \frac{0.08}{12}\right)$$
$$t = \frac{\ln 2}{12 \ln\left(1 + \frac{0.08}{12}\right)} = 8.69 \text{ years}$$

**55.** $5000 = Pe^{(0.09)6}$

$$5000 = Pe^{0.54}$$
$$P = \frac{5000}{e^{0.54}} = \$2914$$

**57.** $20,000 = P\left(1 + \frac{0.12}{12}\right)^{12 \cdot 5}$

$$20,000 = P(1.01)^{60}$$
$$P = \frac{20,000}{(1.01)^{60}} = \$11,009$$

**59.** $x = -\frac{1}{k}(\ln A - \ln y)$

$$kx = -\ln A + \ln y$$
$$kx + \ln A = \ln y$$
$$y = e^{kx + \ln A}$$
$$y = e^{kx} \cdot e^{\ln A}$$
$$y = Ae^{kx}$$

This is the continuous compound interest formula.

**61.** $29 = 3\left(1 + \frac{\frac{r}{100}}{1}\right)^{1 \cdot 50}$

$$\frac{29}{3} = \left(1 + \frac{r}{100}\right)^{50}$$
$$\sqrt[50]{\frac{29}{3}} = 1 + \frac{r}{100}$$
$$\sqrt[50]{\frac{29}{3}} - 1 = \frac{r}{100}$$
$$r = 100\left(\sqrt[50]{\frac{29}{3}} - 1\right) = 4.64\%$$

**Critical Thinking**

**1.** To use the one-to-one property, solve Example 7 as follows:

$$\log_{10}(x^3 - 1) - \log_{10}(x^2 + x + 1) = 1$$
$$\log_{10}\frac{x^3 - 1}{x^2 + x + 1} = \log_{10} 10$$
$$x - 1 = 10$$
$$x = 11$$

**3.** The tangent line which has slope closest to 1 is the tangent line to the graph of $y = e^x$.

**Review Exercises**

**1.** For $b > 1$, $f(x)$ is increasing, concave up, and the $x$-axis is a horizontal asymptote toward the left. For $0 < b < 1$, $f(x)$ is decreasing, concave down, and the $x$-axis is a horizontal asymptote toward the right.

**3.**

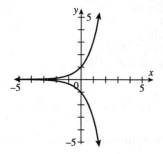

**5.**

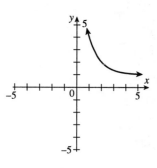

**7.** $8^{3x-2} = 64$

$8^{3x-2} = 8^2$

$3x - 2 = 2$

$3x = 4$

$x = \dfrac{4}{3}$

**9.** $\dfrac{1}{3^{x+2}} = 27$

$3^{-(x+2)} = 3^3$

$-(x+2) = 3$

$-x - 2 = 3$

$-x = 5$

$x = -5$

**11.**

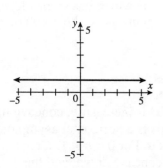

**13.** $x = 3^y$

| $x$ | $\dfrac{1}{9}$ | $\dfrac{1}{3}$ | $1$ | $3$ | $9$ |
|---|---|---|---|---|---|
| $y$ | $-2$ | $-1$ | $0$ | $1$ | $2$ |

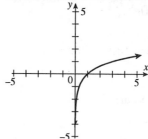

**15.** Domain: all $x > -3$
Asymptote: $x = -3$

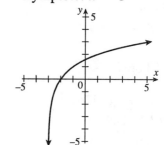

**17.** **(a)** $3^4 = 81$

**(b)** $2^5 = 32$

**19.** $\log_2 64 = y$

$2^y = 64$

$2^y = 2^6$

$y = 6$

**21.** $\log_{\sqrt{5}} x = -4$

$(\sqrt{5})^{-4} = x$

$x = \dfrac{1}{25}$

**23.** $\log_b \dfrac{AB}{C^2} = \log_b A + \log_b B - \log_b C^2$

$= \log_b A + \log_b B - 2\log_b C$

**25.** A common logarithm is a logarithm to base 10.

**27.** $\log_2 75 = \dfrac{\log_{10} 75}{\log_{10} 2} = 6.2288$

**29.** $\log_{12} 1000 = \dfrac{\log_{10} 1000}{\log_{10} 12} = 2.7799$

**31.** $\log_{10}(x^2 - 1) - \log_{10}(x + 1) = 1$

$\log_{10} \dfrac{x^2 - 1}{x + 1} = 1$

$\dfrac{(x+1)(x-1)}{x+1} = 10^1$

$x - 1 = 10$

$x = 11$

**33.** $3\log_{10}(x + 3) = 6$

$\log_{10}(x + 3) = 2$

$x + 3 = 10^2$

$x = 97$

**35.**

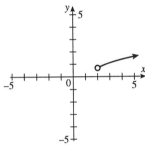

$f(x) = \log(x^2 - 2x) - \log(x - 2)$

$= \log \dfrac{x^2 - 2x}{x - 2}$

$= \log \dfrac{x(x - 2)}{x - 2}$

$= \log x \qquad\qquad x > 2$

**37.** $\ln x$ is the natural logarithm or the logarithm to base $e$.

**39.** The function $f(x) = \ln x$ is an increasing function, concave down, one-to-one, and has a vertical asymptote at $x = 0$.

**41.** $\ln f(x) = \ln \dfrac{2x^2}{x - 4}$

$= \ln 2 + \ln x^2 - \ln(x - 4)$

$= \ln 2 + 2\ln x - \ln(x - 4)$

**43.** $\ln 3 + \ln(x^2 - 1) - \ln(x - 1) = \ln \dfrac{3(x^2 - 1)}{(x - 1)}$

$= \ln \dfrac{3(x - 1)(x + 1)}{x - 1}$

$= \ln 3(x + 1),\ x > 1$

**45.** Domain: $x > \dfrac{2}{3}$

Asymptote: $x = \dfrac{2}{3}$

**47.** **(a)** $e^{\ln(3 - 2x)} = 5x$

$3 - 2x = 5x$

$3 = 7x$

$x = \dfrac{3}{7}$

**(b)** $\ln x - \ln 2 = 1$

$\ln \dfrac{x}{2} = 1$

$\dfrac{x}{2} = e^1$

$x = 2e$

**49.** $\ln(x - 1) + 1 = \ln x$

$\ln(x - 1) - \ln x = -1$

$\ln \dfrac{x - 1}{x} = -1$

$\dfrac{x - 1}{x} = e^{-1}$

$e(x - 1) = x$

$ex - e = x$

$ex - x = e$

$x(e - 1) = e$

$x = \dfrac{e}{e - 1}$

**51.** (i)   (d)
(ii)  (e)
(iii) (f)
(iv)  (a)
(v)   (b)
(vi)  (c)

**53.** $2^x = 132$

$\ln 2^x = \ln 132$

$x \ln 2 = \ln 132$

$x = \dfrac{\ln 132}{\ln 2} = 7.044$

**55.** $5^{x+1} = 145$

$\ln 5^{x+1} = \ln 145$

$(x + 1)\ln 5 = \ln 145$

$x + 1 = \dfrac{\ln 145}{\ln 5}$

$x = \dfrac{\ln 145}{\ln 5} - 1 = 2.092$

**57.** $y = 45e^{(0.06)20}$

$y = 45e^{1.2}$

$y = 149.41$

**59.** $6 = 8e^{k \cdot 30}$

$\dfrac{3}{4} = e^{30k}$

$30k = \ln \dfrac{3}{4}$

$k = \dfrac{1}{30} \ln \dfrac{3}{4}$

$2 = 8e^{\left(\frac{1}{30} \ln \frac{3}{4}\right)t}$

$\dfrac{1}{4} = e^{\frac{t}{30} \ln \frac{3}{4}}$

$\ln \dfrac{1}{4} = \ln e^{\frac{t}{30} \ln \frac{3}{4}}$

$\ln \dfrac{1}{4} = \dfrac{t}{30} \ln \dfrac{3}{4}$

$t = \dfrac{30 \ln \frac{1}{4}}{\ln \frac{3}{4}} = 144.6$ seconds

**61.** $A = Pe^{rt}$

**63.** $A = 5000\left(1 + \dfrac{0.08}{4}\right)^{4 \cdot 3}$

$= 5000(1 + 0.02)^{12}$

$= \$6341.21$

**65.** $10,000 = 5000e^{0.08t}$

$2 = e^{0.08t}$

$0.08t = \ln 2$

$t = \dfrac{\ln 2}{0.08} = 8.664$ years

## Chapter 5 Test: Standard Answer

**1.** (i)   (a)
(ii)  (c)
(iii) (e)
(iv) (h)
(v)  (d)
(vi) (b)

**2.** **(a)** $5^3 = 125$

    **(b)** $\log_{16} 8 = \dfrac{3}{4}$

**3.** **(a)** $81^x = 9$

        $(9^2)^x = 9^1$

        $9^{2x} = 9^1$

        $2x = 1$

        $x = \dfrac{1}{2}$

    **(b)** $e^{\ln(x^2 - x)} = 6$

        $x^2 - x = 6$

        $x^2 - x - 6 = 0$

        $(x - 3)(x + 2) = 0$

        $x - 3 = 0$ or $x + 2 = 0$

        $x = 3$ or $x = -2$

**4.** $\dfrac{1}{2^{x+1}} = 64$

$2^{-(x+1)} = 2^6$

$-x - 1 = 6$

$-x = 7$

$x = -7$

**5.** $80e^{3\ln\left(\frac{1}{2}\right)} = 80e^{\ln\left(\frac{1}{2}\right)^3}$

$= 80 \cdot \left(\dfrac{1}{2}\right)^3 = 80 \cdot \dfrac{1}{8} = 10$

**6. (a)** $\log_b \dfrac{27}{8} = -3$

$b^{-3} = \dfrac{27}{8}$

$(b^{-3})^{-1/3} = \left(\dfrac{27}{8}\right)^{-1/3}$

$b = \dfrac{2}{3}$

**(b)** Let $b = \log_{10} 0.01$. Then

$10^b = 0.01$

$10^b = 10^{-2}$

$b = -2$

**7. (a)** Domain: all real $x$
Range: $y > 0$
**(b)** Increasing for all $x$
**(c)** Concave up for all $x$
**(d)** No $x$-intercept
$y$-intercept: $y = 3^0 = 1$
**(e)** Horizontal asymptote: $y = 0$

**8.** $x = e^{-y}$

$x = \left(\dfrac{1}{e}\right)^y$

$\log_{1/e} x = \log_{1/e} \left(\dfrac{1}{e}\right)^y$

$y = \log_{1/e} x$

$g(x) = \log_{1/e} x$

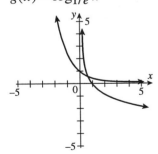

**9.** Domain: all real $x$
Asymptote: $y = -4$

**10.** Domain: all $x > -4$
Vertical asymptote: $x = -4$

**11. (a)** $\log_b x(x^2 + 1)^{10}$

$= \log_b x + \log_b (x^2 + 1)^{10}$

$= \log_b x + 10 \log_b (x^2 + 1)$

**(b)** $\ln \dfrac{x^3}{(x+1)\sqrt{x^2 + 2}}$

$= \ln x^3 - \ln(x+1) - \ln \sqrt{x^2 + 2}$

$= 3 \ln x - \ln(x+1) - \dfrac{1}{2} \ln(x^2 + 2)$

**12. (a)** $\log_7 x + \log_7 2x + \log_7 5$

$= \log_7 (x)(2x)(5) = \log_7 10x^2$

**(b)** $\dfrac{1}{2} \ln x - 2 \ln(x + 2)$

$= \ln \sqrt{x} - \ln(x+2)^2 = \ln \dfrac{\sqrt{x}}{(x+2)^2}$

**13.** $\log_2 75 = \dfrac{\log_{10} 75}{\log_{10} 2} = 6.2288$

**14.** Domain: $x > -2$
Asymptote: $x = -2$
$y$-intercept: $y = \log_2 (0 + 2)$
$= \log_2 2 = 1$
$x$-intercept: $0 = \log_2 (x + 2)$
$2^0 = x + 2$
$x = -1$

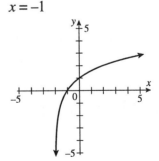

**15.** Domain: all $x$
Asymptote: $y = 0$
$y$-intercept: $y = e^{0+1} = e$

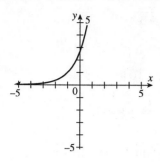

**16.** $\log_{25} x^2 - \log_{25}(2x-5) = \dfrac{1}{2}$

$$\log_{25} \dfrac{x^2}{2x-5} = \dfrac{1}{2}$$

$$\dfrac{x^2}{2x-5} = 25^{1/2}$$

$$x^2 = 5(2x-5)$$

$$x^2 = 10x - 25$$

$$x^2 - 10x + 25 = 0$$

$$(x-5)^2 = 0$$

$$x - 5 = 0$$

$$x = 5$$

**17.** $\log_{10} x + \log_{10}(3x+20) = 2$

$$\log_{10} x(3x+20) = 2$$

$$3x^2 + 20x = 10^2$$

$$3x^2 + 20x - 100 = 0$$

$$(3x-10)(x+10) = 0$$

$$3x - 10 = 0 \text{ or } x + 10 = 0$$

$x = \dfrac{10}{3}$ or $x = -10$, but $x = -10$ is not a

solution, so $x = \dfrac{10}{3}$.

**18.** Shift up 2 units and to the right 1 unit.

**19.** $4^{2x} = 5$

$$\ln 4^{2x} = \ln 5$$

$$2x \ln 4 = \ln 5$$

$$x = \dfrac{\ln 5}{2 \ln 4}$$

$$x = 0.5805$$

**20.** $25 = 50e^{-0.04t}$

$$\dfrac{1}{2} = e^{-0.04t}$$

$$\ln \dfrac{1}{2} = \ln e^{-0.04t}$$

$$\ln \dfrac{1}{2} = -0.04t$$

$$t = \dfrac{\ln \frac{1}{2}}{-0.04}$$

$$t = 17.3 \text{ years}$$

**21.** $A = 2000\left(1 + \dfrac{0.08}{4}\right)^{4 \cdot 6}$

$$= 2000(1 + 0.02)^{24}$$

$$= 2000(1.02)^{24}$$

**22.** $4000 = 2000\left(1 + \dfrac{0.08}{4}\right)^{4t}$

$$2 = (1.02)^{4t}$$

$$\ln 2 = \ln(1.02)^{4t}$$

$$\ln 2 = 4t \ln(1.02)$$

$$t = \dfrac{\ln 2}{4 \ln(1.02)} = 8.75 \text{ years}$$

**23.** $A = 6000e^{(0.07)(10)}$

$$= 6000e^{0.7}$$

$$= \$12{,}083$$

**24.** $y = 120e^{kt}$

$$180 = 120e^{3k}$$

$$\dfrac{3}{2} = e^{3k}$$

$$\ln \dfrac{3}{2} = \ln e^{3k}$$

$$\ln \dfrac{3}{2} = 3k$$

$$k = \dfrac{\ln \frac{3}{2}}{3} = \dfrac{1}{3}\ln \dfrac{3}{2} = 0.1352$$

**25.** $\ln 2 - \ln(1-x) = 1 - \ln(x+1)$

$\ln 2 + \ln(x+1) - \ln(1-x) = 1$

$\ln \dfrac{2(x+1)}{1-x} = 1$

$\dfrac{2x+2}{1-x} = e^1$

$2x + 2 = e(1-x)$

$2x + 2 = e - ex$

$2x + ex = e - 2$

$x(2+e) = e - 2$

$x = \dfrac{e-2}{2+e} = \dfrac{e-2}{e+2}$

## Chapter 5 Test: Multiple Choice

**1.** I is true

II is true

III is also true.

The answer is (d).

**2.** $b^{x^2 - 2x} = 1$

$b^{x^2 - 2x} = b^0$

$x^2 - 2x = 0$

$x(x - 2) = 0$

$x = 0$ or $x = 2$

The answer is (a).

**3.** Since $2^{-1-1} = \dfrac{1}{4}$, $\left(-1, \dfrac{1}{4}\right)$ is on the graph

of $y = f(x) = 2^{x-1}$. The answer is (c).

**4.** The domain is all $x > -3$, so the answer is (e).

**5.** $\log_3 x + \log_3(2x + 51) = 4$

$\log_3 x(2x + 51) = 4$

$2x^2 + 51x = 3^4$

$2x^2 + 51x - 81 = 0$

$(2x - 3)(x + 27) = 0$

$2x - 3 = 0$ or $x + 27 = 0$

$x = \dfrac{3}{2}$ or $x = -27$, but $x = -27$ is not a

solution. The answer is (a).

**6.** $\log_b \dfrac{1}{64} = -\dfrac{3}{2}$

$b^{-3/2} = \dfrac{1}{64}$

$\left(b^{-3/2}\right)^{-2/3} = \left(\dfrac{1}{64}\right)^{-2/3}$

$b = 16$

The answer is (c).

**7.** I is false

II is false

III is also false.

The answer is (e).

**8.** Since all three properties are true, the answer is (d).

**9.** The answer is (b).

**10.** $e^{\ln(2-x)} = 2x$

$2 - x = 2x$

$2 = 3x$

$x = \dfrac{2}{3}$

The answer is (d).

**11.** I is false.

II is false.

III is true.

The answer is (c).

**12.** $\ln(x+1) - 1 = \ln x$

$\ln(x+1) - \ln x = 1$

$\ln \dfrac{x+1}{x} = 1$

$\dfrac{x+1}{x} = e^1$

$x + 1 = ex$

$x - ex = -1$

$x(1 - e) = -1$

$x = -\dfrac{1}{1-e} = \dfrac{1}{e-1}$

The answer is (b).

**13.** $9^{x-1} = 4$

$\ln 9^{x-1} = \ln 4$

$(x-1)\ln 9 = \ln 4$

$x - 1 = \dfrac{\ln 4}{\ln 9}$

$x = \dfrac{\ln 4}{\ln 9} + 1$

The answer is (a).

**14.** $\left(1 + \dfrac{r}{n}\right)^n$ approaches $e^r$, so the answer is (a).

**15.** The answer is (c).

**16.** $y = \log_2 4x = \log_2 4 + \log_2 x = 2 + \log_2 x$, so translate two units up. The answer is (b).

**17.** The answer is (c).

**18.** $30 = 40e^{8k}$

$\dfrac{3}{4} = e^{8k}$

$\ln \dfrac{3}{4} = \ln e^{8k}$

$\ln \dfrac{3}{4} = 8k$

$k = \dfrac{1}{8} \ln \dfrac{3}{4}$

$y = 40e^{\left(\frac{1}{8} \ln \frac{3}{4}\right)x} = 40e^{(x \ln 0.75)/8}$

The answer is (b).

**19.** $A_5 = 3000\left(1 + \dfrac{0.084}{12}\right)^{12 \cdot 5}$

$= 3000(1 + 0.007)^{60}$

$= 3000(1.007)^{60}$

The answer is (c).

**20.** $\log_b Q = \log_b \dfrac{(\sqrt[5]{409})(0.0058)}{7.29}$

$= \log_b \sqrt[5]{409} + \log_b 0.0058 - \log_b 7.29$

$= \dfrac{1}{5} \log_b 409 + \log_b 0.0058 - \log_b 7.29$

The answer is (d).

# Chapter 6: The Trigonometric Functions

**Exercises 6.1**

**1.** $45° = \left(45 \cdot \dfrac{\pi}{180}\right) = \dfrac{\pi}{4}$

**3.** $90° = \left(90 \cdot \dfrac{\pi}{180}\right) = \dfrac{\pi}{2}$

**5.** $270° = \left(270 \cdot \dfrac{\pi}{180}\right) = \dfrac{3\pi}{2}$

**7.** $150° = \left(150 \cdot \dfrac{\pi}{180}\right) = \dfrac{5\pi}{6}$

**9.** $225° = \left(225 \cdot \dfrac{\pi}{180}\right) = \dfrac{5\pi}{4}$

**11.** $210° = \left(210 \cdot \dfrac{\pi}{180}\right) = \dfrac{7\pi}{6}$

**13.** $330° = \left(330 \cdot \dfrac{\pi}{180}\right) = \dfrac{11\pi}{6}$

**15.** $75° = \left(75 \cdot \dfrac{\pi}{180}\right) = \dfrac{5\pi}{12}$

**17.** $100° = \left(100 \cdot \dfrac{\pi}{180}\right) = \dfrac{5\pi}{9} = 1.75$

**19.** $340° = \left(340 \cdot \dfrac{\pi}{180}\right) = \dfrac{17\pi}{9} = 5.93$

**21.** $\pi = \left(\pi \cdot \dfrac{180}{\pi}\right)° = 180°$

**23.** $2\pi = \left(2\pi \cdot \dfrac{180}{\pi}\right)° = 360°$

**25.** $\dfrac{5\pi}{9} = \left(\dfrac{5\pi}{9} \cdot \dfrac{180}{\pi}\right)° = 100°$

**27.** $\dfrac{2\pi}{3} = \left(\dfrac{2\pi}{3} \cdot \dfrac{180}{\pi}\right)° = 120°$

**29.** $\dfrac{5\pi}{4} = \left(\dfrac{5\pi}{4} \cdot \dfrac{180}{\pi}\right)° = 225°$

**31.** $\dfrac{5\pi}{3} = \left(\dfrac{5\pi}{3} \cdot \dfrac{180}{\pi}\right)° = 300°$

**33.** $\dfrac{5\pi}{18} = \left(\dfrac{5\pi}{18} \cdot \dfrac{180}{\pi}\right)° = 50°$

**35.** $\dfrac{\pi}{15} = \left(\dfrac{\pi}{15} \cdot \dfrac{180}{\pi}\right)° = 12° = 12.0°$

**37.** $3 = \left(3 \cdot \dfrac{180}{\pi}\right)° = \left(\dfrac{540}{\pi}\right)° = 171.9°$

**39.** $s = r\theta = 8 \cdot \dfrac{\pi}{4} = 2\pi$

**41.** $\theta = \dfrac{s}{r} = \dfrac{\frac{\pi}{2}}{1} = \dfrac{\pi}{2}$

**43.** $\theta = \dfrac{s}{r} = \dfrac{8\pi}{6} = \dfrac{4\pi}{3}$

**45.** $\theta = 30° = \left(30 \cdot \dfrac{\pi}{180}\right) = \dfrac{\pi}{6}$

$A = \dfrac{1}{2}(12)^2 \cdot \dfrac{\pi}{6} = 12\pi$ square centimeters

**47.** $\theta = 135° = \left(135 \cdot \dfrac{\pi}{180}\right) = \dfrac{3\pi}{4}$

$A = \dfrac{1}{2}(12)^2 \cdot \dfrac{3\pi}{4} = 54\pi$ square centimeters

**49.** $\theta = 315° = \left(315 \cdot \dfrac{\pi}{180}\right) = \dfrac{7\pi}{4}$

$A = \dfrac{1}{2}(12)^2 \cdot \dfrac{7\pi}{4}$
$= 126\pi$ square centimeters

**51.** $\theta = \dfrac{s}{r} = \dfrac{8}{2} = 4$

$A = \dfrac{1}{2}(2)^2 \cdot 4 = 8$ square inches

**53.** $45° = \left(45 \cdot \dfrac{\pi}{180}\right) = \dfrac{\pi}{4}$

$r = \dfrac{s}{\theta} = \dfrac{\frac{\pi}{2}}{\frac{\pi}{4}} = 2$

$A = \dfrac{1}{2}(2)^2 \cdot \dfrac{\pi}{4} = \dfrac{\pi}{2}$ square centimeters

$A = 1.6$ square centimeters

**55.** $\theta = \dfrac{50}{250} = \dfrac{1}{5}$

$\theta = \dfrac{1}{5} = \left(\dfrac{1}{5} \cdot \dfrac{180}{\pi}\right)° = \left(\dfrac{36}{\pi}\right)° = 11°$

**57.** Surface Area $= \dfrac{1}{2}(8)^2 \cdot \dfrac{3\pi}{2} = 48\pi$

$= 150.8$ square inches

**59.** At point $C$, the radius is 15 ft with an angle of $300°$ or $\dfrac{5\pi}{3}$ radians. At points $A$ and $B$, the radius is 5 ft with an angle of $120°$ or $\dfrac{2\pi}{3}$ radians. The amount of grazing area is

$\dfrac{1}{2}(15)^2 \cdot \dfrac{5\pi}{3} + \dfrac{1}{2}(5)^2 \cdot \dfrac{2\pi}{3} + \dfrac{1}{2}(5)^2 \dfrac{2\pi}{3}$

$= \dfrac{1125\pi}{6} + \dfrac{50\pi}{6} + \dfrac{50\pi}{6}$

$= \dfrac{1225\pi}{6}$ square feet

**61. (a)** $r = 4300$ miles

$s = 4300 \cdot 2\pi = 27{,}018$ miles

$t = \dfrac{7}{4}$ hr

$v = \dfrac{27{,}018}{\frac{7}{4}} = 15{,}439$ mph

$15{,}439$ mph $= \dfrac{15{,}439(5280)}{3600}$

$= 22{,}644$ ft/sec

**(b)** $w = \dfrac{2\pi}{\frac{7}{4}} = \dfrac{8\pi}{7}$ rad/hr

**63. (a)** $w = \dfrac{2\pi}{60} = \dfrac{\pi}{30}$ rad/min

**(b)** $w = \dfrac{2\pi}{1} = 2\pi$ rad/hr

**65. (a)** $r = 11$ inches $= \dfrac{11}{12}$ ft

$w = 240 \cdot 2\pi = 480\pi$ rad/min

$v = r \cdot w = \dfrac{11}{12} \cdot 480\pi = 440\pi$ ft/min

**(b)** $\dfrac{5280}{440\pi} = 3.8$ minutes

**67.** $r = \dfrac{1}{4}$ mile, $v = 42$ mph

$w = \dfrac{v}{r} = \dfrac{42}{\frac{1}{4}} = 168$ rad/hr

In $\dfrac{1}{2}$ minute (which is a $\dfrac{1}{120}$ hour):

$\dfrac{1}{120}(168) = 1.4$ rad

**Challenge**

First, drop a perpendicular bisector down from $M$, and call the point where it crosses $AC$, point $N$. Since $M$ is the midpoint of $AB$, and $B$ is directly above $C$, $N$ will be the midpoint of $AC$. So, $AN$ is congruent to $NC$, $MN$ is congruent to itself, and since $MN$ is perpendicular to $AC$, angles $\angle ANM$ and $\angle CNM$ are right angles. Therefore, triangles $\triangle ANM$ and $\triangle CNM$ are congruent, which means that $\angle MCA$ is $60°$ and so $\angle AMC$ must also be $60°$. This means $\angle BMC$ is $120°$ or $\dfrac{2\pi}{3}$ radians. To find the radius, notice that $\triangle AMC$ is an equilateral triangle, so $AC$ has length equal to $AM$, which is half of $AB = 1$. Draw a perpendicular bisector from $M$ to $BC$; call it $R$. Now show $\triangle BMR$ is similar to $\triangle ABC$. Therefore,

$\dfrac{AB}{MB} = \dfrac{AC}{MR} = \dfrac{1}{\frac{1}{2}} = \dfrac{\frac{1}{2}}{MR}$, so $MR$ (the radius)

$= \dfrac{1}{4}$. This is also the radius of the circle since $BC$ is a vertical tangent to the circle. Hence, the area of the shaded region is $\dfrac{1}{2}\left(\dfrac{1}{4}\right)^2 \dfrac{2\pi}{3} = \dfrac{\pi}{48}$.

**Exercises 6.2**

1.  $\sin\dfrac{\pi}{4} = \dfrac{1}{\sqrt{2}}$

3.  $\cos\dfrac{\pi}{3} = \dfrac{1}{2}$

5.  $\tan\dfrac{\pi}{6} = \dfrac{1}{\sqrt{3}}$

7.  $\dfrac{2\pi}{3}$ is in quadrant II.

    $\phi = \dfrac{\pi}{3}$

9.  $\dfrac{5\pi}{4}$ is in quadrant III.

    $\phi = \dfrac{\pi}{4}$

11.  $-\dfrac{13\pi}{3}$ is in quadrant IV.

    $\phi = \dfrac{\pi}{3}$

13.  $\theta = -\dfrac{7\pi}{6}$ is in quadrant II.

    coterminal angle: $\dfrac{5\pi}{6}$

15.  $\theta = \dfrac{11\pi}{3}$ is in quadrant IV.

    coterminal angle: $\dfrac{5\pi}{3}$ or $-\dfrac{\pi}{3}$

17.  $\theta = -\dfrac{17\pi}{4}$ is in quadrant IV.

    coterminal angle: $\dfrac{7\pi}{4}$

19.

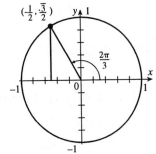

Reference angle: $\dfrac{\pi}{3}$

$\sin\dfrac{2\pi}{3} = \dfrac{\sqrt{3}}{2}$ $\qquad$ $\csc\dfrac{2\pi}{3} = \dfrac{2}{\sqrt{3}}$

$\cos\dfrac{2\pi}{3} = -\dfrac{1}{2}$ $\qquad$ $\sec\dfrac{2\pi}{3} = -2$

$\tan\dfrac{2\pi}{3} = -\sqrt{3}$ $\qquad$ $\cot\dfrac{2\pi}{3} = -\dfrac{1}{\sqrt{3}}$

21.

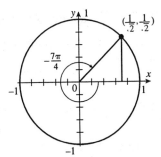

Reference angle: $\dfrac{\pi}{4}$

$\sin\left(-\dfrac{7\pi}{4}\right) = \dfrac{1}{\sqrt{2}}$ $\qquad$ $\csc\left(-\dfrac{7\pi}{4}\right) = \sqrt{2}$

$\cos\left(-\dfrac{7\pi}{4}\right) = \dfrac{1}{\sqrt{2}}$ $\qquad$ $\sec\left(-\dfrac{7\pi}{4}\right) = \sqrt{2}$

$\tan\left(-\dfrac{7\pi}{4}\right) = 1$ $\qquad$ $\cot\left(-\dfrac{7\pi}{4}\right) = 1$

**23.**

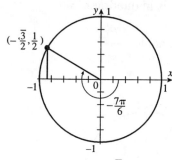

Reference angle: $\dfrac{\pi}{6}$

$$\sin\left(-\dfrac{7\pi}{6}\right)=\dfrac{1}{2} \qquad \csc\left(-\dfrac{7\pi}{6}\right)=2$$

$$\cos\left(-\dfrac{7\pi}{6}\right)=-\dfrac{\sqrt{3}}{2} \qquad \sec\left(-\dfrac{7\pi}{6}\right)=-\dfrac{2}{\sqrt{3}}$$

$$\tan\left(-\dfrac{7\pi}{6}\right)=-\dfrac{1}{\sqrt{3}} \qquad \cot\left(-\dfrac{7\pi}{6}\right)=-\sqrt{3}$$

**25.** $P(0, -1)$

$\sin\theta = -1 \qquad\qquad \csc\theta = -1$

$\cos\theta = 0 \qquad\qquad \sec\theta = \text{undefined}$

$\tan\theta = \text{undefined} \quad \cot\theta = 0$

**27.** $P(0, 1)$

$\sin\theta = 1 \qquad\qquad \csc\theta = 1$

$\cos\theta = 0 \qquad\qquad \sec\theta = \text{undefined}$

$\tan\theta = \text{undefined} \quad \cot\theta = 0$

**29.** $\tan\dfrac{3\pi}{4} = -1$

**31.** $\sin\dfrac{4\pi}{3} = -\dfrac{\sqrt{3}}{2}$

**33.** $\csc(-\pi) = \text{undefined}$

**35.** $\cot\left(\dfrac{7\pi}{2}\right) = 0$

**37.** $\cot(0) = \text{undefined}$

**39.** $\tan 8\pi = 0$

**41.** $\cos 0.23 = 0.9737$

**43.** $\cot 0.95 = 0.7151$

**45.** $\cos\dfrac{19\pi}{4} = -\dfrac{1}{\sqrt{2}}$

**47.** $\tan 1.48 = 10.9834$

**49.** $\tan 220° = 0.8391$

**51.** $\sin 261° = -0.9877$

**53.** $\cot 1200° = -0.5774$

**55.** $\cos(-792.5°) = 0.3007$

**57.** $\sin\theta = 0.4894 \qquad \csc\theta = 2.0434$

$\cos\theta = 0.8721 \qquad \sec\theta = 1.1467$

$\tan\theta = 0.5612 \qquad \cot\theta = 1.7820$

**59.** $\sin\theta = 0.4571 \qquad \csc\theta = 2.1877$

$\cos\theta = -0.8894 \qquad \sec\theta = -1.1243$

$\tan\theta = -0.5139 \qquad \cot\theta = 1.9458$

**61.** $P_1\left(\dfrac{\sqrt{3}}{2}, \dfrac{1}{2}\right); \ P_2\left(\dfrac{1}{\sqrt{2}}, \dfrac{1}{\sqrt{2}}\right);$

$P_3\left(\dfrac{1}{2}, \dfrac{\sqrt{3}}{2}\right); \ P_4\left(-\dfrac{\sqrt{3}}{2}, -\dfrac{1}{2}\right);$

$P_5\left(\dfrac{-1}{\sqrt{2}}, \dfrac{-1}{\sqrt{2}}\right); \ P_6\left(-\dfrac{1}{2}, -\dfrac{\sqrt{3}}{2}\right);$

**63.** (a) $\left(\dfrac{2}{3}\right)^2 + \left(\dfrac{\sqrt{5}}{3}\right)^2 = \dfrac{4}{9} + \dfrac{5}{9} = \dfrac{9}{9} = 1$

(b)

(c) $\sin\theta = \dfrac{\sqrt{5}}{3}; \ \cos\theta = \dfrac{2}{3}; \ \tan\theta = \dfrac{\sqrt{5}}{2}$

$\cot\theta = \dfrac{2}{\sqrt{5}}; \ \sec\theta = \dfrac{3}{2}; \ \csc\theta = \dfrac{3}{\sqrt{5}}$

**65.** $\left(\dfrac{\sqrt{3}}{4},\ y\right)$ in quadrant IV, so $y < 0$.

$$\left(\dfrac{\sqrt{3}}{4}\right)^2 + y^2 = 1$$

$$\dfrac{3}{16} + y^2 = 1$$

$$y^2 = \dfrac{13}{16}$$

$$y = -\dfrac{\sqrt{13}}{4} = -0.9014$$

$$\tan\theta = \dfrac{-0.9014}{\dfrac{\sqrt{3}}{4}} = -2.0817$$

**67.** $\sin\theta = 1$

$$\theta = \dfrac{\pi}{2}$$

**69.** $\sin\theta = 0$

$$\theta = 0,\ \pi$$

**71.** $\sin\theta = \dfrac{1}{2}$

$$\theta = \dfrac{\pi}{6},\ \dfrac{5\pi}{6}$$

**73.** $\cos\theta = -\dfrac{\sqrt{2}}{2}$

$$\theta = \dfrac{3\pi}{4},\ \dfrac{5\pi}{4}$$

**75.** $\tan\theta = 1$; $\sin\theta = -\dfrac{\sqrt{2}}{2}$; $\theta$ is in quadrant III.

$$\cos\theta = \dfrac{\sin\theta}{\tan\theta} = \dfrac{-\dfrac{\sqrt{2}}{2}}{1} = -\dfrac{\sqrt{2}}{2}$$

$$\cot\theta = \dfrac{1}{\tan\theta} = 1$$

$$\sec\theta = \dfrac{1}{\cos\theta} = -\dfrac{2}{\sqrt{2}} = -\sqrt{2}$$

$$\csc\theta = \dfrac{1}{\sin\theta} = -\dfrac{2}{\sqrt{2}} = -\sqrt{2}$$

**77.** $\cot\theta = -\sqrt{3}$; $\sin\theta = -\dfrac{1}{2}$; $\theta$ is in quadrant IV.

$$\cos\theta = \cot\theta\cdot\sin\theta = \dfrac{\sqrt{3}}{2}$$

$$\tan\theta = \dfrac{1}{\cot\theta} = -\dfrac{1}{\sqrt{3}} = -\dfrac{\sqrt{3}}{3}$$

$$\sec\theta = \dfrac{1}{\cos\theta} = \dfrac{2}{\sqrt{3}} = \dfrac{2\sqrt{3}}{3}$$

$$\csc\theta = \dfrac{1}{\sin\theta} = -2$$

**79.** The point on the terminal side of $\theta$ is $P\left(x,\ \dfrac{1}{3}\right)$.

$$x^2 + \left(\dfrac{1}{3}\right)^2 = 1$$

$$x^2 = -\dfrac{1}{9} + 1 = \dfrac{8}{9}$$

$$x = \pm\dfrac{2\sqrt{2}}{3}$$

Since $\theta$ is in quadrant II, $x = -\dfrac{2\sqrt{2}}{3}$.

$$\cos\theta = -\dfrac{2\sqrt{2}}{3}$$

**81.** $\csc\theta = 4$, so $\sin\theta = \dfrac{1}{4}$. The point on the terminal side of $\theta$ is $P\left(x,\ \dfrac{1}{4}\right)$.

$$x^2 + \left(\dfrac{1}{4}\right)^2 = 1$$

$$x^2 = 1 - \dfrac{1}{16} = \dfrac{15}{16}$$

$$x = \dfrac{\sqrt{15}}{4},\ \text{since }\theta\text{ is in quadrant I.}$$

$$\sec\theta = \dfrac{1}{x} = \dfrac{1}{\dfrac{\sqrt{15}}{4}} = \dfrac{4}{\sqrt{15}} = \dfrac{4\sqrt{15}}{15}$$

**83.** $AB = \dfrac{AB}{1} = \dfrac{AB}{OA} = \dfrac{y}{x} = \tan\theta$

$\sec\theta = \dfrac{1}{x} = \dfrac{OP}{x} = \dfrac{OB}{OA} = OB$

## Challenge

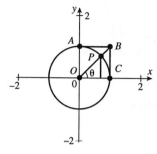

In the figure, $AB$ is tangent to the circle at $A$ and meets the terminal side of angle $\theta$ radians at $B$. Then $C$ is found by dropping a line from $B$ to the $x$-axis. Thus,

$\cot\theta = \dfrac{x}{y} = \dfrac{OC}{BC} = \dfrac{OC}{1} = OC$

$\csc\theta = \dfrac{1}{y} = \dfrac{OP}{y} = \dfrac{OB}{BC} = \dfrac{OB}{1} = OB$.

## Exercises 6.3

**1.**

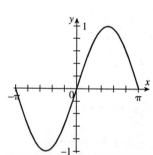

| $x$ | $-\pi$ | $-\dfrac{5\pi}{6}$ | $-\dfrac{2\pi}{3}$ | $-\dfrac{\pi}{2}$ | $-\dfrac{\pi}{3}$ | $-\dfrac{\pi}{6}$ | $0$ |
|---|---|---|---|---|---|---|---|
| $y = \sin x$ | $0$ | $-\dfrac{1}{2}$ | $-\dfrac{\sqrt{3}}{2}$ | $-1$ | $-\dfrac{\sqrt{3}}{2}$ | $-\dfrac{1}{2}$ | $0$ |

## Critical Thinking

**1.** 400 gradients $= 360°$

So, $1° = \dfrac{400}{360} = \dfrac{10}{9}$ grad

and 1 grad $= \dfrac{360}{400} = \left(\dfrac{9}{10}\right)°$

$2\pi$ radians $= 400$ gradients

So, 1 rad $= \dfrac{400}{2\pi} = \dfrac{200}{\pi}$ grad

1 grad $= \dfrac{2\pi}{400} = \dfrac{\pi}{200}$ rad

**3.** When $\theta = \dfrac{\pi}{2}$, the line extending $OP$ will be parallel to the tangent line of the circle at $A$, so they won't intersect. In other words, there will be no point $B$, so no line segment $AB$, and hence no tangent value.

**5.** For $0 < \theta < \dfrac{\pi}{2}$, $\sin\theta$ and $\cos\theta$ are both positive, and the tangent is positive. For $\pi < \theta < \dfrac{3\pi}{2}$, $\sin\theta$ and $\cos\theta$ are both negative, but the tangent is still positive. This is because the coordinates of the points in the third quadrant are the negative of the coordinates of the points in the first quadrant.

**3.** Shift the graph of $f$ to the right $\dfrac{\pi}{2}$ units.

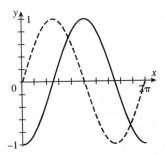

**5.** Shift the graph of $f$ to the right $\dfrac{\pi}{3}$ units.

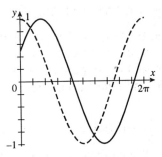

**7.** Shift the graph of $f$ to the left $\pi$ units and multiply the ordinates by 2.

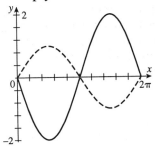

**9.**

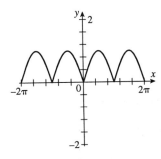

**11.**

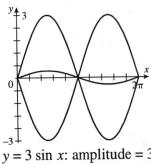

$y = 3 \sin x$: amplitude $= 3$

$y = \dfrac{1}{3} \sin x$: amplitude $= \dfrac{1}{3}$

$y = -3 \sin x$: amplitude $= 3$

**13.**

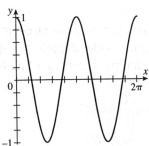

$y = \cos 2x$: amplitude $= 1$

$$\text{period} = \dfrac{2\pi}{2} = \pi$$

**15.**

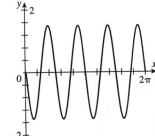

$y = -\dfrac{3}{2} \sin 4x$: amplitude $= \dfrac{3}{2}$

$$\text{period} = \dfrac{2\pi}{4} = \dfrac{\pi}{2}$$

**17.**

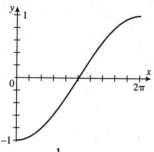

$y = -\cos \dfrac{1}{2}x$: amplitude = 1

$$\text{period} = \dfrac{2\pi}{\frac{1}{2}} = 4\pi$$

**19.** $\text{period} = \dfrac{2\pi}{\frac{1}{4}} = 8\pi$

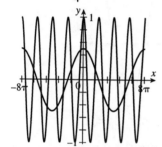

**21.** $\text{period} = \dfrac{2\pi}{\pi} = 2$

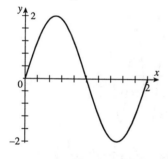

**23.** $a = 1,\ b = 3$
$y = \sin 3x$

**25.** $a = -3,\ b = \dfrac{1}{2}$
$y = -3 \sin \dfrac{1}{2}x$

**27.** amplitude = 5
$\text{period} = \dfrac{2\pi}{3}$
$\text{phase shift} = \dfrac{\pi}{6}$

**29.** amplitude = 2
$\text{period} = \dfrac{2\pi}{2} = \pi$
$\text{phase shift} = -\dfrac{\pi}{2}$

**31.** amplitude = $\dfrac{3}{2}$
$\text{period} = \dfrac{2\pi}{\frac{1}{4}} = 8\pi$
$\text{phase shift} = \dfrac{1}{\frac{1}{4}} = 4$

**33.** $y = \sin(4x - \pi) = \sin 4\left(x - \dfrac{\pi}{4}\right)$
amplitude = 1
$\text{period} = \dfrac{2\pi}{4} = \dfrac{\pi}{2}$
$\text{phase shift} = \dfrac{\pi}{4}$

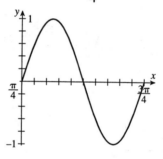

**35.** $y = 2\cos\left(2x - \dfrac{\pi}{2}\right) = 2\cos 2\left(x - \dfrac{\pi}{4}\right)$
amplitude = 2
$\text{period} = \dfrac{2\pi}{2} = \pi$
$\text{phase shift} = \dfrac{\pi}{4}$

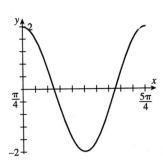

**41.**

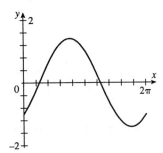

**37.** $y = -\dfrac{5}{2}\sin\left(\dfrac{x}{2} + \dfrac{\pi}{4}\right) = -\dfrac{5}{2}\sin\dfrac{1}{2}\left(x + \dfrac{\pi}{2}\right)$

amplitude $= \dfrac{5}{2}$

period $= \dfrac{2\pi}{\frac{1}{2}} = 4\pi$

phase shift $= -\dfrac{\pi}{2}$

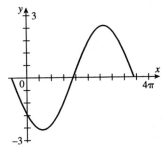

**43.**

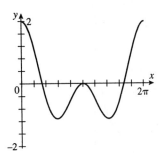

**45.**

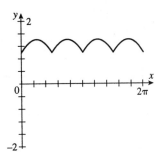

**39.** Shift the graph of $y = \sin\left(\dfrac{x}{2} + \dfrac{\pi}{2}\right)$ three units upward.

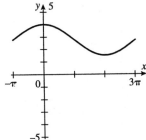

**47.** $(f \circ g)(x) = f(g(x)) = f(\cos x)$
$= 2\cos x + 5$
$(g \circ f)(x) = g(f(x)) = g(2x + 5)$
$= \cos(2x + 5)$

**49.** Let $g(x) = 5x^2, f(x) = \cos x$. Then
$h(x) = (f \circ g)(x) = f(g(x)) = f(5x^2)$
$= \cos(5x^2)$

**51.** Let $h(x) = 1 - 2x, \ g(x) = \sqrt[3]{x}$, and
$f(x) = \cos x$. Then
$F(x) = (f \circ g \circ h)(x) = f(g(h(x)))$
$= f(g(1 - 2x)) = f\left(\sqrt[3]{1 - 2x}\right) = \cos\sqrt[3]{1 - 2x}$

**53.** $f(x+2p) = f((x+p)+p) = f(x+p) = f(x)$

## Challenge

We need to find $g(x) = a\cos(bx + c)$ such that $g(x) = -f(x)$. $g(x)$ and $f(x)$ must have the same period and amplitude, so $|a| = 3$ and $b = 2$. To find $c$, notice that $\cos x$ is the function $\sin x$ shifted left by $\frac{\pi}{2}$ units. Since the period is $\pi$, we need to shift $g(x)$ by $\frac{\pi}{4}$ units to the left. Therefore,

$$g(x) = -3\cos\left(2\left(x - \frac{\pi}{4}\right) - \frac{\pi}{8}\right) = -3\cos\left(2x - \frac{5\pi}{8}\right).$$

## Exercises 6.4

**1.**

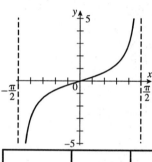

| $x$ | $-1.4$ | $-1.3$ | $-\dfrac{\pi}{3}$ | $-\dfrac{\pi}{4}$ | $-\dfrac{\pi}{6}$ | $0$ |
|---|---|---|---|---|---|---|
| $y = \tan x$ | $-5.8$ | $-3.6$ | $-\sqrt{3}$ | $-1$ | $-\dfrac{\sqrt{3}}{3}$ | $0$ |

**3. (a)** $\cot(-x) = \dfrac{\cos(-x)}{\sin(-x)} = \dfrac{\cos x}{-\sin x} = -\cot x$

**(b)** $\csc(-x) = \dfrac{1}{\sin(-x)} = \dfrac{1}{-\sin x} = -\csc x$

**(c)** $\csc(x + 2\pi) = \dfrac{1}{\sin(x + 2\pi)}$

$= \dfrac{1}{\sin x} = \csc x$

**5.** Shift the graph of $g$ to the left $\dfrac{\pi}{2}$ units.

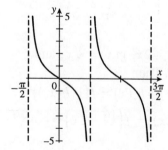

**7.** Shift the graph of $g$ to the right $\dfrac{\pi}{3}$ units.

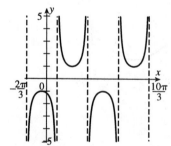

**9.** $y = \cot 3x$

period $= \dfrac{\pi}{3}$

vertical asymptotes: $x = \dfrac{k\pi}{3}$, $k$ an integer

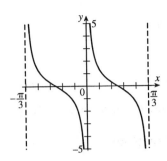

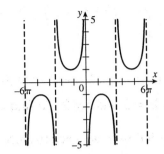

**11.** $y = -2 \cot \dfrac{x}{2}$

period $= \dfrac{\pi}{\frac{1}{2}} = 2\pi$

vertical asymptotes: $x = 2k\pi$, $k$ an integer

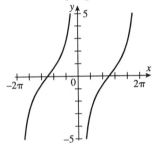

**13.** $y = \sec 4x$

period $= \dfrac{2\pi}{4} = \dfrac{\pi}{2}$

vertical asymptotes: $x = \dfrac{\pi}{8} + \dfrac{k\pi}{4}$,

$k$ an integer

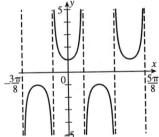

**15.** $y = -\csc \dfrac{1}{3}x$

period $= \dfrac{2\pi}{\frac{1}{3}} = 6\pi$

vertical asymptote: $x = 3k\pi$, $k$ an integer

**17.** $y = 2 \sec \dfrac{3x}{2}$

period $= \dfrac{2\pi}{\frac{3}{2}} = \dfrac{4\pi}{3}$

vertical asymptotes: $x = \dfrac{\pi}{3} + \dfrac{2k\pi}{3}$,

$k$ an integer

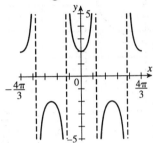

**19.** $y = \dfrac{1}{2} \cot 2\left(x + \dfrac{\pi}{4}\right)$

period $= \dfrac{\pi}{2}$

phase shift $= -\dfrac{\pi}{4}$

**21.** $y = -\csc\left(\dfrac{x}{2} - \dfrac{\pi}{4}\right) = -\csc\dfrac{1}{2}\left(x - \dfrac{\pi}{2}\right)$

$\text{period} = \dfrac{2\pi}{\frac{1}{2}} = 4\pi$

$\text{phase shift} = \dfrac{\pi}{2}$

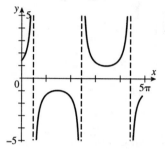

**23.** $a = 1$, $b = \dfrac{2}{3}$, $h = -\dfrac{\pi}{4}$

**25.** $a = \dfrac{1}{2}$, $b = 3$, $h = \dfrac{\pi}{6}$

**27.**

| $x$ | 0.5 | 0.1 | 0.01 | 0.001 | 0.0001 | 0.00001 |
|---|---|---|---|---|---|---|
| $\csc x$ | 2 | 10 | 100 | 1000 | 10,000 | 100,000 |

$\csc x \to \infty$ as $x \to 0^{+}$

**29.**

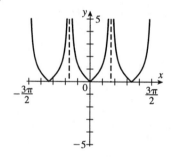

**31.** $(f \circ g)(x) = f\big(g(x)\big) = f(\tan x) = \tan^2 x$

$(g \circ f)(x) = g\big(f(x)\big) = g\left(x^2\right) = \tan x^2$

**33.** $(f \circ g \circ h)(x) = f\big(g(h(x))\big) = f\big(g(\sec x)\big) = f\left(\sqrt{\sec x}\right) = e^{\sqrt{\sec x}}$

$(g \circ f \circ h)(x) = g\big(f(h(x))\big) = g\big(f(\sec x)\big) = g\left(e^{\sec x}\right) = \sqrt{e^{\sec x}}$

$(h \circ g \circ f)(x) = h\big(g(f(x))\big) = h\left(g\left(e^x\right)\right) = h\left(\sqrt{e^x}\right) = \sec\sqrt{e^x}$

**35.** Let $f(x) = \sqrt{x}$, $g(x) = \tan x$, $h(x) = 2x + 1$. Then $F(x) = (f \circ g \circ h)(x) = f\big(g(h(x))\big)$

$= f\big(g(2x + 1)\big) = f\big(\tan(2x + 1)\big) = \sqrt{\tan(2x + 1)}$

**37.** $0 < QP = \sin\theta < \overset{\frown}{AP} = \theta\,AB$

Divide by $\theta$ to get $0 < \dfrac{\sin\theta}{\theta} < 1$

**39.**

| $\theta$ | 1 | 0.5 | 0.25 | 0.1 | 0.01 |
|---|---|---|---|---|---|
| $\dfrac{\sin\theta}{\theta}$ | 0.8415 | 0.9589 | 0.9896 | 0.9983 | 0.999983 |

$\dfrac{\sin\theta}{\theta}$ appears to be getting close to 1 as $\theta$ approaches 0.

**Challenge**

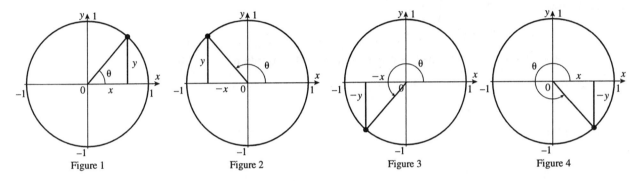

Figure 1    Figure 2    Figure 3    Figure 4

In figure 1, $\cot\theta = \dfrac{x}{y}$. The other angle in the triangle is $\dfrac{\pi}{2} - \theta$, and $\tan\left(\dfrac{\pi}{2} - \theta\right) = \dfrac{x}{y}$. Then

$$\cot\theta = \frac{x}{y} = \tan\left(\frac{\pi}{2} - \theta\right) = -\tan\left(\theta - \frac{\pi}{2}\right)$$

In figure 2, the angle made with the negative $x$-axis is $\pi - \theta$, so the other angle in the triangle is

$\dfrac{\pi}{2} - (\pi - \theta) = \theta - \dfrac{\pi}{2}$. Hence

$$\cot\theta = -\frac{x}{y} = -\tan\left(\theta - \frac{\pi}{2}\right)$$

In figure 3, the angle made with the negative $x$-axis is $\theta - \pi$, so the other angle in the triangle is

$\dfrac{\pi}{2} - (\theta - \pi) = \dfrac{3\pi}{2} - \theta$. Hence

$$\cot\theta = \frac{-x}{-y} = \tan\left(\frac{3\pi}{2} - \theta\right) = \tan\left(\pi + \frac{\pi}{2} - \theta\right) = \tan\left(\frac{\pi}{2} - \theta\right) = -\tan\left(\theta - \frac{\pi}{2}\right)$$

In figure 4, the angle made with the positive $x$-axis is $2\pi - \theta$, so the other angle in the triangle is

$\dfrac{\pi}{2} - (2\pi - \theta) = \theta - \dfrac{3\pi}{2}$.

Hence

$$\cot\theta = \frac{x}{-y} = -\frac{x}{y} = -\tan\left(\theta - \frac{3\pi}{2}\right) = -\tan\left(\theta - \frac{\pi}{2} - \pi\right) = -\tan\left(\theta - \frac{\pi}{2}\right).$$

We don't need to consider the cases $\theta = 0,\ \frac{\pi}{2},\ \pi,\ \frac{3\pi}{2}$ since cot $\theta = 0,\ \pi$ and tan $\theta$ is undefined for $\theta = \frac{\pi}{2}, \frac{3\pi}{2}$.

## Critical Thinking

1. The sine curve on $\left[\frac{\pi}{2},\ \pi\right]$ is a reflection of the sine curve on $\left[0,\ \frac{\pi}{2}\right]$ about the line $x = \frac{\pi}{2}$. Also, the curve on $[\pi, 2\pi]$ is a reflection of the curve on $[0, \pi]$ about the $x$-axis.

3. amplitude $= 1$, period $= \frac{1}{2}\left(\frac{3\pi}{4}\right) = \frac{3\pi}{8}$, phase shift $= 0.$ $y = \sin\frac{16x}{3}$

5. True, since for $-\frac{\pi}{2} < x \le 0$, $|x| = -x$ and $|\tan x| = -\tan x = \tan(-x) = \tan|x|$.

## Exercises 6.5

1. $\arcsin 0 = 0$

3. $\arcsin (-1) = -\frac{\pi}{2}$

5. $\arctan (-1) = -\frac{\pi}{4}$

7. $\arccos\left(-\frac{1}{\sqrt{2}}\right) = \frac{3\pi}{4}$

9. $\arctan 115 = 1.5621$

11. $\arcsin (.5562) = 0.5898$

13. $\arccos (-0.6137) = 2.2315$

15. $\arcsin (-1.0436) = $ undefined

17. $\arctan\frac{7}{11} = 0.5667$

19. $\tan\left(\arctan\frac{x}{2}\right) = \frac{x}{2}$

21. $\cos(\arcsin 0.6208) = 0.7840$

23. $\sin\left(\arctan\frac{1}{12}\right) = 0.0830$

25. $\cos[\arcsin(-1)] = \cos\left(-\frac{\pi}{2}\right) = 0$

27. $\sin\left(\arcsin\frac{\sqrt{3}}{2}\right) = \sin\left(\frac{\pi}{3}\right) = \frac{\sqrt{3}}{2}$

29. $\sin\left(\arccos\frac{13}{12}\right) = $ undefined

31. $\arccos\left(\tan\frac{\pi}{4}\right) = \arccos(1) = 0$

33. $\sin(\arcsin x) = x$ for $-1 \le x \le 1$ since sin and arcsin are inverse functions for the domain $-1 \le x \le 1$.

35. $y = 2 \arcsin x$
    Domain: $-1 \le x \le 1$
    Range: $-\pi \le y \le \pi$

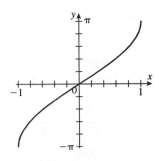

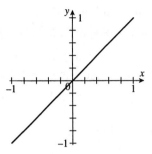

**37.** $y = 2 + \arctan x$
Domain: all real $x$

Range: $2 - \dfrac{\pi}{2} < y < 2 + \dfrac{\pi}{2}$

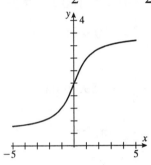

**39.** $y = \dfrac{1}{2} \arcsin 2x$

Domain: $-\dfrac{1}{2} \le x \le \dfrac{1}{2}$

Range: $-\dfrac{\pi}{4} \le y \le \dfrac{\pi}{4}$

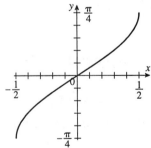

**41.** $y = \sin(\arcsin x)$
Domain: $-1 \le x \le 1$
Range: $-1 \le y \le 1$

**43.** $\tan(\arctan 3 + \arctan 4) = \tan(2.5749)$
$= -0.6364$

**45.** $\cos\left(\arcsin\dfrac{8}{17} - \arccos\dfrac{12}{13}\right) = \cos(0.0952)$
$= 0.9955$

**47.** $\cos\left(2\arcsin\dfrac{1}{3}\right) = \cos(0.6797) = 0.7778$

**49.** $\arcsin(x + 2) = \dfrac{\pi}{6}$

$x + 2 = \sin\dfrac{\pi}{6} = \dfrac{1}{2}$

$x = -\dfrac{3}{2}$

Check: $\arcsin\left(\dfrac{-3}{2} + 2\right) = \arcsin\left(\dfrac{1}{2}\right) = \dfrac{\pi}{6}$

**51.** $\arctan(x^2 + 4x + 3) = -\dfrac{\pi}{4}$

$x^2 + 4x + 3 = \tan\left(-\dfrac{\pi}{4}\right) = -1$

$x^2 + 4x + 4 = 0$
$(x + 2)(x + 2) = 0$
$x + 2 = 0$
$x = -2$

Check: $\arctan\left((-2)^2 + 4(-2) + 3\right)$

$= \arctan(-1) = \dfrac{-\pi}{4}$

**53.** Let $y = \arcsin x$. Then $x = \sin y$
$-x = -\sin y = \sin(-y)$
$-y = \arcsin(-x)$
$-\arcsin x = \arcsin(-x)$

**55.** Let $\alpha$ be the angle subtended by the bottom of the picture to the floor from $P$.

$\tan\alpha = \dfrac{7}{14} = \dfrac{1}{2}$, so $\alpha = \arctan\dfrac{1}{2}$

$\tan(\alpha + \theta) = \dfrac{5+7}{14} = \dfrac{12}{14} = \dfrac{6}{7}$, so

$\alpha + \theta = \arctan\dfrac{6}{7}$. Therefore,

$\theta = \arctan\dfrac{6}{7} - \arctan\dfrac{1}{2} = 0.2450$.

**57.** Let $\theta = \arcsin x$, $-1 \le x \le 1$. Then $\sin\theta = x$ and $\sin^2\theta = x^2$. From Exercise 82, page 381, $\sin^2\theta + \cos^2\theta = 1$ so that $\cos^2\theta = 1 - \sin^2\theta = 1 - x^2$, which gives that $\cos\theta = \pm\sqrt{1-x^2}$. But $\cos\theta \ge 0$ since $-\dfrac{\pi}{2} \le \theta \le \dfrac{\pi}{2}$. Therefore, $\cos\theta = \sqrt{1-x^2}$ and replacing $\theta$ by $\arcsin x$ gives $\cos(\arcsin x) = \sqrt{1-x^2}$.

**59.** $(f \circ g)(x) = f(g(x)) = f(3x+2)$
$= \arcsin(3x+2)$
$(g \circ f)(x) = g(f(x)) = g(\arcsin x)$
$= 3\arcsin x + 2$

**61.** Let $f(x) = \ln x$ and $g(x) = \arccos x$. Then
$(f \circ g)(x) = f(g(x)) = f(\arccos x)$
$= \ln(\arccos x) = h(x)$

### Chapter Review Exercises

**1.** One radian is the measure of a central angle of a circle that is subtended by an arc whose length is equal to the radius of the circle.

**3. (a)** $\dfrac{4\pi}{5} = \left(\dfrac{4\pi}{5} \cdot \dfrac{180}{\pi}\right)^\circ = 144^\circ$

**(b)** $\dfrac{17\pi}{6} = \left(\dfrac{17\pi}{6} \cdot \dfrac{180}{\pi}\right)^\circ = 510^\circ$

**(c)** $\dfrac{1}{5} = \left(\dfrac{1}{5} \cdot \dfrac{180}{\pi}\right)^\circ = \left(\dfrac{36}{\pi}\right)^\circ$

**5.** $\theta = \dfrac{s}{r} = \dfrac{10\pi}{6} = \dfrac{5\pi}{3}$

$\dfrac{5\pi}{3} = \left(\dfrac{5\pi}{3} \cdot \dfrac{180}{\pi}\right)^\circ = 300^\circ$

**7.** angle $= 120^\circ = \left(120 \cdot \dfrac{\pi}{180}\right) = \dfrac{2\pi}{3}$

area $= \dfrac{1}{2}\left(\dfrac{2\pi}{3}\right)10^2 = \dfrac{100\pi}{3}$ cm$^2$

**9.** angle $= 20.5^\circ = \left(20.5 \cdot \dfrac{\pi}{180}\right) = \dfrac{41\pi}{360}$

$s = (30.5)\left(\dfrac{41\pi}{360}\right) = 10.9$ in.

**11. (a)** $v = \dfrac{8400\pi}{\frac{3}{2}} = 17{,}593$ mph

**(b)** $w = \dfrac{2\pi}{\frac{3}{2}} = \dfrac{4\pi}{3}$ rad/hr

**13.** $P\left(-\dfrac{1}{\sqrt{2}}, \dfrac{1}{\sqrt{2}}\right)$

$\sin\theta = \dfrac{1}{\sqrt{2}}$ \qquad $\csc\theta = \sqrt{2}$

$\cos\theta = -\dfrac{1}{\sqrt{2}}$ \qquad $\sec\theta = -\sqrt{2}$

$\tan\theta = -1$ \qquad $\cot\theta = -1$

**15.** $P\left(-\dfrac{1}{2}, -\dfrac{\sqrt{3}}{2}\right)$

$\sin\theta = -\dfrac{\sqrt{3}}{2}$ \qquad $\csc\theta = -\dfrac{2}{\sqrt{3}}$

$\cos\theta = -\dfrac{1}{2}$ \qquad $\sec\theta = -2$

$\tan\theta = \sqrt{3}$ \qquad $\cot\theta = \dfrac{1}{\sqrt{3}}$

**17.** $\sin\theta = 0$ \qquad $\csc\theta =$ undefined
$\cos\theta = 1$ \qquad $\sec\theta = 1$
$\tan\theta = 0$ \qquad $\cot\theta =$ undefined

**19.** $\sin \theta = 1$     $\csc \theta = 1$
$\cos \theta = 0$     $\sec \theta = $ undefined
$\tan \theta = $ undefined     $\cot \theta = 0$

**21.** $\sin \theta = 0.9659$     $\csc \theta = 1.0353$
$\cos \theta = 0.2588$     $\sec \theta = 3.8637$
$\tan \theta = 3.7321$     $\cot \theta = 0.2679$

**23.** $\sin \theta = -0.2419$     $\csc \theta = -4.1336$
$\cos \theta = 0.9703$     $\sec \theta = 1.0306$
$\tan \theta = -0.2493$     $\cot \theta = -4.0108$

**25.** $\sin 215° = -0.5736$

**27.** $\tan 420° = \sqrt{3}$

**29.** $\sec 172.5° = -1.0086$

**31.** $\cot\left(\dfrac{7\pi}{5}\right) = 0.3249$

**33.** $\cos 5\pi = -1$

**35.** $\cot \pi = $ undefined

**37.** $\csc\left(-\dfrac{5\pi}{6}\right) = -2$

**39.** $\sin 200° = -0.3420$

**41.** $\sin \theta = -1$
$\theta = \dfrac{3\pi}{2}$

**43.** $\sec \theta = -\sqrt{2}$
$\theta = \dfrac{3\pi}{4}, \dfrac{5\pi}{4}$

**45.** $\cot \theta = 0$
$\theta = -90°, -270°$

**47.** $\sin \theta = -\dfrac{1}{\sqrt{2}}$
$\theta = -45°, -135°$

**49.** Since $\sin^2 \theta + \cos^2 \theta = 1$,
$$\cos^2 \theta = 1 - \left(-\dfrac{3}{4}\right)^2 = \dfrac{7}{16}$$
$$\cos \theta = \dfrac{\sqrt{7}}{4} \text{ since } \cos \theta > 0$$
$$\tan \theta = \dfrac{\sin \theta}{\cos \theta} = \dfrac{-\frac{3}{4}}{\frac{\sqrt{7}}{4}} = -\dfrac{3}{\sqrt{7}}$$

**51.** Since $-\dfrac{\pi}{2} < \theta < 0$, $\sin \theta < 0$
$$\sin^2 \theta + \cos^2 \theta = 1$$
$$\sin^2 \theta = 1 - \left(\dfrac{3}{8}\right)^2$$
$$\sin^2 \theta = \dfrac{55}{64}$$
$$\sin \theta = -\dfrac{\sqrt{55}}{8} \text{ since } \sin \theta < 0$$

**53.** $\cos \theta = \sin\left(\theta + \dfrac{\pi}{2}\right)$ since the cosine curve
is a sine curve shifted left $\dfrac{\pi}{2}$ units.

**55.** $y = -\dfrac{1}{2}\sin 2x$
amplitude $= \dfrac{1}{2}$
period $= \dfrac{2\pi}{2} = \pi$

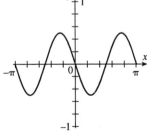

**57.** $a = -\dfrac{3}{2}, b = 3$, sine curve
$$y = -\dfrac{3}{2}\sin 3x$$

**59.** $y = 4\sin\left(x + \dfrac{\pi}{4}\right)$

amplitude $= 4$

period $= \dfrac{2\pi}{1} = 2\pi$

phase shift $= -\dfrac{\pi}{4}$

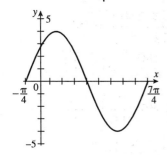

**61.** $y = 4\cos 4\left(x - \dfrac{\pi}{2}\right)$

amplitude $= 4$

period $= \dfrac{2\pi}{4} = \dfrac{\pi}{2}$

phase shift $= \dfrac{\pi}{2}$

**63.** $y = \dfrac{2}{3}\sin\left(2x - \dfrac{\pi}{2}\right) = \dfrac{2}{3}\sin 2\left(x - \dfrac{\pi}{4}\right)$

amplitude $= \dfrac{2}{3}$

period $= \dfrac{2\pi}{2} = \pi$

phase shift $= \dfrac{\pi}{4}$

**65.** $y = \dfrac{1}{3}\cos(\pi - 3x) = \dfrac{1}{3}\cos\left[-3\left(x - \dfrac{\pi}{3}\right)\right]$

amplitude $= \dfrac{1}{3}$

period $= \dfrac{2\pi}{3}$

phase shift $= \dfrac{\pi}{3}$

**67.** Domain: all $x \ne \dfrac{\pi}{2} + k\pi$, $k$ an integer

Range: all $y \ge 1$ and all $y \le -1$

Symmetric about $y$-axis

**69.**

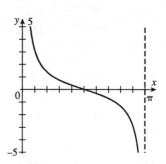

**71.** Decreasing on $\left(0, \dfrac{\pi}{2}\right)$ and $\left(\dfrac{3\pi}{2}, 2\pi\right)$

Increasing on $\left(\dfrac{\pi}{2}, \pi\right)$ and $\left(\pi, \dfrac{3\pi}{2}\right)$

Concave up on $(0, \pi)$

Concave down on $(\pi, 2\pi)$

**73.** $y = \tan 2x$

period $= \dfrac{\pi}{2}$

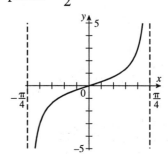

**75.** $y = -2\sec\left(\dfrac{x}{2} - \dfrac{\pi}{8}\right)$

$= -2\sec\dfrac{1}{2}\left(x - \dfrac{\pi}{4}\right)$

Period $= \dfrac{2\pi}{\dfrac{1}{2}} = 4\pi$

phase shift $= \dfrac{\pi}{4}$

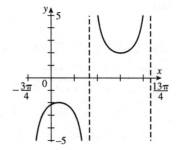

**77.** $y = 3\csc\left(4x - \dfrac{\pi}{2}\right) = 3\csc 4\left(x - \dfrac{\pi}{8}\right)$

period $= \dfrac{2\pi}{4} = \dfrac{\pi}{2}$

phase shift $= \dfrac{\pi}{8}$

asymptote: $x = \dfrac{\pi}{8}$

**79.** Reflect the positive parts of $y = \sec x$ through the $x$-axis.

**81.** $\arctan(-1) = -\dfrac{\pi}{4}$

**83.** $\arccos\dfrac{3}{2} = $ undefined

**85.** $\tan\left(\arcsin\dfrac{5}{13}\right) = \dfrac{5}{12}$

**87.** $\arctan\dfrac{8}{3} = 1.21$ rad

**89.** $\arccos\left(\tan\dfrac{\pi}{4}\right) = \arccos(1) = 0$

**91.** **(a)** $\tan(\arctan 2 - \arctan 5)$
$= \tan(-0.2663) = -0.2727$

   **(b)** $\cos\left(\arcsin\dfrac{5}{8} + \arccos\dfrac{2}{5}\right)$
$= \cos(1.8344) = -0.2606$

**93.** $\arccos(5x + 2) = \dfrac{\pi}{3}$

$5x + 2 = \cos\dfrac{\pi}{3} = \dfrac{1}{2}$

$5x = -\dfrac{3}{2}$

$x = -\dfrac{3}{10} = -0.3$

**95.** Let $\theta = \arccos x$, $-1 \le x \le 1$. Then $\cos\theta = x$, where $0 \le \theta \le \pi$, and $\cos^2\theta = x^2$. Since $\sin^2\theta + \cos^2\theta = 1$, $\sin^2\theta = 1 - \cos^2\theta = 1 - x^2$, which gives $\sin\theta = \pm\sqrt{1 - x^2}$. But $\sin\theta \ge 0$ since

$0 \le \theta \le \pi$. Therefore, $\sin\theta = \sqrt{1 - x^2}$ and substituting arccos $x$ for $\theta$ gives $\sin(\arccos x) = \sqrt{1 - x^2}$.

## Chapter 6 Test: Standard Answer

**1.** **(a)** $\dfrac{5\pi}{12} = \left(\dfrac{5\pi}{12} \cdot \dfrac{180}{\pi}\right)^\circ = 75^\circ$

   **(b)** $\dfrac{3}{2} = \left(\dfrac{3}{2} \cdot \dfrac{180}{\pi}\right)^\circ = \left(\dfrac{270}{\pi}\right)^\circ$

   **(c)** $315^\circ = \left(315 \cdot \dfrac{\pi}{180}\right) = \dfrac{7\pi}{4}$

   **(d)** $20^\circ = \left(20 \cdot \dfrac{\pi}{180}\right) = \dfrac{\pi}{9}$

**2.** Length of base = radius of sector, call it $l$.

$\tan 30^\circ = \dfrac{2}{l}$

$\dfrac{1}{\sqrt{3}} = \dfrac{2}{l}$ or $l = 2\sqrt{3}$

Area of triangle: $\dfrac{1}{2}\left(2\sqrt{3}\right) \cdot 2 = 2\sqrt{3}$

Area of sector: $\dfrac{1}{2}\left(2\sqrt{3}\right)^2 \cdot \dfrac{\pi}{6} = \pi$

Area of shaded part: $2\sqrt{3} - \pi$ square units

**3.** **(a)** $w = \dfrac{246(2\pi)}{1} = 492\pi$ rad/min

   **(b)** $v = rw = \dfrac{13(492\pi)}{12}$
$= 533\pi$ ft/min

   **(c)** In 30 min, the bicycle will travel
$30(533\pi) = 15{,}990\pi$ feet or
$\dfrac{15{,}990\pi}{5280} = \dfrac{533\pi}{176}$ miles

**4.** $120^\circ = \left(120 \cdot \dfrac{\pi}{180}\right) = \dfrac{2\pi}{3}$

$s = 10 \cdot \dfrac{2\pi}{3} = \dfrac{20\pi}{3}$ inches

**5.** (a) $\sin \dfrac{\pi}{3} = \dfrac{\sqrt{3}}{2}$

(b) $\cot \dfrac{3\pi}{2} = 0$

(c) $\sin\left(-\dfrac{3\pi}{4}\right) = -\dfrac{1}{\sqrt{2}}$

**6.** (a) reference angle: $\dfrac{\pi}{4}$

(b) reference angle: $\dfrac{\pi}{6}$

**7.** (a) $\tan \dfrac{17\pi}{6} = -\dfrac{1}{\sqrt{3}}$

(b) $\cos 3\pi = -1$

(c) $\csc\left(-\dfrac{7\pi}{6}\right) = 2$

**8.** $\sec\theta = \dfrac{1}{\cos\theta} = \dfrac{1}{\sin\theta} \cdot \tan\theta$

$= -\dfrac{2}{\sqrt{3}} \cdot \sqrt{3} = -2$

**9.** $\left(-\dfrac{2}{5}\right)^2 + \left(\dfrac{\sqrt{21}}{5}\right)^2 = \dfrac{4}{25} + \dfrac{21}{25} = \dfrac{25}{25} = 1$

$\sin\theta = \dfrac{\sqrt{21}}{5}$ $\qquad \csc\theta = \dfrac{5}{\sqrt{21}}$

$\cos\theta = -\dfrac{2}{5}$ $\qquad \sec\theta = -\dfrac{5}{2}$

$\tan\theta = -\dfrac{\sqrt{21}}{2}$ $\qquad \cot\theta = -\dfrac{2}{\sqrt{21}}$

**10.** (a) Domain: all real numbers
Range: $-1 \le y \le 1$

(b) Increasing on $\left[0, \dfrac{\pi}{2}\right]$ and $\left(\dfrac{3\pi}{2}, 2\pi\right]$

Decreasing on $\left(\dfrac{\pi}{2}, \dfrac{3\pi}{2}\right)$

Concave down on $(0, \pi)$
Concave up on $(\pi, 2\pi)$

**11.** $y = 2 \sin 2x$
amplitude $= 2$

period $= \dfrac{2\pi}{2} = \pi$

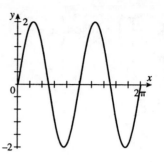

**12.** $y = -\dfrac{1}{2}\sin\left(3x - \dfrac{\pi}{2}\right) = -\dfrac{1}{2}\sin 3\left(x - \dfrac{\pi}{6}\right)$

period $= \dfrac{2\pi}{3}$

amplitude $= \dfrac{1}{2}$

phase shift $= \dfrac{\pi}{6}$

**13.** $y = 2\cos(2x + \pi) = 2\cos 2\left(x + \dfrac{\pi}{2}\right)$

period $= \dfrac{2\pi}{2} = \pi$

amplitude $= 2$

phase shift $= -\dfrac{\pi}{2}$

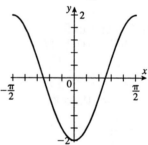

**14.** $a = \dfrac{1}{2}, b = 3$, cosine curve

$y = \dfrac{1}{2}\cos 3x$

**15.** (a) Increasing on $\left(-\dfrac{\pi}{2}, \dfrac{\pi}{2}\right)$

Concave down on $\left(-\dfrac{\pi}{2}, 0\right)$

Concave up on $\left(0, \dfrac{\pi}{2}\right)$

**(b)** Shift the graph of $y = \tan x$ a length of $\frac{\pi}{2}$ to the left.

**16.** $y = \tan 2x$

period $= \dfrac{\pi}{2}$

asymptotes on $[0, \pi]$: $x = \dfrac{\pi}{4}, x = \dfrac{3\pi}{4}$

**17.**

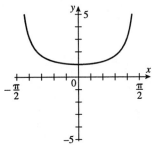

Domain: all $x \neq \dfrac{\pi}{2} + k\pi$, $k$ an integer

Range: all $y \geq 1$ or $y \leq -1$

**18.** $\csc(-x) = \dfrac{1}{\sin(-x)} = \dfrac{1}{-\sin x} = \dfrac{-1}{\sin x}$

$= -\csc x$

**19.** Let $f(x) = x^3$, $g(x) = \csc x$, and $h(x) = 2x$

$f \circ g \circ h = f(g(h(x))) = f(g(2x))$

$= f(\csc 2x) = \csc^3 2x = F(x)$

**20. (a)** Domain: $-1 \leq x \leq 1$

Range: $\dfrac{-\pi}{2} \leq y \leq \dfrac{\pi}{2}$

**(b)** Domain: $-1 \leq x \leq 1$

Range: $0 \leq y \leq \pi$

**21. (a)**

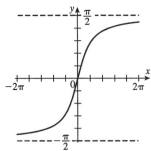

Domain: all real numbers

Range: $-\dfrac{\pi}{2} < y < \dfrac{\pi}{2}$

**(b)** $y = \arctan x$ is symmetric through the origin.

asymptotes: $y = \pm\dfrac{\pi}{2}$

**22. (a)** $\arcsin \dfrac{1}{\sqrt{2}} = \dfrac{\pi}{4}$

**(b)** $\arccos(-1) = \pi$

**(c)** $\arctan\left(-\dfrac{1}{\sqrt{3}}\right) = -\dfrac{\pi}{6}$

**23. (a)** $\arcsin(0.5398) = 0.5702$

**(b)** $\arccos(-0.1896) = 1.7616$

**(c)** $\arctan(-0.2341) = -0.2300$

**24. (a)** $\sin\left(\arccos-\dfrac{1}{\sqrt{2}}\right) = \sin\left(\dfrac{3\pi}{4}\right) = \dfrac{1}{\sqrt{2}}$

**(b)** $\cos\left(2\arcsin\dfrac{1}{2}\right) = \cos\left(2 \cdot \dfrac{\pi}{6}\right) = \dfrac{1}{2}$

$= 0.5$

**(c)** $\arctan(\tan x^2) = x^2$

**25.** $\arcsin(4x + 1) = \dfrac{-\pi}{6}$

$4x + 1 = \sin\left(-\dfrac{\pi}{6}\right) = -\dfrac{1}{2}$

$4x = \dfrac{-3}{2}$

$x = \dfrac{-3}{8}$

## Chapter 6 Test: Multiple Choice

1. $\dfrac{11\pi}{6} = \left(\dfrac{11\pi}{6} \cdot \dfrac{180}{\pi}\right) = 330°$
   The answer is (a).

2. area $= \dfrac{1}{2} \cdot 6^2 \cdot \left(30 \cdot \dfrac{\pi}{180}\right) = 3\pi$ square centimeters
   The answer is (c).

3. $w = 280$ rad/sec, $v = rw = \dfrac{3}{2} \cdot 280$
   $= 420$ ft/sec
   In 60 seconds, the distance traveled is $60 \cdot 420 = 25,200$ ft. The answer is (c).

4. $\tan\left(-\dfrac{3\pi}{4}\right) = 1$, so the answer is (a).

5. I is true, II is true, and III is also true.
   The answer is (d).

6. $\sin\left(-\dfrac{7\pi}{3}\right) = -\dfrac{\sqrt{3}}{2}$, so the answer is (b).

7. All three are true, so the answer is (d).

8. $\left(\dfrac{1}{3}\right)^2 + y^2 = 1, \quad y^2 = 1 - \dfrac{1}{9} = \dfrac{8}{9}$
   $y = -\dfrac{2\sqrt{2}}{3}$ since $y < 0$
   $\tan\theta = \dfrac{\dfrac{-2\sqrt{2}}{3}}{\dfrac{1}{3}} = -2\sqrt{2}$
   The answer is (a).

9. All three are true, so the answer is (d).

10. $y = -\dfrac{1}{2}\sin(3x - \pi) = -\dfrac{1}{2}\sin 3\left(x - \dfrac{\pi}{3}\right)$
    $A = \dfrac{1}{2}, t = \dfrac{\pi}{3}$
    The answer is (a).

11. period $= \dfrac{2\pi}{4} = \dfrac{\pi}{2}$
    The answer is (d).

12. I and III are false, II is true.
    The answer is (b).

13. I and III are true, and II is false, so the correct sequence is T, F, T.
    The answer is (e).

14. Since only I is true, the answer is (c).

15. $\tan\left(\arccos\left(-\dfrac{\sqrt{3}}{2}\right)\right) = \tan\left(\dfrac{5\pi}{6}\right) = -\dfrac{1}{\sqrt{3}}$
    The answer is (b).

16. $a = \dfrac{3}{2}, b = 2$, and $h = \dfrac{\pi}{4}$
    The answer is (d).

17. The two asymptotes are $x = \pm\dfrac{3}{2}\pi$.
    The answer is (b).

18. $\csc\theta = \dfrac{1}{\sin\theta} = \dfrac{1}{\cos\theta} \cdot \dfrac{\cos\theta}{\sin\theta} = \dfrac{1}{\cos\theta} \cdot \dfrac{1}{\tan\theta}$
    $= 2 \cdot \left(-\dfrac{1}{\sqrt{3}}\right) = -\dfrac{2}{\sqrt{3}}$
    The answer is (c).

19. I and III are true, but II is false.
    Hence, the answer is (b).

20. Domain: $-1 \le \dfrac{x}{3} \le 1$ or $-3 \le x \le 3$
    The answer is (c).

# Chapter 7: Right Triangle Trigonometry, Identities, and Equations

## Exercises 7.1

**1.** $\sin\theta = \dfrac{8}{17}$     $\csc\theta = \dfrac{17}{8}$

    $\cos\theta = \dfrac{15}{17}$     $\sec\theta = \dfrac{17}{15}$

    $\tan\theta = \dfrac{8}{15}$     $\cot\theta = \dfrac{15}{8}$

**3.** $\sin\theta = \dfrac{\sqrt{3}}{2}$     $\csc\theta = \dfrac{2}{\sqrt{3}}$

    $\cos\theta = \dfrac{1}{2}$     $\sec\theta = 2$

    $\tan\theta = \sqrt{3}$     $\cot\theta = \dfrac{1}{\sqrt{3}}$

**5.** $\sin\theta = \dfrac{12t}{13t} = \dfrac{12}{13}$     $\csc\theta = \dfrac{13}{12}$

    $\cos\theta = \dfrac{5t}{13t} = \dfrac{5}{13}$     $\sec\theta = \dfrac{13}{5}$

    $\tan\theta = \dfrac{12t}{5t} = \dfrac{12}{5}$     $\cot\theta = \dfrac{5}{12}$

**7.**

$\theta = 45°$

**9.**

$\theta = 60°$

**11.**

$\theta = 30°$

**13.** $c^2 = a^2 + b^2 = 6^2 + 8^2 = 100$
$c = 10$

  (a)   $\sin A = \dfrac{3}{5}$     $\csc A = \dfrac{5}{3}$

        $\cos A = \dfrac{4}{5}$     $\sec A = \dfrac{5}{4}$

        $\tan A = \dfrac{3}{4}$     $\cot A = \dfrac{4}{3}$

  (b)   $\sin B = \dfrac{4}{5}$     $\csc B = \dfrac{5}{4}$

        $\cos B = \dfrac{3}{5}$     $\sec B = \dfrac{5}{3}$

        $\tan B = \dfrac{4}{3}$     $\cot B = \dfrac{3}{4}$

**15.** $c^2 = a^2 + b^2 = 4^2 + 5^2 = 41$
$c = \sqrt{41}$

  (a)   $\sin A = \dfrac{4}{\sqrt{41}}$     $\csc A = \dfrac{\sqrt{41}}{4}$

        $\cos A = \dfrac{5}{\sqrt{41}}$     $\sec A = \dfrac{\sqrt{41}}{5}$

        $\tan A = \dfrac{4}{5}$     $\cot A = \dfrac{5}{4}$

  (b)   $\sin B = \dfrac{5}{\sqrt{41}}$     $\csc B = \dfrac{\sqrt{41}}{5}$

        $\cos B = \dfrac{4}{\sqrt{41}}$     $\sec B = \dfrac{\sqrt{41}}{4}$

        $\tan B = \dfrac{5}{4}$     $\cot B = \dfrac{4}{5}$

**17.** $c^2 = a^2 + b^2 = 1^2 + 2^2 = 5$
$c = \sqrt{5}$

**(a)** $\quad \sin A = \dfrac{1}{\sqrt{5}} \qquad \csc A = \sqrt{5}$

$\qquad \cos A = \dfrac{2}{\sqrt{5}} \qquad \sec A = \dfrac{\sqrt{5}}{2}$

$\qquad \tan A = \dfrac{1}{2} \qquad \cot A = 2$

**(b)** $\quad \sin B = \dfrac{2}{\sqrt{5}} \qquad \csc B = \dfrac{\sqrt{5}}{2}$

$\qquad \cos B = \dfrac{1}{\sqrt{5}} \qquad \sec B = \sqrt{5}$

$\qquad \tan B = 2 \qquad \cot B = \dfrac{1}{2}$

**19.** $\quad b^2 = c^2 - a^2 = 4^2 - 3^2 = 7$
$\qquad b = \sqrt{7}$

**(a)** $\quad \sin A = \dfrac{3}{4} \qquad \csc A = \dfrac{4}{3}$

$\qquad \cos A = \dfrac{\sqrt{7}}{4} \qquad \sec A = \dfrac{4}{\sqrt{7}}$

$\qquad \tan A = \dfrac{3}{\sqrt{7}} \qquad \cot A = \dfrac{\sqrt{7}}{3}$

**(b)** $\quad \sin B = \dfrac{\sqrt{7}}{4} \qquad \csc B = \dfrac{4}{\sqrt{7}}$

$\qquad \cos B = \dfrac{3}{4} \qquad \sec B = \dfrac{4}{3}$

$\qquad \tan B = \dfrac{\sqrt{7}}{3} \qquad \cot B = \dfrac{3}{\sqrt{7}}$

**21.** $\quad a^2 = c^2 - b^2 = 5^2 - 2^2 = 21$
$\qquad b = \sqrt{21}$

**(a)** $\quad \sin A = \dfrac{\sqrt{21}}{5} \qquad \csc A = \dfrac{5}{\sqrt{21}}$

$\qquad \cos A = \dfrac{2}{5} \qquad \sec A = \dfrac{5}{2}$

$\qquad \tan A = \dfrac{\sqrt{21}}{2} \qquad \cot A = \dfrac{2}{\sqrt{21}}$

**(b)** $\quad \sin B = \dfrac{2}{5} \qquad \csc B = \dfrac{5}{2}$

$\qquad \cos B = \dfrac{\sqrt{21}}{5} \qquad \sec B = \dfrac{5}{\sqrt{21}}$

$\qquad \tan B = \dfrac{2}{\sqrt{21}} \qquad \cot B = \dfrac{\sqrt{21}}{2}$

**23.** $\quad b^2 = c^2 - a^2 = 29^2 - 21^2 = 400$
$\qquad b = 20$

**(a)** $\quad \sin A = \dfrac{21}{29} \qquad \csc A = \dfrac{29}{21}$

$\qquad \cos A = \dfrac{20}{29} \qquad \sec A = \dfrac{29}{20}$

$\qquad \tan A = \dfrac{21}{20} \qquad \cot A = \dfrac{20}{21}$

**(b)** $\quad \sin B = \dfrac{20}{29} \qquad \csc B = \dfrac{29}{20}$

$\qquad \cos B = \dfrac{21}{29} \qquad \sec B = \dfrac{29}{21}$

$\qquad \tan B = \dfrac{20}{21} \qquad \cot B = \dfrac{21}{20}$

**25.** $\quad (\tan x)(\cot x) = \dfrac{x}{y} \cdot \dfrac{y}{x} = 1$

**27.** $\quad (\sin x)\left(\dfrac{1}{\csc x}\right) = \dfrac{x}{z} \cdot \dfrac{1}{\frac{z}{x}} = \dfrac{x}{z} \cdot \dfrac{x}{z} = \dfrac{x^2}{z^2}$

**29.** $\quad \sin^2 X + \cos^2 X = \left(\dfrac{x}{z}\right)^2 + \left(\dfrac{y}{z}\right)^2 = \dfrac{x^2}{z^2} + \dfrac{y^2}{z^2}$

$\qquad = \dfrac{x^2 + y^2}{z^2} = \dfrac{z^2}{z^2} = 1$

**31.** $\quad \sec^2 X - \tan^2 X = \left(\dfrac{z}{y}\right)^2 - \left(\dfrac{x}{y}\right)^2 = \dfrac{z^2}{y^2} - \dfrac{x^2}{y^2}$

$\qquad = \dfrac{z^2 - x^2}{y^2} = \dfrac{y^2}{y^2} = 1$

**33.** $\quad a = 3,\ c = 4$
$\qquad b^2 = c^2 - a^2 = 4^2 - 3^2 = 7$
$\qquad b = \sqrt{7}$

$$\cos A = \frac{\sqrt{7}}{4} \qquad \sec A = \frac{4}{\sqrt{7}}$$

$$\csc A = \frac{4}{3} \qquad \tan A = \frac{3}{\sqrt{7}}$$

$$\cot A = \frac{\sqrt{7}}{3}$$

**35.** $a = 9,\ b = 40$

$$c^2 = a^2 + b^2 = 9^2 + 40^2 = 1681$$
$$c = 41$$

$$\sin A = \frac{9}{41} \qquad \csc A = \frac{41}{9}$$

$$\cos A = \frac{40}{41} \qquad \sec A = \frac{41}{40}$$

$$\cot A = \frac{40}{9}$$

**37.** $c = \sqrt{11},\ a = \sqrt{2}$

$$b^2 = c^2 - a^2 = 11 - 2 = 9$$
$$b = 3$$

$$\sin B = \frac{3}{\sqrt{11}} \qquad \csc B = \frac{\sqrt{11}}{3}$$

$$\cos B = \frac{\sqrt{2}}{\sqrt{11}}$$

$$\tan B = \frac{3}{\sqrt{2}} \qquad \cot B = \frac{\sqrt{2}}{3}$$

**39.**

$$\sin\left(\arccos\frac{12}{13}\right) = \frac{5}{13}$$

**41.**

$$\tan\left(\arcsin\frac{3}{4}\right) = \frac{3}{\sqrt{7}}$$

**43.** $(\sin A)(\cos A) = \dfrac{\sqrt{1-x^2}}{1} \cdot \dfrac{x}{1} = x\sqrt{1-x^2}$

**45.** $\dfrac{4\sin^2 A}{\cos A} = \dfrac{4\left(\frac{x}{4}\right)^2}{\frac{\sqrt{16-x^2}}{4}} = \dfrac{16 \cdot \frac{x^2}{16}}{\sqrt{16-x^2}} = \dfrac{x^2}{\sqrt{16-x^2}}$

**47.**

$$(\sin\theta)(\tan\theta) = \frac{x}{2} \cdot \frac{x}{\sqrt{4-x^2}} = \frac{x^2}{2\sqrt{4-x^2}}$$

**49.**

$$\left(\tan^2\theta\right)\left(\cos^2\theta\right) = \left(\frac{x}{4}\right)^2\left(\frac{4}{\sqrt{x^2+16}}\right)^2$$

$$= \frac{16x^2}{16\left(\sqrt{x^2+16}\right)^2} = \frac{x^2}{x^2+16}$$

**51.** Base has length $s$, height has length $\dfrac{\sqrt{3}}{2}s$ by the Pythagorean Theorem. So

$$\text{area} = \frac{1}{2}(s)\left(\frac{\sqrt{3}}{2}s\right) = \frac{\sqrt{3}}{4}s^2$$

**53.** The length of the base is $2y = 2\sqrt{9-x^2}$ since $(x, y)$ lies on $x^2 + y^2 = 9$. The height is also $2\sqrt{9-x^2}$, so

$$\text{Area} = \frac{1}{2}\left(2\sqrt{9-x^2}\right)\left(2\sqrt{9-x^2}\right)$$

$$= 2\left(9-x^2\right)$$

*Chapter 7: Right Triangle Trigonometry, Identities, and Equations*

**55.** Since $AP$ is used to form an equilateral triangle, each side will have length $\frac{x}{3}$ and area $\frac{\sqrt{3}}{4}\left(\frac{x}{3}\right)^2$ by Exercise 51. The diameter of the circle is $\frac{100-x}{\pi}$, so the radius is $\frac{100-x}{2\pi}$. Total area is

$$\frac{\sqrt{3}}{4}\left(\frac{x}{3}\right)^2 + \pi\left(\frac{100-x}{2\pi}\right)^2$$

$$= \frac{\sqrt{3}}{36}x^2 + \frac{(100-x)^2}{4\pi}.$$

Domain: $0 < x < 100$

**57 (a)**

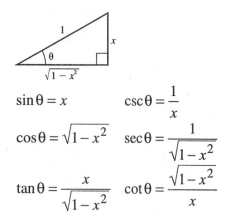

$\sin\theta = x$ $\qquad \csc\theta = \frac{1}{x}$

$\cos\theta = \sqrt{1-x^2}$ $\qquad \sec\theta = \frac{1}{\sqrt{1-x^2}}$

$\tan\theta = \frac{x}{\sqrt{1-x^2}}$ $\qquad \cot\theta = \frac{\sqrt{1-x^2}}{x}$

**(b)**

$\sin\theta = \frac{x}{\sqrt{1+x^2}}$ $\qquad \csc\theta = \frac{\sqrt{1+x^2}}{x}$

$\cos\theta = \frac{1}{\sqrt{1+x^2}}$ $\qquad \sec\theta = \sqrt{1+x^2}$

$\tan\theta = x$ $\qquad \cot\theta = \frac{1}{x}$

**Challenge**

From the figure, $OB = OA = \sqrt{2}$. By the Pythagorean Theorem, $OC = 1$, which makes

$\triangle OBC$ an isosceles right triangle with $\angle BOC$ a $45°$ angle. Hence, $\angle AOB$ is a $135°$ angle and $\triangle AOB$ is also an isosceles triangle. Therefore, $\angle BAO$ is a $22.5°$ angle, and

$$\tan 22.5° = \frac{BC}{AC} = \frac{1}{\sqrt{2}+1} = \frac{1(\sqrt{2}-1)}{(\sqrt{2}+1)(\sqrt{2}-1)}$$

$$\frac{\sqrt{2}-1}{2-1} = \sqrt{2}-1$$

**Exercises 7.2**

**1.** $\tan\frac{\pi}{6} = \frac{x}{25}$

$\frac{1}{\sqrt{3}} = \frac{x}{25}$

$x = \frac{25}{\sqrt{3}}$

**3.** $\sin 45° = \frac{15}{x}$

$\frac{1}{\sqrt{2}} = \frac{15}{x}$

$x = 15\sqrt{2}$

**5.** $\sin\frac{\pi}{3} = \frac{x}{40}$

$\frac{\sqrt{3}}{2} = \frac{x}{40}$

$x = 20\sqrt{3}$

**7.** $\cos\frac{\pi}{6} = \frac{4}{x}$

$\frac{\sqrt{3}}{2} = \frac{4}{x}$

$x = \frac{8}{\sqrt{3}}$

**9.** $\sin\theta = \frac{5}{5\sqrt{2}} = \frac{1}{\sqrt{2}}$

$\theta = \arcsin\frac{1}{\sqrt{2}}$

$\theta = \frac{\pi}{4}$

**11.** $\tan\theta = 6.314$

$\theta = 81.0°$

**13.** $\cot \theta = 0.4592$
$\theta = 65.3°$

**15.** $\cos \theta = 0.9940$
$\theta = 6.3°$

**17.** $\tan 70° = \dfrac{35}{b}$
$b = \dfrac{35}{\tan 70°} = 12.7$

**19.** $\tan 42.3° = \dfrac{b}{20}$
$b = 20 \tan 42.3° = 18.2$

**21.** $\tan B = \dfrac{3}{1}$
$B = 71.6°$

**23.** $\sin B = \dfrac{9}{25}$
$B = 21.1°$

**25.** $\tan 35° = \dfrac{h}{50}$
$h = 50 \tan 35° = 35$ meters

**27.** $\tan 70° = \dfrac{h}{100}$
$h = 100 \tan 70° = 275$ feet

**29.** $\tan 6.7° = \dfrac{225}{d}$
$d = \dfrac{225}{\tan 6.7°} = 1915$ feet

**31.** $\cos \theta = \dfrac{14.5}{82}$
$\theta = 79.8°$

**33.**

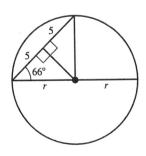

From the figure, $\cos 66° = \dfrac{5}{r}$, so

$$r = \dfrac{5}{\cos 66°} = 12.3.$$

**35.** $\dfrac{1}{4}$ mile = 1320 feet

$\tan 43° = \dfrac{h}{BC}$ and $\tan 23° = \dfrac{h}{BC + 1320}$

So, $BC = \dfrac{h}{\tan 43°}$

and $\tan 23° = \dfrac{h}{\dfrac{h}{\tan 43°} + 1320}$

$\dfrac{h(\tan 23°)}{\tan 43°} + 1320 \tan 23° = h$

$1320 \tan 23° = h - h\left(\dfrac{\tan 23°}{\tan 43°}\right)$

$h = \dfrac{1320 \tan 23°}{1 - \frac{\tan 23°}{\tan 43°}} = 1028$ feet

**37.** In the figure, the radius makes a right angle with the pentagon. The angle made between the radius and a line from the center to one of the closest corners is a 36° angle. So, half of the length of a side of the pentagon is $20 \tan 36° = 14.5$ cm.
The perimeter is 145 cm.

**Challenge**

Notice that $\angle BDA = 145°$, $\angle ADC = 35°$, and $\angle DAC = 55°$. We have that
$\tan 55° = \dfrac{CD}{AC}$ and $\tan 68° = \dfrac{CD + 16}{AC}$

or $AC = \dfrac{CD}{\tan 55°}$ and $AC = \dfrac{CD + 16}{\tan 68°}$

so that $\dfrac{CD}{\tan 55°} = \dfrac{CD + 16}{\tan 68°}$

$CD(\tan 68°) = CD(\tan 55°) + 16 \tan 55°$
$CD(\tan 68° - \tan 55°) = 16 \tan 55°$

$CD = \dfrac{16 \tan 55°}{\tan 68° - \tan 55°} = 21.8$

# Exercises 7.3

**1.** $\dfrac{1}{\csc\theta} = \sin\theta$

**3.** $\csc^2\theta - \cot^2\theta = \csc^2\theta - \left(\csc^2\theta - 1\right) = 1$

**5.** $\sin\theta\cot\theta = \sin\theta \cdot \dfrac{\cos\theta}{\sin\theta} = \cos\theta$

**7.** $\sin^2 39° + \cos^2 39° = 1$

**9.** $\tan^2(3.2°) - \sec^2(3.2°) = -1$

**11.** $(\sin\theta)(\cot\theta)(\sec\theta) = \sin\theta \cdot \dfrac{\cos\theta}{\sin\theta} \cdot \dfrac{1}{\cos\theta} = 1$

**13.** $\cos^2\theta\left(\tan^2\theta + 1\right) = \cos^2\theta \cdot \sec^2\theta$

$= \cos^2\theta \cdot \dfrac{1}{\cos^2\theta} = 1$

**15.** All $\theta$ not coterminal with $\dfrac{\pi}{2}, \dfrac{3\pi}{2}$.

**17.** All $\theta$ not a quadrantal angle.

**19.** All $\theta$ not a quadrantal angle.

**21.** $\dfrac{\tan\theta}{\cot\theta} = \dfrac{\frac{\sin\theta}{\cos\theta}}{\frac{\cos\theta}{\sin\theta}} = \dfrac{\sin^2\theta}{\cos^2\theta}$

$\theta$ not a quadrantal angle.

**23.** $\dfrac{1 - \csc\theta}{\cot\theta} = \dfrac{1 - \frac{1}{\sin\theta}}{\frac{\cos\theta}{\sin\theta}} \cdot \dfrac{\sin\theta}{\sin\theta} = \dfrac{\sin - 1}{\cos\theta}$

$\theta$ not a quadrantal angle.

**25.** $\dfrac{\sec\theta + \csc\theta}{\cos\theta + \sin\theta} = \dfrac{\frac{1}{\cos\theta} + \frac{1}{\sin\theta}}{\cos\theta + \sin\theta} \cdot \dfrac{\sin\theta\cos\theta}{\sin\theta\cos\theta}$

$= \dfrac{\sin\theta + \cos\theta}{(\cos\theta + \sin\theta)\sin\theta\cos\theta} = \dfrac{1}{\sin\theta\cos\theta}$

$\theta$ not a quadrantal angle and $\theta$ not coterminal with $\dfrac{3\pi}{4}, \dfrac{7\pi}{4}$.

**27.** $\cos^2\theta - \sin^2\theta = \left(1 - \sin^2\theta\right) - \sin^2\theta$

$= 1 - 2\sin^2\theta$

**29.** $\dfrac{\cot\theta - \sin\theta}{\csc\theta} = \dfrac{\frac{\cos\theta}{\sin\theta} - \sin\theta}{\frac{1}{\sin\theta}} \cdot \dfrac{\sin\theta}{\sin\theta}$

$= \dfrac{\cos\theta - \sin^2\theta}{1} = \cos\theta - (1 - \cos^2\theta)$

$= \cos^2\theta + \cos\theta - 1$

**31.** $\dfrac{\cos\theta}{\cot\theta} = \cos\theta\left(\dfrac{\sin\theta}{\cos\theta}\right) = \sin\theta$

$\theta$ not a quadrantal angle

**33.** $(\tan\theta - 1)^2 = \tan^2\theta - 2\tan\theta + 1$

$= \sec^2\theta - 2\tan\theta$

$\theta$ not coterminal with $\dfrac{\pi}{2}, \dfrac{3\pi}{2}$.

**35.**

| $\sec\theta - \cos\theta$ | $\sin\theta\tan\theta$ |
|---|---|
| $\dfrac{1}{\cos\theta} - \cos\theta$ | $\sin\theta \cdot \dfrac{\sin\theta}{\cos\theta}$ |
| $\dfrac{1 - \cos^2\theta}{\cos\theta}$ | $\dfrac{\sin^2\theta}{\cos\theta}$ |
| $\dfrac{\sin^2\theta}{\cos\theta}$ | |

$\theta$ not coterminal with $\dfrac{\pi}{2}, \dfrac{3\pi}{2}$.

**37.**

| $\dfrac{\cot\theta - 1}{1 - \tan\theta}$ | $\dfrac{\csc\theta}{\sec\theta}$ |
|---|---|
| $\dfrac{\frac{\cos\theta}{\sin\theta} - 1}{1 - \frac{\sin\theta}{\cos\theta}}$ | $\dfrac{\frac{1}{\sin\theta}}{\frac{1}{\cos\theta}}$ |
| $\dfrac{\cos^2\theta - \sin\theta\cos\theta}{\sin\theta\cos\theta - \sin^2\theta}$ | $\dfrac{\cos\theta}{\sin\theta}$ |
| $\dfrac{\cos\theta(\cos\theta - \sin\theta)}{\sin\theta(\cos\theta - \sin\theta)}$ | |
| $\dfrac{\cos\theta}{\sin\theta}$ | |

$\theta$ is not a quadrantal angle or coterminal with $\frac{\pi}{4}, \frac{5\pi}{4}$.

**39.** $\tan\theta + \cot\theta = \frac{\sin\theta}{\cos\theta} + \frac{\cos\theta}{\sin\theta}$

$= \frac{\sin^2\theta + \cos^2\theta}{\cos\theta\sin\theta}$

$= \frac{1}{\sin\theta\cos\theta}$

**41.** $(\sec\theta + \tan\theta)(1 - \sin\theta)$

$= \left(\frac{1}{\cos\theta} + \frac{\sin\theta}{\cos\theta}\right)(1 - \sin\theta)$

$= \left(\frac{1 + \sin\theta}{\cos\theta}\right)(1 - \sin\theta)$

$= \frac{1 - \sin^2\theta}{\cos\theta}$

$= \frac{\cos^2\theta}{\cos\theta}$

$= \cos\ \theta$

**43.** $(\csc^2\theta - 1)\sin^2\theta = \cot^2\theta\sin^2\theta$

$= \frac{\cos^2\theta}{\sin^2\theta} \cdot \sin^2\theta$

$= \cos^2\theta$

**45.** $\tan^2\theta - \sin^2\theta = \frac{\sin^2\theta}{\cos^2\theta} - \sin^2\theta$

$= \sin^2\theta\left(\frac{1}{\cos^2\theta} - 1\right)$

$= \sin^2\theta\left(\sec^2\theta - 1\right)$

$= \sin^2\theta\tan^2\theta$

**47.** $\frac{1 + \sec\theta}{\csc\theta} = \frac{1 + \frac{1}{\cos\theta}}{\frac{1}{\sin\theta}}$

$= \sin\theta + \frac{\sin\theta}{\cos\theta}$

$= \sin\theta + \tan\theta$

**49.** $\frac{1}{1 + \cos\theta} + \frac{1}{1 - \cos\theta} = \frac{1 - \cos\theta + 1 + \cos\theta}{(1 + \cos\theta)(1 - \cos\theta)}$

$= \frac{2}{1 - \cos^2\theta}$

$= \frac{2}{\sin^2\theta}$

$= 2\csc^2\theta$

**51.** $\frac{1 + \tan^2\theta}{1 + \cot^2\theta} = \frac{\sec^2\theta}{\csc^2\theta}$

$= \frac{\frac{1}{\cos^2\theta}}{\frac{1}{\sin^2\theta}}$

$= \frac{\sin^2\theta}{\cos^2\theta}$

$= \tan^2\theta$

$= \sec^2\theta - 1$

**53.** $\frac{\sec\theta + \csc\theta}{\sec\theta - \csc\theta} = \frac{\frac{1}{\cos\theta} + \frac{1}{\sin\theta}}{\frac{1}{\cos\theta} - \frac{1}{\sin\theta}}$

$= \frac{\sin\theta + \cos\theta}{\sin\theta - \cos\theta}$

**55.** $(\csc\theta - \cot\theta)^2 = \left(\frac{1}{\sin\theta} - \frac{\cos\theta}{\sin\theta}\right)^2$

$= \frac{(1 - \cos\theta)^2}{\sin^2\theta}$

$= \frac{(1 - \cos\theta)^2}{1 - \cos^2\theta}$

$= \frac{(1 - \cos\theta)^2}{(1 + \cos\theta)(1 - \cos\theta)}$

$= \frac{1 - \cos\theta}{1 + \cos\theta}$

**57.** $\left(\sin^2\theta + \cos^2\theta\right)^5 = 1^5$

$= 1$

**59.** 
$$\frac{1}{\cos^2\theta} + \frac{1}{\sin^2\theta} = \frac{\sin^2\theta + \cos^2\theta}{\cos^2\theta\sin^2\theta}$$
$$= \frac{1}{\left(1-\sin^2\theta\right)\sin^2\theta}$$
$$= \frac{1}{\sin^2\theta - \sin^4\theta}$$

**61.** 
$$\frac{\tan^2\theta + 1}{\tan^2\theta} = 1 + \frac{1}{\tan^2\theta}$$
$$= 1 + \cot^2\theta$$
$$= \csc^2\theta$$

**63.** 
$$\frac{\tan\theta}{\sec\theta - 1} = \frac{\tan\theta(\sec\theta + 1)}{(\sec\theta - 1)(\sec\theta + 1)}$$
$$= \frac{\tan\theta(\sec\theta + 1)}{\sec^2\theta - 1}$$
$$= \frac{\tan\theta(\sec\theta + 1)}{\tan^2\theta}$$
$$= \frac{\sec\theta + 1}{\tan\theta}$$

**65.** 
$$\sec^3\theta + \frac{\sin^2\theta}{\cos^3\theta} = \frac{1}{\cos^3\theta} + \frac{\sin^2\theta}{\cos^3\theta}$$
$$= \frac{1 + \sin^2\theta}{\cos^3\theta}$$
$$= \frac{\sin^2\theta + \cos^2\theta + \sin^2\theta}{\cos^3\theta}$$
$$= \frac{\cos^2\theta + 2\sin^2\theta}{\cos^3\theta}$$

**67.** 
$$\frac{1}{(\csc\theta - \sec\theta)^2} = \frac{1}{\left(\frac{1}{\sin\theta} - \frac{1}{\cos\theta}\right)^2}$$
$$= \frac{1}{\left(\frac{\cos\theta - \sin\theta}{\sin\theta\cos\theta}\right)^2}$$
$$= \frac{\sin^2\theta\cos^2\theta}{(\cos\theta - \sin\theta)^2}$$
$$= \frac{\sin^2\theta\cos^2\theta}{\cos^2\theta - 2\cos\theta\sin\theta + \sin^2\theta}$$

$$= \frac{\sin^2\theta\cos^2\theta}{1 - 2\cos\theta\sin\theta}$$
$$= \frac{\sin^2\theta}{\frac{1}{\cos^2\theta} - \frac{2\sin\theta}{\cos\theta}}$$
$$= \frac{\sin^2\theta}{\sec^2\theta - 2\tan\theta}$$

**69.** 
$$\frac{\sec\theta}{\tan\theta - \sin\theta} = \frac{\frac{1}{\cos\theta}}{\frac{\sin\theta}{\cos\theta} - \sin\theta}$$
$$= \frac{1}{\sin\theta - \sin\theta\cos\theta}$$
$$= \frac{1}{\sin\theta(1 - \cos\theta)}$$
$$= \frac{1 + \cos\theta}{\sin\theta(1 - \cos^2\theta)}$$
$$= \frac{1 + \cos\theta}{\sin^3\theta}$$

**71.** 
$$-\ln|\cos x| = \ln|\cos x|^{-1}$$
$$= \ln\frac{1}{|\cos x|}$$
$$= \ln\left|\frac{1}{\cos x}\right|$$
$$= \ln|\sec x|$$

**73.** 
$$-\ln|\sec x - \tan x| = \ln|\sec x - \tan x|^{-1}$$
$$= \ln\frac{1}{|\sec x - \tan x|}$$
$$= \ln\left|\frac{1}{\sec x - \tan x}\right|$$
$$= \ln\left|\frac{\sec x + \tan x}{\sec^2 x - \tan^2 x}\right|$$
$$= \ln|\sec x + \tan x|$$

**75.** 
$$\tan\left(-\frac{\pi}{4}\right) = -1 \neq 1 = \tan\frac{\pi}{4}$$

**77.** 
$$\sin\frac{\pi}{2} = 1 \neq -1 = -\sqrt{1 - \cos^2\frac{\pi}{2}}$$

**79.** $\dfrac{\cot\frac{\pi}{3}-1}{1-\tan\frac{\pi}{3}}=\dfrac{\frac{1}{\sqrt{3}}-1}{1-\sqrt{3}}=\dfrac{\frac{1}{\sqrt{3}}(1-\sqrt{3})}{1-\sqrt{3}}$

$=\dfrac{1}{\sqrt{3}}\neq\sqrt{3}=\dfrac{2}{\frac{2}{\sqrt{3}}}=\dfrac{\sec\frac{\pi}{3}}{\csc\frac{\pi}{3}}$

**81. (a)** $\ln(\sec\theta+\tan\theta)=\ln\left(\dfrac{\sqrt{x^2+9}}{3}+\dfrac{x}{3}\right)$

$=\ln\left(\dfrac{\sqrt{x^2+9}+x}{3}\right)$

**(b)** $F(4)-F(0)$

$=\ln\left(\dfrac{\sqrt{4^2+9}+4}{3}\right)-\ln\left(\dfrac{\sqrt{0+9}+0}{3}\right)$

$=\ln\left(\dfrac{\sqrt{25}+4}{3}\right)-\ln\left(\dfrac{3}{3}\right)$

$=\ln(3)-\ln(1)$

$=\ln 3$

## Challenge

**1.** $\dfrac{\sin\frac{\pi}{4}}{\csc\frac{\pi}{4}}-\dfrac{\cos\frac{\pi}{4}}{\sec\frac{\pi}{4}}=\dfrac{\frac{1}{\sqrt{2}}}{\sqrt{2}}-\dfrac{\frac{1}{\sqrt{2}}}{\sqrt{2}}=\dfrac{1}{2}-\dfrac{1}{2}=0\neq 1$

The equation is not an identity.

**3.** $\dfrac{\sec^2 x-2}{(1+\tan x)^2}=\dfrac{\tan^2 x+1-2}{(1+\tan x)^2}$

$=\dfrac{\tan^2 x-1}{(1+\tan x)^2}$

$=\dfrac{(\tan x-1)(\tan x+1)}{(\tan x+1)^2}$

$=\dfrac{\tan x-1}{\tan x+1}\cdot\dfrac{\cot x}{\cot x}$

$=\dfrac{1-\cot x}{1+\cot x}$

The equation is an identity.

## Critical Thinking

**1.** The two procedures mentioned are equivalent if the sides of a triangle are allowed to be negative quantities in the appropriate quadrants.

**3.** $\tan\alpha=\dfrac{h}{x+d}$ and $\tan\beta=\dfrac{h}{x}$
or $(x+d)\tan\alpha=h$ and $x\tan\beta=h$
Solving both equations for $x$:
$x=\dfrac{h-d\tan\alpha}{\tan\alpha}$ and $x=\dfrac{h}{\tan\beta}$
Therefore,
$\dfrac{h-d\tan\alpha}{\tan\alpha}=\dfrac{h}{\tan\beta}$
$h\tan\beta-d\tan\alpha\tan\beta=h\tan\alpha$
$h\tan\beta-h\tan\alpha=d\tan\alpha\tan\beta$
$h=\dfrac{d\tan\alpha\tan\beta}{\tan\beta-\tan\alpha}$

**5.** Squaring an equation does not necessarily produce an identity, and if it does, the original equation is not necessarily an identity since extraneous solutions may be introduced.

## Exercises 7.4

**1.** $\cos 22°\cos 38°-\sin 22°\sin 38°$
$=\cos(22°+38°)=\cos 60°=\dfrac{1}{2}$

**3.** $\dfrac{\tan 25°-\tan 55°}{1+\tan 25°\tan 55°}=\tan(25°-55°)$
$=\tan(-30°)=-\dfrac{1}{\sqrt{3}}$

**5.** $\sin 75°=\sin(45°+30°)$
$=\sin 45°\cos 30°+\cos 45°\sin 30°$
$=\dfrac{\sqrt{2}}{2}\cdot\dfrac{\sqrt{3}}{2}+\dfrac{\sqrt{2}}{2}\cdot\dfrac{1}{2}=\dfrac{1}{4}\left(\sqrt{6}+\sqrt{2}\right)$
Check: $\sin 75°=.9659$
$\dfrac{1}{4}\left(\sqrt{6}+\sqrt{2}\right)=0.9659$

$\cos 75° = \cos(45° + 30°)$
$= \cos 45° \cos 30° - \sin 45° \sin 30°$
Check: $\cos 75° = .0.2588$

$\dfrac{1}{4}\left(\sqrt{6} - \sqrt{2}\right) = 0.2588$

$= \dfrac{\sqrt{2}}{2} \cdot \dfrac{\sqrt{3}}{2} - \dfrac{\sqrt{2}}{2} \cdot \dfrac{1}{2} = \dfrac{1}{4}\left(\sqrt{6} - \sqrt{2}\right)$

$\tan 75° = \tan(45° + 30°)$

$= \dfrac{\tan 45° + \tan 30°}{1 - \tan 45° \tan 30°} = \dfrac{1 + \frac{1}{\sqrt{3}}}{1 - \frac{1}{\sqrt{3}}} = \dfrac{\sqrt{3} + 1}{\sqrt{3} - 1}$

$= \dfrac{\left(\sqrt{3} + 1\right)\left(\sqrt{3} + 1\right)}{\left(\sqrt{3} - 1\right)\left(\sqrt{3} + 1\right)} = \dfrac{3 + 2\sqrt{3} + 1}{3 - 1} = 2 + \sqrt{3}$

Check: $\tan 75° = 3.7321$

$2 + \sqrt{3} = 3.7321$

7. $\sin\dfrac{\pi}{12} = \sin\left(\dfrac{\pi}{3} - \dfrac{\pi}{4}\right)$

$= \sin\dfrac{\pi}{3} \cos\dfrac{\pi}{4} - \cos\dfrac{\pi}{3} \sin\dfrac{\pi}{4}$

$= \dfrac{\sqrt{3}}{2} \cdot \dfrac{\sqrt{2}}{2} - \dfrac{1}{2} \cdot \dfrac{\sqrt{2}}{2} = \dfrac{1}{4}\left(\sqrt{6} - \sqrt{2}\right)$

Check: $\sin\dfrac{\pi}{12} = 0.2588$

$\dfrac{1}{4}\left(\sqrt{2} - \sqrt{6}\right) = 0.2588$

$\cos\dfrac{\pi}{12} = \cos\left(\dfrac{\pi}{3} - \dfrac{\pi}{4}\right)$

$= \cos\dfrac{\pi}{3} \cos\dfrac{\pi}{4} + \sin\dfrac{\pi}{3}\sin\dfrac{\pi}{4}$

$= \dfrac{1}{2} \cdot \dfrac{\sqrt{2}}{2} + \dfrac{\sqrt{3}}{2} \cdot \dfrac{\sqrt{2}}{2} = \dfrac{1}{4}\left(\sqrt{2} + \sqrt{6}\right)$

Check: $\cos\dfrac{\pi}{12} = 0.9659$

$\dfrac{1}{4}\left(\sqrt{2} + \sqrt{6}\right) = 0.9659$

$\tan\dfrac{\pi}{12} = \tan\left(\dfrac{\pi}{3} - \dfrac{\pi}{4}\right) = \dfrac{\tan\frac{\pi}{3} - \tan\frac{\pi}{4}}{1 + \tan\frac{\pi}{3}\tan\frac{\pi}{4}}$

$= \dfrac{\sqrt{3} - 1}{1 + \sqrt{3}} = \dfrac{\left(\sqrt{3} - 1\right)\left(\sqrt{3} - 1\right)}{\left(\sqrt{3} + 1\right)\left(\sqrt{3} - 1\right)}$

$= \dfrac{3 - 2\sqrt{3} + 1}{3 - 1} = 2 - \sqrt{3}$

Check: $\tan\dfrac{\pi}{12} = 0.2679$

$2 - \sqrt{3} = 0.2679$

9. $\sin 165° = \sin(135° + 30°)$
$= \sin 135° \cos 30° + \cos 135° \sin 30°$

$= \dfrac{\sqrt{2}}{2} \cdot \dfrac{\sqrt{3}}{2} - \dfrac{\sqrt{2}}{2} \cdot \dfrac{1}{2} = \dfrac{1}{4}\left(\sqrt{6} - \sqrt{2}\right)$

Check: $\sin 165° = 0.2588$

$\dfrac{1}{4}\left(\sqrt{6} - \sqrt{2}\right) = 0.2588$

$\cos 165° = \cos(135° + 30°)$
$= \cos 135° \cos 30° - \sin 135° \sin 30°$

$= -\dfrac{\sqrt{2}}{2} \cdot \dfrac{\sqrt{3}}{2} - \dfrac{\sqrt{2}}{2} \cdot \dfrac{1}{2} = -\dfrac{1}{4}\left(\sqrt{6} + \sqrt{2}\right)$

Check: $\cos 165° = -0.9659$

$-\dfrac{1}{4}\left(\sqrt{6} + \sqrt{2}\right) = -0.9659$

$\tan 165° = \tan(135° + 30°)$

$= \dfrac{\tan 135° + \tan 30°}{1 - \tan 135° \tan 30°} = \dfrac{-1 + \frac{1}{\sqrt{3}}}{1 + \frac{1}{\sqrt{3}}} \cdot \dfrac{\sqrt{3}}{\sqrt{3}}$

$= \dfrac{-\sqrt{3} + 1}{\sqrt{3} + 1} \cdot \dfrac{\left(\sqrt{3} - 1\right)}{\left(\sqrt{3} - 1\right)} = \dfrac{-3 + 2\sqrt{3} - 1}{3 - 1}$

$= \sqrt{3} - 2$

Check: $\tan 165° = -0.2679$

$\sqrt{3} - 2 = -0.2679$

11. $\sin\dfrac{17\pi}{12} = \sin\left(\dfrac{7\pi}{6} + \dfrac{\pi}{4}\right)$

$= \sin\dfrac{7\pi}{6} \cos\dfrac{\pi}{4} + \cos\dfrac{7\pi}{6} \sin\dfrac{\pi}{4}$

$= -\dfrac{1}{2} \cdot \dfrac{\sqrt{2}}{2} - \dfrac{\sqrt{3}}{2} \cdot \dfrac{\sqrt{2}}{2} = -\dfrac{1}{4}\left(\sqrt{2} + \sqrt{6}\right)$

Check: $\sin\dfrac{17\pi}{12} = -0.9659$

$-\dfrac{1}{4}\left(\sqrt{2} + \sqrt{6}\right) = -0.9659$

$$\cos\frac{17\pi}{12} = \cos\left(\frac{7\pi}{6} + \frac{\pi}{4}\right)$$

$$= \cos\frac{7\pi}{6}\cos\frac{\pi}{4} - \sin\frac{7\pi}{6}\sin\frac{\pi}{4}$$

$$= -\frac{\sqrt{3}}{2}\cdot\frac{\sqrt{2}}{2} + \frac{1}{2}\cdot\frac{\sqrt{2}}{2} = \frac{1}{4}\left(\sqrt{2} - \sqrt{6}\right)$$

Check: $\cos\dfrac{17\pi}{12} = 0.2588$

$$\frac{1}{4}\left(\sqrt{2} - \sqrt{6}\right) = -0.2588$$

$$\tan\frac{17\pi}{12} = \tan\left(\frac{7\pi}{6} + \frac{\pi}{4}\right) = \frac{\tan\frac{7\pi}{6} + \tan\frac{\pi}{4}}{1 - \tan\frac{7\pi}{6}\tan\frac{\pi}{4}}$$

$$= \frac{\frac{1}{\sqrt{3}} + 1}{1 - \frac{1}{\sqrt{3}}}\cdot\frac{\sqrt{3}}{\sqrt{3}} = \frac{1+\sqrt{3}}{\sqrt{3}-1}\cdot\frac{1+\sqrt{3}}{1+\sqrt{3}}$$

$$= \frac{1 + 2\sqrt{3} + 3}{3 - 1} = 2 + \sqrt{3}$$

Check: $\tan\dfrac{17\pi}{12} = 3.7321$

$2 + \sqrt{3} = 3.7321$

**13.** $\sin(\pi - \theta) = \sin\pi\cos\theta - \cos\pi\sin\theta$
$= 0\cdot\cos\theta - (-1)\sin\theta = \sin\theta$

**15.** $\tan(\pi - \theta) = \dfrac{\tan\pi - \tan\theta}{1 + \tan\pi\tan\theta} = \dfrac{0 - \tan\theta}{1 + 0\cdot\tan\theta}$
$= -\tan\theta$

**17.** $\sin\left(\theta + \dfrac{7\pi}{2}\right) = \sin\theta\cos\dfrac{7\pi}{2} + \cos\theta\sin\dfrac{7\pi}{2}$
$= \sin\theta\cdot 0 + \cos\theta\cdot(-1) = -\cos\theta$

**19.** $\sin\alpha = \dfrac{3}{5},\ \cos\alpha = \dfrac{4}{5},\ \tan\alpha = \dfrac{3}{4}$

$\sin\beta = \dfrac{5}{13},\ \cos\beta = \dfrac{12}{13},\ \tan\beta = \dfrac{5}{12}$

$\sin(\alpha - \beta) = \sin\alpha\cos\beta - \cos\alpha\sin\beta$

$\dfrac{3}{5}\cdot\dfrac{12}{13} - \dfrac{4}{5}\cdot\dfrac{5}{13} = \dfrac{36}{65} - \dfrac{20}{65} = \dfrac{16}{65}$

$$\tan(\alpha + \beta) = \frac{\tan\alpha + \tan\beta}{1 - \tan\alpha\tan\beta} = \frac{\frac{3}{4} + \frac{5}{12}}{1 - \frac{3}{4}\cdot\frac{5}{12}}$$

$$= \frac{3\cdot 12 + 5\cdot 4}{48 - 15} = \frac{56}{33}$$

**21.** $\sin\alpha = -\dfrac{1}{3},\ \cos\alpha = -\dfrac{2\sqrt{2}}{3}$

$\sin\beta = -\dfrac{\sqrt{21}}{5},\ \cos\beta = \dfrac{2}{5}$

$\sin(\alpha + \beta) = \sin\alpha\cos\beta + \cos\alpha\sin\beta$

$$= -\frac{1}{3}\cdot\frac{2}{5} + \left(-\frac{2\sqrt{2}}{3}\right)\left(-\frac{\sqrt{21}}{5}\right) = \frac{2}{15}\left(\sqrt{42} - 1\right)$$

$\cos(\alpha + \beta) = \cos\alpha\cos\beta - \sin\alpha\sin\beta$

$$= -\frac{2\sqrt{2}}{3}\cdot\frac{2}{5} + \frac{1}{3}\left(-\frac{\sqrt{21}}{5}\right)$$

$$= -\frac{1}{15}\left(\sqrt{21} + 4\sqrt{2}\right)$$

**23.** $\sin\alpha = \dfrac{21}{29},\ \cos\alpha = -\dfrac{20}{29},\ \tan\alpha = -\dfrac{21}{20}$

$\sin\beta = -\dfrac{5}{13},\ \cos\beta = \dfrac{12}{13},\ \tan\beta = -\dfrac{5}{12}$

$\sin(\alpha - \beta) = \sin\alpha\cos\beta - \cos\alpha\sin\beta$

$$= \frac{21}{29}\cdot\frac{12}{13} - \left(-\frac{20}{29}\right)\left(-\frac{5}{13}\right) = \frac{152}{377}$$

$$\tan(\alpha + \beta) = \frac{\tan\alpha + \tan\beta}{1 - \tan\alpha\tan\beta}$$

$$= \frac{-\frac{21}{20} - \frac{5}{12}}{1 - \frac{21}{20}\cdot\frac{5}{12}} = \frac{(-21)(12) - 5\cdot 20}{240 - 105}$$

$$= \frac{-352}{135} = -\frac{352}{135}$$

**25.** **(a)** $\cos(191°) = \cos(11° + 180°) = -\cos 11°$
$= -0.9816$

**(b)** $\sin(132°) = \sin(90° + 42°) = \cos 42°$
$= 0.7431$

**(c)** $\tan(173°) = \tan(180° - 7°) = -\tan 7°$
$= -0.1228$

**(d)** $\cos(102°) = \cos(90° + 12°) = -\sin 12°$
$= -0.2079$

**27.** $\cos\left(\theta - \dfrac{\pi}{4}\right) = \cos\theta\cos\dfrac{\pi}{4} + \sin\theta\sin\dfrac{\pi}{4}$

$\qquad = \dfrac{\sqrt{2}}{2}(\cos\theta + \sin\theta)$

**29.** $\cos(\theta + 30°) + \cos(\theta - 30°)$

$\qquad = \cos\theta\cos 30° - \sin\theta\sin 30°$
$\qquad\qquad + \cos\theta\cos 30° + \sin\theta\sin 30°$
$\qquad = 2\cos\theta\cos 30° = \sqrt{3}\cos\theta$

**31.** $\dfrac{\sin\left(\theta + \frac{\pi}{2}\right)}{\cos\left(\theta + \frac{\pi}{2}\right)} = \dfrac{\cos\theta}{-\sin\theta}$

$\qquad = -\cot\theta$

**33.** $\csc(\pi - \theta) = \dfrac{1}{\sin(\pi - \theta)} = \dfrac{1}{\sin\theta} = \csc\theta$

**35.** $\sin(\alpha + \beta) - \sin(\alpha - \beta) = \sin\alpha\cos\beta$

$\qquad\quad + \cos\alpha\sin\beta - \sin\alpha\cos\beta + \cos\alpha\sin\beta$
$\qquad = 2\cos\alpha\sin\beta$

**37.** $\dfrac{\sin(\alpha + \beta)}{\sin(\alpha - \beta)}$

$\qquad = \dfrac{\sin\alpha\cos\beta + \cos\alpha\sin\beta}{\sin\alpha\cos\beta - \cos\alpha\sin\beta} \cdot \dfrac{\frac{1}{\cos\alpha\cos\beta}}{\frac{1}{\cos\alpha\cos\beta}}$

$\qquad = \dfrac{\frac{\sin\alpha\cos\beta}{\cos\alpha\cos\beta} + \frac{\cos\alpha\sin\beta}{\cos\alpha\cos\beta}}{\frac{\sin\alpha\cos\beta}{\cos\alpha\cos\beta} - \frac{\cos\alpha\sin\beta}{\cos\alpha\cos\beta}} = \dfrac{\tan\alpha + \tan\beta}{\tan\alpha - \tan\beta}$

**39.** $2\sin\left(\dfrac{\pi}{4} - \theta\right)\sin\left(\dfrac{\pi}{4} + \theta\right)$

$\qquad = 2\left[\sin\dfrac{\pi}{4}\cos\theta - \cos\dfrac{\pi}{4}\sin\theta\right]$

$\qquad\quad \cdot\left[\sin\dfrac{\pi}{4}\cos\theta + \cos\dfrac{\pi}{4}\sin\theta\right]$

$\qquad = 2\left[\dfrac{\cos\theta}{\sqrt{2}} - \dfrac{\sin\theta}{\sqrt{2}}\right]\left[\dfrac{\cos\theta}{\sqrt{2}} + \dfrac{\sin\theta}{\sqrt{2}}\right]$

$\qquad = 2\left[\dfrac{\cos^2\theta}{2} - \dfrac{\sin^2\theta}{2}\right] = \cos^2\theta - \sin^2\theta$

**41.** $\cos\left(\dfrac{\pi}{2} - x\right) = \sin x$

$\qquad \cos\left(\dfrac{\pi}{2} - \left(\dfrac{\pi}{2} - \theta\right)\right) = \sin\left(\dfrac{\pi}{2} - \theta\right)$

$\qquad \cos\theta = \sin\left(\dfrac{\pi}{2} - \theta\right)$

**43.** $\sin(\alpha - \beta)' = \sin(\alpha + (-\beta))$

$\qquad = \sin\alpha\cos(-\beta) + \cos\alpha\sin(-\beta)$
$\qquad = \sin\alpha\cos\beta - \cos\alpha\sin\beta$

**45.** $\cot(\alpha + \beta) = \dfrac{\cos(\alpha + \beta)}{\sin(\alpha + \beta)}$

$\qquad = \dfrac{\cos\alpha\cos\beta - \sin\alpha\sin\beta}{\sin\alpha\cos\beta + \cos\alpha\sin\beta} \cdot \dfrac{\frac{1}{\sin\alpha\sin\beta}}{\frac{1}{\sin\alpha\sin\beta}}$

$\qquad = \dfrac{\frac{\cos\alpha\cos\beta}{\sin\alpha\sin\beta} - \frac{\sin\alpha\sin\beta}{\sin\alpha\sin\beta}}{\frac{\sin\alpha\cos\beta}{\sin\alpha\sin\beta} + \frac{\cos\alpha\sin\beta}{\sin\alpha\sin\beta}}$

$\qquad = \dfrac{\cot\alpha\cot\beta - 1}{\cot\beta + \cot\alpha} = \dfrac{\cot\alpha\cot\beta - 1}{\cot\alpha + \cot\beta}$

**47.** Since $\sin\theta = -\cos\left(\theta + \dfrac{\pi}{2}\right)$, shift $y = \cos\theta$

by $\dfrac{\pi}{2}$ units to the left and reflect through the

$\theta$-axis.

**49.** $\sin\left(\theta + \dfrac{\pi}{2}k\right) = \sin\theta\cos\dfrac{\pi}{2}k + \cos\theta\sin\dfrac{\pi}{2}k$. If $k$

is even, then $\sin\dfrac{\pi}{2}k = 0$, $\cos\dfrac{\pi}{2}k = \pm 1$,

giving $\sin\left(\theta + \dfrac{\pi}{2}k\right) = \pm\sin\theta$. If $k$ is odd,

then $\cos\dfrac{\pi}{2}k = 0$, $\sin\dfrac{\pi}{2}k = \pm 1$, giving

$\sin\left(\theta + \dfrac{\pi}{2}k\right) = \pm\cos\theta$.

**51.** $y = 3\sin x - 3\cos x$

$\quad = 3\sqrt{2}\left(\cos\left(-\dfrac{\pi}{4}\right)\sin x + \sin\left(-\dfrac{\pi}{4}\right)\cos x\right)$

$\quad = 3\sqrt{2}\sin\left(x - \dfrac{\pi}{4}\right)$

**53.** $f(x) = 2\sin x + 2\sqrt{3}\cos x$

$\quad = 4\left(\cos\dfrac{\pi}{3}\sin x + \sin\dfrac{\pi}{3}\cos x\right) = 4\sin\left(x + \dfrac{\pi}{3}\right)$

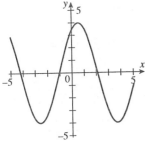

**55.** $\tan(\alpha + \beta) = \dfrac{\tan\alpha + \tan\beta}{1 - \tan\alpha\tan\beta}$

$\quad \dfrac{5}{8} = \dfrac{\frac{2}{8} + \tan\beta}{1 - \frac{2}{8}\tan\beta} \cdot \dfrac{8}{8}$

$\quad \dfrac{5}{8} = \dfrac{2 + 8\tan\beta}{8 - 2\tan\beta}$

$\quad \dfrac{5}{8}(8 - 2\tan\beta) = 2 + 8\tan\beta$

$\quad 5 - \dfrac{5}{4}\tan\beta = 2 + 8\tan\beta$

$\quad -\dfrac{5}{4}\tan\beta - 8\tan\beta = 2 - 5$

$\quad -\dfrac{37}{4}\tan\beta = -3$

$\quad \tan\beta = \dfrac{12}{37}$

**57.** $\tan(\arctan 3 + \arctan 4)$

$\quad = \dfrac{\tan(\arctan 3) + \tan(\arctan 4)}{1 - \tan(\arctan 3)\tan(\arctan 4)}$

$\quad = \dfrac{3 + 4}{1 - 3 \cdot 4} = \dfrac{7}{1 - 12} = -\dfrac{7}{11}$

Check: $\tan(\arctan 3 + \arctan 4) = -0.6364,$

$\quad \dfrac{-7}{11} = -0.6364$

**59.** $\cos\left(\arcsin\dfrac{8}{17} - \arccos\dfrac{12}{13}\right)$

$\quad = \cos\left(\arcsin\dfrac{8}{17}\right)\cos\left(\arccos\dfrac{12}{13}\right)$

$\quad + \sin\left(\arcsin\dfrac{8}{17}\right)\sin\left(\arccos\dfrac{12}{13}\right)$

$\quad = \dfrac{15}{17}\cdot\dfrac{12}{13} + \dfrac{8}{17}\cdot\dfrac{5}{13} = \dfrac{220}{221}$

$\quad \cos\left(\arcsin\dfrac{8}{17} - \arccos\dfrac{12}{13}\right) = 0.9955$

$\quad \dfrac{220}{221} = 0.9955$

Check: $\cos\left(\arcsin\dfrac{8}{17} - \arccos\dfrac{12}{13}\right) = 0.9955$

$\quad \dfrac{220}{221} = 0.9955$

**61. (a)** $\dfrac{f(x+h) - f(x)}{h} = \dfrac{\sin(x+h) - \sin x}{h}$

$\quad = \dfrac{\sin x\cos h + \cos x\sin h - \sin x}{h}$

$\quad = \dfrac{\sin x(\cos h - 1)}{h} + \dfrac{\cos x\sin h}{h}$

$\quad = \left(\dfrac{\cos h - 1}{h}\right)\sin x + \left(\dfrac{\sin h}{h}\right)\cos x$

**(b)** $\dfrac{g(x+h) - g(x)}{h} = \dfrac{\cos(x+h) - \cos x}{h}$

$\quad = \dfrac{\cos x\cos h - \sin x\sin h - \cos x}{h}$

$\quad = \dfrac{\cos x(\cos h - 1)}{h} - \dfrac{\sin x\sin h}{h}$

$\quad = \left(\dfrac{\cos h - 1}{h}\right)\cos x - \left(\dfrac{\sin h}{h}\right)\sin x$

**63.** $\tan\alpha = \text{slope of } l_1 = -\dfrac{3}{2}$

$\quad \tan\beta = \text{slope of } l_2 = -6$

$\quad \tan(\alpha - \beta) = \dfrac{\tan\alpha - \tan\beta}{1 + \tan\alpha\tan\beta} = \dfrac{-\frac{3}{2} + 6}{1 + 9} = \dfrac{9}{20}$

$\quad \alpha - \beta = \arctan\left(\dfrac{9}{20}\right) = 24.2°$

**65.** $\tan\alpha = $ slope of $l_1 = -\dfrac{1}{3}$

$\tan\beta = $ slope of $l_2 = \dfrac{5}{6}$

$\tan(\alpha-\beta) = \dfrac{\tan\alpha - \tan\beta}{1+\tan\alpha\tan\beta} = \dfrac{-\frac{1}{3}-\frac{5}{6}}{1+\left(-\frac{1}{3}\right)\left(\frac{5}{6}\right)}$

$= \dfrac{-6-15}{18-5} = -\dfrac{21}{13}$

$\alpha - \beta = 180° + \arctan\left(-\dfrac{21}{13}\right) = 121.8°$

**Challenge**

$\sin(\alpha+\beta) = \dfrac{e+f}{d} = \dfrac{e}{d} + \dfrac{f}{d}$

$\quad = \dfrac{e}{d} + \dfrac{a}{d} \quad$ since $a = f$

$\quad < \dfrac{b}{d} + \dfrac{a}{d} \quad$ since $b > e$

$\quad < \dfrac{b}{d} + \dfrac{a}{c} \quad$ since $c < d$ so $\dfrac{1}{c} > \dfrac{1}{d}$

$\quad = \sin\beta + \sin\alpha$

$\quad = \sin\alpha + \sin\beta$

Therefore $\sin(\alpha+\beta) < \sin\alpha + \sin\beta$.

**Exercises 7.5**

**1.** $1 - 2\sin^2\dfrac{7\pi}{12} = \cos\left(2\cdot\dfrac{7\pi}{12}\right)$

$= \cos\left(\dfrac{7\pi}{6}\right) = -\dfrac{\sqrt{3}}{2}$

**3.** $\dfrac{6\tan 75°}{1-\tan^2 75°} = 3\cdot\tan(2\cdot 75°)$

$= 3\cdot\tan 150° = -\dfrac{3}{\sqrt{3}} = -\sqrt{3}$

**5.** $\cos 75° = \cos\left(\dfrac{1}{2}\cdot 150°\right) = \sqrt{\dfrac{1+\cos 150°}{2}}$

$= \dfrac{1}{2}\sqrt{2 + 2\cdot\left(\dfrac{-\sqrt{3}}{2}\right)} = \dfrac{1}{2}\sqrt{2-\sqrt{3}}$

**7.** $\tan\dfrac{3\pi}{8} = \tan\left(\dfrac{1}{2}\cdot\dfrac{3\pi}{4}\right) = \dfrac{1-\cos\frac{3\pi}{4}}{\sin\frac{3\pi}{4}}$

$= \dfrac{1+\frac{\sqrt{2}}{2}}{\frac{\sqrt{2}}{2}}\cdot\dfrac{\frac{2}{\sqrt{2}}}{\frac{2}{\sqrt{2}}} = \dfrac{2}{\sqrt{2}} + 1 = \sqrt{2} + 1$

**9. (a)** $\tan 15° = \tan\left(\dfrac{1}{2}\cdot 30°\right) = \dfrac{1-\cos 30°}{\sin 30°}$

$= \dfrac{1-\frac{\sqrt{3}}{2}}{\frac{1}{2}} = 2 - \sqrt{3}$

**(b)** $\tan 15° = 0.2679,\ 2-\sqrt{3} = 0.2679$

**11. (a)** $\tan(22.5°) = \tan\left(\dfrac{1}{2}\cdot 45°\right) = \dfrac{1-\cos 45°}{\sin 45°}$

$= \dfrac{1-\frac{1}{\sqrt{2}}}{\frac{1}{\sqrt{2}}} = \sqrt{2} - 1$

**(b)** $\tan 22.5° = 0.4142,\ \sqrt{2}-1 = 0.4142$

**13.** $\sin\theta = \dfrac{15}{17},\ \cos\theta = -\dfrac{8}{17},\ \tan\theta = -\dfrac{15}{8}$

**(a)** $\sin 2\theta = 2\sin\theta\cos\theta$

$= 2\left(\dfrac{15}{17}\right)\left(-\dfrac{8}{17}\right) = -\dfrac{240}{289}$

**(b)** $\cos 2\theta = 1 - 2\sin^2\theta$

$= 1 - 2\cdot\left(\dfrac{15}{17}\right)^2 = -\dfrac{161}{289}$

**(c)** $\tan 2\theta = \dfrac{2\tan\theta}{1-\tan^2\theta} = \dfrac{2\left(-\frac{15}{8}\right)}{1-\left(-\frac{15}{8}\right)^2}$

$= \dfrac{-240}{64-225} = \dfrac{240}{161}$

**15.** $\tan\theta = -\dfrac{2}{3},\ \sin\theta = -\dfrac{2}{\sqrt{13}},\ \cos\theta = \dfrac{3}{\sqrt{13}}$

$\sin 2\theta = 2\sin\theta\cos\theta = 2\left(-\dfrac{2}{\sqrt{13}}\right)\left(\dfrac{3}{\sqrt{13}}\right)$

$= -\dfrac{12}{13}$

**17.** $9\sin 2\theta = 18\sin\theta\cos\theta = 18\left(\dfrac{x}{3}\right)\left(\dfrac{\sqrt{9-x^2}}{3}\right)$

$= 2x\sqrt{9-x^2}$

**19.** $\sin 15° = \dfrac{b}{10}$ or $b = 10\sin 15°$

$b = 10\sin\left(\dfrac{1}{2}\cdot 30°\right) = 10\cdot\sqrt{\dfrac{1-\cos 30°}{2}}$

$= 10\cdot\sqrt{\dfrac{1-\frac{\sqrt{3}}{2}}{2}} = 10\sqrt{\dfrac{1}{2}-\dfrac{\sqrt{3}}{4}}$

$= \dfrac{10}{2}\sqrt{2-\sqrt{3}} = 5\sqrt{2-\sqrt{3}}$

**21.** $\tan\theta = \dfrac{h}{150}$ and $\tan 2\theta = \dfrac{h}{50}$

$\dfrac{2\tan\theta}{1-\tan^2\theta} = \dfrac{h}{50}$

$\dfrac{2\cdot\frac{h}{150}}{1-\left(\frac{h}{150}\right)^2} = \dfrac{h}{50}$

$\dfrac{300h}{22{,}500-h^2} = \dfrac{h}{50}$

$15{,}000h = 22{,}500h - h^3$

$h^3 - 7500h = 0$

$h\left(h^2 - 7500\right) = 0$

$h^2 = 7500$

$h = \sqrt{7500} = 50\sqrt{3}$

**23.** $\tan\theta = \dfrac{9}{b}$ and $\tan 2\theta = \dfrac{21}{b}$, so

$\dfrac{2\tan\theta}{1-\tan^2\theta} = \dfrac{21}{b}$

$\dfrac{2\left(\frac{9}{b}\right)}{1-\left(\frac{9}{b}\right)^2} = \dfrac{21}{b}$

$\dfrac{18b}{b^2-81} = \dfrac{21}{b}$

$18b^2 = 21b^2 - 1701$

$-3b^2 = -1701$

$b^2 = 567$

$b = 9\sqrt{7}$

**25.** $\cot x\sin 2x = \dfrac{\cos x}{\sin x}\cdot 2\sin x\cos x = 2\cos^2 x$

**27.** $\sec^2\dfrac{x}{2} = \dfrac{1}{\cos^2\frac{x}{2}}$

$= \dfrac{1}{\frac{1+\cos x}{2}}$

$= \dfrac{2}{1+\cos x}$

**29.** $\cos^2\dfrac{x}{2} = \dfrac{1+\cos x}{2}$

$= \dfrac{\tan x + \tan x\cos x}{2\tan x}$

$= \dfrac{\tan x + \sin x}{2\tan x}$

**31.** $\cot\dfrac{x}{2} = \dfrac{1}{\tan\frac{x}{2}}$

$= \dfrac{1}{\frac{\sin x}{1+\cos x}}$

$= \dfrac{1+\cos x}{\sin x}$

$= \dfrac{1}{\sin x} + \dfrac{\cos x}{\sin x}$

$= \csc x + \cot x$

**33.** $(\sin x + \cos x)^2$

$= \sin^2 x + 2\sin x\cos x + \cos^2 x$

$= 1 + 2\sin x\cos x$

$= 1 + \sin 2x$

**35.** $\tan 3x = \tan(2x + x) = \dfrac{\tan 2x + \tan x}{1 - \tan 2x \tan x}$

$= \dfrac{\dfrac{2\tan x}{1-\tan^2 x} + \tan x}{1 - \dfrac{2\tan x}{1-\tan^2 x}\tan x}$

$= \dfrac{2\tan x + \tan x - \tan^3 x}{1 - \tan^2 x - 2\tan^2 x}$

$= \dfrac{3\tan x - \tan^3 x}{1 - 3\tan^2 x}$

**37.** $\cot 2x = \dfrac{1}{\tan 2x} = \dfrac{1}{\dfrac{2\tan x}{1-\tan^2 x}}$

$= \dfrac{1 - \tan^2 x}{2\tan x}$

$= \dfrac{\dfrac{1}{\tan^2 x} - \dfrac{\tan^2 x}{\tan^2 x}}{\dfrac{2\tan x}{\tan^2 x}}$

$= \dfrac{\cot^2 x - 1}{2\cot x}$

**39.** $\csc 2x - \cot 2x = \dfrac{1}{\sin 2x} - \dfrac{\cos 2x}{\sin 2x}$

$= \dfrac{1 - \cos 2x}{\sin 2x}$

$\tan \dfrac{2x}{2}$

$= \tan x$

**41.** $\cos^4 x - \sin^4 x$

$= \left(\cos^2 x - \sin^2 x\right)\left(\cos^2 x + \sin^2 x\right)$

$= \cos^2 x - \sin^2 x$

$= \cos 2x$

**43.** $\cos 4x = \cos 2(2x) = 2\cos^2 2x - 1$

$= 2\left(2\cos^2 x - 1\right)^2 - 1$

$= 2\left(4\cos^4 x - 4\cos^2 x + 1\right) - 1$

$= 8\cos^4 x - 8\cos^2 x + 2 - 1$

$= 8\cos^4 x - 8\cos^2 x + 1$

**45.** $\cos^4 x = \left(\cos^2 x\right)^2 = \left(\dfrac{1+\cos 2x}{2}\right)^2$

$= \dfrac{1}{4}\left(1 + 2\cos 2x + \cos^2 2x\right)$

$= \dfrac{1}{4}\left(1 + 2\cos 2x + \dfrac{1+\cos 4x}{2}\right)$

$= \dfrac{1}{8}\cos 4x + \dfrac{1}{2}\cos 2x + \dfrac{3}{8}$

**47.** $\sin^2 x \cos^2 x = \left(\dfrac{1-\cos 2x}{2}\right)\left(\dfrac{1+\cos 2x}{2}\right)$

$= \dfrac{1}{4}\left(1 - \cos^2 2x\right)$

$= \dfrac{1}{4}\left(1 - \dfrac{1+\cos 4x}{2}\right)$

$= \dfrac{1}{4}\left(\dfrac{1}{2} - \dfrac{1}{2}\cos 4x\right)$

$= \dfrac{1}{8}\left(1 - \cos 4x\right)$

**49.** $a\sin kx + b\cos kx = c \cdot \dfrac{a}{c}\sin kx + c \cdot \dfrac{b}{c}\cos kx = c\cos\theta\sin kx + c\sin\theta\cos kx = c\sin(kx + \theta)$

**51. (a)** $\cos\alpha\cos\beta - \sin\alpha\sin\beta = \cos(\alpha + \beta)$

$\dfrac{\cos\alpha\cos\beta + \sin\alpha\sin\beta = \cos(\alpha - \beta)}{2\cos\alpha\cos\beta \qquad\qquad = \cos(\alpha + \beta) + \cos(\alpha - \beta)}$

$\cos\alpha\cos\beta = \dfrac{1}{2}\left[\cos(\alpha + \beta) + \cos(\alpha - \beta)\right]$

**(b) 1.**

$$\cos\alpha\cos\beta + \sin\alpha\sin\beta = \cos(\alpha - \beta)$$
$$\underline{-\left(\cos\alpha\cos\beta - \sin\alpha\sin\beta = \cos(\alpha + \beta)\right)}$$
$$2\sin\alpha\sin\beta = \cos(\alpha - \beta) - \cos(\alpha + \beta)$$
$$\sin\alpha\sin\beta = \frac{1}{2}\left[\cos(\alpha - \beta) - \cos(\alpha + \beta)\right]$$

**2.**

$$\sin\alpha\cos\beta + \cos\alpha\sin\beta = \sin(\alpha + \beta)$$
$$\underline{\sin\alpha\cos\beta - \cos\alpha\sin\beta = \sin(\alpha - \beta)}$$
$$2\sin\alpha\cos\beta = \sin(\alpha + \beta) + \sin(\alpha - \beta)$$
$$\sin\alpha\cos\beta = \frac{1}{2}\left[\sin(\alpha + \beta) + \sin(\alpha - \beta)\right]$$

**3.**

$$\sin\alpha\cos\beta + \cos\alpha\sin\beta = \sin(\alpha + \beta)$$
$$\underline{-\left(\sin\alpha\cos\beta - \cos\alpha\sin\beta = \sin(\alpha - \beta)\right)}$$
$$2\cos\alpha\sin\beta = \sin(\alpha + \beta) - \sin(\alpha - \beta)$$
$$\cos\alpha\sin\beta = \frac{1}{2}\left[\sin(\alpha + \beta) - \sin(\alpha - \beta)\right]$$

**53.** $2\cos\left(\dfrac{u+v}{2}\right)\cos\left(\dfrac{u-v}{2}\right) = \cos\left(\dfrac{u+v}{2}+\dfrac{u-v}{2}\right) + \cos\left(\dfrac{u+v}{2}-\dfrac{u-v}{2}\right) = \cos u + \cos v$

$2\sin\left(\dfrac{u+v}{2}\right)\sin\left(\dfrac{u-v}{2}\right) = \cos\left(\dfrac{u+v}{2}-\dfrac{u-v}{2}\right) - \cos\left(\dfrac{u+v}{2}+\dfrac{u-v}{2}\right) = \cos v - \cos u$

$2\sin\left(\dfrac{u+v}{2}\right)\cos\left(\dfrac{u-v}{2}\right) = \sin\left(\dfrac{u+v}{2}+\dfrac{u-v}{2}\right) + \sin\left(\dfrac{u+v}{2}-\dfrac{u-v}{2}\right) = \sin u + \sin v$

$2\cos\left(\dfrac{u+v}{2}\right)\sin\left(\dfrac{u-v}{2}\right) = \sin\left(\dfrac{u+v}{2}+\dfrac{u-v}{2}\right) - \sin\left(\dfrac{u+v}{2}-\dfrac{u-v}{2}\right) = \sin u - \sin v$

**55.** $\cos\left(2\arcsin\dfrac{1}{3}\right) = 1 - 2\sin^2\left(\arcsin\dfrac{1}{3}\right) = 1 - 2\left(\dfrac{1}{3}\right)^2 = 1 - \dfrac{2}{9} = \dfrac{7}{9}$

**57.** $\tan\left(2\arctan\dfrac{4}{5}\right) = \dfrac{2\tan\left(\arctan\frac{4}{5}\right)}{1 - \tan^2\left(\arctan\frac{4}{5}\right)} = \dfrac{2\cdot\frac{4}{5}}{1-\left(\frac{4}{5}\right)^2} = \dfrac{40}{25-16} = \dfrac{40}{9}$

**59.** $\sin(2\arctan x) = 2\sin(\arctan x)\cos(\arctan x) = 2\left(\dfrac{x}{\sqrt{1+x^2}}\right)\left(\dfrac{1}{\sqrt{1+x^2}}\right) = \dfrac{2x}{1+x^2}$

**61.** $\tan(2\arctan x) = \dfrac{2\tan(\arctan x)}{1-\tan^2(\arctan x)} = \dfrac{2x}{1-x^2}$

**Chapter 7:** *Right Triangle Trigonometry, Identities, and Equations*

## Challenge

In the figure $\angle PQR = \dfrac{1}{2}\theta$ and $\sin\theta = \dfrac{PR}{PO} = PR$,

and $\cos\theta = \dfrac{RO}{PO} = RO$. So,

$$\tan\frac{\theta}{2} = \frac{PR}{QR} = \frac{PR}{1+RO} = \frac{\sin\theta}{1+\cos\theta}$$

## Critical Thinking

1. In the tangent formulas, all tangents must be defined, so $\alpha$, $\beta$ and $\alpha\pm\beta$ cannot be coterminal with $\dfrac{\pi}{2}$ or $\dfrac{3\pi}{2}$. Also, $\tan\alpha\tan\beta \neq \pm1$. Therefore $\alpha$ and $\beta$ cannot be complementary angles.

3. From Example 8, $\dfrac{a+b}{9} = \tan 30°$ where $a = 9\tan 15°$, so solving this for $b$ yields
$b = 9\tan 30° - 9\tan 15°$
$= 9(\tan 30° - \tan 15°)$.
This is what needs to be evaluated using a calculator.

5. $\cos 15° = \cos\dfrac{1}{2}(30°) = \sqrt{\dfrac{1+\cos 30°}{2}}$

$= \sqrt{\dfrac{1+\frac{\sqrt{3}}{2}}{2}} = \sqrt{\dfrac{1}{2} + \dfrac{\sqrt{3}}{4}} = \dfrac{1}{2}\sqrt{2+\sqrt{3}}$

$\cos 60° = \cos 4(15°) = 8\cos^4 15° - 8\cos^2 15° + 1$

$= 8\left(\dfrac{1}{2}\sqrt{2+\sqrt{3}}\right)^4 - 8\left(\dfrac{1}{2}\sqrt{2+\sqrt{3}}\right)^2 + 1$

$= 8\cdot\dfrac{1}{16}(2+\sqrt{3})^2 - 8\cdot\dfrac{1}{4}(2+\sqrt{3}) + 1$

$= \dfrac{1}{2}(4+4\sqrt{3}+3) - 2(2+\sqrt{3}) + 1$

$= \dfrac{7}{2} + 2\sqrt{3} - 4 - 2\sqrt{3} + 1$

$= \dfrac{1}{2}$

## Exercises 7.6

1. $\cos x = 1$
$x = 2k\pi$

3. $\sin x = \dfrac{1}{2}$
$x = \dfrac{\pi}{6} + 2k\pi \text{ or } \dfrac{5\pi}{6} + 2k\pi$

5. $\sec x = -1$
$x = \pi + 2k\pi$

7. $\sin 2x = \dfrac{-1}{2}$
$2x = \dfrac{7\pi}{6} + 2k\pi \text{ or } 2x = \dfrac{11\pi}{6} + 2k\pi$
$x = \dfrac{7\pi}{12} + k\pi \text{ or } x = \dfrac{11\pi}{12} + k\pi$

9. $\dfrac{1}{2}\sin^2 x = 1$
$\sin^2 x = 2$
$\sin x = \pm\sqrt{2}$
No solutions

11. $\tan 2x = \sqrt{3}$
$2x = \dfrac{\pi}{3} + k\pi \text{ or } 2x = \dfrac{4\pi}{3} + k\pi$
$x = \dfrac{\pi}{6} + \dfrac{k\pi}{2} \text{ or } x = \dfrac{2\pi}{3} + \dfrac{k\pi}{2}$

13. $2\cos(x+1) = -2$
$\cos(x+1) = -1$
$x+1 = \pi + 2k\pi$
$x = \pi + 2k\pi - 1$

15. $\cos 3x = 1$
$3x = 2k\pi$
$x = \dfrac{2k\pi}{3}$

17. $\sqrt{3}\csc\theta = 2$
$\csc\theta = \dfrac{2}{\sqrt{3}} \text{ or } \sin\theta = \dfrac{\sqrt{3}}{2}$
$\theta = 60°,\ 120°$

**19.** $2\sec\theta - 2\sqrt{2} = 0$
$2\sec\theta = 2\sqrt{2}$
$\sec\theta = \sqrt{2}$
$\theta = 45°, \ 315°$

**21.** $\sin^2\theta - \cos^2\theta = 1$
$-\cos 2\theta = 1$
$\cos 2\theta = -1$
$2\theta = 180°, \ 540°$
$\theta = 90°, \ 270°$

**23.** $2\sin\dfrac{\theta}{2} - 1 = 0$
$2\sin\dfrac{\theta}{2} = 1$
$\sin\dfrac{\theta}{2} = \dfrac{1}{2}$
$\dfrac{\theta}{2} = 30°, \ 150°$
$\theta = 60°, \ 300°$

**25.** $-1 + \tan\dfrac{3\theta}{2} = 0$
$\tan\dfrac{3\theta}{2} = 1$
$\dfrac{3\theta}{2} = 45°, \ 225°, \ 405°$
$\theta = 30°, \ 150°, \ 270°$

**27.** $2\tan\theta\cos^2\theta = 1$
$2\dfrac{\sin\theta}{\cos\theta}\cdot\cos^2\theta = 1$
$2\sin\theta\cos\theta = 1$
$\sin 2\theta = 1$
$2\theta = 90°, 450°$
$\theta = 45°, 225°$

**29.** $(\cos x - 1)(2\sin x + 1) = 0$
$\cos x - 1 = 0$ or $2\sin x + 1 = 0$
$\cos x = 1$ or $\sin x = \dfrac{-1}{2}$
$x = 0$ or $x = \dfrac{7\pi}{6}, \ \dfrac{11\pi}{6}$
$x = 0, \ \dfrac{7\pi}{6}, \ \dfrac{11\pi}{6}$

**31.** $\sin x - \sqrt{3}\cos x = 0$
$\sin x = \sqrt{3}\cos x$
$\dfrac{\sin x}{\cos x} = \sqrt{3}$
$\tan x = \sqrt{3}$
$x = \dfrac{\pi}{3}, \ \dfrac{4\pi}{3}$

**33.** $2\cos^2 x - \sqrt{3}\cos x = 0$
$\cos x(2\cos x - \sqrt{3}) = 0$
$\cos x = 0$ or $2\cos x - \sqrt{3} = 0$
$\cos x = \dfrac{\sqrt{3}}{2}$
$x = \dfrac{\pi}{2}, \ \dfrac{3\pi}{2}$ or $x = \dfrac{\pi}{6}, \ \dfrac{11\pi}{6}$
$x = \dfrac{\pi}{2}, \ \dfrac{3\pi}{2}, \ \dfrac{\pi}{6}, \ \dfrac{11\pi}{6}$

**35.** $\sin^2 x + \sin x - 2 = 0$
$(\sin x + 2)(\sin x - 1) = 0$
$\sin x + 2 = 0$ or $\sin x - 1 = 0$
$\sin x = -2$ or $\sin x = 1$
no solutions $\quad x = \dfrac{\pi}{2}$
$x = \dfrac{\pi}{2}$

**37.** $\sin 2x = \cos x$
$2\sin x\cos x = \cos x$
$2\sin x\cos x - \cos x = 0$
$\cos x(2\sin x - 1) = 0$
$\cos x = 0$ or $2\sin x - 1 = 0$
$\sin x = \dfrac{1}{2}$
$x = \dfrac{\pi}{2}, \ \dfrac{3\pi}{2}$ or $x = \dfrac{\pi}{6}, \ \dfrac{5\pi}{6}$
$x = \dfrac{\pi}{6}, \ \dfrac{5\pi}{6}, \ \dfrac{\pi}{2}, \ \dfrac{3\pi}{2}$

**39.** $\sin^2 x + \cos^2 x = 1.5$
$1 = 1.5$
no solutions

**41.** $2\cos^2 x + 9\cos x - 5 = 0$

$(2\cos x - 1)(\cos x + 5) = 0$

$2\cos x - 1 = 0$ or $\cos x = -5$

$\cos x = \dfrac{1}{2}$ \qquad no solutions

$x = \dfrac{\pi}{3}, \dfrac{5\pi}{3}$

**43.** $\cos^2 2x = \cos 2x$

$\cos^2 2x - \cos 2x = 0$

$\cos 2x(\cos 2x - 1) = 0$

$\cos 2x = 0$ or $\cos 2x = 1$

$2x = \dfrac{\pi}{2}, \dfrac{3\pi}{2}, \dfrac{5\pi}{2}, \dfrac{7\pi}{2}$ or $2x = 0, \; 2\pi$

$x = 0, \dfrac{\pi}{4}, \dfrac{3\pi}{4}, \dfrac{5\pi}{4}, \dfrac{7\pi}{4}, \pi$

**45.** $2\sin^4 x - 3\sin^2 x + 1 = 0$

$(2\sin^2 x - 1)(\sin^2 x - 1) = 0$

$2\sin^2 x - 1 = 0$ \qquad or $\sin^2 x - 1 = 0$

$\sin^2 x = \dfrac{1}{2}$ \qquad or $\sin^2 x = 1$

$\sin x = \pm\dfrac{1}{\sqrt{2}}$ \qquad or $\sin x = \pm 1$

$x = \dfrac{\pi}{4}, \dfrac{3\pi}{4}, \dfrac{5\pi}{4}, \dfrac{7\pi}{4}$ or $x = \dfrac{\pi}{2}, \dfrac{3\pi}{2}$

$x = \dfrac{\pi}{4}, \dfrac{\pi}{2}, \dfrac{3\pi}{4}, \dfrac{5\pi}{4}, \dfrac{3\pi}{2}, \dfrac{7\pi}{4}$

**47.** $2\tan x - 1 = \tan^2 x$

$0 = \tan^2 x - 2\tan x + 1$

$0 = (\tan x - 1)^2$

$\tan x - 1 = 0$

$\tan x = 1$

$x = \dfrac{\pi}{4}, \dfrac{5\pi}{4}$

**49.** $3\tan^2 x = 7\sec x - 5$

$3(\sec^2 x - 1) = 7\sec x - 5$

$3\sec^2 x - 7\sec x + 2 = 0$

$(3\sec x - 1)(\sec x - 2) = 0$

$3\sec x - 1 = 0$ or $\sec x - 2 = 0$

$\sec x = \dfrac{1}{3}$ \qquad or $\sec x = 2$

no solutions \qquad $x = \dfrac{\pi}{3}, \dfrac{5\pi}{3}$

$x = \dfrac{\pi}{3}, \dfrac{5\pi}{3}$

**51.** $\cos^2 x - \sin^2 x + \sin x = 1$

$1 - \sin^2 x - \sin^2 x + \sin x = 1$

$-2\sin^2 x + \sin x = 0$

$\sin x(-2\sin x + 1) = 0$

$\sin x = 0$ or $-2\sin x + 1 = 0$

$\sin x = \dfrac{1}{2}$

$x = 0, \pi$ \quad or $x = \dfrac{\pi}{6}, \dfrac{5\pi}{6}$

$x = 0, \dfrac{\pi}{6}, \dfrac{5\pi}{6}, \pi$

**53.** $3\cos 2x + 2\sin^2 x = 2$

$3(1 - 2\sin^2 x) + 2\sin^2 x = 2$

$-4\sin^2 x = -1$

$\sin^2 x = \dfrac{1}{4}$

$\sin x = \pm\dfrac{1}{2}$

$x = \dfrac{\pi}{6}, \dfrac{5\pi}{6}, \dfrac{7\pi}{6}, \dfrac{11\pi}{6}$

**55.** $\sin 2x + \sin 4x = 0$

$\sin 2x + 2\sin 2x \cos 2x = 0$

$\sin 2x(1 + 2\cos 2x) = 0$

$\sin 2x = 0$ or $1 + 2\cos 2x = 0$

$\sin 2x = 0$ or $\cos 2x = -\dfrac{1}{2}$

$2x = 0, \; \pi, \; 2\pi, \; 3\pi$ or

$\qquad\qquad 2x = \dfrac{2\pi}{3}, \dfrac{4\pi}{3}, \dfrac{8\pi}{3}, \dfrac{10\pi}{3}$

$x = 0, \dfrac{\pi}{3}, \dfrac{\pi}{2}, \dfrac{2\pi}{3}, \pi, \dfrac{4\pi}{3}, \dfrac{3\pi}{2}, \dfrac{5\pi}{3}$

**57.** $\sin x + \cos x = 1$

$(\sin x + \cos x)^2 = 1^2$

$\sin^2 x + 2\cos x \sin x + \cos^2 x = 1$

$1 + 2\sin x \cos x = 1$

$2\sin x \cos x = 0$

$\sin x = 0$ or $\cos x = 0$

$x = 0,\ \pi$ or $x = \dfrac{\pi}{2},\ \dfrac{3\pi}{2}$

But, $x = \pi$ or $x = \dfrac{3\pi}{2}$ aren't solutions.

$x = 0,\ \dfrac{\pi}{2}$

**59.** $7\sec x - 15 = 0$

$7\sec x = 15$

$\sec x = \dfrac{15}{7}$ or $\cos x = \dfrac{7}{15}$

$x = 62.2°,\ 297.8°$

**61.** $\sin \dfrac{x}{2} = 0.8259$

$\dfrac{x}{2} = 55.68°,\ 124.32°$

$x = 111.4°,\ 248.6°$

**63.** $4\cot^2 x - 12\cot x + 9 = 0$

$(2\cot x - 3)^2 = 0$

$2\cot x - 3 = 0$

$2\cot x = 3$

$\cot x = \dfrac{3}{2}$ or $\tan x = \dfrac{2}{3}$

$x = 33.7°,\ 213.7°$

**65.** $\cos^2 x - \sin^2 x = -\dfrac{3}{4}$

$\cos 2\theta = -\dfrac{3}{4}$

$2\theta = 138.6°,\ 221.4°,\ 498.6°,\ 581.4°$

$\theta = 69.3°,\ 110.7°,\ 249.3°,\ 290.7°$

**67. (a)** $\cos \alpha = \dfrac{28}{35} = 0.8,$

$\cos \dfrac{\pi}{4} = \dfrac{1}{\sqrt{2}} = 0.7071...$

Since cosine is decreasing for

$0 \le x \le \dfrac{\pi}{2}$, and $\cos \dfrac{\pi}{4} < \cos \alpha$, then

$\alpha < \dfrac{\pi}{4}.$

**(b)** $\cos 2\alpha = 2\cos^2 \alpha - 1$

$= 2(0.8)^2 - 1 = \dfrac{7}{25} = \dfrac{28}{100} = \cos(\alpha + \beta).$

Then, since the cosine is one-to-one

for $0 \le x \le \dfrac{\pi}{2}$, $2\alpha = \alpha + \beta$ or $\alpha = \beta$.

**69.** $y = \sin x$

$y = -\cos x$

$\sin x = -\cos x$

$\dfrac{\sin x}{\cos x} = -1$

$\tan x = -1$

$x = \dfrac{3\pi}{4},\ \dfrac{7\pi}{4}$

For $x = \dfrac{3\pi}{4}, y = \sin \dfrac{3\pi}{4} = \dfrac{\sqrt{2}}{2}$

For $x = \dfrac{7\pi}{4},\ y = \sin \dfrac{7\pi}{4} = -\dfrac{\sqrt{2}}{2}$

$\left( \dfrac{3\pi}{4},\ \dfrac{\sqrt{2}}{2} \right),\ \left( \dfrac{7\pi}{4},\ -\dfrac{\sqrt{2}}{2} \right)$

**Challenge**

For $x = \begin{cases} \dfrac{\pi}{2} + 2k\pi \\ \dfrac{3\pi}{2} + 2k\pi \end{cases}$, we can replace this by

$x = \dfrac{\pi}{2} + k\pi.$

**Chapter 7 Review Exercises**

**1.** $\sin \theta = \dfrac{\text{side opposite}}{\text{hypotenuse}}$

$\cos \theta = \dfrac{\text{side adjacent}}{\text{hypotenuse}}$

$\tan \theta = \dfrac{\text{side opposite}}{\text{side adjacent}}$

$\cot \theta = \dfrac{\text{side adjacent}}{\text{side opposite}}$

$\sec \theta = \dfrac{\text{hypotenuse}}{\text{side adjacent}}$

$\csc \theta = \dfrac{\text{hypotenuse}}{\text{side opposite}}$

**3.** $a = 2, b = 3$

$c^2 = a^2 + b^2 = 2^2 + 3^2 = 13$

$c = \sqrt{13}$

$\cos B = \dfrac{2}{\sqrt{13}}$

**5.** $\sin B = \dfrac{\sqrt{x^2 - 1}}{x}$

$\cos B = \dfrac{1}{x}$

$\tan B = \sqrt{x^2 - 1}$

$\cot B = \dfrac{1}{\sqrt{x^2 - 1}}$

$\sec B = x$

$\csc B = \dfrac{x}{\sqrt{x^2 - 1}}$

**7.**

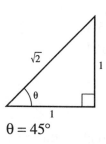

$\theta = 45°$

**9.**

$\theta = 30°$

**11.** $\tan^2 \theta + 1 = \left( \dfrac{\sqrt{25 - x^2}}{x} \right)^2 + 1 = \dfrac{25 - x^2}{x^2} + 1$

$= \dfrac{25 - x^2 + x^2}{x^2} = \dfrac{25}{x^2} = \left( \dfrac{5}{x} \right)^2 = \sec^2 \theta$

**13.** $\sin \dfrac{\pi}{3} = \dfrac{12}{x}$

$\dfrac{\sqrt{3}}{2} = \dfrac{12}{x}$

$x\sqrt{3} = 24$

$x = \dfrac{24}{\sqrt{3}} = \dfrac{24\sqrt{3}}{3} = 8\sqrt{3}$

**15.** $\sin \theta = \dfrac{3}{\sqrt{12}} = \dfrac{3}{2\sqrt{3}} = \dfrac{3\sqrt{3}}{2 \cdot 3} = \dfrac{\sqrt{3}}{2}$

$\theta = \dfrac{\pi}{3}$

**17.** $\sin 73° = 0.9563$

**19.** $\tan 40.1° = 0.8421$

**21.** $\sin \theta = 0.2356$

$\theta = 13.6°$

**23.** $\tan \theta = 1.3452$

$\theta = 53.4°$

**25.** $\sec \theta = 4.3997$

$\theta = 76.9°$

**27.** $\tan 52.8° = \dfrac{20}{b}$

$b = \dfrac{20}{\tan 52.8°} = 15.2$

**29.** $\tan 32° = \dfrac{18}{d}$

$d = \dfrac{18}{\tan 32°} = 29 \text{ m}$

**31.** $\tan 20° = \dfrac{h}{12 + x}$ and $\tan 65° = \dfrac{h}{x}$

$(12 + x)\tan 20° = h$ $\qquad x \tan 65° = h$

$x \tan 20° = h - 12 \tan 20°$ $\qquad x = \dfrac{h}{\tan 65°}$

$x = \dfrac{h - 12 \tan 20°}{\tan 20°}$ and $x = \dfrac{h}{\tan 65°}$

$\dfrac{h - 12 \tan 20°}{\tan 20°} = \dfrac{h}{\tan 65°}$

$h \tan 65° - 12 \tan 20° \tan 65° = h \tan 20°$

$h \tan 65° - h \tan 20° = 12 \tan 20° \tan 65°$

$$h(\tan 65° - \tan 20°) = 12 \tan 20° \tan 65°$$
$$h = \frac{12 \tan 20° \tan 65°}{\tan 65° - \tan 20°}$$
$$h = 5.3$$

**33.** $\cot^2(10°) - \csc^2(10°) = -1$

**35.** $(\sec^2 8° - 1)\cot^2 8° = \tan^2 8° \cdot \cot^2 8° = 1$

**37.**
$$\frac{\sec\theta + \csc\theta}{\tan\theta - \cot\theta} = \frac{\frac{1}{\cos\theta} + \frac{1}{\sin\theta}}{\frac{\sin\theta}{\cos\theta} - \frac{\cos\theta}{\sin\theta}} \cdot \frac{\sin\theta\cos\theta}{\sin\theta\cos\theta}$$
$$= \frac{\sin\theta + \cos\theta}{\sin^2\theta - \cos^2\theta}$$
$$= \frac{\sin\theta + \cos\theta}{(\sin\theta + \cos\theta)(\sin\theta - \cos\theta)}$$
$$= \frac{1}{\sin\theta - \cos\theta}$$

$\theta$ is not a quadrantal angle and $\theta$ is not coterminal with $\dfrac{\pi}{4}, \dfrac{3\pi}{4}, \dfrac{5\pi}{4}, \dfrac{7\pi}{4}$.

**39.** $\sin^2\theta - 2\cos^2\theta = \sin^2\theta - 2(1 - \sin^2\theta)$
$$= \sin^2\theta - 2 + 2\sin^2\theta$$
$$= 3\sin^2\theta - 2$$

**41.** $(1 + \cos\theta)(1 - \cos\theta) = 1 - \cos^2\theta$
$$= \sin^2\theta$$
$$= \frac{1}{\csc^2\theta}$$
$$= \frac{1}{1 + \cot^2\theta}$$

**43.**
$$\frac{\cot\theta\cos\theta}{\csc\theta - 1} = \frac{\frac{\cos\theta}{\sin\theta} \cdot \cos\theta}{\frac{1}{\sin\theta} - 1}$$
$$= \frac{\cos^2\theta}{1 - \sin\theta}$$
$$= \frac{1 - \sin^2\theta}{1 - \sin\theta}$$
$$= \frac{(1 - \sin\theta)(1 + \sin\theta)}{1 - \sin\theta}$$
$$= 1 + \sin\theta$$

**45.**
$$\frac{\cot\theta}{1 + 3\cot^2\theta} = \frac{\frac{\cos\theta}{\sin\theta}}{1 + 3\frac{\cos^2\theta}{\sin^2\theta}} \cdot \frac{\sin^2\theta}{\sin^2\theta}$$
$$= \frac{\cos\theta\sin\theta}{\sin^2\theta + 3\cos^2\theta}$$
$$= \frac{\cos\theta\sin\theta}{\sin^2\theta + \cos^2\theta + 2\cos^2\theta}$$
$$= \frac{\cos\theta\sin\theta}{1 + 2\cos^2\theta} \cdot \frac{\frac{1}{\cos\theta}}{\frac{1}{\cos\theta}}$$
$$= \frac{\sin\theta}{\frac{1}{\cos\theta} + 2\cos\theta}$$
$$= \frac{\sin\theta}{\sec\theta + 2\cos\theta}$$

**47.**
$$\frac{1}{\sec\theta + \tan\theta} = \frac{1}{\sec\theta + \tan\theta} \cdot \frac{\sec\theta - \tan\theta}{\sec\theta - \tan\theta}$$
$$= \frac{\sec\theta - \tan\theta}{\sec^2\theta - \tan^2\theta}$$
$$= \frac{\sec\theta - \tan\theta}{1}$$
$$= \sec\theta - \tan\theta$$

**49.** $\sec\pi = -1$
$$\sqrt{1 + \tan^2\pi} = \sqrt{1 + 0} = \sqrt{1} = 1$$
$$-1 \neq 1$$

**51.** $\cos(6° - \theta)\cos(6° + \theta) - \sin(6° - \theta)\sin(6° + \theta)$
$$= \cos[(6° - \theta) + (6° + \theta)]$$
$$= \cos 12°$$

**53.** $\cos\dfrac{7\pi}{12} = \cos\left(\dfrac{\pi}{3} + \dfrac{\pi}{4}\right)$
$$= \cos\frac{\pi}{3}\cos\frac{\pi}{4} - \sin\frac{\pi}{3}\sin\frac{\pi}{4}$$
$$= \frac{1}{2} \cdot \frac{\sqrt{2}}{2} - \frac{\sqrt{3}}{2} \cdot \frac{\sqrt{2}}{2}$$
$$\frac{1}{4}(\sqrt{2} - \sqrt{6})$$

**55.** From Exercise 54,

$$\tan\alpha = -\frac{3}{4}, \quad \tan\beta = -\frac{12}{5}.$$

$$\tan(\alpha + \beta) = \frac{\tan\alpha + \tan\beta}{1 - \tan\alpha\tan\beta}$$

$$= \frac{-\frac{3}{4} - \frac{12}{5}}{1 - \left(-\frac{3}{4}\right)\left(-\frac{12}{5}\right)} \cdot \frac{20}{20}$$

$$= \frac{-15 - 48}{20 - 36} = \frac{63}{16}$$

**57.** $\cos\left(\frac{\pi}{6} + \theta\right) = \cos\frac{\pi}{6}\cos\theta - \sin\frac{\pi}{6}\sin\theta$

$$= \frac{\sqrt{3}}{2}\cos\theta - \frac{1}{2}\sin\theta$$

$$= \frac{1}{2}(\sqrt{3}\cos\theta - \sin\theta)$$

**59.** $\sin\alpha\cos(\alpha - \beta) - \cos\alpha\sin(\alpha - \beta)$

$= \sin\alpha(\cos\alpha\cos\beta + \sin\alpha\sin\beta)$

$\quad - \cos\alpha(\sin\alpha\cos\beta - \cos\alpha\sin\beta)$

$= \sin\alpha\cos\alpha\cos\beta + \sin^2\alpha\sin\beta$

$\quad - \cos\alpha\sin\alpha\cos\beta + \cos^2\alpha\sin\beta$

$= (\sin^2\alpha + \cos^2\alpha)\sin\beta$

$= \sin\beta$

**61.** $\tan(\alpha + \beta) = \frac{\tan\alpha + \tan\beta}{1 - \tan\alpha\tan\beta}$

$$\frac{10}{6} = \frac{\frac{4}{6} + \tan\beta}{1 - \frac{4}{6}\tan\beta} = \frac{4 + 6\tan\beta}{6 - 4\tan\beta}$$

$$60 - 40\tan\beta = 24 + 36\tan\beta$$

$$-76\tan\beta = -36$$

$$\tan\beta = \frac{9}{19}$$

**63.** $\frac{1}{2}\sin 15° \cos 15° = \frac{1}{4}\sin 30° = \frac{1}{8}$

**65.** $\cos\frac{\pi}{8} = \cos\frac{1}{2}\left(\frac{\pi}{4}\right) = \sqrt{\frac{1 + \cos\frac{\pi}{4}}{2}}$

$$= \sqrt{\frac{1}{2} + \frac{\sqrt{2}}{4}} = \frac{1}{2}\sqrt{2 + \sqrt{2}}$$

**67.** $\tan\theta = -\frac{4}{3}, \quad \sin\theta = \frac{4}{5}, \quad \cos\theta = -\frac{3}{5}$

$$\sin 2\theta = 2\sin\theta\cos\theta = 2\left(\frac{4}{5}\right)\left(-\frac{3}{5}\right) = -\frac{24}{25}$$

$$\cos 2\theta = 2\cos^2\theta - 1 = 2\left(-\frac{3}{5}\right)^2 - 1 = -\frac{7}{25}$$

$$\tan 2\theta = \frac{\sin 2\theta}{\cos 2\theta} = \frac{-\frac{24}{25}}{-\frac{7}{25}} = \frac{24}{7}$$

**69.** **(a)** $\sin 2\theta = 2\sin\theta\cos\theta = 2x\left(\sqrt{1 - x^2}\right)$

$$= 2x\sqrt{1 - x^2}$$

$$\cos 2\theta = 1 - 2\sin^2\theta = 1 - 2x^2$$

**(b)** $\sin 4\theta = 2\sin 2\theta\cos 2\theta$

$$= 4x\sqrt{1 - x^2}\,(1 - 2x^2)$$

**71.** $\dfrac{1 + \cos 2x}{\sin 2x} = \dfrac{1 + 2\cos^2 x - 1}{\sin 2x}$

$$= \frac{2\cos^2 x}{2\sin x\cos x} = \frac{\cos x}{\sin x} = \cot x$$

**73.** $\sin 4x = 2\sin 2x\cos 2x$

$= 2(2\sin x\cos x)(1 - 2\sin^2 x)$

$= 4\sin x\cos x - 8\sin^3 x\cos x$

**75.** $\sec x = \dfrac{2}{\sqrt{3}}$

$$x = \frac{\pi}{6} + 2k\pi \text{ or } x = \frac{11\pi}{6} + 2k\pi$$

**77.** $\tan 3x = -1$

$$3x = \frac{3\pi}{4} + \pi k \text{ or } 3x = \frac{7\pi}{4} + \pi k$$

$$x = \frac{\pi}{4} + \frac{\pi}{3}k \quad \text{or } x = \frac{7\pi}{12} + \frac{\pi}{3}k$$

**79.** $\dfrac{1}{2}\sin\theta - 1 = 0$

$\dfrac{1}{2}\sin\theta = 1$

$\sin\theta = 2$
No solutions

**81.** $2\sin^2\theta\cot\theta = -1$

$2\sin^2\theta\dfrac{\cos\theta}{\sin\theta} = -1$

$2\sin\theta\cos\theta = -1$

$\sin 2\theta = -1$

$2\theta = 270°,\ 630°$

$\theta = 135°,\ 315°$

**83.** $2\cos^2 x - \cos x - 1 = 0$

$(2\cos x + 1)(\cos x - 1) = 0$

$2\cos x + 1 = 0$ or $\cos x - 1 = 0$

$\cos x = \dfrac{-1}{2}$ $\qquad\cos x = 1$

$x = \dfrac{2\pi}{3}, \dfrac{4\pi}{3}$ $\quad$ or $x = 0$

$x = 0,\ \dfrac{2\pi}{3}, \dfrac{4\pi}{3}$

**85.** $\sec x\tan^2 x - 3\sec x = 0$

$\sec x(\tan^2 x - 3) = 0$

$\sec x = 0$ or $\tan^2 x - 3 = 0$

no solutions $\quad\tan^2 x = 3$

$\tan x = \pm\sqrt{3}$

$x = \dfrac{\pi}{3}, \dfrac{2\pi}{3}, \dfrac{4\pi}{3}, \dfrac{5\pi}{3}$

**87.** $5\cos x - 3 = 0$

$5\cos x = 3$

$\cos x = \dfrac{3}{5}$

$x = 53.1°,\ 306.9°$

## Chapter 7 Test: Standard Answer

**1. (a)** $\tan A = \dfrac{30}{16} = \dfrac{15}{8}$

**(b)** $\sin B = \dfrac{16}{34} = \dfrac{8}{17}$

**(c)** $\sec A = \dfrac{34}{16} = \dfrac{17}{8}$

**2.** $c = 3,\ b = 2$

$a^2 = c^2 - b^2 = 3^2 - 2^2 = 9 - 4 = 5$

$a = \sqrt{5}$

$\sin A = \dfrac{\sqrt{5}}{3}$

$\cos A = \dfrac{2}{3}$

$\tan A = \dfrac{\sqrt{5}}{2}$

**3.** $3(\sin^2 A)(\sec A) = 3\left(\dfrac{x}{\sqrt{x^2+9}}\right)^2\left(\dfrac{\sqrt{x^2+9}}{3}\right)$

$\qquad = \dfrac{x^2}{\sqrt{x^2+9}}$

**4.** hypotenuse $= \sqrt{20^2 + 20^2} = \sqrt{400 + 400}$

$= \sqrt{800}$

$= 20\sqrt{2}$

**5.** $\tan 75° = \dfrac{h}{100}$

$h = 100\tan 75°$

$h = 373$ feet

**6.** $\tan 47.8° = \dfrac{h}{120}$

$h = 120\tan 47.8°$

$h = 132.3$ m

**7.** $\cot^2\theta - \cos^2\theta = \dfrac{\cos^2\theta}{\sin^2\theta} - \cos^2\theta$

$= \cos^2\theta\left(\dfrac{1}{\sin^2\theta} - 1\right)$

$= \cos^2\theta\left(\dfrac{1 - \sin^2\theta}{\sin^2\theta}\right)$

$$= \cos^2\theta\left(\frac{\cos^2\theta}{\sin^2\theta}\right)$$

$$= \cos^2\theta \cot^2\theta$$

**8.** $(\sec\theta - \tan\theta)^2 = \left(\dfrac{1}{\cos\theta} - \dfrac{\sin\theta}{\cos\theta}\right)^2$

$$= \frac{(1-\sin\theta)^2}{\cos^2\theta}$$

$$= \frac{(1-\sin\theta)^2}{1-\sin^2\theta}$$

$$= \frac{(1-\sin\theta)^2}{(1-\sin\theta)(1+\sin\theta)}$$

$$= \frac{1-\sin\theta}{1+\sin\theta}$$

**9.** $\dfrac{\tan\theta - \sin\theta}{\csc\theta - \cot\theta} = \dfrac{\frac{\sin\theta}{\cos\theta} - \sin\theta}{\frac{1}{\sin\theta} - \frac{\cos\theta}{\sin\theta}}$

$$= \frac{\sin^2\theta - \sin^2\theta\cos\theta}{\cos\theta - \cos^2\theta}$$

$$= \frac{\sin^2\theta(1-\cos\theta)}{\cos\theta(1-\cos\theta)}$$

$$= \frac{\sin^2\theta}{\cos\theta}$$

$$= \frac{1-\cos^2\theta}{\cos\theta} = \frac{1}{\cos\theta} - \cos\theta = \sec\theta - \cos\theta$$

**10.** $\cos 165° = \cos(135° + 30°)$

$= \cos 135° \cos 30° - \sin 135° \sin 30°$

$$= -\frac{\sqrt{2}}{2}\cdot\frac{\sqrt{3}}{2} - \frac{\sqrt{2}}{2}\cdot\frac{1}{2}$$

$$= -\frac{1}{4}(\sqrt{6} + \sqrt{2})$$

**11.** $\sin\left(-\dfrac{\pi}{12}\right) = \sin\dfrac{1}{2}\left(-\dfrac{\pi}{6}\right)$

$$= -\sqrt{\frac{1-\cos\left(-\frac{\pi}{6}\right)}{2}} = -\sqrt{\frac{1-\frac{\sqrt{3}}{2}}{2}}$$

$$= -\sqrt{\frac{1}{2} - \frac{\sqrt{3}}{4}} = -\frac{1}{2}\sqrt{2-\sqrt{3}}$$

**12.** $\sin\theta = \dfrac{1}{3}, \ \cos\theta = \dfrac{2\sqrt{2}}{3}$

$$\sin 2\theta = 2\sin\theta\cos\theta = 2\left(\frac{1}{3}\right)\left(\frac{2\sqrt{2}}{3}\right) = \frac{4\sqrt{2}}{9}$$

**13.** $\dfrac{\tan 115° - \tan 55°}{1 + \tan 115° \tan 55°} = \tan(115° - 55°)$

$= \tan(60°)$

$= \sqrt{3}$

**14.** $\sin\left(\dfrac{\pi}{4} - \theta\right) = \sin\dfrac{\pi}{4}\cos\theta - \cos\dfrac{\pi}{4}\sin\theta$

$$= \frac{\sqrt{2}}{2}\cos\theta - \frac{\sqrt{2}}{2}\sin\theta$$

$$= \frac{\sqrt{2}}{2}(\cos\theta - \sin\theta)$$

**15.** $\csc 4x = \dfrac{1}{\sin 4x} = \dfrac{1}{2\sin 2x \cos 2x}$

$$= \frac{1}{4\sin x\cos x\cos 2x}$$

$$= \frac{1}{4}\csc x\sec x\sec 2x$$

**16.** $\sin^2\dfrac{x}{2} = \dfrac{1-\cos x}{2}$

$$= \frac{\tan x - \tan x\cos x}{2\tan x}$$

$$= \frac{\tan x - \sin x}{2\tan x}$$

**17.** $\tan\left(\arctan 3 - \arctan\dfrac{4}{9}\right)$

$= \dfrac{\tan(\arctan 3) - \tan\left(\arctan\frac{4}{9}\right)}{1 + \tan(\arctan 3)\tan\left(\arctan\frac{4}{9}\right)}$

$= \dfrac{3 - \frac{4}{9}}{1 + 3\cdot\frac{4}{9}}\cdot\dfrac{9}{9}$

$= \dfrac{27 - 4}{9 + 12}$

$= \dfrac{23}{21}$

**18.** $\cos\left(2\arcsin\dfrac{1}{2}\right) = 1 - 2\sin^2\left(\arcsin\dfrac{1}{2}\right)$

$= 1 - 2\left(\dfrac{1}{2}\right)^2$

$= 1 - \dfrac{1}{2} = \dfrac{1}{2}$

**19.** $\sin^2 x - \cos^2 x + \sin x = 0$

$\sin^2 x - (1 - \sin^2 x) + \sin x = 0$

$2\sin^2 x + \sin x - 1 = 0$
$(2\sin x - 1)(\sin x + 1) = 0$
$2\sin x - 1 = 0$ or $\sin x + 1 = 0$

$\sin x = \dfrac{1}{2}$ or $\sin x = -1$

$x = \dfrac{\pi}{6}, \dfrac{5\pi}{6}$ or $x = \dfrac{3\pi}{2}$

$x = \dfrac{\pi}{6}, \dfrac{5\pi}{6}, \dfrac{3\pi}{2}$

**20.** $\cos 2x = 2\sin^2 x - 2$

$1 - 2\sin^2 x = 2\sin^2 x - 2$

$-4\sin^2 x = -3$

$\sin^2 x = \dfrac{3}{4}$

$\sin x = \pm\dfrac{\sqrt{3}}{2}$

$x = 60°, 120°, 240°, 300°$

**21.** $\sin 2x = 1$

$2x = \dfrac{\pi}{2} + 2k\pi$

$x = \dfrac{\pi}{4} + k\pi$

**22.** $5\cos^2 x - 17x + 6 = 0$
$(5\cos x - 2)(\cos x - 3) = 0$
$5\cos x - 2 = 0$ or $\cos x - 3 = 0$

$\cos x = \dfrac{2}{5}$ or $\cos x = 3$

no solutions

$x = 1.16, 5.12$

**23.** $\sin\alpha = \dfrac{3}{5}, \quad \cos\alpha = -\dfrac{4}{5}$

$\tan\beta = \dfrac{12}{5}, \quad \sin\beta = -\dfrac{12}{13}, \quad \cos\beta = -\dfrac{5}{13}$

$\cos(\alpha + \beta) = \cos\alpha\cos\beta - \sin\alpha\sin\beta$

$= \left(-\dfrac{4}{5}\right)\left(-\dfrac{5}{13}\right) - \left(\dfrac{3}{5}\right)\left(-\dfrac{12}{13}\right)$

$= \dfrac{20}{65} + \dfrac{36}{65} = \dfrac{56}{65}$

**24.** $\tan 15° = \dfrac{8}{b}$

$b = \dfrac{8}{\tan 15°} = \dfrac{8}{\tan\frac{1}{2}(30°)} = \dfrac{8}{\frac{\sin 30°}{1 + \cos 30°}}$

$= \dfrac{8 + 8\cos 30°}{\sin 30°} = \dfrac{8 + 8\cdot\frac{\sqrt{3}}{2}}{\frac{1}{2}}$

$= 16 + 8\sqrt{3} = 8(2 + \sqrt{3})$

**25.** $\tan 2\theta = \dfrac{2\tan\theta}{1 - \tan^2\theta}$

$\dfrac{24}{b} = \dfrac{2\cdot\frac{9}{b}}{1 - \left(\frac{9}{b}\right)^2} = \dfrac{18b}{b^2 - 81}$

$24b^2 - 1944 = 18b^2$

$6b^2 = 1944$

$b^2 = 324$

$b = 18$

## Chapter 7 Test: Multiple Choice

**1.** $a = 2$, $b = 3$, $c^2 = a^2 + b^2 = 2^2 + 3^2 = 13$

$c = \sqrt{13}$

$\sec B = \dfrac{\sqrt{13}}{2}$

The answer is (b).

**2.** $\sin 30° + \cos 60° - \tan 45° = \dfrac{1}{2} + \dfrac{1}{2} - 1 = 0$

The answer is (c).

**3.** $\tan 35° = \dfrac{x}{50}$ or $\tan 55° = \dfrac{50}{x}$

The answer is (b).

**4.** Since all three are true, the answer is (d).

**5.** $\sin^2 \dfrac{\pi}{4} - \cos^2 \dfrac{\pi}{4} = \left(\dfrac{\sqrt{2}}{2}\right)^2 - \left(\dfrac{\sqrt{2}}{2}\right)^2$

$= \dfrac{1}{2} - \dfrac{1}{2} = 0 \neq 1$

The answer is (c).

**6.** $\sec^2 \theta + 2 \sec \theta \tan \theta + \tan^2 \theta$

$= (\sec \theta + \tan \theta)^2$

$= \left(\dfrac{1}{\cos \theta} + \dfrac{\sin \theta}{\cos \theta}\right)^2 = \dfrac{(1 + \sin \theta)^2}{\cos^2 \theta}$

$= \dfrac{(1 + \sin \theta)^2}{1 - \sin^2 \theta} = \dfrac{(1 + \sin \theta)^2}{(1 - \sin \theta)(1 + \sin \theta)}$

$= \dfrac{1 + \sin \theta}{1 - \sin \theta}$

The answer is (b).

**7.** $\cos \dfrac{5\pi}{4} \cos \dfrac{\pi}{2} - \sin \dfrac{5\pi}{4} \sin \dfrac{\pi}{2} = \cos\left(\dfrac{5\pi}{4} + \dfrac{\pi}{2}\right)$

$= \cos \dfrac{7\pi}{4}$

The answer is (a).

**8.** $\sin \alpha = \dfrac{3}{5}$, $\cos \alpha = -\dfrac{4}{5}$

$\sin \beta = -\dfrac{4}{5}$, $\cos \beta = -\dfrac{3}{5}$

$\sin(\alpha - \beta) = \sin \alpha \cos \beta - \cos \alpha \sin \beta$

$= \left(\dfrac{3}{5}\right)\left(-\dfrac{3}{5}\right) - \left(-\dfrac{4}{5}\right)\left(-\dfrac{4}{5}\right)$

$= -\dfrac{9}{25} - \dfrac{16}{25} = -\dfrac{25}{25} = -1$

The answer is (a).

**9.** Only III is true, so the answer is (c).

**10.** $\cos \theta = -\dfrac{5}{13}$, $\sin \theta = \dfrac{12}{13}$

$\sin 2\theta = 2 \sin \theta \cos \theta = 2\left(\dfrac{12}{13}\right)\left(-\dfrac{5}{13}\right) = -\dfrac{120}{169}$

The answer is (d).

**11.** $\sin \dfrac{\pi}{8} = \sin \dfrac{1}{2}\left(\dfrac{\pi}{4}\right) = \sqrt{\dfrac{1 - \cos \frac{\pi}{4}}{2}}$

$= \sqrt{\dfrac{1}{2} - \dfrac{\sqrt{2}}{4}} = \dfrac{1}{2}\sqrt{2 - \sqrt{2}}$

The answer is (c).

**12.** $\dfrac{\sin 2x}{1 - \cos^2 x} = \dfrac{2 \sin x \cos x}{\sin^2 x} = \dfrac{2 \cos x}{\sin x} = 2 \cot x$

The answer is (b).

**13.** $\sin 2x = 1$

$2x = \dfrac{\pi}{2} + 2k\pi$

$x = \dfrac{\pi}{4} + k\pi$

The answer is (a).

**14.** $\sin^2 x = \sin x$

$\sin^2 x - \sin x = 0$

$\sin x(\sin x - 1) = 0$

$\sin x = 0$ or $\sin x = 1$

$x = 0, \dfrac{\pi}{2}, \pi$

The answer is (e).

**15.** $\tan \theta = \dfrac{100}{23}$

$\theta = 77°$

The answer (d).

**16.** $\cot\theta$ must be defined and $\sin\theta \neq \cos\theta$, so $\theta$ must not be coterminal with $0$, $\pi$, $\dfrac{\pi}{4}$ and $\dfrac{5\pi}{4}$.

The answer is (b).

**17.** $\cos\left(\theta + \dfrac{\pi}{3}\right) - \cos\left(\theta - \dfrac{\pi}{3}\right)$

$\quad = \cos\theta\cos\dfrac{\pi}{3} - \sin\theta\sin\dfrac{\pi}{3}$

$\qquad - \cos\theta\cos\dfrac{\pi}{3} - \sin\theta\sin\dfrac{\pi}{3}$

$\quad = -2\sin\theta\sin\dfrac{\pi}{3}$

$\quad = -2\sin\theta \cdot \dfrac{\sqrt{3}}{2}$

$\quad = -\sqrt{3}\sin\theta$

The answer is (d).

**18.** $y = 3\sin x - 3\sqrt{3}\cos x$

$\quad = 6 \cdot \dfrac{1}{2}\sin x - 6 \cdot \dfrac{\sqrt{3}}{2}\cos x$

$\quad = 6 \cdot \cos\left(-\dfrac{\pi}{3}\right)\sin x + 6 \cdot \sin\left(-\dfrac{\pi}{3}\right)\cos x$

$\quad = 6\sin\left(x - \dfrac{\pi}{3}\right)$

$\theta = -\dfrac{\pi}{3}$ and the answer is (c).

**19.** $5\sin^2\theta - 2\sin\theta = 0$

$\sin\theta(5\sin\theta - 2) = 0$

$\sin\theta = 0$ or $\sin\theta = \dfrac{2}{5}$

$\theta = 0°, 180°$ or $\theta = 23.6°, 156.4°$

There are four solutions, so the answer is (a).

**20.** $\sin(2\arccos x)$

$\quad = 2\sin(\arccos x)\cos(\arccos x)$

$\quad = 2\left(\sqrt{1 - x^2}\right)x$

$\quad = 2x\sqrt{1 - x^2}$

The answer is (b).

# Chapter 8: Additional Applications of Trigonometry

**Exercises 8.1**

**1.**  $a^2 = b^2 + c^2 - 2bc\cos A$

$\qquad = 4^2 + 11^2 - 2\cdot 4\cdot 11\cos 60°$
$\qquad = 16 + 121 - 44$
$\qquad = 93$
$a = \sqrt{93}$

**3.**  $a^2 = b^2 + c^2 - 2bc\cos A$

$(5\sqrt{3})^2 = (10\sqrt{3})^2 + 15^2 - 2(10\sqrt{3})(15)\cos A$
$75 = 300 + 225 - 300\sqrt{3}\cos A$
$300\sqrt{3}\cos A = 450$
$\cos A = \dfrac{3}{2\sqrt{3}}\cdot\dfrac{\sqrt{3}}{\sqrt{3}} = \dfrac{\sqrt{3}}{2}$
$A = 30°$

**5.**  $a^2 = b^2 + c^2 - 2bc\cos A$

$\qquad = 8^2 + 13^2 - 2\cdot 8\cdot 13\cos 20°$
$\qquad = 37.5439$
$a = 6.1$
$\cos B = \dfrac{a^2 + c^2 - b^2}{2ac} = \dfrac{(6.1)^2 + 13^2 - 8^2}{2(6.1)(13)}$
$\qquad = 0.8967$
$B = 26.3°$

**7.**  $\cos A = \dfrac{b^2 + c^2 - a^2}{2bc} = \dfrac{5^2 + 13^2 - 12^2}{2\cdot 5\cdot 13}$
$\qquad = 0.3846,\ A = 67.4°$
$\cos C = \dfrac{a^2 + b^2 - c^2}{2ab} = \dfrac{12^2 + 5^2 - 13^2}{2\cdot 12\cdot 5}$
$\qquad = 0$
$C = 90°$

**9.**  $\cos B = \dfrac{a^2 + c^2 - b^2}{2ac} = \dfrac{18^2 + 4^2 - 15^2}{2\cdot 18\cdot 4}$
$\qquad = 0.7986$
$B = 37.0°$
$\cos C = \dfrac{a^2 + b^2 - c^2}{2ab} = \dfrac{18^2 + 15^2 - 4^2}{2\cdot 18\cdot 15}$
$\qquad = 0.9870$
$C = 9.2°$

**11.**  $c^2 = a^2 + b^2 - 2ab\cos C$

$\qquad = 18^2 + 9^2 - 2\cdot 18\cdot 9\cos 30.2°$
$\qquad = 124.9750$
$c = 11.2$
$\cos A = \dfrac{b^2 + c^2 - a^2}{2bc} = \dfrac{9^2 + (11.2)^2 - 18^2}{2(9)(11.2)}$
$\qquad = -0.5831$
$A = 125.7°$

**13.**  $a^2 = b^2 + c^2 - 2bc\cos A$

$\qquad = 2.2^2 + 6.4^2 - 2(2.2)(6.4)\cos 42°$
$\qquad = 24.8730$
$a = 5.0$
$\cos B = \dfrac{a^2 + c^2 - b^2}{2ac} = \dfrac{5^2 + 6.4^2 - 2.2^2}{2(5)(6.4)}$
$\qquad = 0.955$
$B = 17.3°$

**15.**  $(AB)^2 = 180^2 + 120^2 - 2(180)(120)\cos 56.3°$
$\qquad = 22{,}830.7207$
$AB = 151$ meters

**17.**  $(AB)^2 = (3.5)^2 + (2.2)^2 - 2(3.5)(2.2)\cos 112°$
$\qquad = 22.8589$
$AB = 4.8$ miles

**19.** If 2 of the angles sum to 80°, then the remaining 2 angles sum to 280° or 140° for each remaining angle. Let $d$ be the length of a diagonal. Then
$d^2 = 10^2 + 14^2 - 2\cdot 10\cdot 14\cos 140°$
$\qquad = 510.4924$
$d = 22.6$ cm

**21.**  $\angle ACB = 90° + 26° = 116°$
$(AB)^2 = 20^2 + 30^2 - 2\cdot 20\cdot 30\cos 116°$
$\qquad = 1826.0454$
$AB = 42.7$

**23.** The distance from home to second is $90\sqrt{2}$ feet, and the angle that line segment makes with the segment from shortstop to second is $45° + 15° = 60°$. So, if $d$ is the distance from shortstop to home,

$$d^2 = 50^2 + (90\sqrt{2})^2 - 2 \cdot (50)(90\sqrt{2})\cos 60°$$
$$= 12,336.0390$$
$$d = 111 \text{ feet}$$

**25.** After 20 minutes, or $\frac{1}{3}$ hour, the two trains are 18 miles and 20 miles from the station. If $d$ is the distance between the trains, then

$$d^2 = 18^2 + 20^2 - 2 \cdot 18 \cdot 20 \cos 124°$$
$$= 1126.6189$$
$$d = 33.6 \text{ miles}$$

**27.** $K = \frac{1}{2}(25)(18) \sin 30°$
$\qquad = 112.5$

**29.** $K = \frac{1}{2}(8.4)(12.6) \sin 40.5°$
$\qquad = 34.4$

**31.** $K = \frac{1}{2}(8)(6) = 24$

$s = \frac{1}{2}(6 + 8 + 10) = 12$

$K = \sqrt{12(12 - 6)(12 - 8)(12 - 10)}$
$\qquad = \sqrt{12 \cdot 6 \cdot 4 \cdot 2}$
$\qquad = \sqrt{576}$
$\qquad = 24$

**33.** $K = \frac{1}{2}(10)(5) = 25$

$s = \frac{1}{2}(10 + 5\sqrt{2} + 5\sqrt{2}) = 5 + 5\sqrt{2}$

$K = \sqrt{(5 + 5\sqrt{2})(5 + 5\sqrt{2} - 5\sqrt{2})(5 + 5\sqrt{2} - 5\sqrt{2})(5 + 5\sqrt{2} - 10)}$
$\qquad = \sqrt{(5 + 5\sqrt{2})(5)(5)(-5 + 5\sqrt{2})}$
$\qquad = \sqrt{625}$
$\qquad = 25$

**35.** $s = \frac{1}{2}(10 + 13 + 3.4) = 13.2$

$K = \sqrt{(13.2)(13.2 - 10)(13.2 - 13)(13.2 - 3.4)}$
$\qquad = \sqrt{(13.2)(3.2)(0.2)(9.8)}$
$\qquad = \sqrt{82.7904}$
$\qquad = 9.1 \text{ sq. ft}$

**37.** $(AD)^2 = 20^2 + 30^2 - 2 \cdot 20 \cdot 30 \cos 75°$
$\qquad = 989.4171$
$AD = 31.5 \text{ in.}$
$(CD)^2 = 30^2 + 15^2 - 2 \cdot 30 \cdot 15 \cos 40°$
$\qquad = 435.5600$
$CD = 20.9 \text{ in.}$
For $\triangle ABD$, $s = \frac{1}{2}(20 + 30 + 31.5) = 40.75$

$K = \sqrt{(40.75)(40.75 - 20)(40.75 - 30)(40.75 - 31.5)}$
$\qquad = \sqrt{(40.75)(20.75)(10.75)(9.25)}$
$\qquad = \sqrt{84,080.6211}$
$\qquad = 290.0$

For $\triangle BDC$, $s = \frac{1}{2}(30 + 15 + 20.9) = 32.95$

$$K = \sqrt{(32.95)(32.95 - 30)(32.95 - 15)(32.95 - 20.9)}$$
$$= \sqrt{(32.95)(2.95)(17.95)(12.05)}$$
$$= \sqrt{21,024.6577}$$
$$= 145.0$$

area of quadrilateral = 290 + 145 = 435

**39. (a)**

$$\frac{\cos A}{a} + \frac{\cos B}{b} + \frac{\cos C}{c}$$

$$= \frac{1}{a} \cdot \frac{b^2 + c^2 - a^2}{2bc} + \frac{1}{b} \cdot \frac{a^2 + c^2 - b^2}{2ac} + \frac{1}{c} \cdot \frac{a^2 + b^2 - c^2}{2ab}$$

$$= \frac{b^2 + c^2 - a^2 + a^2 + c^2 - b^2 + a^2 + b^2 - c^2}{2abc}$$

$$= \frac{a^2 + b^2 + c^2}{2abc}$$

**(b)**

$$\left[ \frac{\cos A}{a} + \frac{\cos B}{b} + \frac{\cos C}{c} = \frac{a^2 + b^2 + c^2}{2abc} \right] \cdot 2abc$$

$$2(abc)\left[ \frac{\cos A}{a} + \frac{\cos B}{b} + \frac{\cos C}{c} \right] = a^2 + b^2 + c^2$$

$$2(ab\cos C + ac\cos B + bc\cos A) = a^2 + b^2 + c^2$$

## Challenge

**1.** height $= \sqrt{a^2 - \left(\frac{a}{2}\right)^2} = \sqrt{a^2 - \frac{a^2}{4}} = \sqrt{\frac{3}{4}a^2} = \frac{a}{2}\sqrt{3}$

Area $= \frac{1}{2}(a)\left(\frac{a}{2}\sqrt{3}\right) = \frac{\sqrt{3}}{4}a^2$

$s = \frac{1}{2}(a + a + a) = \frac{3}{2}a$

$$k = \sqrt{\left(\frac{3}{2}a\right)\left(\frac{3}{2}a - a\right)\left(\frac{3}{2}a - a\right)\left(\frac{3}{2}a - a\right)}$$

$$= \sqrt{\left(\frac{3}{2}a\right)\left(\frac{a}{2}\right)\left(\frac{a}{2}\right)\left(\frac{a}{2}\right)}$$

$$= \sqrt{\frac{3a^4}{16}} = \frac{\sqrt{3}a^2}{4} = \frac{\sqrt{3}}{4}a^2$$

## Exercises 8.2

**1.** $C = 180° - 120° - 30° = 30°$

$$\frac{a}{\sin 30°} = \frac{54}{\sin 120°}$$

$$a = \frac{54 \sin 30°}{\sin 120°} = \frac{54\left(\frac{1}{2}\right)}{\frac{\sqrt{3}}{2}} = \frac{54}{\sqrt{3}} = 18\sqrt{3}$$

$$\frac{c}{\sin 30°} = \frac{54}{\sin 120°} \text{ so } c = 18\sqrt{3}$$

**3.** $\dfrac{\sin C}{c} = \dfrac{\sin A}{a}$

$$\frac{\sin C}{4\sqrt{3}} = \frac{\sin\frac{2\pi}{3}}{12} = \frac{\frac{\sqrt{3}}{2}}{12}$$

$$\sin C = \frac{1}{2}$$

$$C = \frac{\pi}{6}$$

**5.** $B = 180° - 55° - 25° = 100°$

$$\frac{a}{\sin 25°} = \frac{12}{\sin 100°}$$

$$a = \frac{12 \sin 25°}{\sin 100°} = 5.1$$

$$\frac{c}{\sin 55°} = \frac{12}{\sin 100°}$$

$$c = \frac{12 \sin 55°}{\sin 100°} = 10.0$$

**7.** $C = 180° - 62.2° - 50° = 67.8°$

$$\frac{a}{\sin 62.2°} = \frac{5}{\sin 50°}$$

$$a = \frac{5 \sin 62.2°}{\sin 50°} = 5.8$$

$$\frac{c}{\sin 67.8°} = \frac{5}{\sin 50°}$$

$$c = \frac{5 \sin 67.8°}{\sin 50°} = 6.0$$

**9.** $b\sin A = 10 \sin 32° = 5.3$
$a < b \sin A$
No solutions

**11.** $b \sin A = 19 \sin 30° = 9.5$
$a = b \sin A$
1 triangle

**13.** $b \sin A = 25 \sin 126° = 20.2$
$a < b \sin A$
No solutions

**15.** $\dfrac{\sin B}{15} = \dfrac{\sin 53°}{12}$

$$\sin B = \frac{15 \sin 53°}{12} = 0.9983$$

$B = 86.7°$ or $B = 93.3°$
$C = 180° - 86.7° - 53° = 40.3°$
or $C = 180° - 93.3° - 53° = 33.7°$

$$\frac{c}{\sin 40.3°} = \frac{12}{\sin 53°}$$

$$c = \frac{12 \sin 40.3°}{\sin 53°} = 9.7 \text{ if } B = 86.7°$$

$$\text{or } \frac{c}{\sin 33.7°} = \frac{12}{\sin 53°}$$

$$c = \frac{12 \sin 33.7°}{\sin 53°} = 8.3 \text{ if } B = 93.3°$$

**17.** $\dfrac{\sin B}{25} = \dfrac{\sin 75°}{7}$

$$\sin B = \frac{25 \sin 75°}{7} = 3.45$$

No solution

**19.** $\dfrac{\sin C}{30} = \dfrac{\sin 22.7°}{25}$

$$\sin C = \frac{30 \sin 22.7°}{25} = 0.4631$$

$C = 27.6°$ or $C = 152.4°$

**21.** In the triangle, let $C$ be the largest angle, $A = 15°$, and $B$ be the remaining angle.

Then $\dfrac{\sin B}{80} = \dfrac{\sin 15°}{40}$

$$\sin B = \frac{80 \sin 15°}{40} = 0.5176$$

$B = 31.2°$
$C = 180° - 31.2° - 15° = 133.8°$

**23.** Let $d$ be the distance from the airplane to the airfield. Then

$$\frac{d}{\sin 32°} = \frac{2}{\sin 42°}$$

$$d = \frac{2\sin 32°}{\sin 42°} = 1.6 \text{ miles}$$

**25.** Let $d_1$ be the distance from cottage $A$ to the base of the hill, and let $d_2$ be the distance from cottage $B$ to the base of the hill. Then

$$\frac{d_1}{\sin 74.5°} = \frac{250}{\sin 15.5°}$$

$$d_1 = \frac{250\sin 74.5°}{\sin 15.5°} = 901.5 \text{ feet}$$

$$\frac{d_2}{\sin 60.8°} = \frac{250}{\sin 29.2°}$$

$$d_2 = \frac{250\sin 60.8°}{\sin 29.2°} = 447.3 \text{ feet}$$

distance between cottages = 901.5 – 447.3 = 454 feet

**27.**

$$\frac{BD}{\sin 43°} = \frac{20}{\sin 15°}$$

$$BD = \frac{20\sin 43°}{\sin 15°} = 52.7 \text{ cm}$$

$$\frac{AD}{\sin 12°} = \frac{52.7}{\sin 110°}$$

$$AD = \frac{52.7\sin 12°}{\sin 110°} = 11.7 \text{ cm}$$

**29.** $(AC)^2 = 4^2 + 12^2 - 2(4)(12)\cos 20°$

$$= 69.7895$$

$$AC = 8.4$$

$$\frac{\sin D}{12} = \frac{\sin 20°}{10}$$

$$\sin D = \frac{12\sin 20°}{10} = 0.4104$$

$$D = 24.2°$$

$$\frac{\sin(\angle ACD)}{10} = \frac{\sin 24.2°}{8.4}$$

$$\sin(\angle ACD) = \frac{10\sin 24.2°}{8.4} = 0.4880$$

$$\angle ACD = 29.2°$$

$$\angle DAC = 180° - 24.2° - 29.2° = 126.6°$$

$$\frac{CD}{\sin 126.6°} = \frac{10}{\sin 29.2°}$$

$$CD = \frac{10\sin 126.6°}{\sin 29.2°} = 16 \text{ units}$$

**31.** $\angle ACB = 180° - 52° - 106° = 22°$

$$\frac{BC}{\sin 106°} = \frac{300}{\sin 22°}$$

$$BC = \frac{300\sin 106°}{\sin 22°} = 769.8 \text{ m}$$

$$\angle ADB = 180° - 40° - 123° = 17°$$

$$\frac{AD}{\sin 123°} = \frac{300}{\sin 17°}$$

$$AD = \frac{300\sin 123°}{\sin 17°} = 860.6 \text{ m}$$

Let $E$ be the point where $AD$ and $BC$ intersect. Then

$$\frac{AE}{\sin 52°} = \frac{300}{\sin 88°}$$

$$AE = \frac{300\sin 52°}{\sin 88°} = 236.5 \text{ m}$$

So $ED = 860.6 - 236.5 = 624.1$ m

$$\frac{BE}{\sin 40°} = \frac{300}{\sin 88°}$$

$$BE = \frac{300\sin 40°}{\sin 88°} = 193.0 \text{ m}$$

So $EC = 769.8 - 193.0 = 576.8$ m
Then

$$(CD)^2 = (576.8)^2 + (624.1)^2$$

$$-2 \cdot (576.8)(624.1)\cos 88°$$

$$= 697,073 \text{ m}^2$$

$$CD = 835 \text{ m}$$

**33.** resultant force: $\sqrt{24^2 + 10^2} = \sqrt{676}$
= 26 pounds

$$\tan \theta = \frac{10}{24}$$

$$\theta = 23°$$

**35.** Let $\vec{w}$ be the resultant force. Then

$$|\vec{w}|^2 = 38^2 + 12^2 - 2 \cdot 12 \cdot 38\cos 100°$$

$$= 1746$$

$$|\vec{w}| = 42 \text{ pounds}$$

$$\frac{\sin \theta}{38} = \frac{\sin 100°}{42}$$

$$\sin\theta = \frac{38\sin 100°}{42} = 0.8910$$
$$\theta = 63°$$

**37.** Let $s$ be the ground speed of the airplane. Then
$$s^2 = 30^2 + 210^2 - 2\cdot 30\cdot 210\cos 60°$$
$$= 38{,}700$$
$$s = 197 \text{ mph}$$
If $\theta$ is the angle between the resulting bearing and the original bearing, then
$$\frac{\sin\theta}{30} = \frac{\sin 60°}{197}$$
$$\sin\theta = \frac{30\sin 60°}{197} = 0.1319$$
$$\theta = 7.58°$$
The new bearing is N 22° E.

**39.** $1800\cos 85° = 157$ pounds

**41.** Let $w$ be the weight of the vehicle. Then
$$w\cos 80° = 200$$
$$w = \frac{200}{\cos 80°} = 1150 \text{ pounds}$$

**43.** $K = \dfrac{1}{2}bc\sin A = \dfrac{1}{2}\left(\dfrac{a\sin B}{\sin A}\right)\left(\dfrac{a\sin C}{\sin A}\right)\sin A$
$$= \frac{a^2\sin B\sin C}{2\sin A}$$
$$a^2 = 8^2 + 12^2 - 2\cdot 8\cdot 12\cos 110°$$
$$= 273.7$$
$$a = 16.5 \text{ cm}$$
$$\frac{\sin B}{12} = \frac{\sin 110°}{16.5}$$
$$\sin B = \frac{12\sin 110°}{16.5} = 0.6834$$
$$B = 43.1°$$
$$C = 180° - 110° - 43.1° = 26.9°$$
$$K = \frac{a^2\sin B\sin C}{2\sin A}$$
$$= \frac{(16.5)^2\sin 43.1°\sin 26.9°}{2\sin 110°}$$
$$= 45 \text{ cm}^2$$

**Critical Thinking**

**1.** Let $C = 90°$. Then
$$c^2 = a^2 + b^2 - 2ab\cos 90°$$
or $c^2 = a^2 + b^2$
This is circular reasoning since the Pythagorean Theorem is used to prove the Law of Cosines.

**3.** $AC = \sqrt{a^2 + b^2 - 2ab\cos B}$
$$= \sqrt{a^2 + b^2 - 2ab\cos(\pi - A)}$$
$$= \sqrt{a^2 + b^2 - 2ab[\cos\pi\cos A + \sin\pi\sin A]}$$
$$= \sqrt{a^2 + b^2 + 2ab\cos A}$$

**5.** $\angle ABD = 137°$ and $\angle BDA = 20°$
$$\frac{AD}{\sin 137°} = \frac{0.25}{\sin 20°}$$
$$AD = \frac{0.25\sin 137°}{\sin 20°} = 0.50$$
$$\frac{CD}{\sin 23°} = \frac{0.50}{\sin 90°}$$
$$CD = 0.50\cdot\sin 23° = 0.20 \text{ miles}$$

**Exercises 8.3**

**1.**

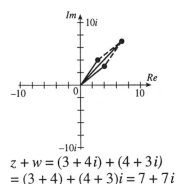

$$z + w = (3 + 4i) + (4 + 3i)$$
$$= (3 + 4) + (4 + 3)i = 7 + 7i$$

**3.**

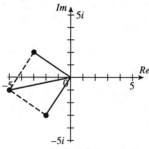

$$z + w = (-3 + 2i) + (-2 - 3i)$$
$$= (-3 - 2) + (2 - 3)i = -5 - i$$

**5.**

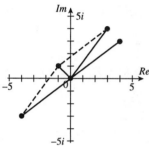

$$z - w = (3 + 4i) - (4 + 3i)$$
$$= (3 - 4) + (4 - 3)i = -1 + i$$

**7.**

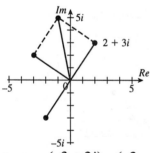

$$z - w = (-3 + 2i) - (-2 - 3i)$$
$$= (-3 + 2) + (2 + 3)i = -1 + 5i$$

**9.**

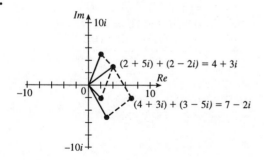

**11.**

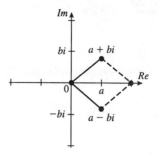

**13.**

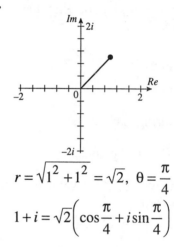

$$r = \sqrt{1^2 + 1^2} = \sqrt{2}, \ \theta = \frac{\pi}{4}$$

$$1 + i = \sqrt{2}\left(\cos\frac{\pi}{4} + i\sin\frac{\pi}{4}\right)$$

**15.**

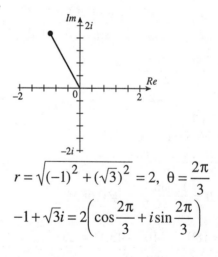

$$r = \sqrt{(-1)^2 + (\sqrt{3})^2} = 2, \ \theta = \frac{2\pi}{3}$$

$$-1 + \sqrt{3}i = 2\left(\cos\frac{2\pi}{3} + i\sin\frac{2\pi}{3}\right)$$

**17.**

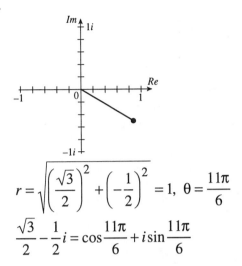

$r = \sqrt{\left(\dfrac{\sqrt{3}}{2}\right)^2 + \left(-\dfrac{1}{2}\right)^2} = 1, \ \theta = \dfrac{11\pi}{6}$

$\dfrac{\sqrt{3}}{2} - \dfrac{1}{2}i = \cos\dfrac{11\pi}{6} + i\sin\dfrac{11\pi}{6}$

**19.**

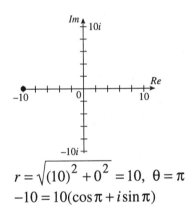

$r = \sqrt{(10)^2 + 0^2} = 10, \ \theta = \pi$

$-10 = 10(\cos\pi + i\sin\pi)$

**21.**

$r = \sqrt{(-10)^2 + (-10)^2} = 10\sqrt{2}, \ \theta = \dfrac{5\pi}{4}$

$-10 - 10i = 10\sqrt{2}\left(\cos\dfrac{5\pi}{4} + i\sin\dfrac{5\pi}{4}\right)$

**23.** $3(\cos 120° + i\sin 120°) = 3\left(-\dfrac{1}{2} + \dfrac{\sqrt{3}}{2}i\right)$

$= -\dfrac{3}{2} + \dfrac{3\sqrt{3}}{2}i$

**25.** $5\left(\cos\dfrac{3\pi}{2} + i\sin\dfrac{3\pi}{2}\right) = 5(0 - i) = -5i$

**27.** $\sqrt{2}(\cos 60° + i\sin 60°) = \sqrt{2}\left(\dfrac{1}{2} + i\dfrac{\sqrt{3}}{2}\right)$

$= \dfrac{\sqrt{2}}{2} + \dfrac{\sqrt{6}}{2}i$

**29.** $4(\cos 315° + i\sin 315°)$

$= 4\left(\dfrac{\sqrt{2}}{2} + i\left(-\dfrac{\sqrt{2}}{2}\right)\right)$

$= 2\sqrt{2} - 2\sqrt{2}i$

**31.** $\cos 350° + i\sin 350° = 0.9848 - 0.1736i$

**33.** $zw = \dfrac{1}{2}(\cos 100° + i\sin 100°)$

$\qquad \cdot 10(\cos 50° + i\sin 50°)$

$= 5(\cos 150° + i\sin 150°)$

$= 5\left(-\dfrac{\sqrt{3}}{2} + \dfrac{1}{2}i\right) = -\dfrac{5\sqrt{3}}{2} + \dfrac{5}{2}i$

$\dfrac{z}{w} = \dfrac{\frac{1}{2}(\cos 100° + i\sin 100°)}{10(\cos 50° + i\sin 50°)}$

$= \dfrac{1}{20}(\cos 50° + i\sin 50°)$

$= \dfrac{1}{20}(0.6428 + i \cdot 0.7660)$

$= 0.0321 + 0.0383i$

$\dfrac{w}{z} = \dfrac{10(\cos 50° + i\sin 50°)}{\frac{1}{2}(\cos 100° + i\sin 100°)}$

$= 20(\cos(-50°) + i\sin(-50°))$

$= 20(0.6428 - 0.7660i)$

$= 12.8558 - 15.3209i$

**35.** $zw = \sqrt{2}\left(\cos\dfrac{5\pi}{4} + i\sin\dfrac{5\pi}{4}\right)$

$\qquad \cdot 8\left(\cos\dfrac{7\pi}{4} + i\sin\dfrac{7\pi}{4}\right)$

$\qquad = 8\sqrt{2}(\cos 3\pi + i\sin 3\pi) = -8\sqrt{2}$

$\dfrac{z}{w} = \dfrac{\sqrt{2}\left(\cos\frac{5\pi}{4} + i\sin\frac{5\pi}{4}\right)}{8\left(\cos\frac{7\pi}{4} + i\sin\frac{7\pi}{4}\right)}$

$\qquad = \dfrac{\sqrt{2}}{8}\left(\cos\left(-\dfrac{\pi}{2}\right) + i\sin\left(-\dfrac{\pi}{2}\right)\right)$

$\qquad = -\dfrac{\sqrt{2}}{8}i$

$\dfrac{w}{z} = \dfrac{8\left(\cos\frac{7\pi}{4} + i\sin\frac{7\pi}{4}\right)}{\sqrt{2}\left(\cos\frac{5\pi}{4} + i\sin\frac{5\pi}{4}\right)}$

$\qquad = \dfrac{8}{\sqrt{2}} \cdot \dfrac{\sqrt{2}}{\sqrt{2}}\left(\cos\dfrac{\pi}{2} + i\sin\dfrac{\pi}{2}\right)$

$\qquad = \dfrac{8\sqrt{2}}{2}i = 4\sqrt{2}i$

**37.** $z = 7i = 7\left(\cos\dfrac{\pi}{2} + i\sin\dfrac{\pi}{2}\right)$

$\qquad w = \sqrt{3} + i = 2\left(\cos\dfrac{\pi}{6} + i\sin\dfrac{\pi}{6}\right)$

$\qquad zw = 14\left(\cos\dfrac{2\pi}{3} + i\sin\dfrac{2\pi}{3}\right)$

$\qquad = 14\left(-\dfrac{1}{2} + \dfrac{\sqrt{3}}{2}i\right) = -7 + 7\sqrt{3}i$

$\qquad \dfrac{z}{w} = \dfrac{7}{2}\left(\cos\dfrac{\pi}{3} + i\sin\dfrac{\pi}{3}\right) = \dfrac{7}{2}\left(\dfrac{1}{2} + \dfrac{\sqrt{3}}{2}i\right)$

$\qquad = \dfrac{7}{4} + \dfrac{7\sqrt{3}}{4}i$

**39.** $z = -2 + 2\sqrt{3}i = 4\left(\cos\dfrac{2\pi}{3} + i\sin\dfrac{2\pi}{3}\right)$

$\qquad w = -2\sqrt{3} - 2i = 4\left(\cos\dfrac{7\pi}{6} + i\sin\dfrac{7\pi}{6}\right)$

$\qquad zw = 16\left(\cos\dfrac{11\pi}{6} + i\sin\dfrac{11\pi}{6}\right)$

$\qquad = 16\left(\dfrac{\sqrt{3}}{2} - \dfrac{1}{2}i\right) = 8\sqrt{3} - 8i$

$\qquad \dfrac{z}{w} = 1\left(\cos\left(-\dfrac{\pi}{2}\right) + i\sin\left(-\dfrac{\pi}{2}\right)\right) = -i$

**41.** $z = -1 + i = \sqrt{2}\left(\cos\dfrac{3\pi}{4} + i\sin\dfrac{3\pi}{4}\right)$

$\qquad w = 1 - i = \sqrt{2}\left(\cos\dfrac{7\pi}{4} + i\sin\dfrac{7\pi}{4}\right)$

$\qquad zw = 2\left(\cos\dfrac{5\pi}{2} + i\sin\dfrac{5\pi}{2}\right) = 2(0 + i) = 2i$

$\qquad \dfrac{z}{w} = 1(\cos(-\pi) + i\sin(-\pi)) = -1$

**43.** $\left[\dfrac{1}{2}(\cos 20° + i\sin 20°)\right]$

$\qquad \cdot \left[\sqrt{2}(\cos 70° + i\sin 70°)\right]$

$\qquad = \dfrac{\sqrt{2}}{2}(\cos 90° + i\sin 90°) = \dfrac{\sqrt{2}}{2}i$

**45.** $\left[\dfrac{2}{3}(\cos 122° + i\sin 122°)\right]$

$\qquad \cdot \left[\dfrac{9}{4}(\cos 77° + i\sin 77°)\right]$

$\qquad = \dfrac{3}{2}(\cos 199° + i\sin 199°)$

$\qquad = \dfrac{3}{2}(-0.9455 - 0.3256i)$

$\qquad = -1.4183 - 0.4884i$

**47.** $|\bar{z}| = |\overline{a + bi}| = |a - bi| = \sqrt{a^2 + (-b)^2}$

$\qquad = \sqrt{a^2 + b^2} = |z|$

**49.** $|zw| = |(a+bi)(c+di)|$

$= |(ac-bd) + (bc+ad)i|$

$= \sqrt{(ac-bd)^2 + (bc+ad)^2}$

$= \sqrt{a^2c^2 - 2abcd + b^2d^2 + b^2c^2 + 2abcd + a^2d^2}$

$= \sqrt{a^2c^2 + b^2d^2 + b^2c^2 + a^2d^2}$

$|z\|w| = |a+bi\|c+di|$

$= \sqrt{a^2+b^2}\sqrt{c^2+d^2}$

$= \sqrt{(a^2+b^2)(c^2+d^2)}$

$= \sqrt{a^2c^2 + b^2d^2 + b^2c^2 + a^2d^2}$

Therefore $|zw| = |z\|w|$.

**51.** $|z-1| = 2$

$|x+yi-1| = 2$

$|x-1+yi|^2 = 4$

$(x-1)^2 + y^2 = 4$

Circle of radius 2 and center $(1, 0)$.

**53.** $|z-(2+3i)| = 1$

$|x+yi-2-3i| = 1$

$|(x-2)+(y-3)i|^2 = 1^2 = 1$

$(x-2)^2 + (y-3)^2 = 1$

Circle of radius 1 and center $(2, 3)$.

**55.** $|z+2| = 8 - |z-2|$

$|x+yi+2| = 8 - |x+yi-2|$

$|x+2+yi| = 8 - |x-2+yi|$

$\left(\sqrt{(x+2)^2 + y^2}\right)^2 = \left(8 - \sqrt{(x-2)^2 + y^2}\right)^2$

$(x+2)^2 + y^2 = 64 - 16\sqrt{(x-2)^2 + y^2} + (x-2)^2 + y^2$

$x^2 + 4x + 4 - x^2 + 4x - 4 - 64 = -16\sqrt{(x-2)^2 + y^2}$

$8x - 64 = -16\sqrt{(x-2)^2 + y^2}$

$\left(-\frac{1}{2}x + 4\right)^2 = \left(\sqrt{(x-2)^2 + y^2}\right)^2$

$\frac{1}{4}x^2 - 4x + 16 = (x-2)^2 + y^2$

$$\frac{1}{4}x^2 - 4x + 16 = x^2 - 4x + 4 + y^2$$

$$-\frac{3}{4}x^2 - y^2 = -12$$

$$\frac{x^2}{16} + \frac{y^2}{12} = 1$$

Ellipse

## Challenge

$$|zw|^2 = (zw)(\overline{zw}) \qquad \text{by Exercise 48}$$
$$= zw\overline{z}\,\overline{w} \qquad\qquad \text{by Exercise 44, p. 233}$$
$$= z\overline{z}w\overline{w}$$
$$= |z|^2|w|^2 \qquad\qquad \text{by Exercise 48}$$

Therefore, $|zw| = |z||w|$.

## Exercise 8.4

**1.** $(\cos 6° + i\sin 6°)^{10} = \cos 60° + i\sin 60°$
$$= \frac{1}{2} + \frac{\sqrt{3}}{2}i$$

**3.** $(\cos 40° + i\sin 40°)^8 = \cos 320° + i\sin 320°$
$$= 0.7660 - 0.6428i$$

**5.** $\left[\frac{1}{2}\left(\cos\frac{\pi}{8} + i\sin\frac{\pi}{8}\right)\right]^6$
$$= \frac{1}{64}\left(\cos\frac{3\pi}{4} + i\sin\frac{3\pi}{4}\right)$$
$$= -\frac{\sqrt{2}}{128} + \frac{\sqrt{2}}{128}i$$

**7.** $(-1-i)^4 = \left[\sqrt{2}\left(\cos\frac{5\pi}{4} + i\sin\frac{5\pi}{4}\right)\right]^4$
$$= 4(\cos 5\pi + i\sin 5\pi) = -4$$

**9.** $(-\sqrt{3} + \sqrt{3}i)^6 = \left[\sqrt{6}\left(\cos\frac{3\pi}{4} + i\sin\frac{3\pi}{4}\right)\right]^6$
$$= 216\left(\cos\frac{9\pi}{2} + i\sin\frac{9\pi}{2}\right) = 216i$$

**11.** $\left(\frac{\sqrt{2}}{2} - \frac{\sqrt{2}}{2}i\right)^{10} = \left(\cos\frac{7\pi}{4} + i\sin\frac{7\pi}{4}\right)^{10}$
$$= \cos\frac{35\pi}{2} + i\sin\frac{35\pi}{2} = -i$$

**13.** $\left(-\frac{\sqrt{3}}{2} + \frac{1}{2}i\right)^{12} = \left(\cos\frac{5\pi}{6} + i\sin\frac{5\pi}{6}\right)^{12}$
$$= \cos 10\pi + i\sin 10\pi = 1$$

**15.** $\sqrt[3]{27}\left(\cos\frac{\pi + 30k\pi}{45} + i\sin\frac{\pi + 30k\pi}{45}\right)$

$k = 0$: $3\left(\cos\dfrac{\pi}{45} + i\sin\dfrac{\pi}{45}\right)$

$k = 1$: $3\left(\cos\dfrac{31\pi}{45} + i\sin\dfrac{31\pi}{45}\right)$

$k = 2$: $3\left(\cos\dfrac{61\pi}{45} + i\sin\dfrac{61\pi}{45}\right)$

**17.** $\sqrt[5]{32}\left(\cos\frac{\pi + 16k\pi}{40} + i\sin\frac{\pi + 16k\pi}{40}\right)$

$k = 0$: $2\left(\cos\dfrac{\pi}{40} + i\sin\dfrac{\pi}{40}\right)$

$k = 1$: $2\left(\cos\dfrac{17\pi}{40} + i\sin\dfrac{17\pi}{40}\right)$

$$k = 2:\ 2\left(\cos\frac{33\pi}{40} + i\sin\frac{33\pi}{40}\right)$$

$$k = 3:\ 2\left(\cos\frac{49\pi}{40} + i\sin\frac{49\pi}{40}\right)$$

$$k = 4:\ 2\left(\cos\frac{65\pi}{40} + i\sin\frac{65\pi}{40}\right)$$

**19.** $16 = 16(\cos 0 + i\sin 0)$

$$\sqrt[4]{16}\left(\cos\frac{2k\pi}{4} + i\sin\frac{2k\pi}{4}\right)$$

$k = 0:\ 2(\cos 0 + i\sin 0) = 2$

$k = 1:\ 2\left(\cos\dfrac{\pi}{2} + i\sin\dfrac{\pi}{2}\right) = 2i$

$k = 2:\ 2(\cos\pi + i\sin\pi) = -2$

$k = 3:\ 2\left(\cos\dfrac{3\pi}{2} + i\sin\dfrac{3\pi}{2}\right) = -2i$

**21.** $-1 = \cos\pi + i\sin\pi$

$$\sqrt[3]{1}\left(\cos\frac{\pi + 2k\pi}{3} + i\sin\frac{\pi + 2k\pi}{3}\right)$$

$k = 0:\ 1\left(\cos\dfrac{\pi}{3} + i\sin\dfrac{\pi}{3}\right) = \dfrac{1}{2} + \dfrac{\sqrt{3}}{2}i$

$k = 1:\ 1(\cos\pi + i\sin\pi) = -1$

$k = 2:\ 1\left(\cos\dfrac{5\pi}{3} + i\sin\dfrac{5\pi}{3}\right) = \dfrac{1}{2} - \dfrac{\sqrt{3}}{2}i$

**23.** $\cos\dfrac{2k\pi}{3} + i\sin\dfrac{2k\pi}{3}$

$k = 0:\ \cos 0 + i\sin 0 = 1$

$k = 1:\ \cos\dfrac{2\pi}{3} + i\sin\dfrac{2\pi}{3} = -\dfrac{1}{2} + \dfrac{\sqrt{3}}{2}i$

$k = 2:\ \cos\dfrac{4\pi}{3} + i\sin\dfrac{4\pi}{3} = -\dfrac{1}{2} - \dfrac{\sqrt{3}}{2}i$

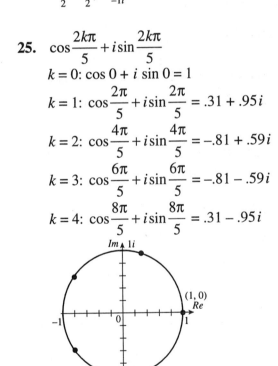

**25.** $\cos\dfrac{2k\pi}{5} + i\sin\dfrac{2k\pi}{5}$

$k = 0:\ \cos 0 + i\sin 0 = 1$

$k = 1:\ \cos\dfrac{2\pi}{5} + i\sin\dfrac{2\pi}{5} = .31 + .95i$

$k = 2:\ \cos\dfrac{4\pi}{5} + i\sin\dfrac{4\pi}{5} = -.81 + .59i$

$k = 3:\ \cos\dfrac{6\pi}{5} + i\sin\dfrac{6\pi}{5} = -.81 - .59i$

$k = 4:\ \cos\dfrac{8\pi}{5} + i\sin\dfrac{8\pi}{5} = .31 - .95i$

**27.** $z^3 = -27 = 27(\cos \pi + i \sin \pi)$

$$z = \sqrt[3]{27}\left(\cos\frac{\pi + 2k\pi}{3} + i\sin\frac{\pi + 2k\pi}{3}\right)$$

$k = 0$: $z = 3\left(\cos\dfrac{\pi}{3} + i\sin\dfrac{\pi}{3}\right) = \dfrac{3}{2} + \dfrac{3\sqrt{3}}{2}i$

$k = 1$: $z = 3(\cos\pi + i\sin\pi) = -3$

$k = 2$: $z = 3\left(\cos\dfrac{5\pi}{3} + i\sin\dfrac{5\pi}{3}\right) = \dfrac{3}{2} - \dfrac{3\sqrt{3}}{2}i$

**29.** $z^3 + 8i = 0$

$$z^3 = -8i = 8\left(\cos\frac{3\pi}{2} + i\sin\frac{3\pi}{2}\right)$$

$$z = \sqrt[3]{8}\left(\cos\frac{3\pi + 4k\pi}{6} + i\sin\frac{3\pi + 4k\pi}{6}\right)$$

$k = 0$: $z = 2\left(\cos\dfrac{\pi}{2} + i\sin\dfrac{\pi}{2}\right) = 2i$

$k = 1$: $z = 2\left(\cos\dfrac{7\pi}{6} + i\sin\dfrac{7\pi}{6}\right) = -\sqrt{3} - i$

$k = 2$: $z = 2\left(\cos\dfrac{11\pi}{6} + i\sin\dfrac{11\pi}{6}\right) = \sqrt{3} - i$

**31.** $z^3 + i = -1$

$$z^3 = -1 - i = \sqrt{2}\left(\cos\frac{5\pi}{4} + i\sin\frac{5\pi}{4}\right)$$

$$z = 2^{1/6}\left(\cos\frac{5\pi + 8k\pi}{12} + i\sin\frac{5\pi + 8k\pi}{12}\right)$$

$k = 0$: $z = 2^{1/6}\left(\cos\dfrac{5\pi}{12} + i\sin\dfrac{5\pi}{12}\right)$

$k = 1$: $z = 2^{1/6}\left(\cos\dfrac{13\pi}{12} + i\sin\dfrac{13\pi}{12}\right)$

$k = 2$: $z = 2^{1/6}\left(\cos\dfrac{21\pi}{12} + i\sin\dfrac{21\pi}{12}\right)$

**33. (a)** For $k = 0$, the formula gives $r^{1/n}\left(\cos\dfrac{\theta}{n} + i\sin\dfrac{\theta}{n}\right)$

For $k = n$, the formula gives $r^{1/n}\left(\cos\dfrac{\theta + 2n\pi}{n} + i\sin\dfrac{\theta + 2n\pi}{n}\right)$

$= r^{1/n}\left(\cos\left(\dfrac{\theta}{n} + 2\pi\right) + i\sin\left(\dfrac{\theta}{n} + 2\pi\right)\right) = r^{1/n}\left(\cos\dfrac{\theta}{n} + i\sin\dfrac{\theta}{n}\right)$

**(b)** For $k = 1$, the formula gives $r^{1/n}\left(\cos\dfrac{\theta + 2\pi}{n} + i\sin\dfrac{\theta + 2\pi}{n}\right)$

For $k = n + 1$, the formula gives $r^{1/n}\left(\cos\dfrac{\theta + 2(n+1)\pi}{n} + i\sin\dfrac{\theta + 2(n+1)\pi}{n}\right)$

$= r^{1/n}\left(\cos\left(\dfrac{\theta + 2\pi}{n} + 2\pi\right) + i\sin\left(\dfrac{\theta + 2\pi}{n} + 2\pi\right)\right) = r^{1/n}\left(\cos\dfrac{\theta + 2\pi}{n} + i\sin\dfrac{\theta + 2\pi}{n}\right)$

**(c)** Use $k$ in the formula to get $r^{1/n}\left(\cos\dfrac{\theta + 2k\pi}{n} + i\sin\dfrac{\theta + 2k\pi}{n}\right)$

Use $k + n$ in the formula to get $r^{1/n}\left(\cos\dfrac{\theta + 2(k+n)\pi}{n} + i\sin\dfrac{\theta + 2(k+n)\pi}{n}\right)$

$= r^{1/n}\left(\cos\left(\dfrac{\theta + 2k\pi}{n} + 2\pi\right) + i\sin\left(\dfrac{\theta + 2k\pi}{n} + 2\pi\right)\right) = r^{1/n}\left(\cos\dfrac{\theta + 2k\pi}{n} + i\sin\dfrac{\theta + 2k\pi}{n}\right)$

**35. (a)**

$$[r(\cos\theta + i\sin\theta)]^{-n} = r^{-n}(\cos\theta + i\sin\theta)^{-n}$$

$$= r^{-n}\left[(\cos\theta + i\sin\theta)^{-1}\right]^{n}$$

$$= r^{-n}(\cos\theta - i\sin\theta)^{n} \text{ by Exercise 34}$$

$$= r^{-n}(\cos(-\theta) + i\sin(-\theta))^{n}$$

$$= r^{-n}(\cos(-n\theta) + i\sin(-n\theta))$$

$$= r^{-n}(\cos n\theta - i\sin n\theta)$$

**(b)** Assuming the rule $z^0 = 1$, the formula holds for $n = 0$. The formula is given in Section 8.4 for positive integers $n$. If in part (a), we let $m = -n$, where $m$ is a negative integer, then

$$[r(\cos\theta + i\sin\theta)]^{m} = [r(\cos\theta + i\sin\theta)]^{-n}$$

$$= r^{-n}(\cos n\theta - i\sin n\theta) \text{ by part (a)}$$

$$= r^{-n}(\cos(-n\theta) - i\sin(-n\theta))$$

$$= r^{m}(\cos(m\theta) - i\sin(m\theta))$$

which shows that the formula holds for negative integers.

**37.**

$$1 + z + z^2 = 1 + \cos\frac{2\pi}{3} + i\sin\frac{2\pi}{3} + \cos\frac{4\pi}{3} + i\sin\frac{4\pi}{3}$$

$$= 1 - \frac{1}{2} + \frac{\sqrt{3}}{2}i - \frac{1}{2} - \frac{\sqrt{3}}{2}i$$

$$= 0$$

## Challenge

Since $z$ is an $n$th root of unity, $z^n = 1$ or

$z^n - 1 = 0$.

This equation factors as

$(z - 1)(z^{n-1} + \cdots + z^2 + z + 1) = 0$.

Since $z \neq 1$ for $n \geq 2$, we have that

$1 + z + z^2 + \cdots + z^{n-1} = 0$.

**Exercises 8.5**

**1.**

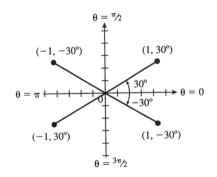

**3.**

**5.**

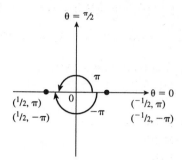

**7. (a)** $\left(1, \dfrac{9\pi}{4}\right)$

**(b)** $\left(1, -\dfrac{7\pi}{4}\right)$

**(c)** $\left(-1, \dfrac{5\pi}{4}\right)$

**(d)** $\left(-1, -\dfrac{3\pi}{4}\right)$

**9. (a)** $\left(3, \dfrac{19\pi}{6}\right)$

**(b)** $\left(3, -\dfrac{5\pi}{6}\right)$

**(c)** $\left(-3, \dfrac{\pi}{6}\right)$

**(d)** $\left(-3, -\dfrac{11\pi}{6}\right)$

**11.** $x = 7\cos\left(-\dfrac{5\pi}{2}\right) = 0$

$y = 7\sin\left(-\dfrac{5\pi}{2}\right) = -7$

$(0, -7)$

**13.** $\left(-\dfrac{3}{2}, \dfrac{5\pi}{2}\right)$

$x = -\dfrac{3}{2}\cos\dfrac{5\pi}{2} = 0$

$y = -\dfrac{3}{2}\sin\dfrac{5\pi}{2} = -\dfrac{3}{2}$

$\left(0, -\dfrac{3}{2}\right)$

**15.** $\left(\dfrac{4}{5}, 405°\right)$

$x = \dfrac{4}{5}\cos 405° = \dfrac{2\sqrt{2}}{5}$

$y = \dfrac{4}{5}\sin 405° = \dfrac{2\sqrt{2}}{5}$

$\left(\dfrac{2\sqrt{2}}{5}, \dfrac{2\sqrt{2}}{5}\right)$

**17.** $r = \sqrt{2^2 + 0^2} = \sqrt{4} = 2$

$\tan\theta = \dfrac{0}{2} = 0$, so $\theta = 0$

since $(x, y)$ is on the positive $x$-axis.

$(r, \theta) = (2, 0)$

**19.** $r = \sqrt{(-2)^2 + 0^2} = \sqrt{4} = 2$

$\tan\theta = \dfrac{0}{-2} = 0$, so $\theta = \pi$ since $(x, y)$

is on the negative $x$-axis.

$(r, \theta) = (2, \pi)$

**21.** $r = \sqrt{(-6)^2 + (-6)^2} = \sqrt{72} = 6\sqrt{2}$

$\tan\theta = \dfrac{-6}{-6} = 1$, so $\theta = \dfrac{5\pi}{4}$ since $(x, y)$

is in quadrant III.

$(r, \theta) = \left(6\sqrt{2}, \dfrac{5\pi}{4}\right)$

**23.** $r = \sqrt{(2\sqrt{2})^2 + (2\sqrt{2})^2} = \sqrt{16} = 4$

$\tan\theta = \dfrac{2\sqrt{2}}{2\sqrt{2}} = 1$, so $\theta = \dfrac{\pi}{4}$ since $(x, y)$ is in quadrant I.

$(r,\ \theta) = \left(4,\ \dfrac{\pi}{4}\right)$

**25.** $r = \sqrt{(\sqrt{3})^2 + (-1)^2} = \sqrt{4} = 2$

$\tan\theta = \dfrac{-1}{\sqrt{3}}$, so $\theta = \dfrac{11\pi}{6}$ since $(x, y)$ is in quadrant IV.

$(r,\ \theta) = \left(2,\ \dfrac{11\pi}{6}\right)$

**27.** $r = -\sqrt{7^2 + 0^2} = -\sqrt{49} = -7$

$\tan\theta = \dfrac{0}{7} = 0$, so $\theta = \pi$

since $(x, y)$ is on the positive $x$-axis and $r$ is negative.

$(r,\ \theta) = (-7,\ \pi)$

**29.** $r = -\sqrt{(-\sqrt{3})^2 + 1^2} = -\sqrt{4} = -2$

$\tan\theta = \dfrac{1}{-\sqrt{3}} = -\dfrac{1}{\sqrt{3}}$, so $\theta = \dfrac{11\pi}{6}$

since $(x, y)$ is in quadrant II and $r$ is negative.

$(r,\ \theta) = \left(-2,\ \dfrac{11\pi}{6}\right)$

**31.** $r = -\sqrt{(\sqrt{3})^2 + (\sqrt{3})^2} = -\sqrt{6}$

$\tan\theta = \dfrac{\sqrt{3}}{\sqrt{3}} = 1$, so $\theta = \dfrac{5\pi}{4}$

since $(x, y)$ is in quadrant I and $r$ is negative.

$(r,\ \theta) = \left(-\sqrt{6},\ \dfrac{5\pi}{4}\right)$

**33.** $r = \sqrt{0^2 + (-10)^2} = \sqrt{100} = 10$

$\tan\theta = \dfrac{-10}{0}$ is undefined, so $\theta = -90°$

since $(x, y)$ is on the negative $y$-axis.

$(r,\ \theta) = (10,\ -90°)$

**35.** $r\cos\theta = 0\cos\theta = 0 = x$
$r\sin\theta = 0\sin\theta = 0 = y$
$(x, y) = (0, 0)$

**37.** Using the polar coordinates $(-r,\ \theta + \pi)$,
$x = -r\cos(\theta + \pi)$ and $y = -r\sin(\theta + \pi)$
since $-r > 0$.
Then
$x = -r\cos(\theta + \pi)$
$= -r(\cos\theta\cos\pi - \sin\theta\sin\pi)$
$= -r(-\cos\theta) = r\cos\theta$
and
$y = -r\sin(\theta + \pi)$
$= -r(\sin\theta\cos\pi + \cos\theta\sin\pi)$
$= -r(-\sin\theta) = r\sin\theta$
$(x,\ y) = (r\cos\theta,\ r\sin\theta)$

**Challenge**

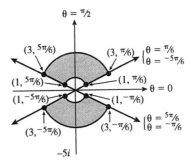

**Critical Thinking**

**1.** Since $z + \bar{z} = x + iy + x - iy = 2x$, then

$\text{Re}\,z = x = \dfrac{1}{2}(z + \bar{z})$. Also

$z - \bar{z} = x + iy - x + iy = 2iy$, so

$\text{Im}\,z = y = \dfrac{1}{2i}(z - \bar{z})$

**3.** The coordinates of the new points are:

$$4\left(\cos\frac{\pi}{6}+i\sin\frac{\pi}{6}\right)=2\sqrt{3}+2i$$

$$4\left(\cos\frac{5\pi}{6}+i\sin\frac{5\pi}{6}\right)=-2\sqrt{3}+2i$$

$$4\left(\cos\frac{3\pi}{2}+i\sin\frac{3\pi}{2}\right)=-4i$$

The points are the cube roots of $64\,i$.

**5.** $x=2\cos15°=2\sqrt{\dfrac{1+\cos30°}{2}}=2\sqrt{\dfrac{1}{2}+\dfrac{\sqrt{3}}{4}}$

$$=\frac{2}{2}\sqrt{2+\sqrt{3}}=\sqrt{2+\sqrt{3}}$$

$$y=2\sin15°=2\sqrt{\dfrac{1-\cos30°}{2}}=2\sqrt{\dfrac{1}{2}-\dfrac{\sqrt{3}}{4}}$$

$$=\frac{2}{2}\sqrt{2-\sqrt{3}}=\sqrt{2-\sqrt{3}}$$

$$(x,y)=\left(\sqrt{2+\sqrt{3}},\ \sqrt{2-\sqrt{3}}\right)$$

## Exercises 8.6

**1.**

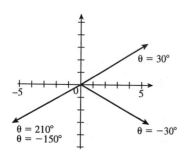

**3.**

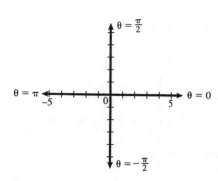

**5.**

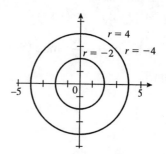

**7.** Since $(0,0°)=(0,-6°)$, and the line passes through $(0,-6°)$ and $(10,-6°)$, the equation is $\theta=-6°$.

**9.** $(\sqrt{3},\ 1)=\left(2,\ \dfrac{\pi}{6}\right)$

$$(-\sqrt{6},\ -\sqrt{2})=\left(-2\sqrt{2},\ \dfrac{\pi}{6}\right)$$

The line is $\theta=\dfrac{\pi}{6}$

**11.**

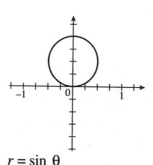

$$r=\sin\theta$$

$$r^2=r\sin\theta$$

$$x^2+y^2=y$$

$$x^2+y^2-y+\frac{1}{4}=\frac{1}{4}$$

$$x^2+\left(y-\frac{1}{2}\right)^2=\left(\frac{1}{2}\right)^2$$

**13.**

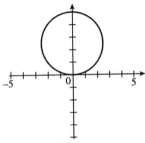

$$r = 5\sin\theta$$
$$r^2 = 5r\sin\theta$$
$$x^2 + y^2 = 5y$$
$$x^2 + y^2 - 5y + \frac{25}{4} = \frac{25}{4}$$
$$x^2 + \left(y - \frac{5}{2}\right)^2 = \left(\frac{5}{2}\right)^2$$

**15.**

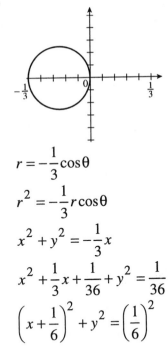

$$r = -\frac{1}{3}\cos\theta$$
$$r^2 = -\frac{1}{3}r\cos\theta$$
$$x^2 + y^2 = -\frac{1}{3}x$$
$$x^2 + \frac{1}{3}x + \frac{1}{36} + y^2 = \frac{1}{36}$$
$$\left(x + \frac{1}{6}\right)^2 + y^2 = \left(\frac{1}{6}\right)^2$$

**17. (a)** $\quad r = 2$
$$r^2 = 4$$
$$x^2 + y^2 = 4$$

**(b)** $\quad r = -2$
$$r^2 = 4$$
$$x^2 + y^2 = 4$$

**(c)** $\quad r = \sqrt{5}$
$$r^2 = 5$$
$$x^2 + y^2 = 5$$

**19.**

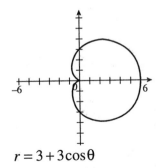

$$r = 3 + 3\cos\theta$$

**21.**

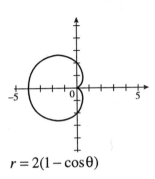

$$r = 2(1 - \cos\theta)$$

**23.** $\quad -2 = r\sin\theta = y$

$$r\sin\theta = -2$$

**25.** $\quad -3 = \cot\theta = \dfrac{1}{\tan\theta}$
$$1 = -3\tan\theta$$
$$\tan\theta = -\frac{1}{3} = \frac{y}{x}$$
$$y = -\frac{1}{3}x$$

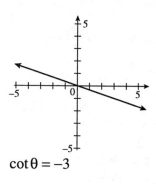

$\cot\theta = -3$

**27.**

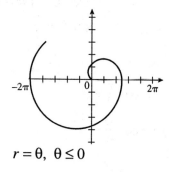

$r = \theta, \ \theta \le 0$

**29.**

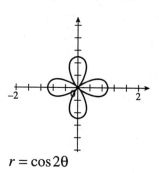

$r = \cos 2\theta$

**31.**

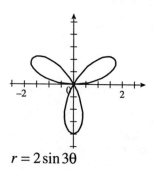

$r = 2\sin 3\theta$

**33.**

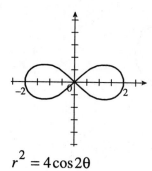

$r^2 = 4\cos 2\theta$

**35.**

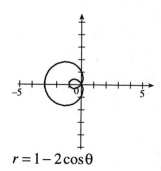

$r = 1 - 2\cos\theta$

**37.** $r = \dfrac{4}{1+\cos\theta}$

$r + r\cos\ \theta = 4$

$r + x = 4$

$r = -x + 4$

$r^2 = (-x+4)^2$

$x^2 + y^2 = x^2 - 8x + 16$

$y^2 = 16 - 8x$

parabola

**39.** $r = -5\sec\ \theta$

$r\cos\ \theta = -5$

$x = -5$

vertical line

**41.** $r = 6\cos\ \theta + 8\sin\ \theta$

$r^2 = 6r\cos\theta + 8r\sin\theta$

$x^2 + y^2 = 6x + 8y$

$x^2 - 6x + 9 + y^2 - 8y + 16 = 9 + 16$

$(x-3)^2 + (y-4)^2 = 25$

circle

**43.** $r(2 \cos \theta + 3 \sin \theta) = 3$
$2r \cos \theta + 3r \sin \theta = 3$
$2x + 3y = 3$
line

**45.** $r \cos\left(\theta - \dfrac{\pi}{3}\right) = \sqrt{3}$

$r\left[\cos\theta\cos\dfrac{\pi}{3} + \sin\theta\sin\dfrac{\pi}{3}\right] = \sqrt{3}$

$\dfrac{1}{2}r\cos\theta + \dfrac{\sqrt{3}}{2}r\sin\theta = \sqrt{3}$

$\dfrac{1}{2}x + \dfrac{\sqrt{3}}{2}y = \sqrt{3}$

$x + \sqrt{3}y = 2\sqrt{3}$
line

**47.**

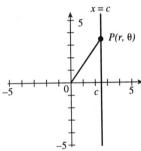

Let $P(r, \theta)$ be a point on the line $x = c$
where $c \neq 0$, and draw $OP$. From the
reference triangle,

$$\cos \theta = \dfrac{x}{r} = \dfrac{c}{r},$$

so that $r \cos \theta = c$. If $c = 0$, then $(r, \theta)$ is
on the ray $\dfrac{\pi}{2}$ so that $r \cos \dfrac{\pi}{2} = 0$.

**49.** $r = 3 \cos \theta, r = 1 + \cos \theta$
$3 \cos \theta = 1 + \cos \theta$
$2 \cos \theta = 1$
$\cos \theta = \dfrac{1}{2}$
$\theta = \dfrac{\pi}{3}, \dfrac{5\pi}{3}$

For $\theta = \dfrac{\pi}{3}, r = 1 + \cos\dfrac{\pi}{3} = \dfrac{3}{2}$

For $\theta = \dfrac{5\pi}{3}, r = 1 + \cos\dfrac{5\pi}{3} = \dfrac{3}{2}$

$\left(\dfrac{3}{2}, \dfrac{\pi}{3}\right), \left(\dfrac{3}{2}, \dfrac{5\pi}{3}\right)$

**51.** $r = \sin 2\theta, r = \cos \theta$
$\sin 2\theta = \cos \theta$
$2 \sin \theta \cos \theta - \cos \theta = 0$
$\cos \theta (2 \sin \theta - 1) = 0$
$\cos \theta = 0 \qquad$ or $\qquad 2\sin \theta - 1 = 0$
$\qquad\qquad\qquad\qquad\qquad\qquad \sin \theta = \dfrac{1}{2}$

$\theta = \dfrac{\pi}{2} \qquad$ or $\qquad \theta = \dfrac{\pi}{6}, \dfrac{5\pi}{6}$

For $\theta = \dfrac{\pi}{2}, r = \cos\dfrac{\pi}{2} = 0$

For $\theta = \dfrac{\pi}{6}, r = \cos\dfrac{\pi}{6} = \dfrac{\sqrt{3}}{2}$

For $\theta = \dfrac{5\pi}{6}, r = \cos\dfrac{5\pi}{6} = -\dfrac{\sqrt{3}}{2}$

$\left(0, \dfrac{\pi}{2}\right), \left(\dfrac{\sqrt{3}}{2}, \dfrac{\pi}{6}\right), \left(-\dfrac{\sqrt{3}}{2}, \dfrac{5\pi}{6}\right)$

**53.** $r \cos \theta = \dfrac{1}{2}, 2r\sin \theta = -\sqrt{3}$

$r = \dfrac{1}{2 \cos \theta}, r = -\dfrac{\sqrt{3}}{2 \sin \theta}$

$\dfrac{1}{2\cos\theta} = -\dfrac{\sqrt{3}}{2\sin\theta}$

$\dfrac{\sin\theta}{\cos\theta} = -\sqrt{3}$

$\tan \theta = -\sqrt{3}$

$\theta = -\dfrac{\pi}{3}$

$r\cos\left(-\dfrac{\pi}{3}\right) = \dfrac{1}{2}$

$\dfrac{1}{2}r = \dfrac{1}{2}$

$r = 1$

$\left(1, -\dfrac{\pi}{3}\right)$

## Challenge

Notice that $\angle POC = \theta - \frac{\pi}{2}$. Then, using the law of cosines

$$2^2 = r^2 + 8^2 - 2 \cdot 8 \cdot r \cos\left(\theta - \frac{\pi}{2}\right)$$

$$4 = r^2 + 64 - 16r \sin\theta$$
$$r^2 = 16r\sin\theta - 60$$

## Chapter 8 Review Exercises

**1.** $a^2 = b^2 + c^2 - 2bc\cos A$

**3.** $c^2 = 5^2 + 8^2 - 2 \cdot 5 \cdot 8\cos 110°$
$= 25 + 64 - 80\cos 110°$
$= 116.3616$
$c = 11$

**5.** $(AB)^2 = 150^2 + 100^2 - 2 \cdot 150 \cdot 100\cos 62°$
$= 18,415.8531$
$AB = 136$ meters

**7.** $K = \frac{1}{2}bc\sin A$

**9.** $K = \sqrt{s(s-a)(s-b)(s-c)}$
where $s = \frac{1}{2}(a + b + c)$

**11.** $\dfrac{a}{\sin A} = \dfrac{b}{\sin B} = \dfrac{c}{\sin C}$

**13.** $\dfrac{a}{\sin 60°} = \dfrac{10}{\sin 45°}$

$a = \dfrac{10\sin 60°}{\sin 45°} = \dfrac{10 \cdot \frac{\sqrt{3}}{2}}{\frac{1}{\sqrt{2}}}$

$a = 5\sqrt{6}$

**15.** $\dfrac{c}{\sin 40°} = \dfrac{250}{\sin 118°}$

$c = \dfrac{250\sin 40°}{\sin 118°} = 182$ meters

**17.** $\dfrac{\sin B}{12} = \dfrac{\sin 40°}{8}$

$\sin B = \dfrac{12\sin 40°}{8} = 0.9642$

$B = 74.6°$ or $B = 105.4°$
For $B = 74.6°$, $C = 65.4°$

$\dfrac{c}{\sin 65.4°} = \dfrac{12}{\sin 74.6°}$

$c = \dfrac{12\sin 65.4°}{\sin 74.6°} = 11.3$

For $B = 105.4°$, $C = 34.6°$

$\dfrac{c}{\sin 34.6°} = \dfrac{12}{\sin 105.4°}$

$c = \dfrac{12\sin 34.6°}{\sin 105.4°} = 7.1$

**19.** Let $s$ be the ground speed. Then
$s^2 = 280^2 + 30^2 - 2 \cdot 280 \cdot 30\cos 130°$
$= 90,098.8318$
$s = 300$ mph
$\dfrac{\sin \alpha}{30} = \dfrac{\sin 130°}{300}$

$\sin \alpha = \dfrac{30\sin 130°}{300} = 0.0766$

$\alpha = 4.4°$
Final bearing: N $44°$ W

**21.** The complex plane is a rectangular coordinate system in which the horizontal is called the real axis and the vertical axis is the imaginary axis.

**23.** $|z| = \sqrt{x^2 + y^2}$

**25.** The trigonometric form of $z = x + iy$ is
$z = r(\cos\theta + i\sin\theta)$
where $r = \sqrt{x^2 + y^2}$ and $\tan\theta = \dfrac{y}{x}$.

**27.** $2(\cos 40° + i\sin 40°)$
$= 2(0.7660 + i \cdot 0.6428)$
$= 1.5321 + 1.2856i$

**29.** If $z_1 = r_1(\cos\theta_1 + i\sin\theta_1)$ and $z_2 = r_2(\cos\theta_2 + i\sin\theta_2)$ then

$$\frac{z_1}{z_2} = \frac{r_1}{r_2}\left(\cos(\theta_1 - \theta_2) + i\sin(\theta_1 - \theta_2)\right)$$

**31.** $\dfrac{z_1}{z_2} = \dfrac{2+2i}{-1-i} = \dfrac{-2(-1-i)}{-1-i} = -2$

$$z_1 = 2 + 2i = 2\sqrt{2}\left(\cos\frac{\pi}{4} + i\sin\frac{\pi}{4}\right)$$

$$z_2 = -1 - i = \sqrt{2}\left(\cos\frac{5\pi}{4} + i\sin\frac{5\pi}{4}\right)$$

$$\frac{z_1}{z_2} = \frac{2\sqrt{2}}{\sqrt{2}}\left[\cos\left(\frac{\pi}{4} - \frac{5\pi}{4}\right) + i\sin\left(\frac{\pi}{4} - \frac{5\pi}{4}\right)\right] = 2(\cos(-\pi) + i\sin(-\pi)) = -2$$

**33.** $(1+i)^6 = \left[\sqrt{2}\left(\cos\frac{\pi}{4} + i\sin\frac{\pi}{4}\right)\right]^6$

$$= (\sqrt{2})^6\left(\cos\frac{3\pi}{2} + i\sin\frac{3\pi}{2}\right)$$

$$= 8(-i) = -8i$$

**35.** The angular difference between the two consecutive $n$th roots of a complex number is $\dfrac{2\pi}{n}$.

**37.** $z^4 + 256 = 0$

$$z^4 = -256 = 256(\cos\pi + i\sin\pi)$$

$$z = \sqrt[4]{256}\left(\cos\frac{\pi + 2k\pi}{4} + i\sin\frac{\pi + 2k\pi}{4}\right)$$

$k = 0$: $4\left(\cos\dfrac{\pi}{4} + i\sin\dfrac{\pi}{4}\right) = 2\sqrt{2} + 2\sqrt{2}i$

$k = 1$:

$$4\left(\cos\frac{3\pi}{4} + i\sin\frac{3\pi}{4}\right) = -2\sqrt{2} + 2\sqrt{2}i$$

$k = 2$:

$$4\left(\cos\frac{5\pi}{4} + i\sin\frac{5\pi}{4}\right) = -2\sqrt{2} - 2\sqrt{2}i$$

$k = 3$: $4\left(\cos\dfrac{7\pi}{4} + i\sin\dfrac{7\pi}{4}\right) = 2\sqrt{2} - 2\sqrt{2}i$

**39.** $\cos\dfrac{2k\pi}{6} + i\sin\dfrac{2k\pi}{6}$ for $k = 0, 1, ..., 5$

$k = 0$: $\cos 0 + i\sin 0 = 1$

$k = 1$: $\cos\dfrac{\pi}{3} + i\sin\dfrac{\pi}{3} = \dfrac{1}{2} + \dfrac{\sqrt{3}}{2}i$

$k = 2$: $\cos\dfrac{2\pi}{3} + i\sin\dfrac{2\pi}{3} = -\dfrac{1}{2} + \dfrac{\sqrt{3}}{2}i$

$k = 3$: $\cos\pi + i\sin\pi = -1$

$k = 4$: $\cos\dfrac{4\pi}{3} + i\sin\dfrac{4\pi}{3} = -\dfrac{1}{2} - \dfrac{\sqrt{3}}{2}i$

$k = 5$: $\cos\dfrac{5\pi}{3} + i\sin\dfrac{5\pi}{3} = \dfrac{1}{2} - \dfrac{\sqrt{3}}{2}i$

**41.**

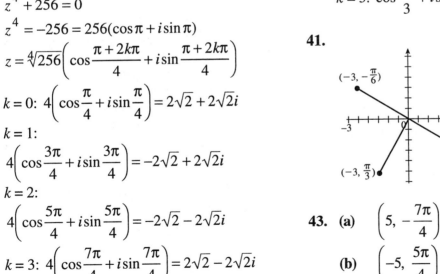

**43. (a)** $\left(5, -\dfrac{7\pi}{4}\right)$

**(b)** $\left(-5, \dfrac{5\pi}{4}\right)$

*Chapter 8: Additional Applications of Trigonometry*

**45.** $(x, y) = (-\sqrt{3}, -1)$

$$r = \sqrt{\left(-\sqrt{3}\right)^2 + (-1)^2} = \sqrt{4} = 2$$

$$\tan\theta = \frac{-1}{-\sqrt{3}} = \frac{1}{\sqrt{3}}, \text{ so } \theta = \frac{7\pi}{6}$$

$$(r, \theta) = \left(2, \frac{7\pi}{6}\right)$$

$$(x, y) = (-2, 2)$$

$$r = \sqrt{(-2)^2 + 2^2} = \sqrt{8} = 2\sqrt{2}$$

$$\tan\theta = \frac{2}{-2} = -1, \text{ so } \theta = \frac{3\pi}{4}$$

$$(r, \theta) = \left(2\sqrt{2}, \frac{3\pi}{4}\right)$$

**47.** The graph of the equation
$$r = r_0$$
is a circle with center 0 and radius $|r_0|$.

**49.** $a = 3$
$$r = -2 \cdot 3\cos\theta$$
$$r = -6\cos\theta$$

**51.**

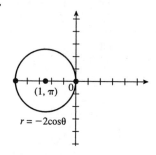

**53.** The complete graph of $r = 4\sin\theta$ is traced out twice as $\theta$ varies from $\theta = 0$ to $\theta = 2\pi$.

**55.** $r = \dfrac{1}{1 + \cos\theta}$

$$r + r\cos\theta = 1$$
$$r + x = 1$$
$$r = -x + 1$$
$$r^2 = (-x + 1)^2$$
$$x^2 + y^2 = x^2 - 2x + 1$$
$$y^2 = -2x + 1$$

$$2x = -y^2 + 1$$
$$x = -\frac{1}{2}y^2 + \frac{1}{2}$$

**57.**

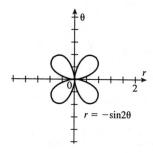

## Chapter 8 Test: Standard Answer

**1. (a)** $a^2 = 8^2 + 12^2 - 2 \cdot 8 \cdot 12 \cos 60°$
$$= 112$$
$$a = \sqrt{112} = 4\sqrt{7}$$
$$60° = \left(60 \cdot \frac{\pi}{180}\right) = \frac{\pi}{3}$$

**(b)** $\dfrac{\sin B}{8} = \dfrac{\sin 60°}{4\sqrt{7}}$

$$\sin B = \frac{8\sin 60°}{4\sqrt{7}}$$

$$= \frac{\sqrt{3}}{\sqrt{7}} \cdot \frac{\sqrt{7}}{\sqrt{7}} = \frac{\sqrt{21}}{7}$$

**2. (a)** $b\sin A = 12\sin 30° = 6$
$a < b\sin A$
no triangles

**(b)** $b\sin A = 8\sin 30° = 4$
$b > a > b\sin A$
two triangles

**3.** $K = \frac{1}{2}bc\sin A$

$$= \frac{1}{2}(20)(12)\sin 60°$$

$$= 120 \cdot \frac{\sqrt{3}}{2} = 60\sqrt{3} \text{ cm}^2$$

**4.** $\dfrac{a}{\sin 45°} = \dfrac{20}{\sin 60°}$

$a = \dfrac{20\sin 45°}{\sin 60°} = \dfrac{10\sqrt{2}}{\frac{\sqrt{3}}{2}}$

$a = \dfrac{20\sqrt{2}}{\sqrt{3}} \cdot \dfrac{\sqrt{3}}{\sqrt{3}} = \dfrac{20\sqrt{6}}{3}$

**5.** $K = \dfrac{1}{2}bc \sin A$

$= \dfrac{1}{2}(20)(34) \sin 18°$

$= 105 \text{ cm}^2$

**6.** $s = \dfrac{1}{2}(7 + 9 + 12) = 14$

$K = \sqrt{14(14-7)(14-9)(14-12)}$

$= \sqrt{14(7)(5)(2)}$

$= 14\sqrt{5} = 31 \text{ cm}^2$

**7.** $\angle BCA = 30°$, so $\angle BAC = 30°$

$\dfrac{x}{\sin 30°} = \dfrac{24}{\sin 120°}$

$x = \dfrac{24\sin 30°}{\sin 120°} = \dfrac{12}{\frac{\sqrt{3}}{2}} = \dfrac{24}{\sqrt{3}}$

$x = \dfrac{24}{\sqrt{3}} \cdot \dfrac{\sqrt{3}}{\sqrt{3}} = 8\sqrt{3}$

**8.** $9^2 = 13^2 + 6^2 - 2 \cdot 13 \cdot 6\cos A$

$-124 = -156 \cos A$

$\cos A = 0.7949$

$A = 37.4°$

$13^2 = 9^2 + 6^2 - 2 \cdot 9 \cdot 6\cos B$

$52 = -108 \cos B$

$\cos B = -0.4815$

$B = 118.8°$

**9.** $\dfrac{\sin B}{16} = \dfrac{\sin 30°}{10}$

$\sin B = \dfrac{16\sin 30°}{10} = \dfrac{4}{5}$

$B = 53.1°$ or $B = 126.9°$

For $B = 53.1°$, $C = 96.9°$

$\dfrac{c}{\sin 96.9°} = \dfrac{10}{\sin 30°}$

$c = \dfrac{10\sin 96.9°}{\sin 30°} = 19.9$

For $B = 126.9°$, $C = 23.1°$

$\dfrac{c}{\sin 23.1°} = \dfrac{10}{\sin 30°}$

$c = \dfrac{10\sin 23.1°}{\sin 30°} = 7.8$

**10.** force $= 2000 \cos 81°$

$= 313$ pounds

**11.** Let $s$ be the speed of the boat. Then

$s^2 = 18^2 + 6^2 - 2 \cdot 6 \cdot 18\cos 50°$

$= 221.1579$

$s = 15$ mph

**12.** $\dfrac{\sin \alpha}{6} = \dfrac{\sin 50}{15}$

$\sin \alpha = \dfrac{6\sin 50}{15} = 0.3064$

$\alpha = 17.8°$

final bearing: N $22°$ W

**13.** $[5(\cos 250° + i\sin 250°)]$

$\cdot [\sqrt{2}(\cos 245° + i\sin 245°)]$

$= 5\sqrt{2}(\cos 495° + i\sin 495°)$

$= 5\sqrt{2}\left(-\dfrac{1}{\sqrt{2}} + \dfrac{1}{\sqrt{2}}i\right) = -5 + 5i$

**14.** $z = \dfrac{\sqrt{3}}{4} + \dfrac{1}{4}i = \dfrac{1}{2}\left(\cos\dfrac{\pi}{6} + i\sin\dfrac{\pi}{6}\right)$

$w = -1 + \sqrt{3}i = 2\left(\cos\dfrac{2\pi}{3} + i\sin\dfrac{2\pi}{3}\right)$

$\dfrac{z}{w} = \dfrac{\frac{1}{2}}{2}\left(\cos\left(-\dfrac{\pi}{2}\right) + i\sin\left(-\dfrac{\pi}{2}\right)\right)$

$= -\dfrac{1}{4}i$

**15.** $z^5 = \left[\dfrac{1}{2}\left(\cos\dfrac{\pi}{6} + i\sin\dfrac{\pi}{6}\right)\right]^5$

$= \dfrac{1}{32}\left(\cos\dfrac{5\pi}{6} + i\sin\dfrac{5\pi}{6}\right)$

$= \dfrac{1}{32}\left(-\dfrac{\sqrt{3}}{2} + \dfrac{1}{2}i\right)$

$= -\dfrac{\sqrt{3}}{64} + \dfrac{1}{64}i$

**16.** $z^3 = 8 = 8(\cos 0 + i\sin 0)$

$z = \sqrt[3]{8}\left(\cos\dfrac{2k\pi}{3} + i\sin\dfrac{2k\pi}{3}\right)$

$k = 0: 2(\cos 0 + i\sin 0) = 2$

$k = 1: 2\left(\cos\dfrac{2\pi}{3} + i\sin\dfrac{2\pi}{3}\right) = -1 + \sqrt{3}i$

$k = 2: 2\left(\cos\dfrac{4\pi}{3} + i\sin\dfrac{4\pi}{3}\right) = -1 - \sqrt{3}i$

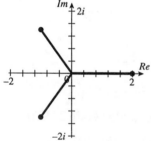

**17.** $z = -8\sqrt{2} + 8\sqrt{2}i = 16\left(\cos\dfrac{3\pi}{4} + i\sin\dfrac{3\pi}{4}\right)$

$z^{1/4} = \sqrt[4]{16}\left(\cos\dfrac{3\pi + 8k\pi}{16} + i\sin\dfrac{3\pi + 8k\pi}{16}\right)$

$k = 0: 2\left(\cos\dfrac{3\pi}{16} + i\sin\dfrac{3\pi}{16}\right)$

$k = 1: 2\left(\cos\dfrac{11\pi}{16} + i\sin\dfrac{11\pi}{16}\right)$

$k = 2: 2\left(\cos\dfrac{19\pi}{16} + i\sin\dfrac{19\pi}{16}\right)$

$k = 3: 2\left(\cos\dfrac{27\pi}{16} + i\sin\dfrac{27\pi}{16}\right)$

**18. (a)** $(-3, -300°)$
    **(b)** $(3, 240°)$

**19. (a)** $x = r\cos\theta = \dfrac{1}{3}\cos\left(-\dfrac{\pi}{6}\right) = \dfrac{\sqrt{3}}{6}$

$y = r\sin\theta = \dfrac{1}{3}\sin\left(-\dfrac{\pi}{6}\right) = -\dfrac{1}{6}$

$(x, y) = \left(\dfrac{\sqrt{3}}{6}, -\dfrac{1}{6}\right)$

    **(b)** $r = \sqrt{\left(-7\sqrt{2}\right)^2 + \left(7\sqrt{2}\right)^2}$

$= \sqrt{196} = 14$

$\tan\theta = \dfrac{7\sqrt{2}}{-7\sqrt{2}} = -1,\ \text{so}\ \theta = \dfrac{3\pi}{4}$

$(r, \theta) = \left(14, \dfrac{3\pi}{4}\right)$

**20.** $a = 3$
$r = -2 \cdot 3 \sin\theta$
$r = -6 \sin\theta$

**21.** $r = 2\sin\theta$
$r^2 = 2r\sin\theta$
$x^2 + y^2 = 2y$
$x^2 + y^2 - 2y + 1 = 1$
$x^2 + (y-1)^2 = 1$
circle

**22.** $r = 4\sin\theta - 3\cos\theta$
$r^2 = 4r\sin\theta - 3r\cos\theta$
$x^2 + y^2 = 4y - 3x$
$x^2 + 3x + \dfrac{9}{4} + y^2 - 4y + 4 = \dfrac{9}{4} + 4$
$\left(x + \dfrac{3}{2}\right)^2 + (y-2)^2 = \dfrac{25}{4}$
circle

**23.**

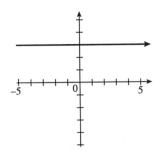

**24.**

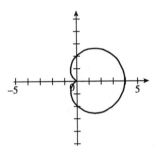

**25.**

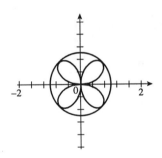

## Chapter 8 Test: Multiple Choice

1. None of the formulas are correct, so the answer is (e).

2. Formula II is correct, so the answer is (b).

3. $\left(\sqrt{106}\right)^2 = 4^2 + \left(5\sqrt{2}\right)^2 - 2 \cdot 4 \cdot 5\sqrt{2}\cos B$
   $106 = 66 - 40\sqrt{2}\cos B$
   $-40\sqrt{2}\cos B = 40$
   $\cos B = -\dfrac{1}{\sqrt{2}}$
   $B = \dfrac{3\pi}{4}$
   The answer is (d).

4. $K = \dfrac{1}{2}bc \sin A = \dfrac{1}{2}(10)(12)\sin 120°$
   $= 30\sqrt{3} \text{ cm}^2$
   The answer is (a).

5. $s = \dfrac{1}{2}(8 + 8 + 12) = 14$
   $K = \sqrt{14(14-8)(14-8)(14-12)}$
   $= \sqrt{14(6)(6)(2)}$
   $= 12\sqrt{7} \text{ cm}^2$
   The answer is (b).

6. $C = 65°$
   $\dfrac{c}{\sin 65°} = \dfrac{40}{\sin 40°}$
   $c = \dfrac{40\sin 65°}{\sin 40°}$
   The answer is (b).

7. $b \sin A = 7 \sin 30° = \dfrac{7}{2}$
   $b \sin A < a < b$
   Two triangles
   The answer is (c).

8. A triangle is not possible if
   $a < b \sin A = 10 \sin 30° = 5$
   The answer is (d).

9. force $= 300 \cos 75°$
   The answer is (a).

10. $1 - i = \sqrt{2}\left(\cos\dfrac{7\pi}{4} + i\sin\dfrac{7\pi}{4}\right)$
    The answer is (c).

11. $z^3 = 2^3(\cos 60° + i\sin 60°)$
    $= 8\left(\dfrac{1}{2} + \dfrac{\sqrt{3}}{2}i\right)$
    $= 4 + 4\sqrt{3}i$
    The answer is (d).

**12.** $\dfrac{z}{w} = \dfrac{8}{24}\left(\cos\left(-\dfrac{\pi}{2}\right) + i\sin\left(-\dfrac{\pi}{2}\right)\right)$

$= \dfrac{1}{3}\left(\cos\dfrac{3\pi}{2} + i\sin\dfrac{3\pi}{2}\right)$

The answer is (c).

**13.** $x = r\cos\theta = -\dfrac{1}{2}\cos\dfrac{\pi}{3} = -\dfrac{1}{4}$

$y = r\sin\theta = -\dfrac{1}{2}\sin\dfrac{\pi}{3} = -\dfrac{\sqrt{3}}{4}$

$(x, y) = \left(-\dfrac{1}{4}, -\dfrac{\sqrt{3}}{4}\right)$

The answer is (a).

**14.** $r = \sqrt{(-3)^2 + (3\sqrt{3})^2} = \sqrt{36} = 6$

$\tan\theta = \dfrac{3\sqrt{3}}{-3} = -\sqrt{3}$, so $\theta = \dfrac{2\pi}{3}$

$(r, \theta) = \left(6, \dfrac{2\pi}{3}\right)$

The answer is (b).

**15.** $-64 = 64(\cos\pi + i\sin\pi)$

Cube roots:

$\sqrt[3]{64}\left(\cos\dfrac{\pi + 2k\pi}{3} + i\sin\dfrac{\pi + 2k\pi}{3}\right)$

$k = 0:\ 4\left(\cos\dfrac{\pi}{3} + i\sin\dfrac{\pi}{3}\right) = 2 + 2\sqrt{3}i$

$k = 1:\ 4(\cos\pi + i\sin\pi) = -4$

$k = 2:\ 4\left(\cos\dfrac{5\pi}{3} + i\sin\dfrac{5\pi}{3}\right) = 2 - 2\sqrt{3}i$

The answer is (d).

**16.** Only I and II are circles, so the answer is (c).

**17.** A cardioid has the form $r = a \pm a\sin\theta$ or $r = a \pm a\cos\theta$. The answer is (a).

**18.** $r = -8\cos\theta$

$r^2 = -8r\cos\theta$

$x^2 + y^2 = -8x$

$x^2 + 8x + 16 + y^2 = 16$

$(x + 4)^2 + y^2 = 16$

The answer is (b).

**19.** $[r(\cos\theta + i\sin\theta)]^n = r^n(\cos n\theta + i\sin n\theta)$

The answer is (c).

**20.** $z^3 + 27i = 0$

$z^3 = -27i = 27\left(\cos\dfrac{3\pi}{2} + i\sin\dfrac{3\pi}{2}\right)$

$z = \sqrt[3]{27}\left(\cos\dfrac{3\pi + 4k\pi}{6} + i\sin\dfrac{3\pi + 4k\pi}{6}\right)$

$k = 0:\ 3\left(\cos\dfrac{\pi}{2} + i\sin\dfrac{\pi}{2}\right) = 3i$

$k = 1:\ 3\left(\cos\dfrac{7\pi}{6} + i\sin\dfrac{7\pi}{6}\right) = -\dfrac{3\sqrt{3}}{2} - \dfrac{3}{2}i$

$k = 2:\ 3\left(\cos\dfrac{11\pi}{6} + i\sin\dfrac{11\pi}{6}\right) = \dfrac{3\sqrt{3}}{2} - \dfrac{3}{2}i$

The answer is (a).

# Chapter 9: Linear Systems, Matrices, and Determinants

**Exercises 9.1**

**1.** $\begin{bmatrix} 1 & 5 & | & -9 \\ 4 & -3 & | & -13 \end{bmatrix}$

$\begin{bmatrix} 1 & 5 & | & -9 \\ 0 & -23 & | & 23 \end{bmatrix} \leftarrow -4 \times \text{(row 1)} + \text{(row 2)}$

$-23y = 23$

$y = -1$

$x + 5(-1) = -9$

$x = -4$

$(-4, -1)$

**3.** $\begin{bmatrix} 3 & 2 & | & 18 \\ 6 & 5 & | & 45 \end{bmatrix}$

$\begin{bmatrix} 3 & 2 & | & 18 \\ 0 & 1 & | & 9 \end{bmatrix} \leftarrow -2 \times \text{(row 1)} + \text{(row 2)}$

$y = 9$

$3x + 2 \cdot 9 = 18$

$3x = 0$

$x = 0$

$(0, 9)$

**5.** $\begin{bmatrix} 4 & -5 & | & -2 \\ 16 & 2 & | & 3 \end{bmatrix}$

$\begin{bmatrix} 4 & -5 & | & -2 \\ 0 & 22 & | & 11 \end{bmatrix} \leftarrow -4 \times \text{(row 1)} + \text{(row 2)}$

$22y = 11$

$y = \dfrac{1}{2}$

$4x - 5\left(\dfrac{1}{2}\right) = -2$

$4x = \dfrac{1}{2}$

$x = \dfrac{1}{8}$

$\left(\dfrac{1}{8}, \dfrac{1}{2}\right)$

**7.** $2x = -8y + 2 \qquad\qquad 2x + 8y = 2$
$4y = x - 1 \qquad \text{or} \qquad -x + 4y = -1$

$\begin{bmatrix} 2 & 8 & | & 2 \\ -1 & 4 & | & -1 \end{bmatrix}$

$$\begin{bmatrix} 1 & 4 & | & 1 \\ -1 & 4 & | & -1 \end{bmatrix} \leftarrow \tfrac{1}{2} \times \text{(row 1)}$$

$$\begin{bmatrix} 1 & 4 & | & 1 \\ 0 & 8 & | & 0 \end{bmatrix} \leftarrow 1 \times \text{(row 1)} + \text{(row 2)}$$

$8y = 0$

$y = 0$

$x + 4 \cdot 0 = 1$

$x = 1$

$(1, 0)$

**9.** $\begin{bmatrix} 30 & 45 & | & 60 \\ 4 & 6 & | & 8 \end{bmatrix}$

$$\begin{bmatrix} 1 & \tfrac{3}{2} & | & 2 \\ 4 & 6 & | & 8 \end{bmatrix} \leftarrow \tfrac{1}{30} \times \text{(row 1)}$$

$$\begin{bmatrix} 1 & \tfrac{3}{2} & | & 2 \\ 0 & 0 & | & 0 \end{bmatrix} \leftarrow -4 \times \text{(row 1)} + \text{(row 2)}$$

Let $y = c$

$x + \dfrac{3}{2} \cdot c = 2$

$x = 2 - \dfrac{3}{2}c$

$\left( 2 - \dfrac{3}{2}c, \ c \right)$

**11.** $\begin{array}{ll} 2x = 8 - 3y & \qquad 2x + 3y = 8 \\ 6x + 9y = 14 & \text{or} \qquad 6x + 9y = 14 \end{array}$

$$\begin{bmatrix} 2 & 3 & | & 8 \\ 6 & 9 & | & 14 \end{bmatrix}$$

$$\begin{bmatrix} 2 & 3 & | & 8 \\ 0 & 0 & | & -10 \end{bmatrix} \leftarrow -3 \times \text{(row 1)} + \text{(row 2)}$$

No solutions

**13.** $\begin{bmatrix} 1 & -2 & 3 & | & -2 \\ -4 & 10 & 2 & | & -2 \\ 3 & 1 & 10 & | & 7 \end{bmatrix}$

$$\begin{bmatrix} 1 & -2 & 3 & | & -2 \\ 0 & 2 & 14 & | & -10 \\ 0 & 7 & 1 & | & 13 \end{bmatrix} \begin{matrix} \leftarrow 4 \times (\text{row 1}) + (\text{row 2}) \\ \leftarrow -3 \times (\text{row 1}) + (\text{row 3}) \end{matrix}$$

$$\begin{bmatrix} 1 & -2 & 3 & | & -2 \\ 0 & 1 & 7 & | & -5 \\ 0 & 7 & 1 & | & 13 \end{bmatrix} \leftarrow \tfrac{1}{2} \times (\text{row } 2)$$

$$\begin{bmatrix} 1 & -2 & 3 & | & -2 \\ 0 & 1 & 7 & | & -5 \\ 0 & 0 & -48 & | & 48 \end{bmatrix} \leftarrow -7 \times (\text{row } 2) + \text{row } 3$$

$-48z = 48$

$z = -1$

$y + 7(-1) = -5$

$y = 2$

$x - 2(2) + 3(-1) = -2$

$x = 5$

$(5, 2, -1)$

**15.** $\begin{bmatrix} -1 & 2 & 3 & | & 11 \\ 2 & -3 & 0 & | & -6 \\ 3 & -3 & 3 & | & 3 \end{bmatrix}$

$\begin{bmatrix} 1 & -2 & -3 & | & -11 \\ 2 & -3 & 0 & | & -6 \\ 3 & -3 & 3 & | & 3 \end{bmatrix} -1 \times (\text{row } 1)$

$\begin{bmatrix} 1 & -2 & -3 & | & -11 \\ 0 & 1 & 6 & | & 16 \\ 0 & 3 & 12 & | & 36 \end{bmatrix} \begin{matrix} \\ \leftarrow -2 \times (\text{row } 1) + (\text{row } 2) \\ \leftarrow -3 \times (\text{row } 1) + (\text{row } 3) \end{matrix}$

$\begin{bmatrix} 1 & -2 & -3 & | & -11 \\ 0 & 1 & 6 & | & 16 \\ 0 & 0 & -6 & | & -12 \end{bmatrix} \begin{matrix} \\ \\ \leftarrow -3 \times (\text{row } 2) + (\text{row } 3) \end{matrix}$

$-6z = -12$

$z = 2$

$y + 6 \cdot 2 = 16$

$y = 4$

$x - 2 \cdot 4 - 3 \cdot 2 = -11$

$x = 3$

$(3, 4, 2)$

**17.** $\begin{bmatrix} 1 & 0 & -2 & | & 5 \\ 0 & 3 & 4 & | & -2 \\ -2 & 3 & 8 & | & 4 \end{bmatrix}$

$\begin{bmatrix} 1 & 0 & -2 & | & 5 \\ 0 & 3 & 4 & | & -2 \\ 0 & 3 & 4 & | & 14 \end{bmatrix} \begin{matrix} \\ \\ \leftarrow 2 \times (\text{row } 1) + (\text{row } 3) \end{matrix}$

$\begin{bmatrix} 1 & 0 & -2 & | & 5 \\ 0 & 3 & 4 & | & -2 \\ 0 & 0 & 0 & | & 16 \end{bmatrix} \begin{matrix} \\ \\ \leftarrow -1 \times (\text{row } 2) + (\text{row } 3) \end{matrix}$

No solutions

**19.**
$$\begin{bmatrix} 1 & -2 & 1 & | & 1 \\ -6 & 1 & 2 & | & -2 \\ -4 & -3 & 4 & | & 0 \end{bmatrix}$$

$$\begin{bmatrix} 1 & -2 & 1 & | & 1 \\ 0 & -11 & 8 & | & 4 \\ 0 & -11 & 8 & | & 4 \end{bmatrix} \begin{matrix} \\ \leftarrow 6\times(\text{row } 1)+(\text{row } 2) \\ \leftarrow 4\times(\text{row } 1)+(\text{row } 3) \end{matrix}$$

$$\begin{bmatrix} 1 & -2 & 1 & | & 1 \\ 0 & -11 & 8 & | & 4 \\ 0 & 0 & 0 & | & 0 \end{bmatrix} \begin{matrix} \\ \\ \leftarrow -1\times(\text{row } 2)+(\text{row } 3) \end{matrix}$$

Let $z = c$

$-11y + 8c = 4$

$-11y = 4 - 8c$

$y = -\dfrac{4}{11} + \dfrac{8}{11}c$

$x - 2\left(-\dfrac{4}{11} + \dfrac{8}{11}c\right) + c = 1$

$x = \dfrac{3}{11} + \dfrac{5}{11}c$

$\left(\dfrac{3}{11} + \dfrac{5}{11}c, \ -\dfrac{4}{11} + \dfrac{8}{11}c, \ c\right)$

**21.**
$$\begin{bmatrix} 1 & -1 & 2 & 2 & | & 0 \\ 2 & 0 & -1 & -3 & | & 0 \\ 0 & 4 & -3 & 1 & | & -2 \\ -3 & 2 & 0 & 4 & | & 1 \end{bmatrix}$$

$$\begin{bmatrix} 1 & -1 & 2 & 2 & | & 0 \\ 0 & 2 & -5 & -7 & | & 0 \\ 0 & 4 & -3 & 1 & | & -2 \\ 0 & -1 & 6 & 10 & | & 1 \end{bmatrix} \begin{matrix} \\ \leftarrow -2\times(\text{row } 1)+(\text{row } 2) \\ \\ \leftarrow 3\times(\text{row } 1)+(\text{row } 4) \end{matrix}$$

$$\begin{bmatrix} 1 & -1 & 2 & 2 & | & 0 \\ 0 & -1 & 6 & 10 & | & 1 \\ 0 & 4 & -3 & 1 & | & -2 \\ 0 & 2 & -5 & -7 & | & 0 \end{bmatrix} \left.\begin{matrix} \\ \\ \\ \end{matrix}\right\} \begin{matrix} \text{Interchange rows} \\ \\ 2 \text{ and } 4 \end{matrix}$$

$$\begin{bmatrix} 1 & -1 & 2 & 2 & | & 0 \\ 0 & -1 & 6 & 10 & | & 1 \\ 0 & 0 & 21 & 41 & | & 2 \\ 0 & 0 & 7 & 13 & | & 2 \end{bmatrix} \begin{matrix} \\ \\ \leftarrow 4\times(\text{row } 2)+(\text{row } 3) \\ \leftarrow 2\times(\text{row } 2)+(\text{row } 4) \end{matrix}$$

$$\begin{bmatrix} 1 & -1 & 2 & 2 & | & 0 \\ 0 & -1 & 6 & 10 & | & 1 \\ 0 & 0 & 7 & 13 & | & 2 \\ 0 & 0 & 21 & 41 & | & 2 \end{bmatrix} \left.\begin{matrix} \\ \\ \\ \end{matrix}\right\} \begin{matrix} \text{Interchange rows} \\ \\ 3 \text{ and } 4 \end{matrix}$$

$$\begin{bmatrix} 1 & -1 & 2 & 2 & 0 \\ 0 & -1 & 6 & 10 & 1 \\ 0 & 0 & 7 & 13 & 2 \\ 0 & 0 & 0 & 2 & -4 \end{bmatrix} \leftarrow -3 \times (\text{row } 3) + (\text{row } 4)$$

$2z = -4 \rightarrow z = -2$

$7y + 13(-2) = 2 \rightarrow 7y = 28 \rightarrow y = 4$

$-x + 6 \cdot 4 + 10(-2) = 1 \rightarrow -x = -3 \rightarrow x = 3$

$w - (3) + 2 \cdot 4 + 2 \cdot (-2) = 0 \rightarrow w = -1$

$(-1, 3, 4, -2)$

**23.**

$$\begin{bmatrix} 1 & 2 & -1 & 0 & 3 & 4 \\ 0 & -1 & 5 & -1 & -1 & 3 \\ 3 & 0 & 6 & 0 & 2 & 1 \\ 2 & 3 & 3 & -1 & 5 & 10 \\ 1 & 0 & 9 & -2 & 1 & 5 \end{bmatrix}$$

$$\begin{bmatrix} 1 & 2 & -1 & 0 & 3 & 4 \\ 0 & -1 & 5 & -1 & -1 & 3 \\ 0 & -6 & 9 & 0 & -7 & -11 \\ 0 & -1 & 5 & -1 & -1 & 2 \\ 0 & -2 & 10 & -2 & -2 & 1 \end{bmatrix} \begin{matrix} \\ \\ \leftarrow -3 \times (\text{row } 1) + (\text{row } 3) \\ \leftarrow -2 \times (\text{row } 1) + (\text{row } 4) \\ \leftarrow -1 \times (\text{row } 1) + (\text{row } 5) \end{matrix}$$

$$\begin{bmatrix} 1 & 2 & -1 & 0 & 3 & 4 \\ 0 & -1 & 5 & -1 & -1 & 3 \\ 0 & 0 & -21 & 6 & -1 & -29 \\ 0 & 0 & 0 & 0 & 0 & -1 \\ 0 & 0 & 0 & 0 & 0 & -5 \end{bmatrix} \begin{matrix} \\ \\ \leftarrow -6 \times (\text{row } 2) + (\text{row } 3) \\ \leftarrow -1 \times (\text{row } 2) + (\text{row } 4) \\ \leftarrow -2 \times (\text{row } 2) + (\text{row } 5) \end{matrix}$$

no solutions

**25.** Let $x$ be the number of one-dollar bills,
$y$ the number of 5 dollar bills, and $z$ the number of 10 dollar bills.

$1x + 5y + 10z = 575$

$x + y + z = 95$

$x + z = 2y + 5$ or $x - 2y + z = 5$

$$\begin{bmatrix} 1 & 5 & 10 & 575 \\ 1 & 1 & 1 & 95 \\ 1 & -2 & 1 & 5 \end{bmatrix}$$

$$\begin{bmatrix} 1 & 5 & 10 & 575 \\ 0 & -4 & -9 & -480 \\ 0 & -7 & -9 & -570 \end{bmatrix} \begin{matrix} \\ \leftarrow -1 \times (\text{row } 1) + (\text{row } 2) \\ \leftarrow -1 \times (\text{row } 1) + (\text{row } 3) \end{matrix}$$

$$\begin{bmatrix} 1 & 5 & 10 & 575 \\ 0 & 1 & \frac{9}{4} & 120 \\ 0 & -7 & -9 & -570 \end{bmatrix} -\frac{1}{4} \times (\text{row } 2)$$

$$\begin{bmatrix} 1 & 5 & 10 & | & 575 \\ 0 & 1 & \frac{9}{4} & | & 120 \\ 0 & 0 & \frac{27}{4} & | & 270 \end{bmatrix} \leftarrow 7 \times (\text{row } 2) + (\text{row } 3)$$

$$\frac{27}{4} z = 270 \;\rightarrow\; z = 40$$

$$y + \frac{9}{4}(40) = 120 \;\rightarrow\; y = 30$$

$$x + 5(30) + 10(40) = 575 \rightarrow x = 25$$

25 ones, 30 fives, and 40 tens

**27.** Let $x$, $y$, and $z$ be the angles from smallest to largest.

$$\begin{array}{lll} x + y + z = 180 & & x + y + z = 180 \\ z = x + y & \text{or} & -x - y + z = 0 \\ 2x = z - 10 & & 2x - z = -10 \end{array}$$

$$\begin{bmatrix} 1 & 1 & 1 & | & 180 \\ -1 & -1 & 1 & | & 0 \\ 2 & 0 & -1 & | & -10 \end{bmatrix}$$

$$\begin{bmatrix} 1 & 1 & 1 & | & 180 \\ 0 & 0 & 2 & | & 180 \\ 0 & -2 & -3 & | & -370 \end{bmatrix} \begin{array}{l} \leftarrow 1 \times (\text{row } 1) + \text{row } 2 \\ \leftarrow -2 \times (\text{row } 1) + \text{row } 3 \end{array}$$

$$\begin{bmatrix} 1 & 1 & 1 & | & 180 \\ 0 & -2 & -3 & | & -370 \\ 0 & 0 & 2 & | & 180 \end{bmatrix} \left. \begin{array}{l} \text{Interchange rows} \\ 2 \text{ and } 3 \end{array} \right.$$

$$2z = 180 \;\rightarrow\; z = 90$$
$$-2y - 3(90) = -370 \;\rightarrow\; -2y = -100 \rightarrow y = 50$$
$$x + 50 + 90 = 180 \;\rightarrow\; x = 40$$
$$40°, 50°, 90°$$

**29.** Let $x$ be the pounds of peanuts, $y$ be pounds of pecans, and $z$ be pounds of Brazil nuts.

$$x + y + z = 50$$
$$2.8x + 4.5y + 5.4z = (4.44)50 = 222$$
$$x + y = z$$

or

$$x + y + z = 50$$
$$2.8x + 4.5y + 5.4z = 222$$
$$x + y - z = 0$$

$$\begin{bmatrix} 1 & 1 & 1 & | & 50 \\ 2.8 & 4.5 & 5.4 & | & 222 \\ 1 & 1 & -1 & | & 0 \end{bmatrix}$$

$$\begin{bmatrix} 1 & 1 & 1 & | & 50 \\ 0 & 1.7 & 2.6 & | & 82 \\ 0 & 0 & -2 & | & -50 \end{bmatrix} \begin{array}{l} \leftarrow -2.8 \times (\text{row } 1) + (\text{row } 2) \\ \leftarrow 1 \times (\text{row } 1) + (\text{row } 3) \end{array}$$

$-2z = -50$

$z = 25$

$1.7y + 2.6(25) = 82$

$1.7y = 17$

$y = 10$

$x + 10 + 25 = 50$

$x = 15$

15 pounds peanuts, 10 pounds pecans, 25 pounds Brazil nuts.

**31.** $35A + 10B + 20C = 900$

$15A + 20B + 15C = 750$

$10A + 10B + 5C = 350$

$$\begin{bmatrix} 35 & 10 & 20 & | & 900 \\ 15 & 20 & 15 & | & 750 \\ 10 & 10 & 5 & | & 350 \end{bmatrix}$$

$$\begin{bmatrix} 10 & 10 & 5 & | & 350 \\ 15 & 20 & 15 & | & 750 \\ 35 & 10 & 20 & | & 900 \end{bmatrix} \begin{matrix} \leftarrow \\ \\ \leftarrow \end{matrix}$$ Interchange rows 1 and 3

$$\begin{bmatrix} 1 & 1 & \frac{1}{2} & | & 35 \\ 15 & 20 & 15 & | & 750 \\ 35 & 10 & 20 & | & 900 \end{bmatrix} \leftarrow \frac{1}{10} \times (\text{row } 1)$$

$$\begin{bmatrix} 1 & 1 & \frac{1}{2} & | & 35 \\ 0 & 5 & \frac{15}{2} & | & 225 \\ 0 & -25 & \frac{5}{2} & | & -325 \end{bmatrix} \begin{matrix} \\ \leftarrow -15 \times (\text{row } 1) + (\text{row } 2) \\ \leftarrow -35 \times (\text{row } 1) + \text{row } 3 \end{matrix}$$

$$\begin{bmatrix} 1 & 1 & \frac{1}{2} & | & 35 \\ 0 & 5 & \frac{15}{2} & | & 225 \\ 0 & 0 & 40 & | & 800 \end{bmatrix} \leftarrow 5 \times (\text{row } 2) + (\text{row } 3)$$

$40C = 800$

$C = 20$

$5B + \dfrac{15}{2}(20) = 225$

$5B = 75$

$B = 15$

$A + 15 + \dfrac{1}{2}(20) = 35$

$A = 10$

10 grams of $A$, 15 grams of $B$, and 20 grams of $C$.

**Challenge**

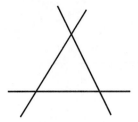

**Exercises 9.2**

**1.** $C + D = \begin{bmatrix} 5 & -1 \\ -3 & 4 \end{bmatrix} + \begin{bmatrix} 1 & 2 \\ 3 & 4 \end{bmatrix} = \begin{bmatrix} 6 & 1 \\ 0 & 8 \end{bmatrix}$

**3.** $A + B$ is undefined.

**5.** $E + 2F = \begin{bmatrix} 1 & -3 & 2 \\ -5 & 2 & 0 \\ 0 & -1 & 3 \end{bmatrix} + 2 \begin{bmatrix} 1 & 1 & 1 \\ -1 & 1 & 1 \\ -1 & -1 & 1 \end{bmatrix} = \begin{bmatrix} 1 & -3 & 2 \\ -5 & 2 & 0 \\ 0 & -1 & 3 \end{bmatrix} + \begin{bmatrix} 2 & 2 & 2 \\ -2 & 2 & 2 \\ -2 & -2 & 2 \end{bmatrix} = \begin{bmatrix} 3 & -1 & 4 \\ -7 & 4 & 2 \\ -2 & -3 & 5 \end{bmatrix}$

**7.** $BA = \begin{bmatrix} 3 & 5 \end{bmatrix} \begin{bmatrix} 1 \\ -2 \end{bmatrix} = [-7]$

**9.** $CD = \begin{bmatrix} 5 & -1 \\ -3 & 4 \end{bmatrix} \begin{bmatrix} 1 & 2 \\ 3 & 4 \end{bmatrix} = \begin{bmatrix} 2 & 6 \\ 9 & 10 \end{bmatrix}$

**11.** $CA = \begin{bmatrix} 5 & -1 \\ -3 & 4 \end{bmatrix} \begin{bmatrix} 1 \\ -2 \end{bmatrix} = \begin{bmatrix} 7 \\ -11 \end{bmatrix}$

**13.** $GE = \begin{bmatrix} 1 & -4 & 0 \\ 2 & -3 & 5 \end{bmatrix} \begin{bmatrix} 1 & -3 & 2 \\ -5 & 2 & 0 \\ 0 & -1 & 3 \end{bmatrix} = \begin{bmatrix} 21 & -11 & 2 \\ 17 & -17 & 19 \end{bmatrix}$

**15.** $GJ$ is undefined.

**17.** $HJ = \begin{bmatrix} 5 & 2 & -3 & 0 \\ 1 & 0 & 3 & -1 \\ 4 & -2 & 0 & 1 \end{bmatrix} \begin{bmatrix} 1 & 0 \\ 0 & -1 \\ 1 & 1 \\ 0 & 0 \end{bmatrix} = \begin{bmatrix} 2 & -5 \\ 4 & 3 \\ 4 & 2 \end{bmatrix}$

**19.** $GH = \begin{bmatrix} 1 & -4 & 0 \\ 2 & -3 & 5 \end{bmatrix} \begin{bmatrix} 5 & 2 & -3 & 0 \\ 1 & 0 & 3 & -1 \\ 4 & -2 & 0 & 1 \end{bmatrix} = \begin{bmatrix} 1 & 2 & -15 & 4 \\ 27 & -6 & -15 & 8 \end{bmatrix}$

**21.** $F\begin{bmatrix} 0 \\ 0 \\ 0 \end{bmatrix} = \begin{bmatrix} 1 & 1 & 1 \\ -1 & 1 & 1 \\ -1 & -1 & 1 \end{bmatrix} \begin{bmatrix} 0 \\ 0 \\ 0 \end{bmatrix} = \begin{bmatrix} 0 \\ 0 \\ 0 \end{bmatrix}$

**23.** $I_3 H = \begin{bmatrix} 1 & 0 & 0 \\ 0 & 1 & 0 \\ 0 & 0 & 1 \end{bmatrix} \begin{bmatrix} 5 & 2 & -3 & -0 \\ 1 & 0 & 3 & -1 \\ 4 & -2 & 0 & 1 \end{bmatrix} = \begin{bmatrix} 5 & 2 & -3 & 0 \\ 1 & 0 & 3 & -1 \\ 4 & -2 & 0 & 1 \end{bmatrix} = H$

**25.** $G(-2F) = \begin{bmatrix} 1 & -4 & 0 \\ 2 & -3 & 5 \end{bmatrix} \begin{bmatrix} -2 & -2 & -2 \\ 2 & -2 & -2 \\ 2 & 2 & -2 \end{bmatrix} = \begin{bmatrix} -10 & 6 & 6 \\ 0 & 12 & -8 \end{bmatrix}$

**27.** $(GF)H = \left( \begin{bmatrix} 1 & -4 & 0 \\ 2 & -3 & 5 \end{bmatrix} \begin{bmatrix} 1 & 1 & 1 \\ -1 & 1 & 1 \\ -1 & -1 & 1 \end{bmatrix} \right) \begin{bmatrix} 5 & 2 & -3 & 0 \\ 1 & 0 & 3 & -1 \\ 4 & -2 & 0 & 1 \end{bmatrix} = \begin{bmatrix} 5 & -3 & -3 \\ 0 & -6 & 4 \end{bmatrix} \begin{bmatrix} 5 & 2 & -3 & 0 \\ 1 & 0 & 3 & -1 \\ 4 & -2 & 0 & 1 \end{bmatrix}$

$= \begin{bmatrix} 10 & 16 & -24 & 0 \\ 10 & -8 & -18 & 10 \end{bmatrix}$

**29.** $B + C = \begin{bmatrix} 0 & 6 \\ 5 & -4 \end{bmatrix} + \begin{bmatrix} 3 & 7 \\ -1 & -2 \end{bmatrix} = \begin{bmatrix} 3 & 13 \\ 4 & -6 \end{bmatrix}$

$C + B = \begin{bmatrix} 3 & 7 \\ -1 & -2 \end{bmatrix} + \begin{bmatrix} 0 & 6 \\ 5 & -4 \end{bmatrix} = \begin{bmatrix} 3 & 13 \\ 4 & -6 \end{bmatrix}$

**31.** $A + (B + C) = \begin{bmatrix} 2 & 1 \\ -4 & 1 \end{bmatrix} + \left( \begin{bmatrix} 0 & 6 \\ 5 & -4 \end{bmatrix} + \begin{bmatrix} 3 & 7 \\ -1 & -2 \end{bmatrix} \right) = \begin{bmatrix} 2 & 1 \\ -4 & 1 \end{bmatrix} + \begin{bmatrix} 3 & 13 \\ 4 & -6 \end{bmatrix} = \begin{bmatrix} 5 & 14 \\ 0 & -5 \end{bmatrix}$

$(A + B) + C = \left( \begin{bmatrix} 2 & 1 \\ -4 & 1 \end{bmatrix} + \begin{bmatrix} 0 & 6 \\ 5 & -4 \end{bmatrix} \right) + \begin{bmatrix} 3 & 7 \\ -1 & -2 \end{bmatrix} = \begin{bmatrix} 2 & 7 \\ 1 & -3 \end{bmatrix} + \begin{bmatrix} 3 & 7 \\ -1 & -2 \end{bmatrix} = \begin{bmatrix} 5 & 14 \\ 0 & -5 \end{bmatrix}$

**33.** $(A + B)C = \left( \begin{bmatrix} 2 & 1 \\ -4 & 1 \end{bmatrix} + \begin{bmatrix} 0 & 6 \\ 5 & -4 \end{bmatrix} \right) \begin{bmatrix} 3 & 7 \\ -1 & -2 \end{bmatrix} = \begin{bmatrix} 2 & 7 \\ 1 & -3 \end{bmatrix} \begin{bmatrix} 3 & 7 \\ -1 & -2 \end{bmatrix} = \begin{bmatrix} -1 & 0 \\ 6 & 13 \end{bmatrix}$

$AC + BC = \begin{bmatrix} 2 & 1 \\ -4 & 1 \end{bmatrix} \begin{bmatrix} 3 & 7 \\ -1 & -2 \end{bmatrix} + \begin{bmatrix} 0 & 6 \\ 5 & -4 \end{bmatrix} \begin{bmatrix} 3 & 7 \\ -1 & -2 \end{bmatrix} = \begin{bmatrix} 5 & 12 \\ -13 & -30 \end{bmatrix} + \begin{bmatrix} -6 & -12 \\ 19 & 43 \end{bmatrix} = \begin{bmatrix} -1 & 0 \\ 6 & 13 \end{bmatrix}$

**35.** $(AB)C = \left( \begin{bmatrix} 1 & -2 \\ 3 & 3 \\ 0 & 4 \end{bmatrix} \begin{bmatrix} -1 & 6 & 8 \\ 2 & 1 & 3 \end{bmatrix} \right) \begin{bmatrix} 5 \\ -2 \\ 1 \end{bmatrix} = \begin{bmatrix} -5 & 4 & 2 \\ 3 & 21 & 33 \\ 8 & 4 & 12 \end{bmatrix} \begin{bmatrix} 5 \\ -2 \\ 1 \end{bmatrix} = \begin{bmatrix} -31 \\ 6 \\ 44 \end{bmatrix}$

$A(BC) = \begin{bmatrix} 1 & -2 \\ 3 & 3 \\ 0 & 4 \end{bmatrix} \left( \begin{bmatrix} -1 & 6 & 8 \\ 2 & 1 & 3 \end{bmatrix} \begin{bmatrix} 5 \\ -2 \\ 1 \end{bmatrix} \right) = \begin{bmatrix} 1 & -2 \\ 3 & 3 \\ 0 & 4 \end{bmatrix} \begin{bmatrix} -9 \\ 11 \end{bmatrix} = \begin{bmatrix} -31 \\ 6 \\ 44 \end{bmatrix}$

**37.** $A(2B) + (2A)C = (2A)B + (2A)C = 2A(B + C) = \begin{bmatrix} 10 & -4 & 2 \end{bmatrix} \left( \begin{bmatrix} 5 & -2 \\ 3 & 3 \\ 0 & 4 \end{bmatrix} + \begin{bmatrix} -1 & 2 \\ 6 & 1 \\ 8 & 3 \end{bmatrix} \right)$

$= \begin{bmatrix} 10 & -4 & 2 \end{bmatrix} \begin{bmatrix} 4 & 0 \\ 9 & 4 \\ 8 & 7 \end{bmatrix} = \begin{bmatrix} 20 & -2 \end{bmatrix}$

**39. (a)** $EA = \begin{bmatrix} 0 & 1 & 0 \\ 1 & 0 & 0 \\ 0 & 0 & 1 \end{bmatrix} \begin{bmatrix} a & b & c \\ d & e & f \\ g & h & i \end{bmatrix} = \begin{bmatrix} d & e & f \\ a & b & c \\ g & h & i \end{bmatrix}$

**(b)** Since $E$ exchanges the first and second rows of $A$ and $E$ is the identity matrix with the first and second rows interchanged, $F = \begin{bmatrix} 0 & 0 & 1 \\ 0 & 1 & 0 \\ 1 & 0 & 0 \end{bmatrix}$, the identity matrix with the first and third rows interchanged.

**41.** $E = \begin{bmatrix} 1 & 0 & 0 \\ 0 & 1 & 0 \\ 3 & 0 & 1 \end{bmatrix}, \quad F = \begin{bmatrix} 1 & 0 & 0 \\ 0 & 1 & 0 \\ 0 & -1 & 1 \end{bmatrix}$

**43. (a)** $\begin{bmatrix} 1 & 1 \\ 0 & 0 \end{bmatrix} \begin{bmatrix} 2 & 3 \\ x & y \end{bmatrix} = \begin{bmatrix} 2+x & 3+y \\ 0 & 0 \end{bmatrix} = \begin{bmatrix} 0 & 0 \\ 0 & 0 \end{bmatrix}$

$\begin{aligned} 2 + x &= 0 & 3 + y &= 0 \\ x &= -2 & y &= -3 \end{aligned}$

**(b)** $\begin{bmatrix} 2 & 3 \\ x & y \end{bmatrix} \begin{bmatrix} 1 & 1 \\ 0 & 0 \end{bmatrix} = \begin{bmatrix} 2 & 2 \\ x & x \end{bmatrix} \neq \begin{bmatrix} 0 & 0 \\ 0 & 0 \end{bmatrix}$

If there were such $x, y$, then $\begin{bmatrix} 2 & 2 \\ x & x \end{bmatrix} = \begin{bmatrix} 0 & 0 \\ 0 & 0 \end{bmatrix}$, which cannot be true since $2 \neq 0$.

**45. (a)** $A^2 = AA = \begin{bmatrix} x & 0 & 0 \\ 0 & y & 0 \\ 0 & 0 & z \end{bmatrix}\begin{bmatrix} x & 0 & 0 \\ 0 & y & 0 \\ 0 & 0 & z \end{bmatrix} = \begin{bmatrix} x^2 & 0 & 0 \\ 0 & y^2 & 0 \\ 0 & 0 & z^2 \end{bmatrix}$

**(b)** $A^3 = AAA = (AA)A = \begin{bmatrix} x^2 & 0 & 0 \\ 0 & y^2 & 0 \\ 0 & 0 & z^2 \end{bmatrix}\begin{bmatrix} x & 0 & 0 \\ 0 & y & 0 \\ 0 & 0 & z \end{bmatrix} = \begin{bmatrix} x^3 & 0 & 0 \\ 0 & y^3 & 0 \\ 0 & 0 & z^3 \end{bmatrix}$

**(c)** $A^n = \begin{bmatrix} x^n & 0 & 0 \\ 0 & y^n & 0 \\ 0 & 0 & z^n \end{bmatrix}$

**47. (a)** $\left(A^t\right)^t = A$

**(b)** $(A+B)^t = \left(\begin{bmatrix} a_1 & a_2 & a_3 \\ a_4 & a_5 & a_6 \end{bmatrix} + \begin{bmatrix} b_1 & b_2 & b_3 \\ b_4 & b_5 & b_6 \end{bmatrix}\right)^t = \begin{bmatrix} a_1 + b_1 & a_2 + b_2 & a_3 + b_3 \\ a_4 + b_4 & a_5 + b_5 & a_6 + b_6 \end{bmatrix}^t$

$= \begin{bmatrix} a_1 + b_1 & a_4 + b_4 \\ a_2 + b_2 & a_5 + b_5 \\ a_3 + b_3 & a_6 + b_6 \end{bmatrix}$

$A^t + B^t = \begin{bmatrix} a_1 & a_4 \\ a_2 & a_5 \\ a_3 & a_6 \end{bmatrix} + \begin{bmatrix} b_1 & b_4 \\ b_2 & b_5 \\ b_3 & b_6 \end{bmatrix} = \begin{bmatrix} a_1 + b_1 & a_4 + b_4 \\ a_2 + b_2 & a_5 + b_5 \\ a_3 + b_3 & a_6 + b_6 \end{bmatrix}$

**49. (a)** 

$$M = \begin{bmatrix} \text{bic.} & \text{row} & \text{tread} \\ 21 & 15 & 34 \\ 19 & 25 & 28 \end{bmatrix} \begin{matrix} \\ \text{Store I} \\ \text{Store II} \end{matrix}$$

$$J = \begin{bmatrix} 28 & 18 & 27 \\ 25 & 17 & 28 \end{bmatrix} \begin{matrix} \text{Store I} \\ \text{Store II} \end{matrix}$$

$$M + J = \begin{bmatrix} 49 & 33 & 61 \\ 44 & 42 & 56 \end{bmatrix}$$ Gives the total sales per item for May and June for each store.

$$M - J = \begin{bmatrix} -7 & -3 & 7 \\ -6 & 8 & 0 \end{bmatrix}$$

A positive entry means the store sold more items in May than in June. If negative, more were sold in June than in May, and if 0, the sales were the same for the two months.

**(b)** July sales $= S - (M+J) = \begin{bmatrix} 67 & 45 & 70 \\ 59 & 62 & 63 \end{bmatrix} - \begin{bmatrix} 49 & 33 & 61 \\ 44 & 42 & 56 \end{bmatrix} = \begin{bmatrix} \text{bic.} & \text{row} & \text{tread} \\ 18 & 12 & 9 \\ 15 & 20 & 7 \end{bmatrix} \begin{matrix} \\ \text{Store I} \\ \text{Store II} \end{matrix}$

**51.** **(a)**
$$\begin{matrix} a & b & c \end{matrix}$$
$$P = \begin{bmatrix} 0 & 1200 & 200 \end{bmatrix}$$

$$PR = \begin{bmatrix} 0 & 1200 & 200 \end{bmatrix} \begin{bmatrix} 2 & 4 & 5 \\ 3 & 5 & 3 \\ 3 & 6 & 4 \end{bmatrix} = \begin{matrix} m & p & w \\ [4200 & 7200 & 4400] \end{matrix}$$

**(b)** $(PR)C = P(RC) = \begin{bmatrix} 0 & 1200 & 200 \end{bmatrix} \begin{bmatrix} 12.15 & 12.00 \\ 12.20 & 12.25 \\ 14.20 & 14.15 \end{bmatrix} = \begin{bmatrix} 17,480 & 17,530 \end{bmatrix}$

$17,530 - $17,480 = $50$, so it is more economical to buy from supplier $s_1$.

## Challenge

To transform $A$ into $I_2$, first add twice the first row to the second, then multiply the entire matrix by $\frac{1}{2}$.

So, let $E_1 = \begin{bmatrix} 1 & 0 \\ 2 & 1 \end{bmatrix}$ and $E_2 = \begin{bmatrix} \frac{1}{2} & 0 \\ 0 & \frac{1}{2} \end{bmatrix}$. Then $E_2 E_1 A = \begin{bmatrix} \frac{1}{2} & 0 \\ 0 & \frac{1}{2} \end{bmatrix} \begin{bmatrix} 1 & 0 \\ 2 & 1 \end{bmatrix} \begin{bmatrix} 2 & 0 \\ -4 & 2 \end{bmatrix}$

$= \begin{bmatrix} \frac{1}{2} & 0 \\ 0 & \frac{1}{2} \end{bmatrix} \begin{bmatrix} 2 & 0 \\ 0 & 2 \end{bmatrix} = \begin{bmatrix} 1 & 0 \\ 0 & 1 \end{bmatrix}$.

## Exercises 9.3

**1.** $\begin{bmatrix} 4 & -1 & | & 1 & 0 \\ 2 & 0 & | & 0 & 1 \end{bmatrix}$

$\begin{bmatrix} 2 & 0 & | & 0 & 1 \\ 4 & -1 & | & 1 & 0 \end{bmatrix}$ ← Interchange rows 1 and 2

$\begin{bmatrix} 1 & 0 & | & 0 & \frac{1}{2} \\ 4 & -1 & | & 1 & 0 \end{bmatrix}$ ← $\frac{1}{2} \times R_1$

$\begin{bmatrix} 1 & 0 & | & 0 & \frac{1}{2} \\ 0 & -1 & | & 1 & -2 \end{bmatrix}$ ← $-4 \times R_1 + R_2$

$\begin{bmatrix} 1 & 0 & | & 0 & \frac{1}{2} \\ 0 & 1 & | & -1 & 2 \end{bmatrix}$ ← $-1 \times R_2$

$A^{-1} = \begin{bmatrix} 0 & \frac{1}{2} \\ -1 & 2 \end{bmatrix}$

**3.** $\begin{bmatrix} \frac{1}{3} & -\frac{4}{3} & | & 1 & 0 \\ -2 & 8 & | & 0 & 1 \end{bmatrix}$

$\begin{bmatrix} 1 & -4 & | & 3 & 0 \\ -2 & 8 & | & 0 & 1 \end{bmatrix}$ ← $3 \times R_1$

$$\begin{bmatrix} 1 & -4 & 3 & 0 \\ 0 & 0 & 6 & 1 \end{bmatrix} \leftarrow 2 \times R_1 + R_2$$

$A^{-1}$ does not exist.

**5.** $\begin{bmatrix} 2 & -5 & 1 & 0 \\ -3 & 4 & 0 & 1 \end{bmatrix}$

$\begin{bmatrix} 1 & -\frac{5}{2} & \frac{1}{2} & 0 \\ -3 & 4 & 0 & 1 \end{bmatrix} \leftarrow \frac{1}{2} \times R_1$

$\begin{bmatrix} 1 & -\frac{5}{2} & \frac{1}{2} & 0 \\ 0 & -\frac{7}{2} & \frac{3}{2} & 1 \end{bmatrix} \leftarrow 3 \times R_1 + R_2$

$\begin{bmatrix} 1 & -\frac{5}{2} & \frac{1}{2} & 0 \\ 0 & 1 & -\frac{3}{7} & -\frac{2}{7} \end{bmatrix} \leftarrow -\frac{2}{7} \times R_2$

$\begin{bmatrix} 1 & 0 & -\frac{4}{7} & -\frac{5}{7} \\ 0 & 1 & -\frac{3}{7} & -\frac{2}{7} \end{bmatrix} \leftarrow \frac{5}{2} \times R_2 + R_1$

$A^{-1} = \begin{bmatrix} -\frac{4}{7} & -\frac{5}{7} \\ -\frac{3}{7} & -\frac{2}{7} \end{bmatrix}$

**7.** $\begin{bmatrix} \frac{1}{3} & \frac{2}{3} & 1 & 0 \\ \frac{2}{3} & \frac{1}{3} & 0 & 1 \end{bmatrix}$

$\begin{bmatrix} 1 & 2 & 3 & 0 \\ \frac{2}{3} & \frac{1}{3} & 0 & 1 \end{bmatrix} \leftarrow 3 \times R_1$

$\begin{bmatrix} 1 & 2 & 3 & 0 \\ 0 & -1 & -2 & 1 \end{bmatrix} \leftarrow -\frac{2}{3} \times R_1 + R_2$

$\begin{bmatrix} 1 & 2 & 3 & 0 \\ 0 & 1 & 2 & -1 \end{bmatrix} \leftarrow -1 \times R_2$

$\begin{bmatrix} 1 & 0 & -1 & 2 \\ 0 & 1 & 2 & -1 \end{bmatrix} \leftarrow -2 \times R_2 + R_1$

$A^{-1} = \begin{bmatrix} -1 & 2 \\ 2 & -1 \end{bmatrix}$

**9.** $\begin{bmatrix} 0 & a & 1 & 0 \\ b & 0 & 0 & 1 \end{bmatrix}$

$\begin{bmatrix} b & 0 & 0 & 1 \\ 0 & a & 1 & 0 \end{bmatrix} \left.\right\} \begin{matrix} \text{Interchange} \\ R_1 \text{ and } R_2 \end{matrix}$

$$\begin{bmatrix} 1 & 0 & | & 0 & \frac{1}{b} \\ 0 & 1 & | & \frac{1}{a} & 0 \end{bmatrix} \begin{matrix} \leftarrow \frac{1}{b} \times R_1 \\ \leftarrow \frac{1}{a} \times R_2 \end{matrix}$$

$$A^{-1} = \begin{bmatrix} 0 & \frac{1}{b} \\ \frac{1}{a} & 0 \end{bmatrix}$$

**11.**
$$\begin{bmatrix} 0 & 1 & 0 & | & 1 & 0 & 0 \\ 1 & 0 & 0 & | & 0 & 1 & 0 \\ 0 & 0 & 1 & | & 0 & 0 & 1 \end{bmatrix}$$

$$\begin{bmatrix} 1 & 0 & 0 & | & 0 & 1 & 0 \\ 0 & 1 & 0 & | & 1 & 0 & 0 \\ 0 & 0 & 1 & | & 0 & 0 & 1 \end{bmatrix} \begin{matrix} \text{Interchange} \\ R_1 \text{ and } R_2 \end{matrix}$$

$$A^{-1} = \begin{bmatrix} 0 & 1 & 0 \\ 1 & 0 & 0 \\ 0 & 0 & 1 \end{bmatrix} = A$$

**13.**
$$\begin{bmatrix} 1 & 0 & 2 & | & 1 & 0 & 0 \\ 2 & -1 & 0 & | & 0 & 1 & 0 \\ 0 & 3 & 4 & | & 0 & 0 & 1 \end{bmatrix}$$

$$\begin{bmatrix} 1 & 0 & 2 & | & 1 & 0 & 0 \\ 0 & -1 & -4 & | & -2 & 1 & 0 \\ 0 & 3 & 4 & | & 0 & 0 & 1 \end{bmatrix} \leftarrow -2 \times R_1 + R_2$$

$$\begin{bmatrix} 1 & 0 & 2 & | & 1 & 0 & 0 \\ 0 & 1 & 4 & | & 2 & -1 & 0 \\ 0 & 3 & 4 & | & 0 & 0 & 1 \end{bmatrix} \leftarrow -1 \times R_2$$

$$\begin{bmatrix} 1 & 0 & 2 & | & 1 & 0 & 0 \\ 0 & 1 & 4 & | & 2 & -1 & 0 \\ 0 & 0 & -8 & | & -6 & 3 & 1 \end{bmatrix} \leftarrow -3 \times R_2 + R_3$$

$$\begin{bmatrix} 1 & 0 & 2 & | & 1 & 0 & 0 \\ 0 & 1 & 4 & | & 2 & -1 & 0 \\ 0 & 0 & 1 & | & \frac{3}{4} & -\frac{3}{8} & -\frac{1}{8} \end{bmatrix} \leftarrow -\frac{1}{8} \times R_3$$

$$\begin{bmatrix} 1 & 0 & 0 & | & -\frac{1}{2} & \frac{3}{4} & \frac{1}{4} \\ 0 & 1 & 0 & | & -1 & \frac{1}{2} & \frac{1}{2} \\ 0 & 0 & 1 & | & \frac{3}{4} & -\frac{3}{8} & -\frac{1}{8} \end{bmatrix} \begin{matrix} \leftarrow -2 \times R_3 + R_1 \\ \leftarrow -4 \times R_3 + R_2 \end{matrix}$$

$$A^{-1} = \begin{bmatrix} -\frac{1}{2} & \frac{3}{4} & \frac{1}{4} \\ -1 & \frac{1}{2} & \frac{1}{2} \\ \frac{3}{4} & -\frac{3}{8} & -\frac{1}{8} \end{bmatrix}$$

**15.**
$$\left[\begin{array}{ccc|ccc} 4 & -3 & 1 & 1 & 0 & 0 \\ 0 & -1 & 9 & 0 & 1 & 0 \\ -2 & 1 & 4 & 0 & 0 & 1 \end{array}\right]$$

$$\left[\begin{array}{ccc|ccc} 1 & -\frac{3}{4} & \frac{1}{4} & \frac{1}{4} & 0 & 0 \\ 0 & -1 & 9 & 0 & 1 & 0 \\ -2 & 1 & 4 & 0 & 0 & 1 \end{array}\right] \leftarrow \frac{1}{4} \times R_1$$

$$\left[\begin{array}{ccc|ccc} 1 & -\frac{3}{4} & \frac{1}{4} & \frac{1}{4} & 0 & 0 \\ 0 & -1 & 9 & 0 & 1 & 0 \\ 0 & -\frac{1}{2} & \frac{9}{2} & \frac{1}{2} & 0 & 1 \end{array}\right] \leftarrow 2 \times R_1 + R_3$$

$$\left[\begin{array}{ccc|ccc} 1 & -\frac{3}{4} & \frac{1}{4} & \frac{1}{4} & 0 & 0 \\ 0 & -1 & 9 & 0 & 1 & 0 \\ 0 & 0 & 0 & \frac{1}{2} & -\frac{1}{2} & 1 \end{array}\right] \leftarrow -\frac{1}{2} R_2 + R_3$$

$A^{-1}$ does not exist.

**17.**
$$\left[\begin{array}{ccc|ccc} -11 & 2 & 2 & 1 & 0 & 0 \\ -4 & 0 & 1 & 0 & 1 & 0 \\ 6 & -1 & -1 & 0 & 0 & 1 \end{array}\right]$$

$$\left[\begin{array}{ccc|ccc} 1 & 0 & 0 & 1 & 0 & 2 \\ -4 & 0 & 1 & 0 & 1 & 0 \\ 6 & -1 & -1 & 0 & 0 & 1 \end{array}\right] \leftarrow 2 \times R_3 + R_1$$

$$\left[\begin{array}{ccc|ccc} 1 & 0 & 0 & 1 & 0 & 2 \\ 0 & 0 & 1 & 4 & 1 & 8 \\ 0 & -1 & -1 & -6 & 0 & -11 \end{array}\right] \begin{array}{l} \leftarrow 4 \times R_1 + R_2 \\ \\ \leftarrow -6 \times R_1 + R_2 \end{array}$$

$$\left[\begin{array}{ccc|ccc} 1 & 0 & 0 & 1 & 0 & 2 \\ 0 & -1 & -1 & -6 & 0 & -11 \\ 0 & 0 & 1 & 4 & 1 & 8 \end{array}\right] \left.\begin{array}{l} \\ \\ \end{array}\right\} \begin{array}{l} \text{Interchange} \\ R_2 \text{ and } R_3 \end{array}$$

$$\left[\begin{array}{ccc|ccc} 1 & 0 & 0 & 1 & 0 & 2 \\ 0 & 1 & 1 & 6 & 0 & 11 \\ 0 & 0 & 1 & 4 & 1 & 8 \end{array}\right] \leftarrow -1 \times R_2$$

$$\begin{bmatrix} 1 & 0 & 0 & | & 1 & 0 & 2 \\ 0 & 1 & 0 & | & 2 & -1 & 3 \\ 0 & 0 & 1 & | & 4 & 1 & 8 \end{bmatrix} \leftarrow -1 \times R_3 + R_2$$

$$A^{-1} = \begin{bmatrix} 1 & 0 & 2 \\ 2 & -1 & 3 \\ 4 & 1 & 8 \end{bmatrix}$$

**19.**
$$\begin{bmatrix} 1 & 1 & 0 & 2 & | & 1 & 0 & 0 & 0 \\ -1 & 0 & 2 & -1 & | & 0 & 1 & 0 & 0 \\ 0 & 2 & 0 & -2 & | & 0 & 0 & 1 & 0 \\ 2 & 0 & 0 & 5 & | & 0 & 0 & 0 & 1 \end{bmatrix}$$

$$\begin{bmatrix} 1 & 1 & 0 & 2 & | & 1 & 0 & 0 & 0 \\ 0 & 1 & 2 & 1 & | & 1 & 1 & 0 & 0 \\ 0 & 2 & 0 & -2 & | & 0 & 0 & 1 & 0 \\ 0 & -2 & 0 & 1 & | & -2 & 0 & 0 & 1 \end{bmatrix} \begin{matrix} \\ \leftarrow 1 \times R_1 + R_2 \\ \\ \leftarrow -2 \times R_1 + R_4 \end{matrix}$$

$$\begin{bmatrix} 1 & 0 & -2 & 1 & | & 0 & -1 & 0 & 0 \\ 0 & 1 & 2 & 1 & | & 1 & 1 & 0 & 0 \\ 0 & 0 & -4 & -4 & | & -2 & -2 & 1 & 0 \\ 0 & 0 & 4 & 3 & | & 0 & 2 & 0 & 1 \end{bmatrix} \begin{matrix} \leftarrow -1 \times R_2 + R_1 \\ \\ \leftarrow -2 \times R_2 + R_3 \\ \leftarrow 2 \times R_2 + R_4 \end{matrix}$$

$$\begin{bmatrix} 1 & 0 & -2 & 1 & | & 0 & -1 & 0 & 0 \\ 0 & 1 & 2 & 1 & | & 1 & 1 & 0 & 0 \\ 0 & 0 & 1 & 1 & | & \frac{1}{2} & \frac{1}{2} & -\frac{1}{4} & 0 \\ 0 & 0 & 4 & 3 & | & 0 & 2 & 0 & 1 \end{bmatrix} \begin{matrix} \\ \\ \leftarrow -\frac{1}{4} \times R_3 \\ \\ \end{matrix}$$

$$\begin{bmatrix} 1 & 0 & 0 & 3 & | & 1 & 0 & -\frac{1}{2} & 0 \\ 0 & 1 & 0 & -1 & | & 0 & 0 & \frac{1}{2} & 0 \\ 0 & 0 & 1 & 1 & | & \frac{1}{2} & \frac{1}{2} & -\frac{1}{4} & 0 \\ 0 & 0 & 0 & -1 & | & -2 & 0 & 1 & 1 \end{bmatrix} \begin{matrix} \leftarrow 2 \times R_3 + R_1 \\ \leftarrow -2 \times R_3 + R_2 \\ \\ \leftarrow -4 \times R_3 + R_4 \end{matrix}$$

$$\begin{bmatrix} 1 & 0 & 0 & 3 & | & 1 & 0 & -\frac{1}{2} & 0 \\ 0 & 1 & 0 & -1 & | & 0 & 0 & \frac{1}{2} & 0 \\ 0 & 0 & 1 & 1 & | & \frac{1}{2} & \frac{1}{2} & -\frac{1}{4} & 0 \\ 0 & 0 & 0 & 1 & | & 2 & 0 & -1 & -1 \end{bmatrix} \begin{matrix} \\ \\ \\ \leftarrow -1 \times R_4 \end{matrix}$$

$$\begin{bmatrix} 1 & 0 & 0 & 0 & | & -5 & 0 & \frac{5}{2} & 3 \\ 0 & 1 & 0 & 0 & | & 2 & 0 & -\frac{1}{2} & -1 \\ 0 & 0 & 1 & 0 & | & -\frac{3}{2} & \frac{1}{2} & \frac{3}{4} & 1 \\ 0 & 0 & 0 & 1 & | & 2 & 0 & -1 & -1 \end{bmatrix} \begin{matrix} \leftarrow -3 \times R_4 + R_1 \\ \leftarrow 1 \times R_4 + R_2 \\ \leftarrow -1 \times R_4 + R_3 \\ \\ \end{matrix}$$

**21.** $AX = C$

$$\begin{bmatrix} 3 & 1 \\ 2 & -2 \end{bmatrix}\begin{bmatrix} x \\ y \end{bmatrix} = \begin{bmatrix} 9 \\ 14 \end{bmatrix}$$

$3x + y = 9$
$2x - 2y = 14$

**23.** $A = \begin{bmatrix} 2 & -5 \\ -3 & 4 \end{bmatrix}$, $A^{-1} = \begin{bmatrix} -\frac{4}{7} & -\frac{5}{7} \\ -\frac{3}{7} & -\frac{2}{7} \end{bmatrix}$ (Exercise 5)

$$\begin{bmatrix} x \\ y \end{bmatrix} = \begin{bmatrix} -\frac{4}{7} & -\frac{5}{7} \\ -\frac{3}{7} & -\frac{2}{7} \end{bmatrix}\begin{bmatrix} 7 \\ -14 \end{bmatrix} = \begin{bmatrix} 6 \\ 1 \end{bmatrix}$$

$(6, 1)$

**25.** $A = \begin{bmatrix} \frac{1}{3} & \frac{2}{3} \\ \frac{2}{3} & \frac{1}{3} \end{bmatrix}$, $A^{-1} = \begin{bmatrix} -1 & 2 \\ 2 & -1 \end{bmatrix}$, (Exercise 7)

$$\begin{bmatrix} x \\ y \end{bmatrix} = \begin{bmatrix} -1 & 2 \\ 2 & -1 \end{bmatrix}\begin{bmatrix} -8 \\ 5 \end{bmatrix} = \begin{bmatrix} 18 \\ -21 \end{bmatrix}$$

$(18, -21)$

**27.** $A = \begin{bmatrix} 1 & 0 & 2 \\ 2 & -1 & 0 \\ 0 & 3 & 4 \end{bmatrix}$, $A^{-1} = \begin{bmatrix} -\frac{1}{2} & \frac{3}{4} & \frac{1}{4} \\ -1 & \frac{1}{2} & \frac{1}{2} \\ \frac{3}{4} & -\frac{3}{8} & -\frac{1}{8} \end{bmatrix}$, (Exercise 13)

$$\begin{bmatrix} x \\ y \\ z \end{bmatrix} = \begin{bmatrix} -\frac{1}{2} & \frac{3}{4} & \frac{1}{4} \\ -1 & \frac{1}{2} & \frac{1}{2} \\ \frac{3}{4} & -\frac{3}{8} & -\frac{1}{8} \end{bmatrix}\begin{bmatrix} 4 \\ -8 \\ 0 \end{bmatrix} = \begin{bmatrix} -8 \\ -8 \\ 6 \end{bmatrix}$$

$(-8, -8, 6)$

**29.** $A = \begin{bmatrix} 1 & 1 & -1 \\ 1 & -1 & -1 \\ -1 & -1 & -1 \end{bmatrix}$, $A^{-1} = \begin{bmatrix} 0 & \frac{1}{2} & -\frac{1}{2} \\ \frac{1}{2} & -\frac{1}{2} & 0 \\ -\frac{1}{2} & 0 & -\frac{1}{2} \end{bmatrix}$, (Exercise 14)

$$\begin{bmatrix} x \\ y \\ z \end{bmatrix} = \begin{bmatrix} 0 & \frac{1}{2} & -\frac{1}{2} \\ \frac{1}{2} & -\frac{1}{2} & 0 \\ -\frac{1}{2} & 0 & -\frac{1}{2} \end{bmatrix}\begin{bmatrix} 1 \\ 2 \\ 3 \end{bmatrix} = \begin{bmatrix} -\frac{1}{2} \\ -\frac{1}{2} \\ -2 \end{bmatrix}$$

$\left(-\dfrac{1}{2}, -\dfrac{1}{2}, -2\right)$

**31.** $A = \begin{bmatrix} 8 & -13 & 2 \\ -4 & 7 & -1 \\ 3 & -5 & 1 \end{bmatrix}$, $A^{-1} = \begin{bmatrix} 2 & 3 & -1 \\ 1 & 2 & 0 \\ -1 & 1 & 4 \end{bmatrix}$, (Exercise 16)

$\begin{bmatrix} x \\ y \\ z \end{bmatrix} = \begin{bmatrix} 2 & 3 & -1 \\ 1 & 2 & 0 \\ -1 & 1 & 4 \end{bmatrix} \begin{bmatrix} 1 \\ 3 \\ -2 \end{bmatrix} = \begin{bmatrix} 13 \\ 7 \\ -6 \end{bmatrix}$

$(13, 7, -6)$

**33.** $A = \begin{bmatrix} 1 & 1 & 0 & 2 \\ -1 & 0 & 2 & -1 \\ 0 & 2 & 0 & -2 \\ 2 & 0 & 0 & 5 \end{bmatrix}$, $A^{-1} = \begin{bmatrix} -5 & 0 & \frac{5}{2} & 3 \\ 2 & 0 & -\frac{1}{2} & -1 \\ -\frac{3}{2} & \frac{1}{2} & \frac{3}{4} & 1 \\ 2 & 0 & -1 & -1 \end{bmatrix}$, (Exercise 19)

$\begin{bmatrix} w \\ x \\ y \\ z \end{bmatrix} = \begin{bmatrix} -5 & 0 & \frac{5}{2} & 3 \\ 2 & 0 & -\frac{1}{2} & -1 \\ -\frac{3}{2} & \frac{1}{2} & \frac{3}{4} & 1 \\ 2 & 0 & -1 & -1 \end{bmatrix} \begin{bmatrix} 2 \\ -6 \\ 0 \\ 8 \end{bmatrix} = \begin{bmatrix} 14 \\ -4 \\ 2 \\ -4 \end{bmatrix}$

$(14, -4, 2, -4)$

**35.** $AB = \begin{bmatrix} 2 & 1 \\ 0 & -4 \end{bmatrix} \begin{bmatrix} 0 & 2 \\ 1 & -2 \end{bmatrix} = \begin{bmatrix} 1 & 2 \\ -4 & 8 \end{bmatrix}$

$(AB)^{-1} = \begin{bmatrix} \frac{1}{2} & -\frac{1}{8} \\ \frac{1}{4} & \frac{1}{16} \end{bmatrix}$, $A^{-1} = \begin{bmatrix} \frac{1}{2} & \frac{1}{8} \\ 0 & -\frac{1}{4} \end{bmatrix}$, $B^{-1} = \begin{bmatrix} 1 & 1 \\ \frac{1}{2} & 0 \end{bmatrix}$

$B^{-1}A^{-1} = \begin{bmatrix} 1 & 1 \\ \frac{1}{2} & 0 \end{bmatrix} \begin{bmatrix} \frac{1}{2} & \frac{1}{8} \\ 0 & -\frac{1}{4} \end{bmatrix} = \begin{bmatrix} \frac{1}{2} & -\frac{1}{8} \\ \frac{1}{4} & \frac{1}{16} \end{bmatrix}$

**37. (a)** $AB = I$
$C(AB) = CI$
$(CA)B = C$
$IB = C$
$B = C$

**(b)** The inverse of a matrix is unique.

**39. (a)** $(A^{-1})^2 = \begin{bmatrix} \frac{1}{2} & \frac{1}{8} \\ 0 & -\frac{1}{4} \end{bmatrix} \begin{bmatrix} \frac{1}{2} & \frac{1}{8} \\ 0 & -\frac{1}{4} \end{bmatrix} = \begin{bmatrix} \frac{1}{4} & \frac{1}{32} \\ 0 & \frac{1}{16} \end{bmatrix}$

$A^2 = \begin{bmatrix} 2 & 1 \\ 0 & -4 \end{bmatrix} \begin{bmatrix} 2 & 1 \\ 0 & -4 \end{bmatrix} = \begin{bmatrix} 4 & -2 \\ 0 & 16 \end{bmatrix}$

$(A^2)^{-1} = \begin{bmatrix} \frac{1}{4} & \frac{1}{32} \\ 0 & \frac{1}{16} \end{bmatrix}$

**(b)** $(A^2)^{-1} = (AA)^{-1} = A^{-1}A^{-1}$ by Exercise 38b
$$= (A^{-1})^2$$

**(c)** $(A^n)^{-1} = (A^{-1})^n$

## Challenge

Notice that $A^2 + 5A = I$ and $A(A + 5I) = I$. So, $A$ has an inverse and $A^{-1} = A + 5I$

## Critical Thinking

1. It is necessary to verify both distributive properties for matrices since matrices don't satisfy the commutative property of multiplication.

3. If $A$ is invertible, then if $AB = AC$, it follows that $B = C$ since
$$AB = AC$$
$$\left(A^{-1}A\right)B = \left(A^{-1}A\right)C$$
$$IB = IC$$
$$B = C$$

5. Notice that $A^{-1} = \begin{bmatrix} \frac{1}{a} & 0 \\ 0 & \frac{1}{b} \end{bmatrix}$. By Exercise 39c, $(A^n)^{-1} = (A^{-1})^n = \begin{bmatrix} \frac{1}{a} & 0 \\ 0 & \frac{1}{b} \end{bmatrix}^n = \begin{bmatrix} \frac{1}{a^n} & 0 \\ 0 & \frac{1}{b^n} \end{bmatrix}$

by Section 9.2 Exercise 45c.

## Exercises 9.4

1. $\begin{vmatrix} 5 & -1 \\ -3 & 4 \end{vmatrix} = 5 \cdot 4 - (-3)(-1) = 17$

3. $\begin{vmatrix} 17 & -3 \\ 20 & 2 \end{vmatrix} = 17 \cdot 2 - (20)(-3) = 94$

5. $\begin{vmatrix} 10 & 5 \\ 6 & -3 \end{vmatrix} = 10 \cdot (-3) - 6 \cdot 5 = -60$

7. $\begin{vmatrix} 16 & 0 \\ -9 & 0 \end{vmatrix} = 16 \cdot 0 - (-9) \cdot 0 = 0$

**9.** $x = \dfrac{\begin{vmatrix} 15 & 9 \\ 18 & 12 \end{vmatrix}}{\begin{vmatrix} 3 & 9 \\ 6 & 12 \end{vmatrix}} = \dfrac{15 \cdot 12 - 18 \cdot 9}{3 \cdot 12 - 6 \cdot 9} = \dfrac{18}{-18} = -1$

$y = \dfrac{\begin{vmatrix} 3 & 15 \\ 6 & 18 \end{vmatrix}}{\begin{vmatrix} 3 & 9 \\ 6 & 12 \end{vmatrix}} = \dfrac{3 \cdot 18 - 6 \cdot 15}{-18} = \dfrac{-36}{-18} = 2$

$(-1, 2)$

**11.** $x = \dfrac{\begin{vmatrix} 8 & 10 \\ 15 & -9 \end{vmatrix}}{\begin{vmatrix} -4 & 10 \\ 11 & -9 \end{vmatrix}} = \dfrac{8(-9) - 15 \cdot 10}{(-4)(-9) - 11 \cdot 10} = \dfrac{-222}{-74} = 3$

$y = \dfrac{\begin{vmatrix} -4 & 8 \\ 11 & 15 \end{vmatrix}}{\begin{vmatrix} -4 & 10 \\ 11 & -9 \end{vmatrix}} = \dfrac{-4 \cdot 15 - 11 \cdot 8}{-74} = \dfrac{-148}{-74} = 2$

$(3, 2)$

**13.** $x = \dfrac{\begin{vmatrix} 3 & 2 \\ -1 & 3 \end{vmatrix}}{\begin{vmatrix} 5 & 2 \\ 2 & 3 \end{vmatrix}} = \dfrac{3 \cdot 3 - (-1) \cdot 2}{5 \cdot 3 - 2 \cdot 2} = \dfrac{11}{11} = 1$

$y = \dfrac{\begin{vmatrix} 5 & 3 \\ 2 & -1 \end{vmatrix}}{\begin{vmatrix} 5 & 2 \\ 2 & 3 \end{vmatrix}} = \dfrac{5(-1) - 2 \cdot 3}{11} = \dfrac{-11}{11} = -1$

$(1, -1)$

**15.** $x = \dfrac{\begin{vmatrix} 13 & \frac{3}{8} \\ -42 & -\frac{9}{4} \end{vmatrix}}{\begin{vmatrix} \frac{1}{3} & \frac{3}{8} \\ 1 & -\frac{9}{4} \end{vmatrix}} = \dfrac{13\left(-\frac{9}{4}\right) - (-42) \cdot \frac{3}{8}}{\frac{1}{3}\left(-\frac{9}{4}\right) - 1 \cdot \frac{3}{8}} = \dfrac{-\frac{54}{4}}{-\frac{9}{8}} = 12$

$y = \dfrac{\begin{vmatrix} \frac{1}{3} & 13 \\ 1 & -42 \end{vmatrix}}{\begin{vmatrix} \frac{1}{3} & \frac{3}{8} \\ 1 & -\frac{9}{4} \end{vmatrix}} = \dfrac{\frac{1}{3}(-42) - 1 \cdot 13}{-\frac{9}{8}} = \dfrac{-27}{-\frac{9}{8}} = 24$

$(12, 24)$

**17.** $x = \dfrac{\begin{vmatrix} 20 & 1 \\ 0 & 1 \end{vmatrix}}{\begin{vmatrix} 3 & 1 \\ -1 & 1 \end{vmatrix}} = \dfrac{20 \cdot 1 - 0 \cdot 1}{3 \cdot 1 - (-1) \cdot 1} = \dfrac{20}{4} = 5$

$y = \dfrac{\begin{vmatrix} 3 & 20 \\ -1 & 0 \end{vmatrix}}{\begin{vmatrix} 3 & 1 \\ -1 & 1 \end{vmatrix}} = \dfrac{3 \cdot 0 - (-1) \cdot 20}{4} = \dfrac{20}{4} = 5$

$(5, 5)$

**19.** $x = \dfrac{\begin{vmatrix} -\frac{1}{2} & -\frac{2}{7} \\ \frac{31}{6} & -\frac{1}{2} \end{vmatrix}}{\begin{vmatrix} \frac{1}{2} & -\frac{2}{7} \\ -\frac{1}{3} & -\frac{1}{2} \end{vmatrix}} = \dfrac{\left(-\frac{1}{2}\right)\left(-\frac{1}{2}\right) - \left(\frac{31}{6}\right)\left(-\frac{2}{7}\right)}{\frac{1}{2}\left(-\frac{1}{2}\right) - \left(-\frac{1}{3}\right)\left(-\frac{2}{7}\right)} = \dfrac{\frac{145}{84}}{-\frac{29}{84}} = -5$

$y = \dfrac{\begin{vmatrix} \frac{1}{2} & -\frac{1}{2} \\ -\frac{1}{3} & \frac{31}{6} \end{vmatrix}}{\begin{vmatrix} \frac{1}{2} & -\frac{2}{7} \\ -\frac{1}{3} & -\frac{1}{2} \end{vmatrix}} = \dfrac{\left(\frac{1}{2}\right)\left(\frac{31}{6}\right) - \left(-\frac{1}{3}\right)\left(-\frac{1}{2}\right)}{-\frac{29}{84}} = \dfrac{\frac{29}{12}}{-\frac{29}{84}} = -7$

$(-5, -7)$

**21.** $x = \dfrac{\begin{vmatrix} 12 & -4 \\ 3 & 2 \end{vmatrix}}{\begin{vmatrix} 9 & -4 \\ 3 & 2 \end{vmatrix}} = \dfrac{12 \cdot 2 - 3(-4)}{9 \cdot 2 - 3(-4)} = \dfrac{36}{30} = \dfrac{6}{5}$

$y = \dfrac{\begin{vmatrix} 9 & 12 \\ 3 & 3 \end{vmatrix}}{\begin{vmatrix} 9 & -4 \\ 3 & 2 \end{vmatrix}} = \dfrac{9 \cdot 3 - 3 \cdot 12}{30} = \dfrac{-9}{30} = -\dfrac{3}{10}$

$\left( \dfrac{6}{5}, -\dfrac{3}{10} \right)$

**23.** $\begin{vmatrix} 5 & -2 \\ -15 & 6 \end{vmatrix} = 5 \cdot 6 - (-15)(-2) = 0$

$\left[ \begin{array}{cc|c} 5 & -2 & 3 \\ -15 & 6 & -4 \end{array} \right]$

$\left[ \begin{array}{cc|c} 5 & -2 & 3 \\ 0 & 0 & 5 \end{array} \right] \leftarrow 3 \times R_1 + R_2$ Inconsistent

**25.** $\begin{vmatrix} 16 & -4 \\ 12 & -3 \end{vmatrix} = 16(-3) - 12(-4) = 0$

$\left[ \begin{array}{cc|c} 16 & -4 & 20 \\ 12 & -3 & 15 \end{array} \right]$

$\left[ \begin{array}{cc|c} 16 & -4 & 20 \\ 0 & 0 & 0 \end{array} \right] \leftarrow -\dfrac{3}{4} \times R_1 + R_2$

Dependent

**27.** $\begin{vmatrix} 3 & -5 \\ 6 & -10 \end{vmatrix} = 3(-10) - 6(-5) = 0$

$\left[ \begin{array}{cc|c} 3 & -5 & -10 \\ 6 & -10 & -25 \end{array} \right]$

$\left[ \begin{array}{cc|c} 3 & -5 & -10 \\ 0 & 0 & -5 \end{array} \right] \leftarrow -2 \times R_1 + R_2$

Inconsistent

**29.** $\begin{vmatrix} x & 2 \\ 5 & 3 \end{vmatrix} = 8$

$x \cdot 3 - 5 \cdot 2 = 8$

$3x = 18$

$x = 6$

**31.** $\begin{vmatrix} -2 & 4 \\ x & 3 \end{vmatrix} = -1$

$-2 \cdot 3 - x \cdot 4 = -1$

$-4x = 5$

$x = -\dfrac{5}{4}$

**33.** $\begin{vmatrix} x & y \\ 2 & 4 \end{vmatrix} = 5, \quad \begin{vmatrix} 1 & y \\ -1 & x \end{vmatrix} = -\dfrac{1}{2}$ is equivalent to

$4x - 2y = 5, \; x + y = -\dfrac{1}{2}$

$x = \dfrac{\begin{vmatrix} 5 & -2 \\ -\frac{1}{2} & 1 \end{vmatrix}}{\begin{vmatrix} 4 & -2 \\ 1 & 1 \end{vmatrix}} = \dfrac{5 \cdot 1 - \left(-\frac{1}{2}\right)(-2)}{4 \cdot 1 - 1(-2)} = \dfrac{4}{6} = \dfrac{2}{3}$

$y = \dfrac{\begin{vmatrix} 4 & 5 \\ 1 & -\frac{1}{2} \end{vmatrix}}{\begin{vmatrix} 4 & -2 \\ 1 & 1 \end{vmatrix}} = \dfrac{4\left(-\frac{1}{2}\right) - 1 \cdot 5}{6} = \dfrac{-7}{6} = -\dfrac{7}{6}$

$\left(\dfrac{2}{3}, -\dfrac{7}{6}\right)$

**35.** $\begin{vmatrix} a_1 & b_1 \\ a_2 & b_2 \end{vmatrix} = a_1 b_2 - a_2 b_1 = a_1 b_2 - b_1 a_2 = \begin{vmatrix} a_1 & a_2 \\ b_1 & b_2 \end{vmatrix}$

**37.** Let $b_1 = ka_1$ and $b_2 = ka_2$; then

$\begin{vmatrix} a_1 & b_1 \\ a_2 & b_2 \end{vmatrix} = \begin{vmatrix} a_1 & ka_1 \\ a_2 & ka_2 \end{vmatrix} = \begin{vmatrix} a_1 & a_2 \\ ka_1 & ka_2 \end{vmatrix}$ (by Exercise 35) $= 0$ (by Exercise 36)

**39.** $\begin{vmatrix} 27 & 3 \\ 105 & -75 \end{vmatrix} = 3\begin{vmatrix} 9 & 1 \\ 105 & -75 \end{vmatrix} = 3 \cdot 15 \begin{vmatrix} 9 & 1 \\ 7 & -5 \end{vmatrix} = 45\begin{vmatrix} 9 & 1 \\ 7 & -5 \end{vmatrix} = -2340$

$\begin{vmatrix} 27 & 3 \\ 105 & -75 \end{vmatrix} = 3\begin{vmatrix} 9 & 1 \\ 105 & -75 \end{vmatrix} = 3 \cdot 3\begin{vmatrix} 3 & 1 \\ 35 & -75 \end{vmatrix} = 3 \cdot 3 \cdot 5\begin{vmatrix} 3 & 1 \\ 7 & -15 \end{vmatrix} = -2340$

**41.** $\begin{vmatrix} a_1 + kb_1 & b_1 \\ a_2 + kb_2 & b_2 \end{vmatrix} = \begin{vmatrix} a_1 & b_1 \\ a_2 & b_2 \end{vmatrix} + \begin{vmatrix} kb_1 & b_1 \\ kb_2 & b_2 \end{vmatrix}$ (by Exercise 40) $= \begin{vmatrix} a_1 & b_1 \\ a_2 & b_2 \end{vmatrix} + 0$ (by Exercise 37) $= \begin{vmatrix} a_1 & b_1 \\ a_2 & b_2 \end{vmatrix}$

**43.** $\begin{vmatrix} 12 & -42 \\ -6 & 27 \end{vmatrix} = 6\begin{vmatrix} 2 & -42 \\ -1 & 27 \end{vmatrix} = 12\begin{vmatrix} 1 & -21 \\ -1 & 27 \end{vmatrix} = 12\begin{vmatrix} 1 & -21 \\ 0 & 6 \end{vmatrix} = 12(6) = 72$

**45.** Let $x$ be the rate walking and $y$ be the rate riding. Then

$$\frac{1}{2}x + \frac{1}{2}y = 7$$
$$\frac{2}{3}x + \frac{1}{3}y = 6$$

$$x = \frac{\begin{vmatrix} 7 & \frac{1}{2} \\ 6 & \frac{1}{3} \end{vmatrix}}{\begin{vmatrix} \frac{1}{2} & \frac{1}{2} \\ \frac{2}{3} & \frac{1}{3} \end{vmatrix}} = \frac{7\left(\frac{1}{3}\right) - 6 \cdot \frac{1}{2}}{\frac{1}{2} \cdot \frac{1}{3} - \frac{2}{3} \cdot \frac{1}{2}} = \frac{-\frac{2}{3}}{-\frac{1}{6}} = 4$$

$$y = \frac{\begin{vmatrix} \frac{1}{2} & 7 \\ \frac{2}{3} & 6 \end{vmatrix}}{\begin{vmatrix} \frac{1}{2} & \frac{1}{2} \\ \frac{2}{3} & \frac{1}{3} \end{vmatrix}} = \frac{\frac{1}{2} \cdot 6 - \frac{2}{3} \cdot 7}{-\frac{1}{6}} = \frac{-\frac{5}{3}}{-\frac{1}{6}} = 10$$

Karin walked at 4 mph and rode at 10 mph.

## Exercises 9.5

**1. (a)** $|A| = 6\begin{vmatrix} -9 & 4 \\ 5 & 1 \end{vmatrix} - 0\begin{vmatrix} -2 & -1 \\ 5 & 1 \end{vmatrix} + (-3)\begin{vmatrix} -2 & -1 \\ -9 & 4 \end{vmatrix}$

$6(-29) - 3(-17) = -123$

**(b)** $|A| = -3\begin{vmatrix} -2 & -1 \\ -9 & 4 \end{vmatrix} - 5\begin{vmatrix} 6 & -1 \\ 0 & 4 \end{vmatrix} + 1\begin{vmatrix} 6 & -2 \\ 0 & -9 \end{vmatrix}$

$-3(-17) - 5(24) + (-54) = -123$

**(c)** $\begin{vmatrix} 6 & -2 & -1 \\ 0 & -9 & 4 \\ -3 & 5 & 1 \end{vmatrix}\begin{matrix} 6 & -2 \\ 0 & -9 \\ -3 & 5 \end{matrix} = -54 + 24 + 0 - (-27) - 120 - 0 = -123$

**3. (a)** $(-1)^{1+1}(-5)\begin{vmatrix} 7 & 4 \\ -6 & -2 \end{vmatrix} + (-1)^{1+2}(-2)\begin{vmatrix} -3 & 4 \\ 1 & -2 \end{vmatrix} + (-1)^{1+3}(1)\begin{vmatrix} -3 & 7 \\ 1 & -6 \end{vmatrix}$

$= -5(10) + 2(2) + (11) = -35$

**(b)** $(-1)^{2+1}(-3)\begin{vmatrix} -2 & 1 \\ -6 & -2 \end{vmatrix} + (-1)^{2+2}(7)\begin{vmatrix} -5 & 1 \\ 1 & -2 \end{vmatrix} + (-1)^{2+3}(4)\begin{vmatrix} -5 & -2 \\ 1 & -6 \end{vmatrix}$

$= 3(10) + 7(9) - 4(32) = -35$

**(c)** $(-1)^{1+3}(1)\begin{vmatrix} -3 & 7 \\ 1 & -6 \end{vmatrix} + (-1)^{2+3}(4)\begin{vmatrix} -5 & -2 \\ 1 & -6 \end{vmatrix} + (-1)^{3+3}(-2)\begin{vmatrix} -5 & -2 \\ -3 & 7 \end{vmatrix}$

$= 11 - 4(32) - 2(-41) = -35$

**5.** $\begin{vmatrix} a_1 & b_1 & c_1 \\ a_2 & b_2 & c_2 \\ a_3 & b_3 & c_3 \end{vmatrix} = a_1b_2c_3 + a_2b_3c_1 + a_3b_1c_2 - a_1b_3c_2 - a_2b_1c_3 - a_3b_2c_1 = \begin{vmatrix} a_1 & a_2 & a_3 \\ b_1 & b_2 & b_3 \\ c_1 & c_2 & c_3 \end{vmatrix}$

**7.**
$$\begin{vmatrix} c_1 & b_1 & a_1 \\ c_2 & b_2 & a_2 \\ c_3 & b_3 & a_3 \end{vmatrix} = c_1 b_2 a_3 + c_2 b_3 a_1 + c_3 b_1 a_2 - c_1 b_3 a_2 - c_2 b_1 a_3 - c_3 b_2 a_1$$

$$= a_3 b_2 c_1 + a_1 b_3 c_2 + a_2 b_1 c_3 - a_2 b_3 c_1 - a_3 b_1 c_2 - a_1 b_2 c_3$$

$$= a_1 b_3 c_2 + a_2 b_1 c_3 + a_3 b_2 c_1 - a_1 b_2 c_3 - a_2 b_3 c_1 - a_3 b_1 c_2$$

$$= -(a_1 b_2 c_3 + a_2 b_3 c_1 + a_3 b_1 c_2 - a_1 b_3 c_2 - a_2 b_1 c_3 - a_3 b_2 c_1)$$

$$= -\begin{vmatrix} a_1 & b_1 & c_1 \\ a_2 & b_2 & c_2 \\ a_3 & b_3 & c_3 \end{vmatrix}$$

**9.**
$$\begin{vmatrix} a_1 + kb_1 & b_1 & c_1 \\ a_2 + kb_2 & b_2 & c_2 \\ a_3 + kb_3 & b_3 & c_3 \end{vmatrix} = (a_1 + kb_1)\begin{vmatrix} b_2 & c_2 \\ b_3 & c_3 \end{vmatrix} - (a_2 + kb_2)\begin{vmatrix} b_1 & c_1 \\ b_3 & c_3 \end{vmatrix} + (a_3 + kb_3)\begin{vmatrix} b_1 & c_1 \\ b_2 & c_2 \end{vmatrix}$$

$$= \left( a_1 \begin{vmatrix} b_2 & c_2 \\ b_3 & c_3 \end{vmatrix} - a_2 \begin{vmatrix} b_1 & c_1 \\ b_3 & c_3 \end{vmatrix} + a_3 \begin{vmatrix} b_1 & c_1 \\ b_2 & c_2 \end{vmatrix} \right) + kb_1 \begin{vmatrix} b_2 & c_2 \\ b_3 & c_3 \end{vmatrix} - kb_2 \begin{vmatrix} b_1 & c_1 \\ b_3 & c_3 \end{vmatrix} + kb_3 \begin{vmatrix} b_1 & c_1 \\ b_2 & c_2 \end{vmatrix}$$

$$= \begin{vmatrix} a_1 & b_1 & c_1 \\ a_2 & b_2 & c_2 \\ a_3 & b_3 & c_3 \end{vmatrix} + k(b_1 b_2 c_3 - b_1 b_3 c_2 - b_1 b_2 c_3 + b_2 b_3 c_1 + b_1 b_3 c_2 - b_2 b_3 c_1)$$

$$= \begin{vmatrix} a_1 & b_1 & c_1 \\ a_2 & b_2 & c_2 \\ a_3 & b_3 & c_3 \end{vmatrix}$$

**11.**
$$\begin{vmatrix} 7 & 2 & -14 \\ 0 & 6 & 0 \\ -3 & 1 & 6 \end{vmatrix} = 0 \text{ since the third column is } -2 \text{ times the first column.}$$

**13.**
$$\begin{vmatrix} 8 & -10 & 2 \\ 4 & 25 & -1 \\ 2 & 10 & 0 \end{vmatrix} = 2\begin{vmatrix} -10 & 2 \\ 25 & -1 \end{vmatrix} - 10\begin{vmatrix} 8 & 2 \\ 4 & -1 \end{vmatrix} + 0\begin{vmatrix} 8 & -10 \\ 4 & 25 \end{vmatrix}$$

$$= 2(-40) - 10(-16) = 80$$

$$20\begin{vmatrix} 2 & -1 & 1 \\ 2 & 5 & -1 \\ 1 & 2 & 0 \end{vmatrix} = 20\left[ 1 \cdot \begin{vmatrix} -1 & 1 \\ 5 & -1 \end{vmatrix} - 2\begin{vmatrix} 2 & 1 \\ 2 & -1 \end{vmatrix} + 0\begin{vmatrix} 2 & -1 \\ 2 & 5 \end{vmatrix} \right]$$

$$= 20[-4 - 2(-4)]$$

$$= 20(4) = 80$$

**15.** 
$$\begin{vmatrix} 5 & -4 & 3 \\ -6 & 6 & 2 \\ -7 & 3 & 4 \end{vmatrix} = \begin{vmatrix} 14 & -13 & 3 \\ 0 & 0 & 2 \\ 5 & -9 & 4 \end{vmatrix} = -2\begin{vmatrix} 14 & -13 \\ 5 & -9 \end{vmatrix}$$
$$= -2[14(-9) - 5(-13)]$$
$$= -2(-61)$$
$$= 122$$

**17.** **(a)** 0 since column 3 is a multiple of column 1
  **(b)** 0 since each of the six terms in the simplified form on page 591 has a zero factor.
  **(c)** 0 since row 1 is a multiple of row 3
  **(d)** 0 since row 2 is a multiple of row 3

**19.** 
$$\begin{vmatrix} 2 & 2 & -1 \\ -1 & 3 & -3 \\ 1 & 2 & 3 \end{vmatrix} = \begin{vmatrix} 2 & 2 & -1 \\ -1 & 3 & -3 \\ 0 & 5 & 0 \end{vmatrix} = -5\begin{vmatrix} 2 & -1 \\ -1 & -3 \end{vmatrix} = -5(-6-1) = 35$$

**21.** 
$$\begin{vmatrix} 1 & -3 & 2 \\ -5 & 2 & 0 \\ 4 & -1 & 3 \end{vmatrix} = 2\begin{vmatrix} -5 & 2 \\ 4 & -1 \end{vmatrix} + 3\begin{vmatrix} 1 & -3 \\ -5 & 2 \end{vmatrix} = 2(5-8) + 3(2-15)$$
$$= -6 - 39 = -45$$

**23.** 
$$\begin{vmatrix} 1 & 1 & 1 \\ -1 & 1 & 1 \\ -1 & -1 & 1 \end{vmatrix} = \begin{vmatrix} 1 & 1 & 1 \\ 0 & 2 & 2 \\ 0 & 0 & 2 \end{vmatrix} = 1 \cdot \begin{vmatrix} 2 & 2 \\ 0 & 2 \end{vmatrix} = 1 \cdot (4-0) = 4$$

**25.** 
$$\begin{vmatrix} 3 & -2 & 1 & 5 \\ 0 & 4 & -3 & 1 \\ -1 & 0 & 6 & 2 \\ 1 & -5 & 0 & -4 \end{vmatrix} = \begin{vmatrix} 0 & 13 & 1 & 17 \\ 0 & 4 & -3 & 1 \\ 0 & -5 & 6 & -2 \\ 1 & -5 & 0 & -4 \end{vmatrix} = -\begin{vmatrix} 13 & 1 & 17 \\ 4 & -3 & 1 \\ -5 & 6 & -2 \end{vmatrix}$$
$$= -\begin{vmatrix} -55 & 52 & 17 \\ 0 & 0 & 1 \\ 3 & 0 & -2 \end{vmatrix} = -\begin{vmatrix} -55 & 52 \\ 3 & 0 \end{vmatrix} = 0 - 156 = -156$$

**27.** 
$$\begin{vmatrix} 2 & -1 & 3 & 0 \\ 1 & 0 & 5 & -3 \\ 0 & 2 & -4 & 6 \\ -5 & 3 & 0 & 1 \end{vmatrix} = 2\begin{vmatrix} 0 & 5 & -3 \\ 2 & -4 & 6 \\ 3 & 0 & 1 \end{vmatrix} - (-1)\begin{vmatrix} 1 & 5 & -3 \\ 0 & -4 & 6 \\ -5 & 0 & 1 \end{vmatrix} + 3\begin{vmatrix} 1 & 0 & -3 \\ 0 & 2 & 6 \\ -5 & 3 & 1 \end{vmatrix} - 0\begin{vmatrix} 1 & 0 & 5 \\ 0 & 2 & -4 \\ -5 & 3 & 0 \end{vmatrix}$$
$$= 2\begin{vmatrix} 9 & 5 & -3 \\ -16 & -4 & 6 \\ 0 & 0 & 1 \end{vmatrix} + \begin{vmatrix} -14 & 5 & -3 \\ 30 & -4 & 6 \\ 0 & 0 & 1 \end{vmatrix} + 3\begin{vmatrix} 1 & 0 & -3 \\ 0 & 2 & 6 \\ 0 & 3 & -14 \end{vmatrix}$$

$$= 2 \begin{vmatrix} 9 & 5 \\ -16 & -4 \end{vmatrix} + \begin{vmatrix} -14 & 5 \\ 30 & -4 \end{vmatrix} + 3 \begin{vmatrix} 2 & 6 \\ 3 & -14 \end{vmatrix}$$

$$= 2(-36 + 80) + (56 - 150) + 3(-28 - 18)$$

$$= 88 - 94 - 138 = -144$$

**29.** $\begin{vmatrix} -1 & x & -1 \\ x & -3 & 0 \\ -3 & 5 & -1 \end{vmatrix} = 0$

$$\begin{vmatrix} -1 & x & -1 \\ x & -3 & 0 \\ -2 & 5-x & 0 \end{vmatrix} = 0$$

$$- \begin{vmatrix} x & -3 \\ -2 & 5-x \end{vmatrix} = 0$$

$$-(5x - x^2 - 6) = 0$$

$$x^2 - 5x + 6 = 0$$

$$(x - 2)(x - 3) = 0$$

$$x = 2 \text{ or } x = 3$$

**31.** $\begin{vmatrix} 5-x & 0 & -2 \\ 4 & -1-x & 3 \\ 2 & 0 & 1-x \end{vmatrix} = 0$

$$(-1-x) \begin{vmatrix} 5-x & -2 \\ 2 & 1-x \end{vmatrix} = 0$$

$$(-1-x)[(5-x)(1-x) + 4] = 0$$

$$-(x+1)[x^2 - 6x + 9] = 0$$

$$-(x+1)(x-3)^2 = 0$$

$$x + 1 = 0 \text{ or } x - 3 = 0$$

$$x = -1 \text{ or } x = 3$$

**33.** $D_x = \begin{vmatrix} 2 & -1 & 4 \\ 0 & 3 & -7 \\ 12 & -4 & 4 \end{vmatrix} = \begin{vmatrix} 2 & -1 & 4 \\ 0 & 3 & -7 \\ 0 & 2 & -20 \end{vmatrix} = 2 \begin{vmatrix} 3 & -7 \\ 2 & -20 \end{vmatrix}$

$$= 2(-60 + 14) = -92$$

**35.** $D = -23$ (Exercise 32)

$$\frac{D_y}{D} = \frac{\begin{vmatrix} 3 & 2 & 4 \\ -5 & 0 & -7 \\ 7 & 12 & 4 \end{vmatrix}}{-23} = -\frac{1}{23}\begin{vmatrix} 3 & 2 & 4 \\ -5 & 0 & -7 \\ -11 & 0 & -20 \end{vmatrix} = \frac{2}{23}\begin{vmatrix} -5 & -7 \\ -11 & -20 \end{vmatrix}$$

$$= \frac{2}{23}(100 - 77) = 2$$

**37.** $D = \begin{vmatrix} 1 & 2 & 3 \\ 3 & -1 & 0 \\ -4 & 0 & 1 \end{vmatrix} = \begin{vmatrix} 13 & 2 & 3 \\ 3 & -1 & 0 \\ 0 & 0 & 1 \end{vmatrix} = \begin{vmatrix} 13 & 2 \\ 3 & -1 \end{vmatrix} = -13 - 6 = -19$

$$x = \frac{\begin{vmatrix} 5 & 2 & 3 \\ -3 & -1 & 0 \\ 6 & 0 & 1 \end{vmatrix}}{-19} = \frac{\begin{vmatrix} -13 & 2 & 0 \\ -3 & -1 & 0 \\ 6 & 0 & 1 \end{vmatrix}}{-19} = \frac{\begin{vmatrix} -13 & 2 \\ -3 & -1 \end{vmatrix}}{-19} = \frac{13 + 6}{-19} = -1$$

$$y = \frac{\begin{vmatrix} 1 & 5 & 3 \\ 3 & -3 & 0 \\ -4 & 6 & 1 \end{vmatrix}}{-19} = \frac{\begin{vmatrix} 1 & 6 & 3 \\ 3 & 0 & 0 \\ -4 & 2 & 1 \end{vmatrix}}{-19} = \frac{-3\begin{vmatrix} 6 & 3 \\ 2 & 1 \end{vmatrix}}{-19} = \frac{3}{19}(6 - 6) = 0$$

$$z = \frac{\begin{vmatrix} 1 & 2 & 5 \\ 3 & -1 & -3 \\ -4 & 0 & 6 \end{vmatrix}}{-19} = \frac{\begin{vmatrix} 7 & 0 & -1 \\ 3 & -1 & -3 \\ -4 & 0 & 6 \end{vmatrix}}{-19} = \frac{-\begin{vmatrix} 7 & -1 \\ -4 & 6 \end{vmatrix}}{-19} = \frac{42 - 4}{19} = 2$$

$(-1, 0, 2)$

**39.** $D = \begin{vmatrix} 2 & 1 & 0 \\ 3 & 0 & -2 \\ 0 & -3 & 8 \end{vmatrix} = \begin{vmatrix} 0 & 1 & 0 \\ 3 & 0 & -2 \\ 6 & -3 & 8 \end{vmatrix} = -\begin{vmatrix} 3 & -2 \\ 6 & 8 \end{vmatrix} = -(24 + 12) = -36$

$$x = \frac{\begin{vmatrix} 5 & 1 & 0 \\ -7 & 0 & -2 \\ -5 & -3 & 8 \end{vmatrix}}{-36} = \frac{\begin{vmatrix} 5 & 1 & 0 \\ -7 & 0 & -2 \\ 10 & 0 & 8 \end{vmatrix}}{-36} = \frac{-\begin{vmatrix} -7 & -2 \\ 10 & 8 \end{vmatrix}}{-36} = \frac{-56 + 20}{36} = -1$$

$$y = \frac{\begin{vmatrix} 2 & 5 & 0 \\ 3 & -7 & -2 \\ 0 & -5 & 8 \end{vmatrix}}{-36} = \frac{\begin{vmatrix} 2 & 5 & 0 \\ 3 & -7 & -2 \\ 12 & -33 & 0 \end{vmatrix}}{-36} = \frac{2\begin{vmatrix} 2 & 5 \\ 12 & -33 \end{vmatrix}}{-36} = \frac{-66 - 60}{-18} = 7$$

$$z = \frac{\begin{vmatrix} 2 & 1 & 5 \\ 3 & 0 & -7 \\ 0 & -3 & -5 \end{vmatrix}}{-36} = \frac{\begin{vmatrix} 2 & 1 & 5 \\ 3 & 0 & -7 \\ 6 & 0 & 10 \end{vmatrix}}{-36} = \frac{-\begin{vmatrix} 3 & -7 \\ 6 & 10 \end{vmatrix}}{-36} = \frac{30 + 42}{36} = 2$$

$(-1, 7, 2)$

**41.** $D = \begin{vmatrix} 6 & 3 & -4 \\ \frac{3}{2} & 1 & -4 \\ 3 & -1 & 8 \end{vmatrix} = \begin{vmatrix} 6 & 3 & -4 \\ -\frac{9}{2} & -2 & 0 \\ 15 & 5 & 0 \end{vmatrix} = -4 \begin{vmatrix} -\frac{9}{2} & -2 \\ 15 & 5 \end{vmatrix} = -4\left(-\frac{9}{2} \cdot 5 - 15(-2)\right) = -4\left(\frac{15}{2}\right) = -30$

$$x = \frac{\begin{vmatrix} 5 & 3 & -4 \\ 0 & 1 & -4 \\ 5 & -1 & 8 \end{vmatrix}}{-30} = \frac{\begin{vmatrix} 5 & 3 & -4 \\ 0 & 1 & -4 \\ 0 & -4 & 12 \end{vmatrix}}{-30} = \frac{5\begin{vmatrix} 1 & -4 \\ -4 & 12 \end{vmatrix}}{-30} = -\frac{12 - 16}{6} = \frac{2}{3}$$

$$y = \frac{\begin{vmatrix} 6 & 5 & -4 \\ \frac{3}{2} & 0 & -4 \\ 3 & 5 & 8 \end{vmatrix}}{-30} = \frac{\begin{vmatrix} 6 & 5 & -4 \\ \frac{3}{2} & 0 & -4 \\ -3 & 0 & 12 \end{vmatrix}}{-30} = \frac{-5\begin{vmatrix} \frac{3}{2} & -4 \\ -3 & 12 \end{vmatrix}}{-30} = \frac{18 - 12}{6} = 1$$

$$z = \frac{\begin{vmatrix} 6 & 3 & 5 \\ \frac{3}{2} & 1 & 0 \\ 3 & -1 & 5 \end{vmatrix}}{-30} = \frac{\begin{vmatrix} 6 & 3 & 5 \\ \frac{3}{2} & 1 & 0 \\ -3 & -4 & 0 \end{vmatrix}}{-30} = \frac{5\begin{vmatrix} \frac{3}{2} & 1 \\ -3 & -4 \end{vmatrix}}{-30} = -\frac{-6 + 3}{6} = \frac{1}{2}$$

$\left(\frac{2}{3}, 1, \frac{1}{2}\right)$

**43.** **(i)** The property in Exercise 8

    **(ii)** $y$ times the second column added to the first (Exercise 9)

    **(iii)** $z$ times the third column added to the first (Exercise 9)

    **(iv)** Substituting for the given values of $d_1$, $d_2$, $d_3$ in the general system on page 596.

    **(v)** Divide by $D$.

**45.** $\left|A^{-1}\right|\left|A\right| = \left|A^{-1}A\right|$ by the given property $= |I| = 1$

Then, $\left|A^{-1}\right| = \dfrac{1}{|A|}$ by dividing by $|A|$

$$|A| = \begin{vmatrix} 2 & 0 & -1 \\ -1 & 2 & 1 \\ 3 & -2 & -4 \end{vmatrix} = \begin{vmatrix} 2 & 0 & -1 \\ -1 & 2 & 1 \\ 2 & 0 & -3 \end{vmatrix} = 2\begin{vmatrix} 2 & -1 \\ 2 & -3 \end{vmatrix} = 2(-6 + 2) = -8$$

$$\left|A^{-1}\right| = \begin{vmatrix} \frac{3}{4} & -\frac{1}{4} & -\frac{1}{4} \\ \frac{1}{8} & \frac{5}{8} & \frac{1}{8} \\ \frac{1}{2} & -\frac{1}{2} & -\frac{1}{2} \end{vmatrix} = \begin{vmatrix} \frac{3}{4} & \frac{1}{2} & \frac{1}{2} \\ \frac{1}{8} & \frac{3}{4} & \frac{1}{4} \\ \frac{1}{2} & 0 & 0 \end{vmatrix} = \frac{1}{2}\begin{vmatrix} \frac{1}{2} & \frac{1}{2} \\ \frac{3}{4} & \frac{1}{4} \end{vmatrix} = \frac{1}{2}\left(\frac{1}{8} - \frac{3}{8}\right) = -\frac{1}{8} = \frac{1}{|A|}$$

## Challenge

$$0 = \begin{vmatrix} 0 & 1 & 1 \\ x-a & a & c \\ y-b & b & d \end{vmatrix} = \begin{vmatrix} 0 & 0 & 1 \\ x-a & a-c & c \\ y-b & b-d & d \end{vmatrix} = +1 \begin{vmatrix} x-a & a-c \\ y-b & b-d \end{vmatrix} = (x-a)(b-d) - (y-b)(a-c)$$

so $\dfrac{y-b}{a-c} = (b-d)(x-a); \quad y-b = \dfrac{b-d}{a-c}(x-a)$

## Exercises 9.6

1. $y \le -2x + 1$

3. $y \le 1$

5.

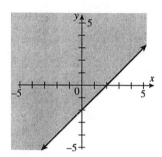

7.

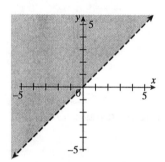

9.

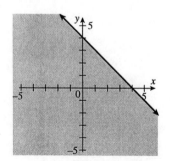

**11.**

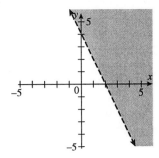

**13.**

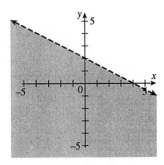

**15.**

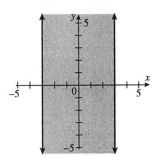

**17.**

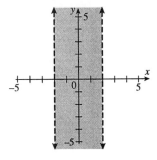

**19.**

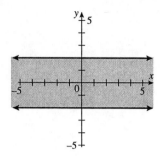

**21.**

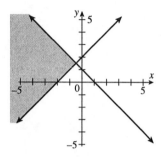

**23.**

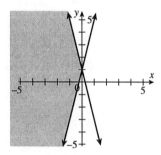

**25.**

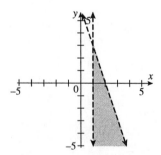

**27.**

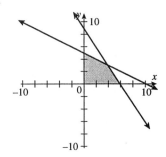

**29.**

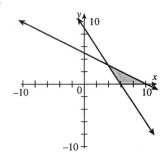

**31.**

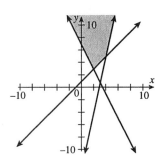

**33.**

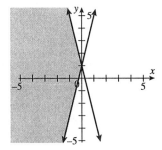

**35.**

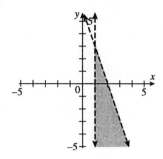

**37.** $2x + y < 6$
$3x - 4y \le 12$

**39.** $x + y \ge 4$
$x - y \le 8$
$y \le 4$

**41.** $y < 4$
$x - y \ge 8$
$x + y > 4$

**43.** $y \ge 4$
$x + y > 4$
$x - y < 8$

**45.** $y < 4$
$x + y \le 4$
$x - y < 8$

**47.** Let $x$ be the number of basketballs and $y$ be the number of uniforms. Then
$x \ge 4$
$y \ge 8$
$30x + 40y \le 600$  or  $3x + 4y \le 60$

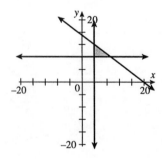

**49.** Let $x$ be the number of style 1 sold and let $y$ be the number of style 2 sold. Then

$x \geq 60$

$x \geq \dfrac{3}{2}y$  or  $y \leq \dfrac{2}{3}x$

$10x + 15y \leq 3000$  or  $2x + 3y \leq 600$

Because $x$ and $y$ are numbers of jeans, $x \geq 0$ and $y \geq 0$ are implied constraints.

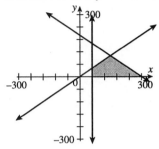

## Challenge

$(4, 8)$, $(4, 9)$, $(4, 10)$, $(4, 11)$, $(4, 12)$,
$(5, 8)$, $(5, 9)$, $(5, 10)$, $(5, 11)$,
$(6, 8)$, $(6, 9)$, $(6, 10)$,
$(7, 8)$, $(7, 9)$
$(8, 8)$, $(8, 9)$
$(9, 8)$

## Critical Thinking

1. The determinant of the coefficients of a linear system must be non-zero for the system to have a unique solution.

3. To the left:

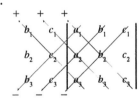

Above:

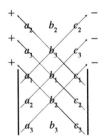

Below:

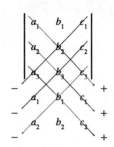

**5.**

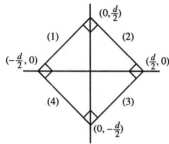

$$y \le x + \frac{d}{2}$$

$$y \le -x + \frac{d}{2}$$

$$y \ge x - \frac{d}{2}$$

$$y \ge -x - \frac{d}{2}$$

## Exercises 9.7

**1. (a)**

| Vertices | $k = 6x + 2y$ |
|----------|---------------|
| (4, 7)   | 38            |
| (8, 5)   | 58            |
| (2, 3)   | 18            |
| (3, 1)   | 20            |

maximum: 58
minimum: 18

**(b)**

| Vertices | $k = 2x + 5y$ |
|----------|---------------|
| (4, 7)   | 43            |
| (8, 5)   | 41            |
| (2, 3)   | 19            |
| (3, 1)   | 11            |

maximum: 43
minimum: 11

**3. (a)**

| Vertices | $k = 4x + 8y$ |
|----------|---------------|
| (3,6)    | 60            |
| (11, 2)  | 60            |
| (3, 0)   | 12            |
| (0, 3)   | 24            |

maximum: 60
minimum: 12

**(b)**

| Vertices | $k = 3x + 3y$ |
|----------|---------------|
| (3, 6)   | 27            |
| (11, 2)  | 39            |
| (3, 0)   | 9             |
| (0, 3)   | 9             |

maximum: 39
minimum: 9

**5. (a)**

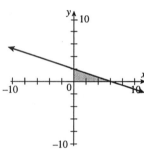

**(b)** (0, 0)
(6, 0)
(0, 2)

**(c)**

| Vertices | $p = x + y$ |
|----------|-------------|
| (0, 0)   | 0           |
| (6, 0)   | 6           |
| (0, 2)   | 2           |

maximum: 6
minimum: 0

**(d)**

| Vertices | $q = 6x + 10y$ |
|----------|----------------|
| (0, 0)   | 0              |
| (6, 0)   | 36             |
| (0, 2)   | 20             |

maximum: 36
minimum: 0

**(e)**

| Vertices | $r = 2x + 9y$ |
|----------|---------------|
| (0, 0)   | 0             |
| (6, 0)   | 12            |
| (0, 2)   | 18            |

maximum: 18
minimum: 0

**7. (a)**

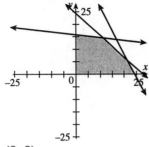

**(b)**  (0, 0)
(0, 16)
(12, 14)
(22, 6)
(24, 0)

**(c)**

| Vertices | $p = x + y$ |
|----------|-------------|
| (0, 0)   | 0           |
| (0, 16)  | 16          |
| (12, 14) | 26          |
| (22, 6)  | 28          |
| (24, 0)  | 24          |

maximum: 28
minimum: 0

**(d)**

| Vertices | $q = 6x + 10y$ |
|----------|----------------|
| (0, 0)   | 0              |
| (0, 16)  | 160            |
| (12, 14) | 212            |
| (22, 6)  | 192            |
| (24, 0)  | 144            |

maximum: 212
minimum: 0

**(e)**

| Vertices | $r = 2x + 9y$ |
|----------|---------------|
| (0, 0)   | 0             |
| (0, 16)  | 144           |
| (12, 14) | 150           |
| (22, 6)  | 98            |
| (24, 0)  | 48            |

maximum: 150
minimum: 0

**9. (a)**

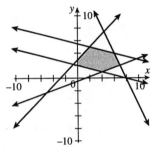

**(b)** (0, 2)
(0, 3)
(2, 5)
(6, 4)
(7, 2)
(4, 1)

**(c)**

| Vertices | $p = x + y$ |
|----------|-------------|
| (0, 2)   | 2           |
| (0, 3)   | 3           |
| (2, 5)   | 7           |
| (6, 4)   | 10          |
| (7, 2)   | 9           |
| (4, 1)   | 5           |

maximum: 10
minimum: 2

**(d)**

| Vertices | $q = 6x + 10y$ |
|----------|----------------|
| (0, 2)   | 20             |
| (0, 3)   | 30             |
| (2, 5)   | 62             |
| (6, 4)   | 76             |
| (7, 2)   | 62             |
| (4, 1)   | 34             |

maximum: 76
minimum: 20

**(e)**

| Vertices | $r = 2x + 9y$ |
|----------|---------------|
| (0, 2)   | 18            |
| (0, 3)   | 27            |
| (2, 5)   | 49            |
| (6, 4)   | 48            |
| (7, 2)   | 32            |
| (4, 1)   | 17            |

maximum: 49
minimum: 17

**11.**

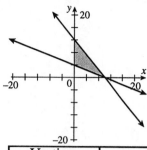

**(a)**

| Vertices | $p = x + 3y$ |
|----------|--------------|
| $(10, 0)$ | 10 |
| $(0, 4)$ | 12 |
| $(0, 12)$ | 36 |

maximum: 36
minimum: 10

**(b)**

| Vertices | $q = 4x + 3y$ |
|----------|---------------|
| $(10, 0)$ | 40 |
| $(0, 4)$ | 12 |
| $(0, 12)$ | 36 |

maximum: 40
minimum: 12

**(c)**

| Vertices | $r = \frac{1}{2}x + \frac{1}{4}y$ |
|----------|-----------------------------------|
| $(10, 0)$ | 5 |
| $(0, 4)$ | 1 |
| $(0, 12)$ | 3 |

maximum: 5
minimum: 1

**13.**

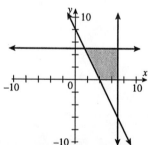

**(a)**

| Vertices | $p = 2x + 2y$ |
|----------|---------------|
| $(4, 0)$ | 8 |
| $(7, 0)$ | 14 |
| $(7, 5)$ | 24 |
| $\left(\frac{3}{2}, 5\right)$ | 13 |

maximum: 24
minimum: 8

**(b)**

| Vertices | $q = 3x + y$ |
|---|---|
| $(4, 0)$ | 12 |
| $(7, 0)$ | 21 |
| $(7, 5)$ | 26 |
| $\left(\dfrac{3}{2}, 5\right)$ | $\dfrac{19}{2}$ |

maximum: 26

minimum: $\dfrac{19}{2} = 9.5$

**(a)**

| Vertices | $r = x + \dfrac{1}{5}y$ |
|---|---|
| $(4, 0)$ | 4 |
| $(7, 0)$ | 7 |
| $(7, 5)$ | 8 |
| $\left(\dfrac{3}{2}, 5\right)$ | $\dfrac{5}{2}$ |

maximum: 8

minimum: $\dfrac{5}{2} = 2.5$

**15.**

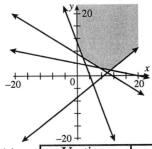

**(a)**

| Vertices | $p = x + y$ |
|---|---|
| $(0, 11)$ | 11 |
| $(2, 6)$ | 8 |
| $\left(\dfrac{11}{2}, 3\right)$ | $\dfrac{17}{2}$ |
| $(11, 2)$ | 13 |

no maximum

minimum: 8

**(b)**

| Vertices | $q = 6x + 10y$ |
|---|---|
| $(0, 11)$ | 110 |
| $(2, 6)$ | 72 |
| $\left(\dfrac{11}{2}, 3\right)$ | 63 |
| $(11, 2)$ | 86 |

no maximum

minimum: 63

**(c)**

| Vertices | $r = 2x + y$ |
|---|---|
| $(0, 11)$ | 11 |
| $(2, 6)$ | 10 |
| $\left(\dfrac{11}{2}, 3\right)$ | 14 |
| $(11, 2)$ | 24 |

no maximum
minimum: 10

**17. (a)** $x \geq 0$ and $y \geq 0$ because a negative number of either model is not possible. $\dfrac{3}{2}x + y$ is the amount of time that machine $M_1$ works per day, and $\dfrac{3}{2}x + y \leq 12$ means that $M_1$ works at most 12 hours daily. The remaining inequalities are the constraints for machines $M_2$ and $M_3$; the explanations are similar, as for $M_1$.

**(b)** $p = 5x + 8y$

**(c)**

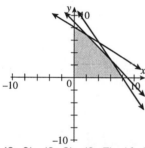

$(0, 0), (0, 8), (2, 7), (6, 3), (8, 0)$

**(d)**

| Vertices | $p = 5x + 8y$ |
|---|---|
| $(0, 0)$ | 0 |
| $(0, 8)$ | 64 |
| $(2, 7)$ | 66 |
| $(6, 3)$ | 54 |
| $(8, 0)$ | 40 |

maximum = $66 when $x = 2$, $y = 7$

**(e)**

| Vertices | $p = 8x + 5y$ |
|---|---|
| $(0, 0)$ | 0 |
| $(0, 8)$ | 40 |
| $(2, 7)$ | 51 |
| $(6, 3)$ | 63 |
| $(8, 0)$ | 64 |

maximum = $64 when $x = 8$, $y = 0$

**19.**

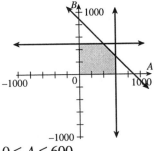

$0 \leq A \leq 600$
$0 \leq B \leq 500$
$A + B \leq 900$

| Vertices | $p = 100A + 120B$ |
|:--------:|:-----------------:|
| $(0, 0)$ | 0 |
| $(600, 0)$ | 60,000 |
| $(0, 500)$ | 60,000 |
| $(600, 300)$ | 96,000 |
| $(400, 500)$ | 100,000 |

400 of model $A$ and 500 of model $B$ for a profit of $100,000.

## Chapter 9 Review Exercises

**1.** Eliminate one of the variables resulting in a system of two equations in two variables that we know how to solve.

**3.** $2x - 2y + z = -7$ or $z = -7 - 2x + 2y$
$3x + y + 2z = -2$
$5x + 3y - 3z = -7$

$3x + y + 2(-7 - 2x + 2y) = -2$ or $\quad -x + 5y = 12$
$5x + 3y - 3(-7 - 2x + 2y) = -7$ or $\quad 11x - 3y = -28$

$x = 5y - 12$
$11(5y - 12) - 3y = -28$
$52y = 104$
$y = 2$
$x = 5 \cdot 2 - 12 = -2$
$z = -7 - 2(-2) + 2 \cdot 2 = 1$
$(-2, 2, 1)$

**5.** $\begin{bmatrix} 1 & 2 & 3 & 5 \\ -4 & 0 & 1 & 6 \\ 3 & -1 & 0 & -3 \end{bmatrix}$

$\begin{bmatrix} 1 & 2 & 3 & 5 \\ 0 & 8 & 13 & 26 \\ 0 & -7 & -9 & -18 \end{bmatrix} \begin{matrix} \\ \leftarrow 4 \times R_1 + R_2 \\ \leftarrow -3 \times R_1 + R_3 \end{matrix}$

$$\begin{bmatrix} 1 & 2 & 3 & 5 \\ 0 & 1 & \frac{13}{8} & \frac{13}{4} \\ 0 & -7 & -9 & -18 \end{bmatrix} \leftarrow \frac{1}{8} \times R_2$$

$$\begin{bmatrix} 1 & 2 & 3 & 5 \\ 0 & 1 & \frac{13}{8} & \frac{13}{4} \\ 0 & 0 & \frac{19}{8} & \frac{19}{4} \end{bmatrix} \leftarrow 7 \times R_2 + R_3$$

$\frac{19}{8}z = \frac{19}{4}$ or $z = 2$

$y + \frac{13}{8} \cdot 2 = \frac{13}{4}$ or $y = 0$

$x + 2 \cdot 0 + 3 \cdot 2 = 5$ or $x = -1$

$(-1, 0, 2)$

7. 
$$\begin{bmatrix} 2 & 1 & 0 & 5 \\ -3 & 0 & 2 & 7 \\ 0 & 3 & -8 & 5 \end{bmatrix}$$

$$\begin{bmatrix} 1 & \frac{1}{2} & 0 & \frac{5}{2} \\ -3 & 0 & 2 & 7 \\ 0 & 3 & -8 & 5 \end{bmatrix} \leftarrow \frac{1}{2} \times R_1$$

$$\begin{bmatrix} 1 & \frac{1}{2} & 0 & \frac{5}{2} \\ 0 & \frac{3}{2} & 2 & \frac{29}{2} \\ 0 & 3 & -8 & 5 \end{bmatrix} \leftarrow 3 \times R_1 + R_2$$

$$\begin{bmatrix} 1 & \frac{1}{2} & 0 & \frac{5}{2} \\ 0 & 1 & \frac{4}{3} & \frac{29}{3} \\ 0 & 3 & -8 & 5 \end{bmatrix} \leftarrow \frac{2}{3} \times R_2$$

$$\begin{bmatrix} 1 & \frac{1}{2} & 0 & \frac{5}{2} \\ 0 & 1 & \frac{4}{3} & \frac{29}{3} \\ 0 & 0 & -12 & -24 \end{bmatrix} \leftarrow -3 \times R_2 + R_3$$

$-12z = -24$ or $z = 2$

$y + \frac{4}{3}(2) = \frac{29}{3}$ or $y = 7$

$x + \frac{1}{2}(7) + 0 = \frac{5}{2}$ or $x = -1$

$(-1, 7, 2)$

**9.**
$$\begin{bmatrix} 1 & 2 & -1 & 3 \\ 2 & -3 & 3 & 0 \\ 0 & 1 & -2 & 6 \end{bmatrix}$$

$$\begin{bmatrix} 1 & 2 & -1 & 3 \\ 0 & -7 & 5 & -6 \\ 0 & 1 & -2 & 6 \end{bmatrix} \leftarrow -2 \times R_1 + R_2$$

$$\begin{bmatrix} 1 & 2 & -1 & 3 \\ 0 & 1 & -2 & 6 \\ 0 & -7 & 5 & -6 \end{bmatrix} \left.\begin{array}{l} \\ \\ \end{array}\right\} \begin{array}{l} \text{Interchange} \\ R_2 \text{ and } R_3 \end{array}$$

$$\begin{bmatrix} 1 & 2 & -1 & 3 \\ 0 & 1 & -2 & 6 \\ 0 & 0 & -9 & 36 \end{bmatrix} \leftarrow 7 \times R_2 + R_3$$

$9z = -36$ or $z = -4$
$y - 2(-4) = 6$ or $y = -2$
$x + 2(-2) - (-4) = 3$ or $x = 3$
$(3, -2, -4)$

**11. (a)** $D + C = \begin{bmatrix} -3 & 7 \\ 1 & -1 \end{bmatrix} + \begin{bmatrix} 4 & 1 \\ -2 & 3 \end{bmatrix} = \begin{bmatrix} 1 & 8 \\ -1 & 2 \end{bmatrix}$

**(b)** $B - A$ is undefined.

**13.** $CD = \begin{bmatrix} 4 & 1 \\ -2 & 3 \end{bmatrix}\begin{bmatrix} -3 & 7 \\ 1 & -1 \end{bmatrix} = \begin{bmatrix} -11 & 27 \\ 9 & -17 \end{bmatrix}$

**15.** $AB = \begin{bmatrix} 3 \\ -2 \\ 1 \end{bmatrix}\begin{bmatrix} -5 & 6 & 4 \end{bmatrix} = \begin{bmatrix} -15 & 18 & 12 \\ 10 & -12 & -8 \\ -5 & 6 & 4 \end{bmatrix}$

**17.** $HD$ is undefined.

**19.** $(HA)B = \left(\begin{bmatrix} 5 & 4 & 3 \\ 2 & 1 & 0 \end{bmatrix}\begin{bmatrix} 3 \\ -2 \\ 1 \end{bmatrix}\right)\begin{bmatrix} -5 & 6 & 4 \end{bmatrix}$

$= \begin{bmatrix} 10 \\ 4 \end{bmatrix}\begin{bmatrix} -5 & 6 & 4 \end{bmatrix} = \begin{bmatrix} -50 & 60 & 40 \\ -20 & 24 & 16 \end{bmatrix}$

**21.** $(-F)(2E) = \begin{bmatrix} -1 & -2 & -3 \\ -1 & 3 & -2 \\ -1 & -4 & 5 \end{bmatrix}\begin{bmatrix} 4 & 0 & -8 \\ -2 & 12 & 4 \\ 0 & 6 & -4 \end{bmatrix} = \begin{bmatrix} 0 & -42 & 12 \\ -10 & 24 & 28 \\ 4 & -18 & -28 \end{bmatrix}$

**23.** $G\begin{bmatrix} 0 \\ 0 \\ 0 \end{bmatrix} = \begin{bmatrix} 4 & 1 & 5 \\ -2 & 0 & 2 \\ 0 & 3 & -3 \\ 1 & -1 & 0 \end{bmatrix}\begin{bmatrix} 0 \\ 0 \\ 0 \end{bmatrix} = \begin{bmatrix} 0 \\ 0 \\ 0 \\ 0 \end{bmatrix}$

**25.** $H(C+D)$ is undefined.

**27.** $E = \begin{bmatrix} 1 & 2 \\ 0 & 1 \end{bmatrix}$

**29.** $\left(\dfrac{3}{2}D\right)H + D\left(-\dfrac{7}{2}H\right) = \dfrac{3}{2}(DH) - \dfrac{7}{2}(DH) = (-2D)H$

$= \begin{bmatrix} 6 & -14 \\ -2 & 2 \end{bmatrix}\begin{bmatrix} 5 & 4 & 3 \\ 2 & 1 & 0 \end{bmatrix} = \begin{bmatrix} 2 & 10 & 18 \\ -6 & -6 & -6 \end{bmatrix}$

**31.** $\begin{bmatrix} 0 & 4 & | & 1 & 0 \\ -5 & -8 & | & 0 & 1 \end{bmatrix}$

$\begin{bmatrix} -5 & -8 & | & 0 & 1 \\ 0 & 4 & | & 1 & 0 \end{bmatrix} \left.\begin{matrix} \\ \\ \end{matrix}\right\}$ Interchange $R_1$ and $R_2$

$\begin{bmatrix} 1 & \frac{8}{5} & | & 0 & -\frac{1}{5} \\ 0 & 1 & | & \frac{1}{4} & 0 \end{bmatrix} \begin{matrix} \leftarrow -\frac{1}{5} \times R_1 \\ \leftarrow \frac{1}{4} \times R_2 \end{matrix}$

$\begin{bmatrix} 1 & 0 & | & -\frac{2}{5} & -\frac{1}{5} \\ 0 & 1 & | & \frac{1}{4} & 0 \end{bmatrix} \leftarrow -\frac{8}{5} \times R_2 + R_1$

$A^{-1} = \begin{bmatrix} -\frac{2}{5} & -\frac{1}{5} \\ \frac{1}{4} & 0 \end{bmatrix}$

**33.** $\begin{bmatrix} \frac{3}{4} & -\frac{6}{5} & | & 1 & 0 \\ -\frac{3}{8} & \frac{3}{5} & | & 0 & 1 \end{bmatrix}$

$\begin{bmatrix} \frac{3}{4} & -\frac{6}{5} & | & 1 & 0 \\ 0 & 0 & | & \frac{1}{2} & 1 \end{bmatrix} \leftarrow \frac{1}{2} \times R_1 + R_2$

No inverse

**35.**
$$\begin{bmatrix} 1 & 2 & 3 & | & 1 & 0 & 0 \\ 4 & 5 & 6 & | & 0 & 1 & 0 \\ 7 & 8 & 9 & | & 0 & 0 & 1 \end{bmatrix}$$

$$\begin{bmatrix} 1 & 2 & 3 & | & 1 & 0 & 0 \\ 0 & -3 & -6 & | & -4 & 1 & 0 \\ 0 & -6 & -12 & | & -7 & 0 & 1 \end{bmatrix} \begin{matrix} \\ \leftarrow -4 \times R_1 + R_2 \\ \leftarrow -7 \times R_1 + R_3 \end{matrix}$$

$$\begin{bmatrix} 1 & 2 & 3 & | & 1 & 0 & 0 \\ 0 & -3 & -6 & | & -4 & 1 & 0 \\ 0 & 0 & 0 & | & 1 & -2 & 1 \end{bmatrix} \begin{matrix} \\ \\ \leftarrow -2 \times R_2 + R_3 \end{matrix}$$

$A^{-1}$ does not exist.

**37.**
$$\begin{bmatrix} 1 & 0 & 0 & 0 & | & 1 & 0 & 0 & 0 \\ 0 & 0 & 0 & 1 & | & 0 & 1 & 0 & 0 \\ 0 & 0 & 1 & 0 & | & 0 & 0 & 1 & 0 \\ 0 & 1 & 0 & 0 & | & 0 & 0 & 0 & 1 \end{bmatrix}$$

$$\begin{bmatrix} 1 & 0 & 0 & 0 & | & 1 & 0 & 0 & 0 \\ 0 & 1 & 0 & 0 & | & 0 & 0 & 0 & 1 \\ 0 & 0 & 1 & 0 & | & 0 & 0 & 1 & 0 \\ 0 & 0 & 0 & 1 & | & 0 & 1 & 0 & 0 \end{bmatrix} \begin{matrix} \\ \text{Interchange} \\ \\ R_2 \text{ and } R_4 \end{matrix}$$

$$A^{-1} = \begin{bmatrix} 1 & 0 & 0 & 0 \\ 0 & 0 & 0 & 1 \\ 0 & 0 & 1 & 0 \\ 0 & 1 & 0 & 0 \end{bmatrix}$$

**39.**
$$\begin{bmatrix} 1 & -1 & 0 & 1 & | & 1 & 0 & 0 & 0 \\ -1 & 1 & 0 & 1 & | & 0 & 1 & 0 & 0 \\ -1 & 1 & 1 & -1 & | & 0 & 0 & 1 & 0 \\ 1 & 1 & 0 & 1 & | & 0 & 0 & 0 & 1 \end{bmatrix}$$

$$\begin{bmatrix} 1 & -1 & 0 & 1 & | & 1 & 0 & 0 & 0 \\ 0 & 0 & 0 & 2 & | & 1 & 1 & 0 & 0 \\ 0 & 0 & 1 & 0 & | & 1 & 0 & 1 & 0 \\ 0 & 2 & 0 & 0 & | & -1 & 0 & 0 & 1 \end{bmatrix} \begin{matrix} \\ \leftarrow 1 \times R_1 + R_2 \\ \leftarrow 1 \times R_1 + R_3 \\ \leftarrow -1 \times R_1 + R_4 \end{matrix}$$

$$\begin{bmatrix} 1 & -1 & 0 & 1 & | & 1 & 0 & 0 & 0 \\ 0 & 2 & 0 & 0 & | & -1 & 0 & 0 & 1 \\ 0 & 0 & 1 & 0 & | & 1 & 0 & 1 & 0 \\ 0 & 0 & 0 & 2 & | & 1 & 1 & 0 & 0 \end{bmatrix} \begin{matrix} \\ \text{Interchange} \\ \\ R_2 \text{ and } R_4 \end{matrix}$$

$$\begin{bmatrix} 1 & -1 & 0 & 1 & | & 1 & 0 & 0 & 0 \\ 0 & 1 & 0 & 0 & | & -\frac{1}{2} & 0 & 0 & \frac{1}{2} \\ 0 & 0 & 1 & 0 & | & 1 & 0 & 1 & 0 \\ 0 & 0 & 0 & 1 & | & \frac{1}{2} & \frac{1}{2} & 0 & 0 \end{bmatrix} \begin{matrix} \\ \leftarrow \frac{1}{2} \times R_2 \\ \\ \leftarrow \frac{1}{2} \times R_4 \end{matrix}$$

$$\begin{bmatrix} 1 & 0 & 0 & 1 & | & \frac{1}{2} & 0 & 0 & \frac{1}{2} \\ 0 & 1 & 0 & 0 & | & -\frac{1}{2} & 0 & 0 & \frac{1}{2} \\ 0 & 0 & 1 & 0 & | & 1 & 0 & 1 & 0 \\ 0 & 0 & 0 & 1 & | & \frac{1}{2} & \frac{1}{2} & 0 & 0 \end{bmatrix} \begin{matrix} \leftarrow 1 \times R_2 + R_1 \\ \\ \\ \end{matrix}$$

$$\begin{bmatrix} 1 & 0 & 0 & 0 & | & 0 & -\frac{1}{2} & 0 & \frac{1}{2} \\ 0 & 1 & 0 & 0 & | & -\frac{1}{2} & 0 & 0 & \frac{1}{2} \\ 0 & 0 & 1 & 0 & | & 1 & 0 & 1 & 0 \\ 0 & 0 & 0 & 1 & | & \frac{1}{2} & \frac{1}{2} & 0 & 0 \end{bmatrix} \begin{matrix} \leftarrow -1 \times R_4 + R_1 \\ \\ \\ \end{matrix}$$

$$A^{-1} = \begin{bmatrix} 0 & -\frac{1}{2} & 0 & \frac{1}{2} \\ -\frac{1}{2} & 0 & 0 & \frac{1}{2} \\ 1 & 0 & 1 & 0 \\ \frac{1}{2} & \frac{1}{2} & 0 & 0 \end{bmatrix}$$

41. $A = \begin{bmatrix} 0 & 2 & -1 \\ -1 & -1 & 1 \\ 1 & -2 & 1 \end{bmatrix}$, $A^{-1} = \begin{bmatrix} 1 & 0 & 1 \\ 2 & 1 & 1 \\ 3 & 2 & 2 \end{bmatrix}$ by Exercise 34

$$\begin{bmatrix} x \\ y \\ z \end{bmatrix} = \begin{bmatrix} 1 & 0 & 1 \\ 2 & 1 & 1 \\ 3 & 2 & 2 \end{bmatrix} \begin{bmatrix} -1 \\ 9 \\ -4 \end{bmatrix} = \begin{bmatrix} -5 \\ 3 \\ 7 \end{bmatrix}$$
$(-5, 3, 7)$

43. $A = \begin{bmatrix} 1 & -1 & 0 & 1 \\ -1 & 1 & 0 & 1 \\ -1 & 1 & 1 & -1 \\ 1 & 1 & 0 & 1 \end{bmatrix}$, $A^{-1} = \begin{bmatrix} 0 & -\frac{1}{2} & 0 & \frac{1}{2} \\ -\frac{1}{2} & 0 & 0 & \frac{1}{2} \\ 1 & 0 & 1 & 0 \\ \frac{1}{2} & \frac{1}{2} & 0 & 0 \end{bmatrix}$ by Exercise 39.

$$\begin{bmatrix} x \\ y \\ z \\ w \end{bmatrix} = \begin{bmatrix} 0 & -\frac{1}{2} & 0 & \frac{1}{2} \\ -\frac{1}{2} & 0 & 0 & \frac{1}{2} \\ 1 & 0 & 1 & 0 \\ \frac{1}{2} & \frac{1}{2} & 0 & 0 \end{bmatrix} \begin{bmatrix} 5 \\ 9 \\ -1 \\ 6 \end{bmatrix} = \begin{bmatrix} -\frac{3}{2} \\ \frac{1}{2} \\ 4 \\ 7 \end{bmatrix}$$
$$\left( -\frac{3}{2}, \frac{1}{2}, 4, 7 \right)$$

**45.** $\begin{vmatrix} -\frac{1}{2} & \frac{2}{3} \\ 9 & -6 \end{vmatrix} = \left(-\frac{1}{2}\right)(-6) - 9 \cdot \frac{2}{3} = 3 - 6 = -3$

**47.** $\begin{vmatrix} x & x-1 \\ x-2 & x-3 \end{vmatrix} = x(x-3) - (x-2)(x-1) = x^2 - 3x - (x^2 - 3x + 2) = -2$

**49.** $x = \dfrac{\begin{vmatrix} 16 & 8 \\ 103 & -5 \end{vmatrix}}{\begin{vmatrix} -3 & 8 \\ 16 & -5 \end{vmatrix}} = \dfrac{-80 - 824}{15 - 128} = \dfrac{-904}{-113} = 8$

$y = \dfrac{\begin{vmatrix} -3 & 16 \\ 16 & 103 \end{vmatrix}}{\begin{vmatrix} -3 & 8 \\ 16 & -5 \end{vmatrix}} = \dfrac{-309 - 256}{-113} = \dfrac{-565}{-113} = 5$

$(8, 5)$

**51.** $\begin{vmatrix} x & y \\ 7 & 3 \end{vmatrix} = -4, \quad \begin{vmatrix} -4 & 9 \\ -y & x \end{vmatrix} = 6$ is equivalent to

$3x - 7y = -4, \ -4x + 9y = 6$

$\begin{bmatrix} 3 & -7 & | & -4 \\ -4 & 9 & | & 6 \end{bmatrix}$

$\begin{bmatrix} 1 & -\frac{7}{3} & | & -\frac{4}{3} \\ -4 & 9 & | & 6 \end{bmatrix} \leftarrow \frac{1}{3} \times R_1$

$\begin{bmatrix} 1 & -\frac{7}{3} & | & -\frac{4}{3} \\ 0 & -\frac{1}{3} & | & \frac{2}{3} \end{bmatrix} \leftarrow 4 \times R_1 + R_2$

$-\frac{1}{3}y = \frac{2}{3}$ or $y = -2$

$x - \frac{7}{3}(-2) = -\frac{4}{3}$ or $x = -6$

$(-6, -2)$

**53.** $\begin{vmatrix} 2 & -1 & 4 \\ 0 & 3 & -2 \\ 5 & 1 & 6 \end{vmatrix} = 2\begin{vmatrix} 3 & -2 \\ 1 & 6 \end{vmatrix} - 0\begin{vmatrix} -1 & 4 \\ 1 & 6 \end{vmatrix} + 5\begin{vmatrix} -1 & 4 \\ 3 & -2 \end{vmatrix}$

$= 2(18 + 2) + 5(2 - 12) = -10$

**55.**
$$\begin{vmatrix} 2 & -1 & 4 \\ 0 & 3 & -2 \\ 5 & 1 & 6 \end{vmatrix} = \begin{vmatrix} 2 & -1 & 4 \\ 6 & 0 & 10 \\ 7 & 0 & 10 \end{vmatrix} = \begin{vmatrix} 6 & 10 \\ 7 & 10 \end{vmatrix}$$
$$= 60 - 70 = -10$$

**57.** $(-1)^{1+1}(-1)\begin{vmatrix} 6 & 1 \\ 5 & -3 \end{vmatrix} + (-1)^{1+2}(3)\begin{vmatrix} 4 & 1 \\ 8 & -3 \end{vmatrix} + (-1)^{1+3}(2)\begin{vmatrix} 4 & 6 \\ 8 & 5 \end{vmatrix}$
$$= -(-18 - 5) - 3(-12 - 8) + 2(20 - 48) = 27$$

**59.**
$$\begin{vmatrix} 3 & -6 & 7 \\ 1 & -2 & 4 \\ -6 & 12 & 9 \end{vmatrix} = \begin{vmatrix} 3 & 0 & 7 \\ 1 & 0 & 4 \\ -6 & 0 & 9 \end{vmatrix} = 0$$

**61.**
$$\begin{vmatrix} 3 & 0 & 0 \\ -3 & 3 & 0 \\ 4 & -4 & 4 \end{vmatrix} = 3\begin{vmatrix} 3 & 0 \\ -4 & 4 \end{vmatrix} = 3(12 - 0) = 36$$

**63.**
$$\begin{vmatrix} \frac{1}{2} & 0 & 0 \\ 0 & \frac{1}{2} & 0 \\ 0 & 0 & \frac{1}{2} \end{vmatrix} = \frac{1}{2}\begin{vmatrix} \frac{1}{2} & 0 \\ 0 & \frac{1}{2} \end{vmatrix} = \frac{1}{2}\left(\frac{1}{4} - 0\right) = \frac{1}{8}$$

**65.**
$$\begin{vmatrix} -3 & 4 & 0 & 2 \\ 6 & 0 & 1 & 0 \\ -9 & -2 & 5 & 6 \\ 3 & -5 & 1 & -4 \end{vmatrix} = \begin{vmatrix} -3 & 4 & 0 & 2 \\ 0 & 0 & 1 & 0 \\ -39 & -2 & 5 & 6 \\ -3 & -5 & 1 & -4 \end{vmatrix} = -\begin{vmatrix} -3 & 4 & 2 \\ -39 & -2 & 6 \\ -3 & -5 & -4 \end{vmatrix}$$
$$= -\begin{vmatrix} -3 & 4 & 2 \\ -30 & -14 & 0 \\ -9 & 3 & 0 \end{vmatrix} = -2\begin{vmatrix} -30 & -14 \\ -9 & 3 \end{vmatrix} = -2(-90 - 126) = 432$$

**67.** $D = \begin{vmatrix} 1 & 1 & 2 \\ 3 & -2 & 1 \\ 2 & 5 & 3 \end{vmatrix} = \begin{vmatrix} 1 & 0 & 0 \\ 3 & -5 & -5 \\ 2 & 3 & -1 \end{vmatrix} = \begin{vmatrix} -5 & -5 \\ 3 & -1 \end{vmatrix} = 5 + 15 = 20$

$D_x = \begin{vmatrix} 7 & 1 & 2 \\ 6 & -2 & 1 \\ 11 & 5 & 3 \end{vmatrix} = \begin{vmatrix} 7 & 1 & 2 \\ 20 & 0 & 5 \\ -24 & 0 & -7 \end{vmatrix} = -\begin{vmatrix} 20 & 5 \\ -24 & -7 \end{vmatrix} = -(-140 + 120) = 20$

$D_y = \begin{vmatrix} 1 & 7 & 2 \\ 3 & 6 & 1 \\ 2 & 11 & 3 \end{vmatrix} = \begin{vmatrix} 1 & 0 & 0 \\ 3 & -15 & -5 \\ 2 & -3 & -1 \end{vmatrix} = \begin{vmatrix} -15 & -5 \\ -3 & -1 \end{vmatrix} = 15 - 15 = 0$

$$D_z = \begin{vmatrix} 1 & 1 & 7 \\ 3 & -2 & 6 \\ 2 & 5 & 11 \end{vmatrix} = \begin{vmatrix} 1 & 0 & 0 \\ 3 & -5 & -15 \\ 2 & 3 & -3 \end{vmatrix} = \begin{vmatrix} -5 & -15 \\ 3 & -3 \end{vmatrix} = 15 + 45 = 60$$

$$x = \frac{D_x}{D} = \frac{20}{20} = 1; \quad y = \frac{D_y}{D} = \frac{0}{20} = 0; \quad z = \frac{D_z}{D} = \frac{60}{20} = 3$$

69. Multiplying matrix $A$ by the scalar 3 gives the matrix each of whose elements are 3 times each of the elements of $A$. Then, $n$ applications of Exercise 8, page 598 gives the stated result.

71.

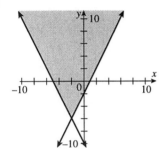

73.

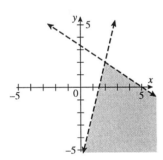

75.

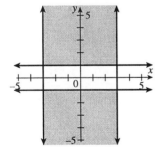

**77.**

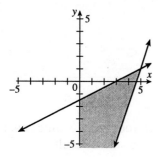

**79.**

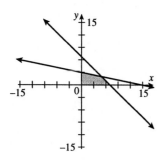

**81.** $y \le \frac{1}{5}x + 8$

$y \ge x$

$y \ge -3x + 8$

**83.** $y \le -3x + 8$

$y \le x$

$y \ge 0$

**85.** $y \ge \frac{1}{5}x + 8$

$y \le x$

**87.** $x + y \le 600$

$x \le \frac{2}{3}y$ or $y \ge \frac{3}{2}x$

$x \ge 150$

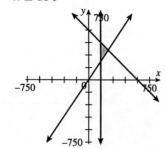

**89. (a)**

| Vertices | $k = 3x + 5y$ |
|----------|---------------|
| (0, 2)   | 10            |
| (2, 0)   | 6             |
| (1, 5)   | 28            |
| (3, 7)   | 44            |

minimum: 6
no maximum

**(b)**

| Vertices | $k = 7x + 4y$ |
|----------|---------------|
| (0, 2)   | 8             |
| (2, 0)   | 14            |
| (1, 5)   | 27            |
| (3, 7)   | 49            |

minimum: 8
no maximum

**91.** $4x - y \geq 7$
$x - 2y \geq -7$
$x + 3y \leq 23$
$2x + y \leq 21$
$y \geq 1$

**93. (a)**

| Vertices | $p = 10x + 7y$ |
|----------|----------------|
| (0, 0)   | 0              |
| (9, 0)   | 90             |
| (0, 6)   | 42             |
| (5, 5)   | 85             |

maximum: 90
minimum: 0

**(b)**

| Vertices | $q = 7x + 4y$ |
|----------|---------------|
| (0, 0)   | 0             |
| (9, 0)   | 63            |
| (0, 6)   | 24            |
| (5, 5)   | 55            |

maximum: 63
minimum: 0

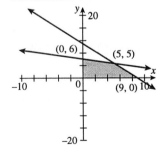

**95.**

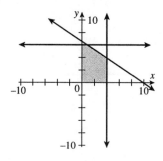

**(a)**

| Vertices | $p = 100x + 500y$ |
|----------|------------------|
| (0, 0)   | 0                |
| (4, 0)   | 400              |
| (0, 6)   | 3000             |
| (4, 4)   | 2400             |
| (1, 6)   | 3100             |

maximum: 3100
minimum: 0

**(b)**

| Vertices | $q = 200x + 200y$ |
|----------|------------------|
| (0, 0)   | 0                |
| (4, 0)   | 800              |
| (0, 6)   | 1200             |
| (4, 4)   | 1600             |
| (1, 6)   | 1400             |

maximum: 1600
minimum: 0

**97.**

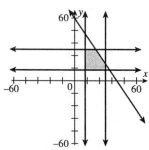

Let $x$ be the number of loaves of wheat bread and let $y$ be the number of loaves of rye bread. Then
$10 \le x \le 30$
$10 \le y \le 30$
$3x + 2y \le 120$

| Vertices | $p = 0.9x + 0.8y$ |
|----------|------------------|
| (10, 10) | 17               |
| (10, 30) | 33               |
| (30, 10) | 35               |
| (30, 15) | 39               |
| (20, 30) | 42               |

A maximum profit of $42 is obtained for 20 loaves of wheat bread and 30 loaves of rye bread.

# Chapter 9 Test: Standard Answer

**1.** $\begin{bmatrix} -6 & 3 & | & 9 \\ 10 & -5 & | & -12 \end{bmatrix}$

$\begin{bmatrix} 1 & -\frac{1}{2} & | & -\frac{3}{2} \\ 10 & -5 & | & -12 \end{bmatrix} \leftarrow -\frac{1}{6} \times R_1$

$\begin{bmatrix} 1 & -\frac{1}{2} & | & -\frac{3}{2} \\ 0 & 0 & | & 3 \end{bmatrix} \leftarrow -10 \times R_1 + R_2$

no solutions

**2.** $\begin{bmatrix} 2 & -1 & 2 & | & -3 \\ 1 & 4 & -3 & | & 18 \\ -4 & 2 & -3 & | & 0 \end{bmatrix}$

$\begin{bmatrix} 1 & 4 & -3 & | & 18 \\ 2 & -1 & 2 & | & -3 \\ -4 & 2 & -3 & | & 0 \end{bmatrix}$ } Interchange $R_1$ and $R_2$

$\begin{bmatrix} 1 & 4 & -3 & | & 18 \\ 0 & -9 & 8 & | & -39 \\ 0 & 18 & -15 & | & 72 \end{bmatrix} \begin{matrix} \\ \leftarrow -2 \times R_1 + R_2 \\ \leftarrow 4 \times R_1 + R_3 \end{matrix}$

$\begin{bmatrix} 1 & 4 & -3 & | & 18 \\ 0 & -9 & 8 & | & -39 \\ 0 & 0 & 1 & | & -6 \end{bmatrix} \leftarrow 2 \times R_2 + R_3$

$z = -6$

$-9y + 8(-6) = -39$

$-9y = 9$

$y = -1$

$x + 4(-1) - 3(-6) = 18$

$x = 4$

$(4, -1, -6)$

**3.** $\begin{bmatrix} 1 & 1 & 2 & | & 3 \\ 3 & 0 & 2 & | & 2 \\ -1 & 2 & 2 & | & 4 \end{bmatrix}$

$\begin{bmatrix} 1 & 1 & 2 & | & 3 \\ 0 & -3 & -4 & | & -7 \\ 0 & 3 & 4 & | & 7 \end{bmatrix} \begin{matrix} \\ \leftarrow -3 \times R_1 + R_2 \\ \leftarrow 1 \times R_1 + R_3 \end{matrix}$

$\begin{bmatrix} 1 & 1 & 2 & | & 3 \\ 0 & -3 & -4 & | & -7 \\ 0 & 0 & 0 & | & 0 \end{bmatrix} \leftarrow 1 \times R_2 + R_3$

Let $z = c$,

$$-3y - 4 \cdot c = -7$$

$$-3y = -7 + 4c$$

$$y = -\frac{4}{3}c + \frac{7}{3}$$

$$x + \left(-\frac{4}{3}c + \frac{7}{3}\right) + 2c = 3$$

$$x = -\frac{2}{3}c + \frac{2}{3}$$

$$\left(-\frac{2}{3}c + \frac{2}{3}, \ -\frac{4}{3}c + \frac{7}{3}, \ c\right)$$

4. (a) $\begin{bmatrix} -6 & 3 \\ 10 & -5 \end{bmatrix}\begin{bmatrix} x \\ y \end{bmatrix} = \begin{bmatrix} 9 \\ -12 \end{bmatrix}$

(b) $\begin{bmatrix} 2 & -1 & 2 \\ 1 & 4 & -3 \\ -4 & 2 & -3 \end{bmatrix}\begin{bmatrix} x \\ y \\ z \end{bmatrix} = \begin{bmatrix} -3 \\ 18 \\ 0 \end{bmatrix}$

5.  $25x + 35y + 40z = 550$
    $20x + 15y + 20z = 300$
    $15x + 25y + 20z = 350$

$$\left[\begin{array}{ccc|c} 25 & 35 & 40 & 550 \\ 20 & 15 & 20 & 300 \\ 15 & 25 & 20 & 350 \end{array}\right]$$

$$\left[\begin{array}{ccc|c} 1 & \frac{7}{5} & \frac{8}{5} & 22 \\ 20 & 15 & 20 & 300 \\ 15 & 25 & 20 & 350 \end{array}\right] \leftarrow \frac{1}{25} \times R_1$$

$$\left[\begin{array}{ccc|c} 1 & \frac{7}{5} & \frac{8}{5} & 22 \\ 0 & -13 & -12 & -140 \\ 0 & 4 & -4 & 20 \end{array}\right] \begin{array}{l} \\ \leftarrow -20 \times R_1 + R_2 \\ \leftarrow -15 \times R_1 + R_3 \end{array}$$

$$\left[\begin{array}{ccc|c} 1 & \frac{7}{5} & \frac{8}{5} & 22 \\ 0 & 4 & -4 & 20 \\ 0 & -13 & -12 & -140 \end{array}\right] \left.\begin{array}{l} \\ \end{array}\right\} \begin{array}{l} \text{Interchange} \\ R_2 \ \text{and} \ R_3 \end{array}$$

$$\left[\begin{array}{ccc|c} 1 & \frac{7}{5} & \frac{8}{5} & 22 \\ 0 & 1 & -1 & 5 \\ 0 & -13 & -12 & -140 \end{array}\right] \leftarrow \frac{1}{4} \times R_2$$

$$\begin{bmatrix} 1 & \frac{7}{5} & \frac{8}{5} & 22 \\ 0 & 1 & -1 & 5 \\ 0 & 0 & -25 & -75 \end{bmatrix} \leftarrow 13 \times R_2 + R_3$$

$-25z = -75$

$z = 3$

$y - 3 = 5$

$y = 8$

$x + \dfrac{7}{5} \cdot 8 + \dfrac{8}{5} \cdot 3 = 22$

$x = 6$

6 oz. of I; 8 oz. of II; 3 oz. of III.

**6. (a)** $B + G = \begin{bmatrix} 1 & 0 & -1 \\ 0 & -1 & 1 \\ -1 & 0 & 11 \end{bmatrix} + \begin{bmatrix} 0 & -1 & 1 \\ 6 & 2 & 3 \\ -5 & -5 & -8 \end{bmatrix} = \begin{bmatrix} 1 & -1 & 0 \\ 6 & 1 & 4 \\ -6 & -5 & 3 \end{bmatrix}$

**(b)** $A - B$ does not exist.

**7.** $5E - 2F = 5\begin{bmatrix} 3 & 0 \\ 1 & 1 \end{bmatrix} - 2\begin{bmatrix} -5 & 7 \\ 4 & -9 \end{bmatrix}$

$= \begin{bmatrix} 15 & 0 \\ 5 & 5 \end{bmatrix} - \begin{bmatrix} -10 & 14 \\ 8 & -18 \end{bmatrix} = \begin{bmatrix} 25 & -14 \\ -3 & 23 \end{bmatrix}$

**8.** $AB = \begin{bmatrix} 2 & 0 & 3 \\ -1 & 4 & 9 \end{bmatrix}\begin{bmatrix} 1 & 0 & -1 \\ 0 & -1 & 1 \\ -1 & 0 & 11 \end{bmatrix} = \begin{bmatrix} -1 & 0 & 31 \\ -10 & -4 & 104 \end{bmatrix}$

**9. (a)** $m_{32} = \begin{bmatrix} 7 & 8 & 9 \end{bmatrix}\begin{bmatrix} 0 \\ -1 \\ 0 \end{bmatrix} = -8$

**(b)** $|E| = \begin{vmatrix} 3 & 0 \\ 1 & 1 \end{vmatrix} = 3$

$|F| = \begin{vmatrix} -5 & 7 \\ 4 & -9 \end{vmatrix} = 45 - 28 = 17$

$(|E|)(|F|) = 3 \cdot 17 = 51$

**10. (a)** $AD = \begin{bmatrix} 2 & 0 & 3 \\ -1 & 4 & 9 \end{bmatrix}\begin{bmatrix} -1 & -1 \\ 2 & 2 \\ -3 & -3 \end{bmatrix} = \begin{bmatrix} -11 & -11 \\ -18 & -18 \end{bmatrix}$

**(b)** $A(EF)$ does not exist.

**11.** $E^3 = \begin{bmatrix} 3 & 0 \\ 1 & 1 \end{bmatrix}\begin{bmatrix} 3 & 0 \\ 1 & 1 \end{bmatrix}\begin{bmatrix} 3 & 0 \\ 1 & 1 \end{bmatrix} = \begin{bmatrix} 3 & 0 \\ 1 & 1 \end{bmatrix}\begin{bmatrix} 9 & 0 \\ 4 & 1 \end{bmatrix} = \begin{bmatrix} 27 & 0 \\ 13 & 1 \end{bmatrix}$

**12.** $HJ = \begin{bmatrix} -1 \\ 1 \\ 2 \end{bmatrix}\begin{bmatrix} 2 & -2 & 4 \end{bmatrix} = \begin{bmatrix} -2 & 2 & -4 \\ 2 & -2 & 4 \\ 4 & -4 & 8 \end{bmatrix}$

**13.** **(a)** 4 by 3
  **(b)** does not exist

**14.** $\left(A + \dfrac{1}{2}B\right)(C - 2D) = \left([11 \quad 7 \quad 2] + [-1 \quad 3 \quad -1]\right)\left(\begin{bmatrix} -5 \\ -2 \\ 4 \end{bmatrix} - \begin{bmatrix} -16 \\ 10 \\ 2 \end{bmatrix}\right)$

$= [10 \quad 10 \quad 1]\begin{bmatrix} 11 \\ -12 \\ 2 \end{bmatrix} = [-8]$

**15.** $\begin{bmatrix} 3 & 1 & | & 1 & 0 \\ 2 & -2 & | & 0 & 1 \end{bmatrix}$

$\begin{bmatrix} 1 & \frac{1}{3} & | & \frac{1}{3} & 0 \\ 2 & -2 & | & 0 & 1 \end{bmatrix} \leftarrow \frac{1}{3} \times R_1$

$\begin{bmatrix} 1 & \frac{1}{3} & | & \frac{1}{3} & 0 \\ 0 & -\frac{8}{3} & | & -\frac{2}{3} & 1 \end{bmatrix} \leftarrow -2 \times R_1 + R_2$

$\begin{bmatrix} 1 & \frac{1}{3} & | & \frac{1}{3} & 0 \\ 0 & 1 & | & \frac{1}{4} & -\frac{3}{8} \end{bmatrix} \leftarrow -\frac{3}{8} \times R_2$

$\begin{bmatrix} 1 & 0 & | & \frac{1}{4} & \frac{1}{8} \\ 0 & 1 & | & \frac{1}{4} & -\frac{3}{8} \end{bmatrix} \leftarrow -\frac{1}{3} \times R_2 + R_1$

So $A^{-1} = \begin{bmatrix} \frac{1}{4} & \frac{1}{8} \\ \frac{1}{4} & -\frac{3}{8} \end{bmatrix}$

$\begin{bmatrix} x \\ y \end{bmatrix} = \begin{bmatrix} \frac{1}{4} & \frac{1}{8} \\ \frac{1}{4} & -\frac{3}{8} \end{bmatrix}\begin{bmatrix} 9 \\ 14 \end{bmatrix} = \begin{bmatrix} 4 \\ -3 \end{bmatrix}$

$(4, -3)$

**16.**
$$\begin{bmatrix} 1 & 0 & -2 & 1 & 0 & 0 \\ 0 & 1 & 0 & 0 & 1 & 0 \\ 3 & 2 & 0 & 0 & 0 & 1 \end{bmatrix}$$

$$\begin{bmatrix} 1 & 0 & -2 & 1 & 0 & 0 \\ 0 & 1 & 0 & 0 & 1 & 0 \\ 0 & 2 & 6 & -3 & 0 & 1 \end{bmatrix} \leftarrow -3 \times R_1 + R_3$$

$$\begin{bmatrix} 1 & 0 & -2 & 1 & 0 & 0 \\ 0 & 1 & 0 & 0 & 1 & 0 \\ 0 & 0 & 6 & -3 & -2 & 1 \end{bmatrix} \leftarrow -2 \times R_2 + R_3$$

$$\begin{bmatrix} 1 & 0 & -2 & 1 & 0 & 0 \\ 0 & 1 & 0 & 0 & 1 & 0 \\ 0 & 0 & 1 & -\frac{1}{2} & -\frac{1}{3} & \frac{1}{6} \end{bmatrix} \leftarrow \frac{1}{6} \times R_3$$

$$\begin{bmatrix} 1 & 0 & 0 & 0 & -\frac{2}{3} & \frac{1}{3} \\ 0 & 1 & 0 & 0 & 1 & 0 \\ 0 & 0 & 1 & -\frac{1}{2} & -\frac{1}{3} & \frac{1}{6} \end{bmatrix} \leftarrow 2 \times R_3 + R_1$$

$$A^{-1} = \begin{bmatrix} 0 & -\frac{2}{3} & \frac{1}{3} \\ 0 & 1 & 0 \\ -\frac{1}{2} & -\frac{1}{3} & \frac{1}{6} \end{bmatrix}$$

**17.**
$$\begin{bmatrix} 1 & 0 & 2 \\ 2 & -1 & 3 \\ 4 & 1 & 8 \end{bmatrix} \begin{bmatrix} 3 \\ -1 \\ -2 \end{bmatrix} = \begin{bmatrix} -1 \\ 1 \\ -5 \end{bmatrix} = \begin{bmatrix} x \\ y \\ z \end{bmatrix}$$
$(-1, 1, -5)$

**18.**
$$\begin{vmatrix} -2 & 4 & 3 \\ 6 & -1 & -4 \\ 5 & 2 & 0 \end{vmatrix} = \begin{vmatrix} 22 & 0 & -13 \\ 6 & -1 & -4 \\ 17 & 0 & -8 \end{vmatrix} = -\begin{vmatrix} 22 & -13 \\ 17 & -8 \end{vmatrix}$$
$-(-176 + 221) = -45$

**19.**
$$\begin{vmatrix} 4 & 0 & -2 & 6 \\ -3 & 6 & 0 & -7 \\ -1 & 3 & -4 & 0 \\ 2 & 9 & -9 & 4 \end{vmatrix} = \begin{vmatrix} 0 & 0 & -2 & 0 \\ -3 & 6 & 0 & -7 \\ -9 & 3 & -4 & -12 \\ -16 & 9 & -9 & -23 \end{vmatrix} = -2\begin{vmatrix} -3 & 6 & -7 \\ -9 & 3 & -12 \\ -16 & 9 & -23 \end{vmatrix}$$

$$= -2\begin{vmatrix} 15 & 0 & 17 \\ -9 & 3 & -12 \\ 11 & 0 & 13 \end{vmatrix} = -6\begin{vmatrix} 15 & 17 \\ 11 & 13 \end{vmatrix} = -6(195 - 187) = -48$$

**20.** $x = \dfrac{\begin{vmatrix} 15 & -2 \\ -8 & 2 \end{vmatrix}}{\begin{vmatrix} 4 & -2 \\ 3 & 2 \end{vmatrix}} = \dfrac{30-16}{8+6} = \dfrac{14}{14} = 1$

$y = \dfrac{\begin{vmatrix} 4 & 15 \\ 3 & -8 \end{vmatrix}}{\begin{vmatrix} 4 & -2 \\ 3 & 2 \end{vmatrix}} = \dfrac{-32-45}{14} = -\dfrac{77}{14} = -\dfrac{11}{2}$

$\left(1, -\dfrac{11}{2}\right)$

**21.** $D = \begin{vmatrix} 2 & 1 & -1 \\ 1 & -2 & 1 \\ 3 & -1 & -2 \end{vmatrix} = \begin{vmatrix} 3 & -1 & 0 \\ 1 & -2 & 1 \\ 5 & -5 & 0 \end{vmatrix} = -1 \cdot \begin{vmatrix} 3 & -1 \\ 5 & -5 \end{vmatrix} = -(-15+5) = 10$

$D_x = \begin{vmatrix} -3 & 1 & -1 \\ 8 & -2 & 1 \\ -1 & -1 & -2 \end{vmatrix} = \begin{vmatrix} -3 & 4 & 5 \\ 8 & -10 & -15 \\ -1 & 0 & 0 \end{vmatrix} = -\begin{vmatrix} 4 & 5 \\ -10 & -15 \end{vmatrix} = -(-60+50) = 10$

$D_y = \begin{vmatrix} 2 & -3 & -1 \\ 1 & 8 & 1 \\ 3 & -1 & -2 \end{vmatrix} = \begin{vmatrix} 3 & 5 & 0 \\ 1 & 8 & 1 \\ 5 & 15 & 0 \end{vmatrix} = -\begin{vmatrix} 3 & 5 \\ 5 & 15 \end{vmatrix} = -(45-25) = -20$

$D_z = \begin{vmatrix} 2 & 1 & -3 \\ 1 & -2 & 8 \\ 3 & -1 & -1 \end{vmatrix} = \begin{vmatrix} 2 & 5 & -19 \\ 1 & 0 & 0 \\ 3 & 5 & -25 \end{vmatrix} = -\begin{vmatrix} 5 & -19 \\ 5 & -25 \end{vmatrix} = -(-125+95) = 30$

$x = \dfrac{D_x}{D} = \dfrac{10}{10} = 1; \quad y = \dfrac{D_y}{D} = \dfrac{-20}{10} = -2; \quad z = \dfrac{D_z}{D} = \dfrac{30}{10} = 3$

$(1, -2, 3)$

**22.** Let $x$ = square feet used for sofas, and $y$ = square feet used for easy chairs. Then

$0 \le x \le 800$

$y \ge 300$

$x + y \le 1200$

$x \ge \dfrac{3}{2}y$ or $y \le \dfrac{2}{3}x$

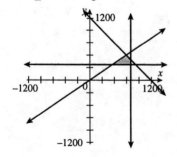

**23.**

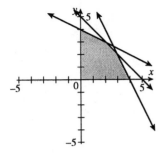

**24.**

| Vertices | $k = 3x + 2y$ |
|----------|---------------|
| (0, 0)   | 0             |
| (0, 4)   | 8             |
| (2, 3)   | 12            |
| (3, 2)   | 13            |
| (4, 0)   | 12            |

maximum: 13

**25.**

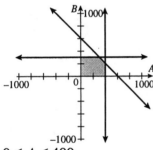

$0 \le A \le 400$

$0 \le B \le 300$

$A + B \le 600$

| Vertices   | $p = 80A + 100B$ |
|------------|-------------------|
| (0, 0)     | 0                 |
| (0, 300)   | 30,000            |
| (400, 0)   | 32,000            |
| (400, 200) | 52,000            |
| (300, 300) | 54,000            |

Produce 300 of model A and 300 of model B for a profit of $54,000.

## Chapter 9 Test: Multiple Choice

**1.** All three are allowed, so the answer is (d).

**2.** $\begin{vmatrix} 2 & 5 \\ 7 & 3 \end{vmatrix}$ is not equal to any of the four determinants, so the answer is (e).

**3.** $9A - 2A = 7A = 7\begin{bmatrix} 3 & -5 \\ -2 & 6 \\ 0 & 4 \end{bmatrix} = \begin{bmatrix} 21 & -35 \\ -14 & 42 \\ 0 & 28 \end{bmatrix}$

The answer is (c).

**4.** $AB = \begin{bmatrix} -2 & 1 & 5 \\ 3 & 0 & 2 \end{bmatrix} \begin{bmatrix} 4 & -2 \\ 3 & 4 \\ 0 & -1 \end{bmatrix} = \begin{bmatrix} -5 & 3 \\ 12 & -8 \end{bmatrix}$

The answer is (a).

**5.** The system has an infinite number of solutions, so the answer is (b).

**6.** Since $\begin{bmatrix} -1 & 2 \\ 2 & -3 \end{bmatrix}\begin{bmatrix} 3 & 2 \\ 2 & 1 \end{bmatrix} = \begin{bmatrix} 1 & 0 \\ 0 & 1 \end{bmatrix}$, the inverse of $\begin{bmatrix} -1 & 2 \\ 2 & -3 \end{bmatrix}$ is $\begin{bmatrix} 3 & 2 \\ 2 & 1 \end{bmatrix}$. The answer is (d).

**7.** The only possible product is $ACB$. The answer is (d).

**8.** $\begin{bmatrix} -2 & 4 \\ 6 & -8 \end{bmatrix}\begin{bmatrix} w & x \\ y & z \end{bmatrix} = \begin{bmatrix} 1 & 0 \\ 0 & 1 \end{bmatrix}$

$-2w + 4y = 1$

$-2x + 4z = 0$ or $x = 2z$

$6w - 8y = 0$

$6x - 8z = 1$

$6(2z) - 8z = 1$

$4z = 1$

$z = \dfrac{1}{4}$

The answer is (c).

**9.** (b)

**10.** (a)

**11.** $\begin{vmatrix} 1 & -2 & 3 \\ -1 & 0 & 1 \\ -2 & 3 & -1 \end{vmatrix} = \begin{vmatrix} 1 & -2 & 4 \\ -1 & 0 & 0 \\ -2 & 3 & -3 \end{vmatrix} = \begin{vmatrix} -2 & 4 \\ 3 & -3 \end{vmatrix} = 6 - 12 = -6$

The answer is (c).

**12.**

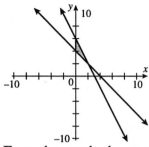

From the graph, the answer is (a).

**13.** The numerator is $\begin{vmatrix} 2 & 5 & -3 \\ -1 & 1 & 1 \\ 3 & -3 & 2 \end{vmatrix}$, so the answer is (e).

**14.**
$$\left[\begin{array}{ccc|ccc} 1 & 0 & 2 & 1 & 0 & 0 \\ 0 & 2 & -1 & 0 & 1 & 0 \\ 2 & 5 & 2 & 0 & 0 & 1 \end{array}\right]$$

$$\left[\begin{array}{ccc|ccc} 1 & 0 & 2 & 1 & 0 & 0 \\ 0 & 1 & -\frac{1}{2} & 0 & \frac{1}{2} & 0 \\ 0 & 5 & -2 & -2 & 0 & 1 \end{array}\right] \begin{array}{l} \leftarrow \frac{1}{2} \times R_2 \\ \leftarrow -2 \times R_1 + R_3 \end{array}$$

$$\left[\begin{array}{ccc|ccc} 1 & 0 & 2 & 1 & 0 & 0 \\ 0 & 1 & -\frac{1}{2} & 0 & \frac{1}{2} & 0 \\ 0 & 0 & \frac{1}{2} & -2 & -\frac{5}{2} & 1 \end{array}\right] -5 \times R_2 + R_3$$

$$\left[\begin{array}{ccc|ccc} 1 & 0 & 2 & 1 & 0 & 0 \\ 0 & 1 & -\frac{1}{2} & 0 & \frac{1}{2} & 0 \\ 0 & 0 & 1 & -4 & -5 & 2 \end{array}\right] \leftarrow 2 \times R_3$$

$$\left[\begin{array}{ccc|ccc} 1 & 0 & 0 & 9 & 10 & -4 \\ 0 & 1 & 0 & -2 & -2 & 1 \\ 0 & 0 & 1 & -4 & -5 & 2 \end{array}\right] \begin{array}{l} \leftarrow -2 \times R_3 + R_1 \\ \leftarrow \frac{1}{2} \times R_3 + R_2 \end{array}$$

The answer is (b).

**15.**

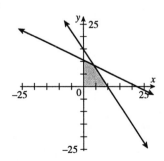

| Vertices | $k = 4x + 3y$ |
|----------|---------------|
| (0, 0) | 0 |
| (10, 0) | 40 |
| (0, 11) | 33 |
| (4, 9) | 43 |

Since the maximum is 43, the answer is (c).

**16.** $z = \dfrac{\begin{vmatrix} 2 & -3 & 11 \\ 3 & 1 & 9 \\ -1 & 3 & -1 \end{vmatrix}}{\begin{vmatrix} 2 & -3 & 1 \\ 3 & 1 & -1 \\ -1 & 3 & -2 \end{vmatrix}} = \dfrac{\begin{vmatrix} 0 & 3 & 9 \\ 0 & 10 & 6 \\ -1 & 3 & -1 \end{vmatrix}}{\begin{vmatrix} 0 & 3 & -3 \\ 0 & 10 & -7 \\ -1 & 3 & -2 \end{vmatrix}} = \dfrac{-\begin{vmatrix} 3 & 9 \\ 10 & 6 \end{vmatrix}}{-\begin{vmatrix} 3 & -3 \\ 10 & -7 \end{vmatrix}} = \dfrac{18 - 90}{-21 + 30} = \dfrac{-72}{9} = -8$

The answer is (d).

**17.** $P = \begin{bmatrix} 200 & 350 & 300 \end{bmatrix}$

$C = \begin{bmatrix} 8.50 & 8 \\ 10 & 10.50 \\ 12.50 & 12 \end{bmatrix}$

$PC = \begin{bmatrix} 200 & 350 & 300 \end{bmatrix} \begin{bmatrix} 8.50 & 8 \\ 10 & 10.50 \\ 12.50 & 12 \end{bmatrix} = \begin{bmatrix} 8950 & 8875 \end{bmatrix}$

The answer is (d).

**18.** Adding 3 times row 1 to row 3 does not change the determinant. However, interchanging two rows changes the sign of the determinant, so the answer is (b).

**19.** Since $\left| A^t \right| = |A| = 2$, the answer is (a).

**20.** $(I - A)^2 = I \cdot I - I \cdot A - A \cdot I + A \cdot A$

$= I - A - A + A^2$
$= I - 2A + A$
$= I - A$

The answer is (c).

# Chapter 10: The Conic Sections

## Exercises 10.1

**1.** $\dfrac{x^2}{25} + \dfrac{y^2}{16} = 1$

$\dfrac{x^2}{5^2} + \dfrac{y^2}{4^2} = 1$

Center: $(0, 0)$
Vertices: $(\pm 5, 0)$
$4^2 = 5^2 - c^2$
$c^2 = 9$
$c = 3$
Foci: $(\pm 3, 0)$

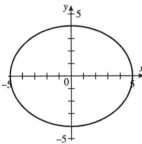

**3.** $\dfrac{x^2}{16} + \dfrac{y^2}{9} = 1$

$\dfrac{x^2}{4^2} + \dfrac{y^2}{3^2} = 1$

$3^2 = 4^2 - c^2$
$c^2 = 7$
$c = \sqrt{7}$
Center: $(0, 0)$
Vertices: $(\pm 4, 0)$
Foci: $(\pm \sqrt{7}, 0)$

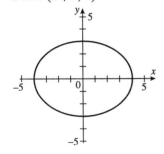

**5.** $\dfrac{x^2}{4} + \dfrac{y^2}{36} = 1$

$4 = 36 - c^2$
$c^2 = 32$
$c = 4\sqrt{2}$
Center: $(0, 0)$
Vertices: $(0, \pm 6)$
Foci: $(0, \pm 4\sqrt{2})$

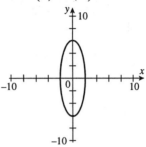

**7.** $9x^2 + y^2 = 9$

$\dfrac{x^2}{1} + \dfrac{y^2}{9} = 1$

$1 = 9 - c^2$
$c^2 = 8$
$c = 2\sqrt{2}$
Center: $(0, 0)$
Vertices: $(0, \pm 3)$
Foci: $(0, \pm 2\sqrt{2})$

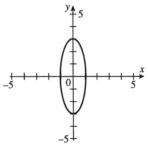

**9.** $4x^2 + 9y^2 = 36$

$\dfrac{x^2}{9} + \dfrac{y^2}{4} = 1$

$4 = 9 - c^2$
$c^2 = 5$
$c = \sqrt{5}$

Center: (0, 0)
Vertices: (±3, 0)
Foci: (±√5, 0)

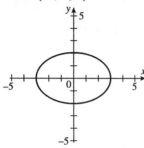

**11.** $25x^2 + 9y^2 = 225$

$$\frac{x^2}{9} + \frac{y^2}{25} = 1$$

$9 = 25 - c^2$

$c^2 = 16$

$c = 4$

Center: (0, 0)
Vertices: (0, ±5)
Foci: (0, ±4)

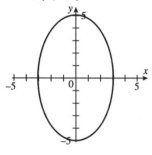

**13.** $\frac{(x-1)^2}{9} + \frac{(y+2)^2}{4} = 1$

$4 = 9 - c^2$

$c^2 = 5$

$c = \sqrt{5}$

Center: (1, –2)
Vertices: (–2, –2), (4, –2)
Foci: $(1 \pm \sqrt{5}, -2)$

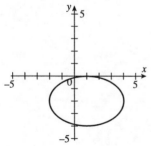

**15.** $\frac{(x+2)^2}{4} + \frac{(y-3)^2}{9} = 1$

$4 = 9 - c^2$

$c^2 = 5$

$c = \sqrt{5}$

Center: (–2, 3)
Vertices: (–2, 0), (–2, 6)
Foci: $(-2, 3 \pm \sqrt{5})$

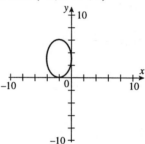

**17.** Center: (0, 0)

$a = 3,\ b = 2$

$$\frac{x^2}{3^2} + \frac{y^2}{2^2} = 1 \text{ or } \frac{x^2}{9} + \frac{y^2}{4} = 1$$

**19.** Center: (0, 0)

$a = 6,\ b = 3$

$$\frac{x^2}{6^2} + \frac{y^2}{3^2} = 1 \text{ or } \frac{x^2}{36} + \frac{y^2}{9} = 1$$

**21.** Center: (–2, 0)

$a = 5,\ b = 3$

$$\frac{[x-(-2)]^2}{5^2} + \frac{(y-0)^2}{3^2} = 1$$

$$\frac{(x+2)^2}{25} + \frac{y^2}{9} = 1$$

**23.** $a = \frac{1}{2}(10) = 5,\ b = \frac{1}{2}(6) = 3$

$$\frac{x^2}{5^2} + \frac{y^2}{3^2} = 1 \text{ or } \frac{x^2}{25} + \frac{y^2}{9} = 1$$

**25.** Ellipse is horizontal. $b = \frac{1}{2}(8) = 4$, $c = 4$.

$$b^2 = a^2 - c^2 \text{ or } 16 = a^2 - 16$$
$$a^2 = 32$$
$$\frac{(x-2)^2}{32} + \frac{(y-3)^2}{16} = 1$$

**27.** Center: $(0, 0)$
Ellipse is vertical.
$a = 5$, $c = 3$
$b^2 = a^2 - c^2 = 25 - 9$
$b^2 = 16$
$$\frac{x^2}{16} + \frac{y^2}{25} = 1$$

**29.** Center: $(0, 0)$
$a = 8$, $b = 4$
Ellipse is horizontal.
$$\frac{x^2}{8^2} + \frac{y^2}{4^2} = 1 \text{ or } \frac{x^2}{64} + \frac{y^2}{16} = 1$$

**31.** Center: $(6, 1)$
Ellipse is horizontal.
$a = 3$, $b = 2$
$$\frac{(x-6)^2}{3^2} + \frac{(y-1)^2}{2^2} = 1 \text{ or }$$
$$\frac{(x-6)^2}{9} + \frac{(y-1)^2}{4} = 1$$

**33.** Center: $(0, 0)$
$c = 4$, $e = \dfrac{c}{a}$
$\dfrac{2}{3} = \dfrac{4}{a}$, so $a = 6$
$b^2 = a^2 - c^2 = 36 - 16 = 20$
$$\frac{x^2}{36} + \frac{y^2}{20} = 1$$

**35.** Center: $(-1, 0)$
$a = 10$, $\dfrac{4}{5} = \dfrac{c}{10}$
$c = 8$
$b^2 = a^2 - c^2 = 100 - 64 = 36$
$$\frac{(x+1)^2}{36} + \frac{y^2}{100} = 1$$

**37.** $25x^2 + y^2 - 12y = -11$
$25x^2 + y^2 - 12y + 36 = -11 + 36$
$25x^2 + (y-6)^2 = 25$
$$\frac{x^2}{1} + \frac{(y-6)^2}{25} = 1$$
Center: $(0, 6)$
Vertices: $(0, 1)$, $(0, 11)$
$b^2 = a^2 - c^2$
$c^2 = 24$
$c = 2\sqrt{6}$
Foci: $(0, 6 \pm 2\sqrt{6})$

**39.** $4x^2 + 24x + 13y^2 - 26y = 3$
$4(x^2 + 6x) + 13(y^2 - 2y) = 3$
$4(x^2 + 6x + 9) + 13(y^2 - 2y + 1)$
$= 3 + 36 + 13$
$4(x+3)^2 + 13(y-1)^2 = 52$
$$\frac{(x+3)^2}{13} + \frac{(y-1)^2}{4} = 1$$
$b^2 = a^2 - c^2$
$4 = 13 - c^2$
$c^2 = 9$
$c = 3$
Center: $(-3, 1)$
Vertices: $(-3 \pm \sqrt{13}, 1)$
Foci: $(-6, 1)$, $(0, 1)$

**41. (a)** $a - c = 4130$
$a + c = 4580$
$2a = 8710$ or $a = 4355$
$4355 + c = 4580$ or $c = 225$
$b^2 = (4355)^2 - (225)^2$
$$\frac{x^2}{b^2} + \frac{y^2}{(4355)^2} = 1$$

**(b)** $b = \sqrt{18,915,400} = 4349$ miles
$$\frac{x^2}{(4349)^2} + \frac{y^2}{(4355)^2} = 1$$

**43.** $a = \frac{1}{2}(3.365 \times 10^9) = 1.6825 \times 10^9$
$e = \dfrac{c}{a}$
$$0.967 = \frac{c}{1.6825 \times 10^9}$$

$c = 1.627 \times 10^9$

$a - c = 56{,}000{,}000$

The closest the comet gets to the sun is 56 million miles.

45. (a) $a = \frac{1}{2}(60) = 30$, $b = 20$

$$\frac{x^2}{30^2} + \frac{y^2}{20^2} = 1 \text{ or } \frac{x^2}{900} + \frac{y^2}{400} = 1$$

When $x = \pm 20$:

$$\frac{400}{900} + \frac{y^2}{400} = 1$$

$$\frac{y^2}{400} = 1 - \frac{400}{900}$$

$$y^2 = 400\left(1 - \frac{400}{900}\right)$$

$$y^2 = \frac{400}{900}(900 - 400)$$

$$y = \frac{2}{3}\sqrt{900 - 400} = \frac{20}{3}\sqrt{5}$$

(b) $y = 14.9$ feet.

## Challenge Problem

Let $A$ be a point on the circle, $(x_1, y_1)$, so that $x_1^2 + y_1^2 = 9$. Then $B$ is the point $(0, y_1)$. The distance $BA$ is equal to $x$, so $BP = \frac{2}{3}x_1$. Hence, $P$ is the point $\left(\frac{2}{3}x_1, y_1\right)$. Notice the $P$ satisfies the equation $\frac{x^2}{4} + \frac{y^2}{9} = 1$, since

$$\frac{\left(\frac{2}{3}x_1\right)^2}{4} + \frac{y_1^2}{9} = \frac{\frac{4}{9}x_1^2}{4} + \frac{y_1^2}{9} = \frac{x_1^2 + y_1^2}{9} = \frac{9}{9} = 1.$$

Hence, the set of all points $P$ determines the ellipse $\frac{x^2}{4} + \frac{y^2}{9} = 1$.

1.

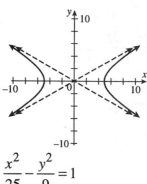

$$\frac{x^2}{25} - \frac{y^2}{9} = 1$$

$9 = c^2 - 25$

$c^2 = 34$

$c = \sqrt{34}$

Center: $(0, 0)$

Vertices: $(\pm 5, 0)$

Foci: $(\pm\sqrt{34}, 0)$

Asymptotes: $y = \pm\frac{3}{5}x$

3.

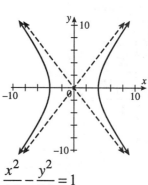

$$\frac{x^2}{16} - \frac{y^2}{25} = 1$$

$25 = c^2 - 16$

$c^2 = 41$

$c = \sqrt{41}$

Center: $(0, 0)$

Vertices: $(\pm 4, 0)$

Foci: $(\pm\sqrt{41}, 0)$

Asymptotes: $y = \pm\frac{5}{4}x$

**5.**

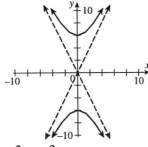

$$\frac{y^2}{36} - \frac{x^2}{9} = 1$$
$9 = c^2 - 36$
$c^2 = 45$
$c = 3\sqrt{5}$
Center: $(0, 0)$
Vertices: $(0, \pm 6)$
Foci: $(0, \pm 3\sqrt{5})$
Asymptotes: $y = \pm\frac{6}{3}x$ or $y = \pm 2x$

**7.**

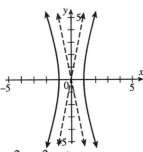

$9x^2 - y^2 = 9$
$$x^2 - \frac{y^2}{9} = 1$$
$9 = c^2 - 1$
$c^2 = 10$
$c = \sqrt{10}$
Center: $(0, 0)$
Vertices: $(\pm 1, 0)$
Foci: $(\pm\sqrt{10}, 0)$
Asymptotes: $y = \pm 3x$

**9.**

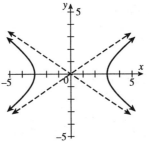

$4x^2 - 9y^2 = 36$
$$\frac{x^2}{9} - \frac{y^2}{4} = 1$$
$4 = c^2 - 9$
$c^2 = 13$
$c = \sqrt{13}$
Center: $(0, 0)$
Vertices: $(\pm 3, 0)$
Foci: $(\pm\sqrt{13}, 0)$
Asymptotes: $y = \pm\frac{2}{3}x$

**11.**

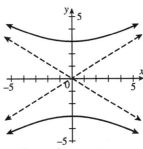

$25y^2 - 9x^2 = 225$
$$\frac{y^2}{9} - \frac{x^2}{25} = 1$$
$25 = c^2 - 9$
$c^2 = 34$
$c = \sqrt{34}$
Center: $(0, 0)$
Vertices: $(0, \pm 3)$
Foci: $(0, \pm\sqrt{34})$
Asymptotes: $y = \pm\frac{3}{5}x$

**13.**

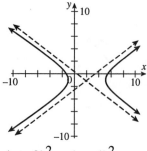

$$\frac{(x-2)^2}{9} - \frac{(y+1)^2}{4} = 1$$

$4 = c^2 - 9$

$c^2 = 13$

$c = \sqrt{13}$

Center: $(2, -1)$

Vertices: $(-1, -1)$, $(5, -1)$

Foci: $(2 \pm \sqrt{13}, -1)$

Asymptotes: $y + 1 = \pm\frac{2}{3}(x - 2)$

**15.**

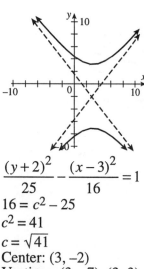

$$\frac{(y+2)^2}{25} - \frac{(x-3)^2}{16} = 1$$

$16 = c^2 - 25$

$c^2 = 41$

$c = \sqrt{41}$

Center: $(3, -2)$

Vertices: $(3, -7)$, $(3, 3)$

Foci: $(3, -2 \pm \sqrt{41})$

Asymptotes: $y + 2 = \pm\frac{5}{4}(x - 3)$

**17.** $a = 3$, $c = \sqrt{17}$

$b^2 = c^2 - a^2 = 17 - 9 = 8$

$b = 2\sqrt{2}$

Center: $(0, 0)$

$$\frac{x^2}{9} - \frac{y^2}{8} = 1$$

**19.** Center: $(2, 3)$

$a = 3$

$\frac{b}{a} = \frac{2}{3}$, $b = \frac{2}{3}(a)$, $b = 2$

$$\frac{(x-2)^2}{9} - \frac{(y-3)^2}{4} = 1$$

**21.** $c = 6$, $a = 4$

$b^2 = c^2 - a^2 = 36 - 16 = 20$

$$\frac{x^2}{16} - \frac{y^2}{20} = 1$$

**23.** $a = \frac{1}{2}(6) = 3$

$b^2 = c^2 - a^2 = 16 - 9 = 7$

$$\frac{(y-3)^2}{9} - \frac{(x+2)^2}{7} = 1$$

**25.** $a = 4$, $\frac{b}{a} = \frac{1}{2}$, $b = \frac{1}{2}a = 2$

$$\frac{x^2}{16} - \frac{y^2}{4} = 1$$

**27.** Center: $(0, 0)$

$c = 17$

$\frac{a}{b} = \frac{8}{15}$

$a = \frac{8}{15}b$

$b^2 = c^2 - a^2$

$b^2 = 289 - \frac{64}{225}b^2$

$\frac{289b^2}{225} = 289$

$b^2 = 225$

$b = 15$

$$a = \frac{8}{15}(15) = 8$$

$$\frac{y^2}{64} - \frac{x^2}{225} = 1$$

**29.** $9 = c^2 - 16$
$c^2 = 25$
$c = 5$
Center: $(-1, 3)$
Vertices: $(-5, 3)$, $(3, 3)$
Foci: $(-6, 3)$, $(4, 3)$

**31.** $36 = c^2 - 64$
$c^2 = 100$
$c = 10$
Center: $(3, -1)$
Vertices: $(3, -9)$, $(3, 7)$
Foci: $(3, -11)$, $(3, 9)$

**33.** $4x^2 - 8x - 9y^2 - 36y = 68$
$4(x^2 - 2x) - 9(y^2 + 4y) = 68$
$4(x^2 - 2x + 1) - 9(y^2 + 4y + 4)$
$= 68 + 4 - 36$
$4(x - 1)^2 - 9(y + 2)^2 = 36$
$$\frac{(x-1)^2}{9} - \frac{(y+2)^2}{4} = 1$$
$4 = c^2 - 9$
$c^2 = 13$
$c = \sqrt{13}$
Center: $(1, -2)$
Vertices: $(-2, -2)$, $(4, -2)$
Foci: $(1 \pm \sqrt{13}, -2)$

**35.** $y^2 + 4y - 4x^2 + 8x = 4$
$(y^2 + 4y) - 4(x^2 - 2x) = 4$
$(y^2 + 4y + 4) - 4(x^2 - 2x + 1) = 4 + 4 - 4$
$(y + 2)^2 - 4(x - 1)^2 = 4$
$$\frac{(y+2)^2}{4} - (x - 1)^2 = 1$$
$1 = c^2 - 4$
$c^2 = 5$
$c = \sqrt{5}$
Center: $(1, -2)$
Vertices: $(1, -4)$, $(1, 0)$
Foci: $(1, -2 \pm \sqrt{5})$

**37.** Center: $(-2, 4)$
$c = 2\sqrt{34}$
$$\frac{b}{a} = \frac{3}{5} \text{ or } b = \frac{3}{5}a$$
$b^2 = c^2 - a^2$
$$\frac{9}{25}a^2 = 136 - a^2$$
$$\frac{34}{25}a^2 = 136$$
$a^2 = 100$
$a = 10$
$$b = \frac{3}{5}(10) = 6$$
$$\frac{(x+2)^2}{100} - \frac{(y-4)^2}{36} = 1$$

**39. (a)** Let $(x, y)$ be the location of the boat. The distance to one station is
$\sqrt{x^2 + y^2}$, and the distance to the other station is $\sqrt{(x-2)^2 + y^2}$.
Since time $= \dfrac{\text{distance}}{\text{rate}}$,
$$\frac{\sqrt{x^2 + y^2}}{1100} - \frac{\sqrt{(x-2)^2 + y^2}}{1100} = 2.7.$$
Thus,
$$\sqrt{x^2 + y^2} - \sqrt{(x-2)^2 + y^2}$$
$= 2.7(1100)$ feet
Since the difference in the distances is constant, $(x, y)$ lies on a hyperbola. The center is at $(1, 0)$, and $c = 1$.
$$2a = \frac{2.7(1100)}{5280} = \frac{9}{16} \text{ miles}$$
$$a = \frac{9}{32}$$
$$b^2 = c^2 - a^2 = 1 - \frac{81}{1024} = \frac{943}{1024}$$
$$\frac{(x-1)^2}{\frac{81}{1024}} - \frac{y^2}{\frac{943}{1024}} = 1 \text{ or}$$
$$\frac{1024(x-1)^2}{81} - \frac{1024y^2}{943} = 1$$

**(b)** $y = 1, x > 1$

$$\frac{1024(x-1)^2}{81} - \frac{1024}{943} = 1$$

$$\frac{1024(x-1)^2}{81} = \frac{1967}{943}$$

$(x-1)^2 = 0.1650$

$x - 1 = \pm 0.406$

$x = 1.406$ or $0.594$

Since $x > 1$, $x = 1.406$

$(1.406, 1)$ is the location of the boat.

## Challenge Problem

The amount of time it takes for the sound to reach the northern vehicle is

$$\frac{5280\sqrt{(2-0)^2 + (0.53-1.5)^2}}{1100} = 10.67 \text{ seconds.}$$

The amount of time it takes for the sound to reach the southern vehicle is

$$\frac{5280\sqrt{(2-0)^2 + (0.53+1.5)^2}}{1100} = 13.68 \text{ seconds.}$$

Since the sound takes 3.01 seconds more to reach the southern vehicle, this confirms the answer. The extra 0.01 second is due to round-off error.

## Exercises 10.3

**1.** $x^2 = \frac{1}{4}y$

$4p = \frac{1}{4}$

$p = \frac{1}{16}$

Focus: $\left(0, \frac{1}{16}\right)$

Directrix: $y = -\frac{1}{16}$

**3.** $x^2 = \frac{1}{2}y$

$4p = \frac{1}{2}$

$p = \frac{1}{8}$

Focus: $\left(0, \frac{1}{8}\right)$

Directrix: $y = -\frac{1}{8}$

**5.** $y^2 = 2x$

$4p = 2$

$p = \frac{1}{2}$

Focus: $\left(\frac{1}{2}, 0\right)$

Directrix: $x = -\frac{1}{2}$

**7.** $y = -4x^2$

$x^2 = -\frac{1}{4}y$

$4p = -\frac{1}{4}$

$p = -\frac{1}{16}$

Focus: $\left(0, -\frac{1}{16}\right)$

Directrix: $y = \frac{1}{16}$

**9.** $-9(x-2) = 6y^2$

$y^2 = -\frac{3}{2}(x-2)$

$4p = -\frac{3}{2}$

$p = -\frac{3}{8}$

Focus: $\left(\frac{13}{8}, 0\right)$

Directrix: $x = \frac{19}{8}$

**11.** $(x-4)^2 = -\dfrac{1}{3}(y+5)$

$4p = -\dfrac{1}{3}$

$p = -\dfrac{1}{12}$

Focus: $\left(4, \ -\dfrac{61}{12}\right)$

Directrix: $y = -\dfrac{59}{12}$

**13.**

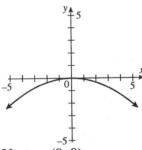

Vertex: $(0, 0)$

$p = -3$

$x^2 = 4(-3)y$

$x^2 = -12y$

**15.**

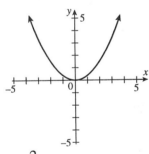

$p = \dfrac{2}{3}$

$x^2 = 4\left(\dfrac{2}{3}\right)y$

$x^2 = \dfrac{8}{3}y$

**17.**

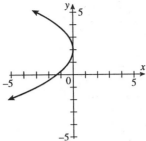

Vertex: $(0, 2)$

$p = -\dfrac{3}{4}$

$(y-2)^2 = 4\left(-\dfrac{3}{4}\right)x$

$(y-2)^2 = -3x$

**19.**

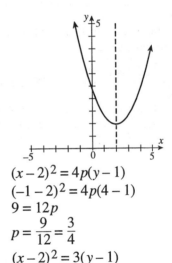

$(x-2)^2 = 4p(y-1)$

$(-1-2)^2 = 4p(4-1)$

$9 = 12p$

$p = \dfrac{9}{12} = \dfrac{3}{4}$

$(x-2)^2 = 3(y-1)$

**21.** $p = \dfrac{3}{2}$

$x^2 = 4\left(\dfrac{3}{2}\right)(y+3)$

$x^2 = 6(y+3)$

**23.** $(x-3)^2 = 4p(y-9)$

$(-3)^2 = 4p(-9)$

$p = -\dfrac{1}{4}$

$(x-3)^2 = -(y-9)$

**25.** $p = 2$

$(x-2)^2 = 8(y+5)$

$y + 5 = \frac{1}{8}(x-2)^2$

$y = \frac{1}{8}(x-2)^2 - 5$

$y = \frac{1}{8}(x^2 - 4x + 4) - 5 = \frac{1}{8}x^2 - \frac{1}{2}x - \frac{9}{2}$

**27.** $4p = 4$

$p = 1$

Vertex: $(2, -5)$

Axis: $x = 2$

Focus: $(2, -4)$

Directrix: $y = -6$

**29.** $y = \frac{1}{4}x^2 - x + 4$

$x^2 - 4x + 16 = 4y$

$x^2 - 4x = 4y - 16$

$x^2 - 4x + 4 = 4y - 16 + 4$

$(x-2)^2 = 4(1)(y-3)$

Vertex: $(2, 3)$

Focus: $(2, 4)$

Directrix: $y = 2$

**31.** $4x^2 - 4x - 28y - 83 = 0$

$4x^2 - 4x = 28y + 83$

$x^2 - x = 7y + \frac{83}{4}$

$x^2 - x + \frac{1}{4} = 7y + 21$

$\left(x - \frac{1}{2}\right)^2 = 7(y+3)$

$\left(x - \frac{1}{2}\right)^2 = 4\left(\frac{7}{4}\right)(y+3)$

Vertex: $\left(\frac{1}{2}, -3\right)$

Focus: $\left(\frac{1}{2}, -\frac{5}{4}\right)$

Directrix: $y = -\frac{19}{4}$

**33.** $-p = -\frac{9}{2} - (-3)$

$p = \frac{3}{2}$

Focus: $\left(-\frac{3}{2}, -2\right)$

Axis: $y = -2$

$(y+2)^2 = 4\left(\frac{3}{2}\right)(x+3)$

$(y+2)^2 = 6(x+3)$

**35.** $x = ay^2 + by + c$

$6 = a + b + c$

$-8 = 4a + 2b + c$

$-18 = 9a - 3b + c$

$$\begin{bmatrix} 1 & 1 & 1 & : & 6 \\ 4 & 2 & 1 & : & -8 \\ 9 & -3 & 1 & : & -18 \end{bmatrix}$$

$$\begin{bmatrix} 1 & 1 & 1 & : & 6 \\ 0 & -2 & -3 & : & -32 \\ 0 & -12 & -8 & : & -72 \end{bmatrix} \begin{matrix} \\ \leftarrow -4 \times R_1 + R_2 \\ \leftarrow -9 \times R_1 + R_3 \end{matrix}$$

$$\begin{bmatrix} 1 & 1 & 1 & : & 6 \\ 0 & -2 & -3 & : & -32 \\ 0 & 0 & 10 & : & 120 \end{bmatrix} \begin{matrix} \\ \\ \leftarrow -6 \times R_2 + R_3 \end{matrix}$$

$10c = 120$

$c = 12$

$-2b - 3 \cdot 12 = -32$

$b = -2$

$a + (-2) + 12 = 6$

$a = -4$

$x = -4y^2 - 2y + 12$

**37.** $4p = -9$

$p = -\frac{9}{4}$

Vertex: $(4, 0)$

Axis: $y = 0$

Focus: $\left(\frac{7}{4}, 0\right)$

Directrix: $x = \frac{25}{4}$

**39.** $x^2 = \dfrac{9}{2}y$

$4p = \dfrac{9}{2}$

$p = \dfrac{9}{8}$

The receiver is $\dfrac{9}{8}$ feet from the bottom of the dish antenna. Since the diameter is 8 feet, the circumference is $8\pi$ feet.

**41.** Vertex: $(0, 10)$
$x^2 = 4p(y - 10)$
$400^2 = 4p(170 - 10)$
$4p = 1000$
$p = 250$
$x^2 = 1000(y - 10)$
When $x = \pm300$,
$90,000 = 1000(y - 10)$
$y - 10 = 90$
$y = 100$
The height of the cable is 100 feet.

**43.** Ellipse
Center: $(0, 0)$
Vertices: $(\pm5, 0)$
$16 = 25 - c^2$
$c^2 = 9$
$c = 3$
Foci: $(\pm3, 0)$

**45.** Hyperbola
Center: $(0, 0)$
Vertices: $(\pm6, 0)$
$25 = c^2 - 36$
$c^2 = 61$
$c = \sqrt{61}$
Foci: $(\pm\sqrt{61}, 0)$
Asymptotes: $y = \pm\dfrac{5}{6}x$

**47.** Circle
Center: $(0, 0)$
Radius: 4

**49.** $16y^2 - 9x^2 = 144$

$\dfrac{y^2}{9} - \dfrac{x^2}{16} = 1$

Hyperbola

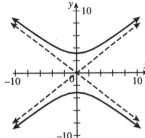

**51.** $\dfrac{(x-1)^2}{64} + \dfrac{(y-2)^2}{36} = 1$

Ellipse

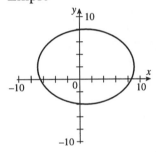

**53.** $16(y-3)^2 - 9(x+2)^2 = -144$

$\dfrac{(x+2)^2}{16} - \dfrac{(y-3)^2}{9} = 1$

Hyperbola

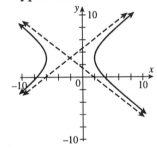

**55.** $x^2 + y^2 - 2x + 4y + 1 = 0$
$x^2 - 2x + y^2 + 4y = -1$
$x^2 - 2x + 1 + y^2 + 4y + 4 = -1 + 1 + 4$
$(x-1)^2 + (y+2)^2 = 4$
Circle

**57.** $x^2 + 4y^2 + 2x - 3 = 0$

$x^2 + 2x + 4y^2 = 3$

$x^2 + 2x + 1 + 4y^2 = 3 + 1$

$(x + 1)^2 + 4y^2 = 4$

$$\frac{(x+1)^2}{4} + y^2 = 1$$

Ellipse

**59.** $9x^2 + 18x - 16y^2 + 96y = 279$

$9(x^2 + 2x) - 16(y^2 - 6y) = 279$

$9(x^2 + 2x + 1) - 16(y^2 - 6y + 9)$
$= 279 + 9 - 144$

$9(x + 1)^2 - 16(y - 3)^2 = 144$

$$\frac{(x+1)^2}{16} - \frac{(y-3)^2}{9} = 1$$

Hyperbola

**61.** $y^2 + 10y = 6x - 1$

$y^2 + 10y + 25 = 6x - 1 + 25$

$(y + 5)^2 = 6x + 24$

$(y + 5)^2 = 6(x + 4)$

Parabola

## Challenge Problem

Vertex: $(0, 0)$

$4p = 8$ or $p = 2$

Focus: $(2, 0)$

$VF = 2$. For any point $A = (x, y)$ on the parabola,

$$AF = \sqrt{(x-2)^2 + y^2} = \sqrt{\left(\frac{y^2}{8} - 2\right)^2 + y^2}$$

$$= \frac{1}{8}\sqrt{(y^2 - 16)^2 + 64y^2}$$

$$= \frac{1}{8}\sqrt{y^4 - 32y^2 + 64y^2 + 256}$$

$$= \frac{1}{8}\sqrt{y^4 + 32y^2 + 256} = \frac{1}{8}\sqrt{(y^2 + 16)^2}$$

$$= \frac{y^2 + 16}{8} = \frac{y^2}{8} + 2 \geq 2$$

Since $AF \geq VF$, the vertex on the parabola is the closest to the focus.

## Critical Thinking

**1.**

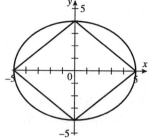

The four lines are:

$y = \frac{4}{5}(x - 5)$    $\frac{1}{5}x - \frac{1}{4}y = 1$

$y = \frac{4}{5}(x + 5)$    $-\frac{1}{5}x + \frac{1}{4}y = 1$

or

$y = -\frac{4}{5}(x - 5)$    $\frac{1}{5}x + \frac{1}{4}y = 1$

$y = -\frac{4}{5}(x + 5)$    $-\frac{1}{5}x - \frac{1}{4}y = 1$

We can combine these four equations into one as follows:

$$\pm\frac{1}{5}x \pm \frac{1}{4}y = 1 \text{ or } \left|\frac{x}{5}\right| + \left|\frac{y}{4}\right| = 1$$

**3.** You can obtain one graph from another by rotating it 90°.

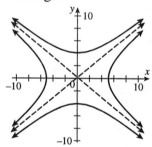

**5.** Consider the line $m$ which is a distance of $OB$ below the line $l$. Then the distance from $P$ to the line $m$ is the same as the distance from $P$ to $O$. Hence, the set of all points $P$ are equidistant from $O$ and $m$, and form a parabola. ($m$ is the directrix and $O$ the focus by construction.)

## Exercises 10.4

**1.**

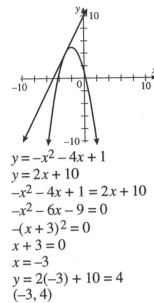

$y = -x^2 - 4x + 1$
$y = 2x + 10$
$-x^2 - 4x + 1 = 2x + 10$
$-x^2 - 6x - 9 = 0$
$-(x + 3)^2 = 0$
$x + 3 = 0$
$x = -3$
$y = 2(-3) + 10 = 4$
$(-3, 4)$

**3.**

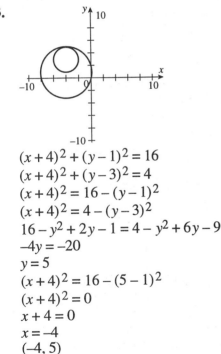

$(x + 4)^2 + (y - 1)^2 = 16$
$(x + 4)^2 + (y - 3)^2 = 4$
$(x + 4)^2 = 16 - (y - 1)^2$
$(x + 4)^2 = 4 - (y - 3)^2$
$16 - y^2 + 2y - 1 = 4 - y^2 + 6y - 9$
$-4y = -20$
$y = 5$
$(x + 4)^2 = 16 - (5 - 1)^2$
$(x + 4)^2 = 0$
$x + 4 = 0$
$x = -4$
$(-4, 5)$

**5.**

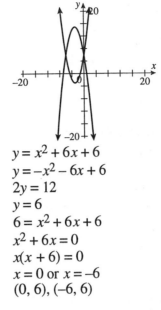

$y = x^2 + 6x + 6$
$y = -x^2 - 6x + 6$
$2y = 12$
$y = 6$
$6 = x^2 + 6x + 6$
$x^2 + 6x = 0$
$x(x + 6) = 0$
$x = 0 \text{ or } x = -6$
$(0, 6), (-6, 6)$

**7.**  $y = (x + 1)^2$
$y = (x - 1)^2$
$(x + 1)^2 = (x - 1)^2$
$x^2 + 2x + 1 = x^2 - 2x + 1$
$4x = 0$
$x = 0$
$y = (0 + 1)^2 = 1$
$(0, 1)$

**9.**  $y = x^2$
$y = -x^2 + 8x - 16$
$x^2 = -x^2 + 8x - 16$
$2x^2 - 8x + 16 = 0$
$2(x^2 - 4x + 8) = 0$
$x = \dfrac{4 \pm \sqrt{16 - 32}}{2}$, no solutions.

**11.**  $7x + 3y = 42$
$y = -3x^2 - 12x - 15$
$3y = 42 - 7x \text{ or } y = 14 - \dfrac{7}{3}x$
$14 - \dfrac{7}{3}x = -3x^2 - 12x - 15$
$3x^2 + \dfrac{29}{3}x + 29 = 0$
$9x^2 + 29x + 87 = 0$
$x = \dfrac{-29 \pm \sqrt{841 - 3132}}{18}$, no solutions.

**13.** $y - x = 0$
$(x-2)^2 + (y+5)^2 = 25$
$y = x$
$(x-2)^2 + (x+5)^2 = 25$
$x^2 - 4x + 4 + x^2 + 10x + 25 = 25$
$2x^2 + 6x + 4 = 0$
$2(x^2 + 3x + 2) = 0$
$2(x+2)(x+1) = 0$
$x + 2 = 0$ or $x + 1 = 0$
$x = -2$ or $x = -1$
$(-2, -2), (-1, -1)$

**15.** $x - 2y^2 = 0$
$x^2 - y^2 = 3$
$x = 2y^2$
$(2y^2)^2 - y^2 = 3$
$4y^4 - y^2 - 3 = 0$
$(4y^2 + 3)(y^2 - 1) = 0$

$4y^2 + 3 = 0 \qquad y^2 - 1 = 0$
$y^2 = -\dfrac{3}{4} \qquad y^2 = 1$

Not possible $\qquad y = \pm 1$
For $y = \pm 1$, $x = 2(\pm 1)^2 = 2$
$(2, -1), (2, 1)$

**17.** $4x^2 + y^2 = 4$
$2x - y = 2$
$y = 2x - 2$
$4x^2 + (2x-2)^2 = 4$
$4x^2 + 4x^2 - 8x + 4 = 4$
$8x^2 - 8x = 0$
$8x(x-1) = 0$
$x = 0$ or $x = 1$
For $x = 0$, $y = 2 \cdot 0 - 2 = -2$
For $x = 1$, $y = 2 \cdot 1 - 2 = 0$
$(0, -2), (1, 0)$

**19.** $4x^2 - 9y^2 = 36$
$9x^2 + 4y^2 = 36$
$x^2 = 9 + \dfrac{9}{4}y^2$

$9\left(9 + \dfrac{9}{4}y^2\right) + 4y^2 = 36$

$81 + \dfrac{81y^2}{4} + 4y^2 = 36$

$\dfrac{97}{4}y^2 = -45$
No solutions.

**21.** $2x^2 + y^2 = 11$
$x^2 - 2y^2 = -2$
$x^2 = 2y^2 - 2$
$2(2y^2 - 2) + y^2 = 11$
$4y^2 - 4 + y^2 = 11$
$5y^2 = 15$
$y^2 = 3$
$y = \pm\sqrt{3}$
For $y = \pm\sqrt{3}$, $y^2 = 3$ and $x^2 = 2 \cdot 3 - 2 = 4$
$x = \pm 2$
$(2, \sqrt{3}), (2, -\sqrt{3}), (-2, \sqrt{3}), (-2, -\sqrt{3})$

**23.** $y = -x^2 + 3$
$x^2 + \dfrac{(y-2)^2}{9} = 1$
$x^2 = -y + 3$
$-y + 3 + \dfrac{(y-2)^2}{9} = 1$
$-9y + 27 + y^2 - 4y + 4 = 9$
$y^2 - 13y + 22 = 0$
$(y-2)(y-11) = 0$
$y - 2 = 0$ or $y - 11 = 0$
$y = 2$ or $y = 11$
For $y = 11$, $x^2 = -11 + 3 = -8$,
No solutions.
For $y = 2$, $x^2 = -2 + 3 = 1$,
$x = \pm 1$
$(\pm 1, 2)$

**25.** $(x-1)^2 + y^2 = 1$
$x^2 + (y-1)^2 = 1$
$(x-1)^2 + y^2 = x^2 + (y-1)^2$
$x^2 - 2x + 1 + y^2 = x^2 + y^2 - 2y + 1$
$-2x = -2y$
$x = y$
$(x-1)^2 + x^2 = 1$
$x^2 - 2x + 1 + x^2 = 1$
$2x^2 - 2x = 0$
$2x(x-1) = 0$
$x = 0$ or $x = 1$
For $x = 0$, $y = 0$

For $x = 1$, $y = 1$
$(0, 0)$, $(1, 1)$

**27.** $y = \frac{1}{3}(x-3)^2 - 3$

$x^2 - 6x + y^2 + 2y = -6$

$y = \frac{1}{3}(x-3)^2 - 3$ or $\quad 3y = (x-3)^2 - 9$

$\qquad\qquad\qquad\qquad (x-3)^2 = 3y + 9$

$x^2 - 6x + y^2 + 2y = -6$ or

$\qquad\qquad x^2 - 6x + 9 = -y^2 - 2y - 6 + 9$

$\qquad\qquad\quad (x-3)^2 = -y^2 - 2y + 3$

$-y^2 - 2y + 3 = 3y + 9$

$-y^2 - 5y - 6 = 0$

$-(y+2)(y+3) = 0$

$y + 2 = 0$ or $y + 3 = 0$

$y = -2$ or $y = -3$

For $y = -2$, $(x-3)^2 = 3(-2) + 9 = 3$

$\qquad\qquad\quad x - 3 = \pm\sqrt{3}$

$\qquad\qquad\quad x = 3 \pm \sqrt{3}$

For $y = -3$, $(x-3)^2 = 3(-3) + 9 = 0$

$\qquad\qquad\quad x - 3 = 0$

$\qquad\qquad\quad x = 3$

$(3, -3)$, $(3 + \sqrt{3}, -2)$, $(3 - \sqrt{3}, -2)$

**29.**

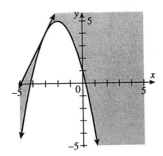

**31.**

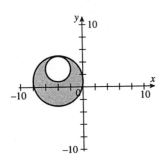

**33.**

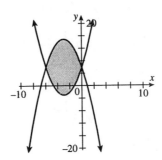

**35.**

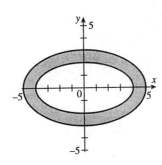

**37.**

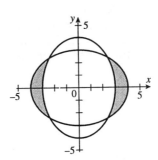

**39.**

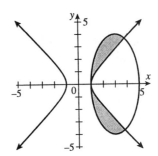

*Chapter 10: The Conic Sections*

**41.** $2x + 2y = 17$ or $x = \dfrac{17}{2} - y$

$xy + \dfrac{1}{2}x\left(\dfrac{1}{3}y\right) = 20 \quad xy + \dfrac{1}{6}xy = 20$

$\dfrac{7}{6}xy = 20$

$xy = \dfrac{120}{7}$

$\left(\dfrac{17}{2} - y\right)y = \dfrac{120}{7}$

$\dfrac{17}{2}y - y^2 = \dfrac{120}{7}$

$14y^2 - 119y + 240 = 0$

$y = \dfrac{119 \pm \sqrt{119^2 - 4(14)(240)}}{28}$

$= 5.2 \text{ or } 3.3$

For $y = 5.2$, $\quad x = \dfrac{120}{y(7)} = 3.3$

For $y = 3.3$, $\quad x = \dfrac{120}{y(7)} = 5.2$

Since $y > x$, $y = 5.2$ and $x = 3.3$. The base of the rectangle is 3.3 feet, the height is 5.2 feet, and the altitude of the triangle is 1.7 feet.

**43.** Area common to both ellipses:
$x^2 + 4y^2 \le 4$
$y^2 + 4x^2 \le 4$
Area inside the horizontal ellipse and outside the vertical:
$x^2 + 4y^2 \le 4$
$y^2 + 4x^2 \ge 4$
Area inside the vertical ellipse and outside the horizontal:
$x^2 + 4y^2 \ge 4$
$y^2 + 4x^2 \le 4$

## Challenge Problem

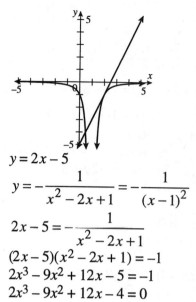

$y = 2x - 5$

$y = -\dfrac{1}{x^2 - 2x + 1} = -\dfrac{1}{(x-1)^2}$

$2x - 5 = -\dfrac{1}{x^2 - 2x + 1}$

$(2x - 5)(x^2 - 2x + 1) = -1$

$2x^3 - 9x^2 + 12x - 5 = -1$

$2x^3 - 9x^2 + 12x - 4 = 0$

Possible rational roots: $\pm 1, \pm\dfrac{1}{2}, \pm 2, \pm 4$.

$$
\begin{array}{r|rrrr}
2 & 2 & -9 & +12 & -4 \\
  &   & +4 & -10 & +4 \\
\hline
  & 2 & -5 & +2 & +0
\end{array}
$$

$2x^3 - 9x^2 + 12x - 4 = 0$

$(x - 2)(2x^2 - 5x + 2) = 0$

$(x - 2)(2x - 1)(x - 2) = 0$

$x - 2 = 0$ or $2x - 1 = 0$

$x = 2$ or $x = \dfrac{1}{2}$

For $x = 2$, $y = 2 \cdot 2 - 5 = -1$

For $x = \dfrac{1}{2}$, $y = 2 \cdot \dfrac{1}{2} - 5 = -4$

$(2, -1)$, $\left(\dfrac{1}{2}, -4\right)$

## Chapter 10 Review Exercises

**1.** An ellipse is the set of all points in a plane such that the sum of the distance from two points, called the foci, is a constant.

**3.**

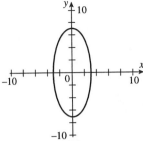

Center: (0, 0)
Vertices: (0, ±7)
Foci: $9 = 49 - c^2$
$\qquad c^2 = 40$
$\qquad c = 2\sqrt{10}$
$\qquad (0, \pm 2\sqrt{10})$

**5.**

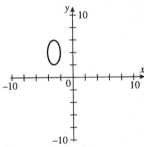

Center: (−3, 4)
Vertices: (−3, 2), (−3, 6)
Foci: $1 = 4 - c^2$
$\qquad c^2 = 3$
$\qquad c = \sqrt{3}$
$\qquad (-3, 4 \pm \sqrt{3})$

**7.**

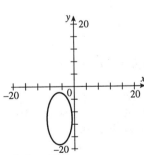

Center: (−5, −10)
Vertices: (−5, −18), (−5, −2)
Foci: $16 = 64 - c^2$
$\qquad c^2 = 48$
$\qquad c = 4\sqrt{3}$
$\qquad (-5, -10 \pm 4\sqrt{3})$

**9.** $4x^2 + 24x + 9y^2 - 18y = -9$
$\quad 4(x^2 + 6x) + 9(y^2 - 2y) = -9$
$\quad 4(x^2 + 6x + 9) + 9(y^2 - 2y + 1)$
$\quad = -9 + 36 + 9$
$\quad 4(x + 3)^2 + 9(y - 1)^2 = 36$
$$\frac{(x+3)^2}{9} + \frac{(y-1)^2}{4} = 1$$
Center: (−3, 1)
Vertices: (0, 1), (−6, 1)
Foci: $4 = 9 - c^2$
$\qquad c^2 = 5$
$\qquad c = \sqrt{5}$
$\qquad (-3 \pm \sqrt{5}, 1)$

**11.** $7y^2 + 10x^2 - 100x - 42y = -243$
$\quad 10(x^2 - 10x) + 7(y^2 - 6y) = -243$
$\quad 10(x^2 - 10x + 25) + 7(y^2 - 6y + 9)$
$\quad = -243 + 250 + 63$
$\quad 10(x - 5)^2 + 7(y - 3)^2 = 70$
$$\frac{(x-5)^2}{7} + \frac{(y-3)^2}{10} = 1$$
Center: (5, 3)
Vertices: $(5, 3 \pm \sqrt{10})$
Foci: $7 = 10 - c^2$
$\qquad c^2 = 3$
$\qquad c = \sqrt{3}$
$\qquad (5, 3 \pm \sqrt{3})$

**13.** Center: (0, 0)
$\quad a = 7, c = 6$
$\quad b^2 = a^2 - c^2 = 49 - 36 = 13$
$$\frac{x^2}{49} + \frac{y^2}{13} = 1$$

**15.** $a = 5$
$$\frac{(x-4)^2}{25} + \frac{(y-6)^2}{b^2} = 1$$
For $(x, y) = (0, 8)$
$$\frac{(-4)^2}{25} + \frac{(8-6)^2}{b^2} = 1$$
$$\frac{16}{25} + \frac{4}{b^2} = 1$$
$\quad 16b^2 + 100 = 25b^2$
$\quad 100 = 9b^2$

$$b^2 = \frac{100}{9}$$

$$\frac{(x-4)^2}{25} + \frac{(y-6)^2}{\frac{100}{9}} = 1 \text{ or}$$

$$\frac{(x-4)^2}{25} + \frac{9(y-6)^2}{100} = 1$$

**17.** Center: (0, 0)

$$c = 6, \frac{3}{4} = \frac{6}{a}, a = 8$$

$$b^2 = a^2 - c^2 = 64 - 36 = 28$$

$$\frac{x^2}{28} + \frac{y^2}{64} = 1$$

**19.** The set of such points is an ellipse, with center at (−1, 1).
$c = 2$ and $2a = 6$, so $a = 3$.
$$b^2 = a^2 - c^2 = 9 - 4 = 5$$
$$\frac{(x+1)^2}{9} + \frac{(y-1)^2}{5} = 1$$

**21.** Center: (0, 0)
$a - c = 4{,}400$
$a + c = 6000$
$2a = 10{,}400$ or $a = 5200$
$c = 6000 - 5200 = 800$
$$b^2 = a^2 - c^2 = (5200)^2 - (800)^2$$
$$\frac{x^2}{(5200)^2 - (800)^2} + \frac{y^2}{(5200)^2} = 1$$

**23.** A hyperbola is the set of all points in the plane such that the difference of the distances from two fixed points, called the foci, is a constant.

**25.**

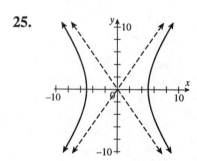

Center: (0, 0)
Vertices: (±5, 0)
Foci:  $c^2 = a^2 + b^2 = 25 + 49 = 74$
$$c = \sqrt{74}$$
$$(\pm\sqrt{74}, 0)$$
Asymptotes: $y = \pm\frac{7}{5}x$

**27.**

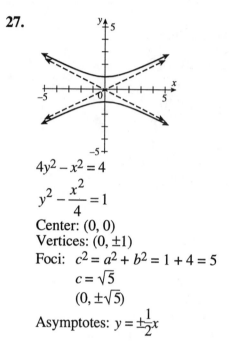

$$4y^2 - x^2 = 4$$
$$y^2 - \frac{x^2}{4} = 1$$
Center: (0, 0)
Vertices: (0, ±1)
Foci:  $c^2 = a^2 + b^2 = 1 + 4 = 5$
$$c = \sqrt{5}$$
$$(0, \pm\sqrt{5})$$
Asymptotes: $y = \pm\frac{1}{2}x$

**29.**

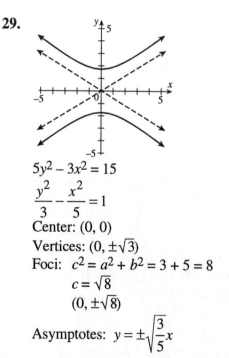

$$5y^2 - 3x^2 = 15$$
$$\frac{y^2}{3} - \frac{x^2}{5} = 1$$
Center: (0, 0)
Vertices: $(0, \pm\sqrt{3})$
Foci:  $c^2 = a^2 + b^2 = 3 + 5 = 8$
$$c = \sqrt{8}$$
$$(0, \pm\sqrt{8})$$

Asymptotes: $y = \pm\sqrt{\frac{3}{5}}x$

**31.**

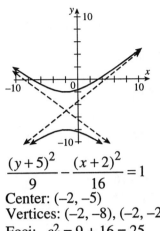

$$\frac{(y+5)^2}{9} - \frac{(x+2)^2}{16} = 1$$

Center: $(-2, -5)$

Vertices: $(-2, -8), (-2, -2)$

Foci: $c^2 = 9 + 16 = 25$

$\quad c = 5$

$\quad (-2, -10), (-2, 0)$

Asymptotes: $y + 5 = \pm\frac{3}{4}(x + 2)$

**33.**

$$49(x + 1)^2 - 4(y - 5)^2 = 196$$

$$\frac{(x+1)^2}{4} - \frac{(y-5)^2}{49} = 1$$

Center: $(-1, 5)$

Vertices: $(-3, 5), (1, 5)$

Foci: $c^2 = 4 + 49 = 53$

$\quad c = \sqrt{53}$

$\quad (-1 \pm \sqrt{53}, 5)$

Asymptotes: $y - 5 = \pm\frac{7}{2}(x + 1)$

**35.** $9y^2 - 36y - 4x^2 - 8x = 4$

$9(y^2 - 4y) - 4(x^2 + 2x) = 4$

$9(y^2 - 4y + 4) - 4(x^2 + 2x + 1) = 4 + 36 - 4$

$9(y - 2)^2 - 4(x + 1)^2 = 36$

$$\frac{(y-2)^2}{4} - \frac{(x+1)^2}{9} = 1$$

Center: $(-1, 2)$

Vertices: $(-1, 0), (-1, 4)$

Foci: $c^2 = 4 + 9 = 13$

$\quad c = \sqrt{13}$

$\quad (-1, 2 \pm \sqrt{13})$

Asymptotes: $y - 2 = \pm\frac{2}{3}(x + 1)$

**37.** $5y^2 - 20y - 4x^2 + 12x - 9 = 0$

$5(y^2 - 4y) - 4(x^2 - 3x) = 9$

$5(y^2 - 4y + 4) - 4\left(x^2 - 3x + \frac{9}{4}\right)$

$\quad = 9 + 20 - 9$

$5(y - 2)^2 - 4\left(x - \frac{3}{2}\right)^2 = 20$

$$\frac{(y-2)^2}{4} - \frac{\left(x - \frac{3}{2}\right)^2}{5} = 1$$

Center: $\left(\frac{3}{2}, 2\right)$

Vertices: $\left(\frac{3}{2}, 0\right), \left(\frac{3}{2}, 4\right)$

Foci: $c^2 = 4 + 5 = 9$

$\quad c = 3$

$\quad \left(\frac{3}{2}, -1\right), \left(\frac{3}{2}, 5\right)$

Asymptotes: $y - 2 = \pm\frac{2}{\sqrt{5}}\left(x - \frac{3}{2}\right)$

**39.** $a = \sqrt{10},\ c = 5$

$b^2 = c^2 - a^2 = 25 - 10 = 15$

$$\frac{y^2}{10} - \frac{x^2}{15} = 1$$

**41.** $a = \frac{1}{2}(8) = 4$

$b^2 = c^2 - a^2 = 20 - 16 = 4$

$$\frac{(y-2)^2}{16} - \frac{(x+3)^2}{4} = 1$$

**43.** Center: $(4, -2)$

$c = \sqrt{13}$

$\frac{a}{b} = \frac{2}{3}$ or $a = \frac{2}{3}b$

$c^2 = a^2 + b^2$

$$13 = a^2 + b^2$$
$$13 = \frac{4}{9}b^2 + b^2$$
$$13 = \frac{13}{9}b^2$$
$$b^2 = 9$$
$$b = 3, \ a = \frac{2}{3}(3) = 2$$
$$\frac{(y+2)^2}{4} - \frac{(x-4)^2}{9} = 1$$

**45.** From Exercise 44, $P$ satisfies
$$\frac{144y^2}{25} - \frac{144x^2}{119} = 1$$
Also,
$$\sqrt{(x-3)^2 + (y+1)^2} = \sqrt{x^2 + (y+1)^2}$$
$$(x-3)^2 + (y+1)^2 = x^2 + (y+1)^2$$
$$x^2 - 6x + 9 = x^2$$
$$-6x + 9 = 0$$
$$-6x = -9$$
$$x = \frac{3}{2}$$

$$\frac{144y^2}{25} - \frac{144\left(\frac{9}{4}\right)}{119} = 1$$
$$y^2 = 0.6463$$
$$y = \pm 0.804$$
$y = -0.804$ since $(x, y)$ is closer to $(0, -1)$ than $(0, 1)$. $P = (1.5, -0.804)$

**47.** $x^2 = 6y$
Vertex: $(0, 0)$
$4p = 6$ or $p = \frac{3}{2}$
Focus: $\left(0, \frac{3}{2}\right)$
Directrix: $y = -\frac{3}{2}$

**49.** $x^2 = -2y$
$4p = -2$ or $p = -\frac{1}{2}$
Vertex: $(0, 0)$
Focus: $\left(0, -\frac{1}{2}\right)$
Directrix: $y = \frac{1}{2}$

**51.** $y - 3 = 2x^2$
$x^2 = \frac{1}{2}(y - 3)$
$4p = \frac{1}{2}$ or $p = \frac{1}{8}$
Vertex: $(0, 3)$
Focus: $\left(0, \frac{25}{8}\right)$
Directrix: $y = \frac{23}{8}$

**53.** $y = 12x^2 - 24x + 16$
$12(x^2 - 2x) = y - 16$
$12(x^2 - 2x + 1) = y - 16 + 12$
$12(x - 1)^2 = y - 4$
$(x - 1)^2 = \frac{1}{12}(y - 4)$
$4p = \frac{1}{12}$ or $p = \frac{1}{48}$
Vertex: $(1, 4)$
Focus: $\left(1, \frac{193}{48}\right)$
Directrix: $y = \frac{191}{48}$

**55.** $6y^2 + 36y + 5x + 29 = 0$
$6(y^2 + 6y) = -5x - 29$
$6(y^2 + 6y + 9) = -5x - 29 + 54$
$6(y + 3)^2 = -5x + 25$
$(y + 3)^2 = -\frac{5}{6}(x - 5)$
$4p = -\frac{5}{6}$ or $p = -\frac{5}{24}$
Vertex: $(5, -3)$
Focus: $\left(\frac{115}{24}, -3\right)$
Directrix: $x = \frac{125}{24}$

**57.** Vertex: $(0, 8)$
$p = -1$
$x^2 = -4(y - 8)$

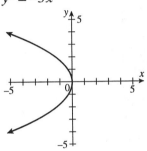

**59.** $p = -\dfrac{3}{4}$

$y^2 = -4\left(\dfrac{3}{4}\right)x$

$y^2 = -3x$

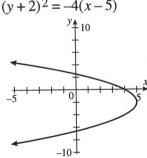

**61.** Vertex: $(5, -2)$
$p = -1$
$(y + 2)^2 = -4(x - 5)$

**63.** $p = 3$
$(x - 6)^2 = 12(y + 1)$

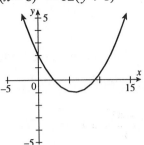

**65.** $(y - 6)^2 = 4p(x + 2)$
at $(-10, 2)$,
$(2 - 6)^2 = 4p(-10 + 2)$
$16 = -32p$
$p = -\dfrac{1}{2}$
$(y - 6)^2 = -2(x + 2)$

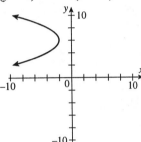

**67.** $-1 = a + b + c$
$35 = 9a + 3b + c$
$20 = 4a - 2b + c$

$$\begin{bmatrix} 1 & 1 & 1 & : & -1 \\ 9 & 3 & 1 & : & 35 \\ 4 & -2 & 1 & : & 20 \end{bmatrix}$$

$$\begin{bmatrix} 1 & 1 & 1 & : & -1 \\ 0 & -6 & -8 & : & 44 \\ 0 & -6 & -3 & : & 24 \end{bmatrix} \begin{matrix} \\ \leftarrow -9 \times R_1 + R_2 \\ \leftarrow -4 \times R_1 + R_3 \end{matrix}$$

$$\begin{bmatrix} 1 & 1 & 1 & : & -1 \\ 0 & -6 & -8 & : & 44 \\ 0 & 0 & 5 & : & -20 \end{bmatrix} \begin{matrix} \\ \\ \leftarrow -1 \times R_2 + R_3 \end{matrix}$$

$5c = -20$ or $c = -4$
$-6b - 8(-4) = 44$
$-6b = 12$
$b = -2$
$a + (-2) + (-4) = -1$
$a = 5$
$x = 5y^2 - 2y - 4$

**69.** $y = \frac{1}{14}x^2$ or $x^2 = 14y$

$4p = 14$

$p = \frac{7}{2} = 3.5$

The receiver is 3.5 feet from the bottom of the antenna. The depth is $\frac{25}{14}$ or $1\frac{11}{14}$ feet.

**71.** $\frac{x^2}{9} + \frac{y^2}{25} = 1$

Ellipse

Center: (0, 0)

Vertices: (0, ±5)

$9 = 25 - c^2$

$c^2 = 16$

$c = 4$

Foci: (0, ±4)

**73.** $y^2 = 6(x + 2)$

$4p = 6,\ p = \frac{3}{2}$

Parabola

Vertex: (−2, 0)

Focus: $\left(-\frac{1}{2},\ 0\right)$

Directrix: $x = -\frac{7}{2}$

**75.** $\frac{y^2}{16} - \frac{(x+3)^2}{9} = 1$

Hyperbola

Center: (−3, 0)

Vertices: (−3, ±4)

Foci:  $c^2 = 9 + 16 = 25$

$c = 5$

(−3, ±5)

Asymptotes: $y = \pm\frac{4}{3}(x + 3)$

**77.** $y = \frac{1}{8}x^2 + 3$

$y - 3 = \frac{1}{8}x^2$

$x^2 = 8(y - 3)$

Parabola

**79.** $9x^2 - 36x + 4y^2 - 24y + 36 = 0$

$9(x^2 - 4x) + 4(y^2 - 6y) = -36$

$9(x^2 - 4x + 4) + 4(y^2 - 6y + 9)$

$= -36 + 36 + 36$

$9(x - 2)^2 + 4(y - 3)^2 = 36$

$\frac{(x-2)^2}{4} + \frac{(y-3)^2}{9} = 1$

Ellipse

**81.**

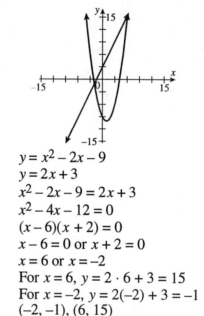

$y = x^2 - 2x - 9$

$y = 2x + 3$

$x^2 - 2x - 9 = 2x + 3$

$x^2 - 4x - 12 = 0$

$(x - 6)(x + 2) = 0$

$x - 6 = 0$ or $x + 2 = 0$

$x = 6$ or $x = -2$

For $x = 6$, $y = 2 \cdot 6 + 3 = 15$

For $x = -2$, $y = 2(-2) + 3 = -1$

(−2, −1), (6, 15)

**83.**

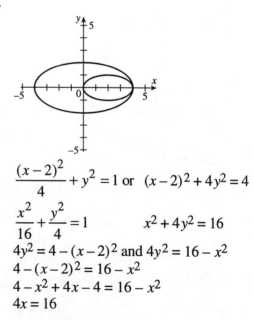

$\frac{(x-2)^2}{4} + y^2 = 1$ or $(x - 2)^2 + 4y^2 = 4$

$\frac{x^2}{16} + \frac{y^2}{4} = 1$ \qquad $x^2 + 4y^2 = 16$

$4y^2 = 4 - (x - 2)^2$ and $4y^2 = 16 - x^2$

$4 - (x - 2)^2 = 16 - x^2$

$4 - x^2 + 4x - 4 = 16 - x^2$

$4x = 16$

$x = 4$
$4^2 + 4y^2 = 16$
$4y^2 = 0$
$y = 0$
$(4, 0)$

**85.**

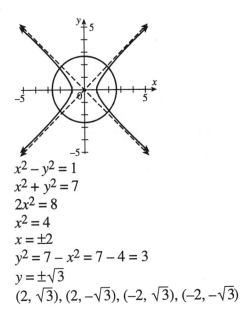

$x^2 - y^2 = 1$
$x^2 + y^2 = 7$
$2x^2 = 8$
$x^2 = 4$
$x = \pm 2$
$y^2 = 7 - x^2 = 7 - 4 = 3$
$y = \pm\sqrt{3}$
$(2, \sqrt{3}), (2, -\sqrt{3}), (-2, \sqrt{3}), (-2, -\sqrt{3})$

**87.** $y = -x^2 + 8x - 10$
$y + 2x = 6$ or $y = -2x + 6$
$-x^2 + 8x - 10 = -2x + 6$
$-x^2 + 10x - 16 = 0$
$-(x - 8)(x - 2) = 0$
$x - 8 = 0$ or $x - 2 = 0$
$x = 8$ or $x = 2$
For $x = 8$, $y = -2 \cdot 8 + 6 = -10$
For $x = 2$, $y = -2 \cdot 2 + 6 = 2$
$(2, 2), (8, -10)$

**89.** $\dfrac{x^2}{5} - \dfrac{y^2}{9} = 1$
$5y - 6x = 0$
$y = \dfrac{6}{5}x$

$\dfrac{x^2}{5} - \dfrac{\left(\frac{6}{5}x\right)^2}{9} = 1$

$\dfrac{x^2}{5} - \dfrac{36x^2}{225} = 1$

$\dfrac{9x^2}{225} = 1$

$\dfrac{x^2}{25} = 1$

$x^2 = 25$
$x = \pm 5$

For $x = 5$, $y = \dfrac{6}{5}(5) = 6$

For $x = -5$, $y = \dfrac{6}{5}(-5) = -6$

$(5, 6), (-5, -6)$

**91.** $(x - 4)^2 + (y - 6)^2 = 4$ or
$(x - 4)^2 = 4 - (y - 6)^2$
$y = -\dfrac{1}{2}(x - 4)^2 + 8$ or $(x - 4)^2 = -2y + 16$
$4 - (y - 6)^2 = -2y + 16$
$4 - y^2 + 12y - 36 = -2y + 16$
$-y^2 + 14y - 48 = 0$
$-(y - 6)(y - 8) = 0$
$y - 6 = 0$ or $y - 8 = 0$
$y = 6$ or $y = 8$
For $y = 6$, $(x - 4)^2 = -2 \cdot 6 + 16 = 4$
$\qquad\qquad x - 4 = \pm 2$
$\qquad\qquad x = 6$ or $x = 2$
For $y = 8$, $(x - 4)^2 = -2 \cdot 8 + 16 = 0$
$\qquad\qquad x - 4 = 0$
$\qquad\qquad x = 4$
$(4, 8), (2, 6), (6, 6)$

**93.** $3x^2 + y^2 = 12$ or $\quad 3x^2 + y^2 = 12$
$\quad x^2 - 3y^2 = 4 \qquad\quad -3x^2 + 9y^2 = -12$
$\qquad\qquad\qquad\qquad\qquad\quad 10y^2 = 0$
$\qquad\qquad\qquad\qquad\qquad\quad y = 0$

$3x^2 = 12$
$x^2 = 4$
$x = \pm 2$
$(\pm 2, 0)$

**95.** $x^2 + 4y^2 = 4$
$4x^2 + y^2 = 4$
$-3x^2 + 3y^2 = 0$
$3y^2 = 3x^2$
$y^2 = x^2$
$y = \pm x$
$x^2 + 4x^2 = 4$
$5x^2 = 4$

$$x^2 = \frac{4}{5}$$

$$x = \pm\sqrt{\frac{4}{5}}$$

$$\left(\pm\sqrt{\frac{4}{5}}, \ \pm\sqrt{\frac{4}{5}}\right)$$

**97.** $y = x^2 - 4x + 4 = (x-2)^2$
$(x-2)^2 + (y-4)^2 = 16$
$y + (y-4)^2 = 16$
$y + y^2 - 8y + 16 = 16$
$y^2 - 7y = 0$
$y(y - 7) = 0$
$y = 0$ or $y = 7$
For $y = 0$, $\quad (x-2)^2 = 0$
$\qquad\qquad\qquad x = 2$
For $y = 7$, $\quad (x-2)^2 = 7$
$\qquad\qquad\quad x - 2 = \pm\sqrt{7}$
$\qquad\qquad\qquad x = 2 \pm \sqrt{7}$
$(2, 0), (2 \pm \sqrt{7}, 7)$

**99.** $x^2 - 12x + y^2 + 4y = -15$
$x + 7y = 17$ or $x = -7y + 17$
$(-7y + 17)^2 - 12(-7y + 17) + y^2 + 4y$
$= -15$
$49y^2 - 238y + 289 + 84y - 204 + y^2 + 4y$
$= -15$
$50y^2 - 150y + 100 = 0$
$50(y - 1)(y - 2) = 0$
$y - 1 = 0$ or $y - 2 = 0$
$y = 1$ or $y = 2$
For $y = 1$, $x = -7 \cdot 1 + 17 = 10$
For $y = 2$, $x = -7 \cdot 2 + 17 = 3$
$(3, 2), (10, 1)$

**101.**

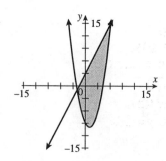

**103.**

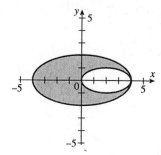

**105.**

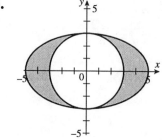

## Chapter 10 Test: Standard Answer

**1.**

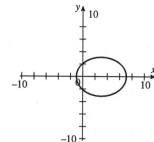

Center: $(3, 0)$
Vertices: $(-1, 0), (7, 0)$
Foci: $\quad c^2 = 16 - 9 = 7$
$\qquad\quad c = \sqrt{7}$
$\qquad\quad (3 \pm \sqrt{7}, 0)$

**2.**

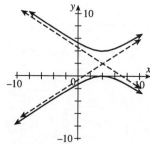

Center: $(4, 2)$
Vertices: $(4, 0), (4, 4)$

Foci: $c^2 = 4 + 9 = 13$

$\qquad c = \sqrt{13}$

$\qquad (4, 2 \pm \sqrt{13})$

Asymptotes: $y - 2 = \pm\dfrac{2}{3}(x - 4)$

**3.**

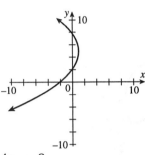

$4p = -8$

$p = -2$

Vertex: $(1, 5)$

Focus: $(-1, 5)$

Directrix: $x = 3$

**4.** $a = 4, b = 3$

$\dfrac{x^2}{16} + \dfrac{y^2}{9} = 1$

**5.** Center: $(-1, 2)$

$a = 3, c = \sqrt{5}$

$b^2 = a^2 - c^2 = 9 - 5 = 4$

$\dfrac{(x+1)^2}{4} + \dfrac{(y-2)^2}{9} = 1$

**6.** $a = 6, \dfrac{2}{3} = \dfrac{c}{6}, c = 4$

$b^2 = a^2 - c^2 = 36 - 16 = 20$

$\dfrac{(x+2)^2}{36} + \dfrac{(y-4)^2}{20} = 1$

**7.** $a = 6, c = 8$

$b^2 = c^2 - a^2 = 64 - 36 = 28$

$\dfrac{x^2}{36} - \dfrac{y^2}{28} = 1$

**8.** $c = 5, a = \dfrac{1}{2}(8) = 4$

$b^2 = c^2 - a^2 = 25 - 16 = 9$

$\dfrac{(y+5)^2}{16} - \dfrac{(x-4)^2}{9} = 1$

**9.** $c = \sqrt{40}, \dfrac{b}{a} = \dfrac{1}{3}, a = 3b$

$b^2 = c^2 - a^2 = 40 - 9b^2$

$10b^2 = 40$

$b^2 = 4$

$\dfrac{x^2}{36} - \dfrac{(y-1)^2}{4} = 1$

**10.** $p = -\dfrac{2}{3}$

$x^2 = -4\left(\dfrac{2}{3}\right)y$

$x^2 = -\dfrac{8}{3}y$

**11.** $p = 3$

$(y + 3)^2 = 4 \cdot 3(x + 5)$

$(y + 3)^2 = 12(x + 5)$

**12.** $(x - 2)^2 = 4p(y + 1)$

$(5 - 2)^2 = 4p(8 + 1)$

$9 = 36p$

$p = \dfrac{1}{4}$

$(x - 2)^2 = (y + 1)$

**13.** $y^2 - 9x^2 = 9$

$\dfrac{y^2}{9} - x^2 = 1$

Hyperbola

Center: $(0, 0)$

Vertices: $(0, \pm3)$

Foci: $c^2 = 9 + 1 = 10$

$\qquad c = \sqrt{10}$

$\qquad (0, \pm\sqrt{10})$

Asymptotes: $y = \pm3x$

**14.** $9(x+4)^2 + 16(y-1)^2 = 144$

$$\frac{(x+4)^2}{16} + \frac{(y-1)^2}{9} = 1$$

Ellipse
Center: $(-4, 1)$
Vertices: $(-8, 1), (0, 1)$
Foci: $c^2 = 16 - 9 = 7$
$\qquad c = \sqrt{7}$
$\qquad (-4 \pm \sqrt{7}, 1)$

**15.** $(x+5)^2 = -12(y-5)$
Parabola
Vertex: $(-5, 5)$
Focus: $4p = -12$
$\qquad p = -3$
$\qquad (-5, 2)$
Directrix: $y = 8$

**16.** $4x^2 - 16x - 9y^2 = 20$
$4(x^2 - 4x) - 9y^2 = 20$
$4(x^2 - 4x + 4) - 9y^2 = 20 + 16$
$4(x-2)^2 - 9y^2 = 36$

$$\frac{(x-2)^2}{9} - \frac{y^2}{4} = 1$$

Hyperbola

**17.** $y = 2x^2 - 4x + 9$
$2(x^2 - 2x) = y - 9$
$2(x^2 - 2x + 1) = y - 9 + 2$
$2(x-1)^2 = y - 7$
$(x-1)^2 = \frac{1}{2}(y - 7)$

Parabola

**18.** $9x^2 - 18x + 4y^2 + 16y = 11$
$9(x^2 - 2x) + 4(y^2 + 4y) = 11$
$9(x^2 - 2x + 1) + 4(y^2 + 4y + 4)$
$\quad = 11 + 9 + 16$
$9(x-1)^2 + 4(y+2)^2 = 36$

$$\frac{(x-1)^2}{4} + \frac{(y+2)^2}{9} = 1$$

Ellipse

**19.** $4x^2 + 16x - y^2 + 6y = -3$
$4(x^2 + 4x) - (y^2 - 6y) = -3$
$4(x^2 + 4x + 4) - (y^2 - 6y + 9) = -3 + 16 - 9$
$4(x+2)^2 - (y-3)^2 = 4$

$$\frac{(x+2)^2}{1} - \frac{(y-3)^2}{4} = 1$$

Hyperbola

**20.**

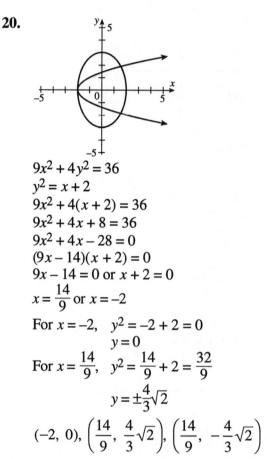

$9x^2 + 4y^2 = 36$
$y^2 = x + 2$
$9x^2 + 4(x + 2) = 36$
$9x^2 + 4x + 8 = 36$
$9x^2 + 4x - 28 = 0$
$(9x - 14)(x + 2) = 0$
$9x - 14 = 0$ or $x + 2 = 0$
$x = \dfrac{14}{9}$ or $x = -2$

For $x = -2$, $\quad y^2 = -2 + 2 = 0$
$\qquad\qquad\qquad y = 0$

For $x = \dfrac{14}{9}$, $\quad y^2 = \dfrac{14}{9} + 2 = \dfrac{32}{9}$

$\qquad\qquad\qquad y = \pm\dfrac{4}{3}\sqrt{2}$

$(-2, 0), \left(\dfrac{14}{9}, \dfrac{4}{3}\sqrt{2}\right), \left(\dfrac{14}{9}, -\dfrac{4}{3}\sqrt{2}\right)$

**21.** $2x^2 - 3y^2 = 15$ or $\quad -6x^2 + 9y^2 = -45$
$\quad\ 3x^2 + 2y^2 = 29 \qquad\quad 6x^2 + 4y^2 = 58$
$\qquad\qquad\qquad\qquad\qquad\quad 13y^2 = 13$
$\qquad\qquad\qquad\qquad\qquad\qquad y^2 = 1$
$\qquad\qquad\qquad\qquad\qquad\qquad y = \pm 1$

$2x^2 - 3(1) = 15$
$2x^2 = 18$
$x^2 = 9$
$x = \pm 3$
$(\pm 3, \pm 1)$

**22.**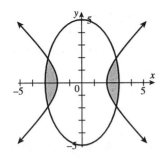

**23.** $a = 25$, $b = 20$.
The equation of the ellipse is
$$\frac{x^2}{625} + \frac{y^2}{400} = 1$$
For $x = \pm 15$,
$$\frac{225}{625} + \frac{y^2}{400} = 1$$
$$\frac{y^2}{400} = 1 - \frac{225}{625} = \frac{400}{625} = \frac{16}{25}$$
$$y^2 = 256$$
$$y = \pm 16$$
The clearance is 16 feet.

**24.** $2a = 8000 + 800 + 200 = 9000$
$a = 4500$
$c = 4500 - 4000 - 200 = 300$
$$e = \frac{300}{4500} = \frac{1}{15}$$

**25.** Let $s_1$ be the distance from $(x, y)$ to $A$ and let $s_2$ be the distance from $(x, y)$ to $B$.
Then, $\dfrac{s_2}{1100} - \dfrac{s_1}{1100} = 3$ or
$s_2 - s_1 = 3300$ feet. Since the difference in the distances is a constant the equation will be a hyperbola with center $(0, 0)$ and $c = 1.5$. Also,
$$2a = \frac{3300}{5280} = \frac{5}{8}$$
$$a = \frac{5}{16}$$
$$\frac{x^2}{\frac{25}{256}} - \frac{y^2}{\frac{9}{4} - \frac{25}{256}} = 1$$

$$\frac{x^2}{\frac{25}{256}} - \frac{y^2}{\frac{551}{256}} = 1$$
$$\frac{256x^2}{25} - \frac{256y^2}{551} = 1$$

## Chapter 10 Test: Multiple Choice

**1.** $c = 3$, $a = 5$, Center: $(0, 0)$
$b^2 = a^2 - c^2 = 25 - 9 = 16$
$b = 4$
$$\frac{x^2}{16} + \frac{y^2}{25} = 1$$
The answer is (c).

**2.** $\dfrac{x^2}{10} - \dfrac{y^2}{15} = 1$ is a hyperbola with foci
$(\pm 5, 0)$ since $c^2 = 10 + 15 = 25$.
The answer is (b).

**3.** $y^2 = -\dfrac{1}{2}x$
$4p = -\dfrac{1}{2}$
$p = -\dfrac{1}{8}$
Focus: $\left(-\dfrac{1}{8}, 0\right)$
Directrix: $x = \dfrac{1}{8}$
The answer is (a).

**4.** $c^2 = a^2 - b^2 = 36 - 4 = 32$
The foci are $(\pm\sqrt{32}, 0)$, so the answer is (d).

**5.** $25y^2 - 4x^2 = 100$
$$\frac{y^2}{4} - \frac{x^2}{25} = 1$$
Asymptotes: $y = \pm\dfrac{2}{5}x$
The answer is (b).

**6.** $a = 4$, $b = 3$,
$c^2 = a^2 - b^2 = 16 - 9 = 7$
$c = \sqrt{7}$

$$e = \frac{\sqrt{7}}{4}$$

The answer is (e).

7. Center: $(1, 0)$, $p = -1$

$$y^2 = -4(x-1)$$

The answer is (a).

8. $b = \frac{1}{2}(4) = 2$, $c = 1$

$$a^2 = b^2 + c^2 = 4 + 1 = 5$$

$$\frac{(x+3)^2}{5} + \frac{(y-2)^2}{4} = 1$$

The answer is (c).

9. $a = 3$, $c = \sqrt{13}$

$$b^2 = c^2 - a^2 = 13 - 9 = 4$$

$$\frac{(x+2)^2}{9} - \frac{(y-1)^2}{4} = 1$$

The answer is (d).

10. Center: $(-4, 2)$

Vertices: $(-4, 2 \pm \sqrt{7})$

The answer is (b).

11. The vertex is $\left(2, -\frac{7}{2}\right)$, so

the answer is (b).

12. $y^2 - 2x^2 = 16$

$y - x = 2$ or $y = x + 2$

$(x+2)^2 - 2x^2 = 16$

$x^2 + 4x + 4 - 2x^2 = 16$

$-x^2 + 4x - 12 = 0$

$$x = \frac{-4 \pm \sqrt{16 - 48}}{-2}$$

No solutions.

The answer is (a).

13. $a = 3$, $b = 2$, Center: $(2, -1)$

$$c^2 = a^2 - b^2 = 9 - 4 = 5$$

$$c = \sqrt{5}$$

Foci: $(2 \pm \sqrt{5}, -1)$

The answer is (a).

14. $x^2 - 10x + 13 = -12y$

$x^2 - 10x + 25 = -12y - 13 + 25$

$(x-5)^2 = -12(y-1)$

Vertex: $(5, 1)$

$4p = -12$, $p = -3$

Focus: $(5, -2)$

The answer is (c).

15. $x^2 - 4y^2 + 6x + 16y - 11 = 0$

$x^2 + 6x - 4(y^2 - 4y) = 11$

$x^2 + 6x + 9 - 4(y^2 - 4y + 4) = 11 + 9 - 16$

$(x+3)^2 - 4(y-2)^2 = 4$

$$\frac{(x+3)^2}{4} - \frac{(y-2)^2}{1} = 1$$

This is a hyperbola with a horizontal transverse axis.

The answer is (d).

16. Vertex: $(5, 2)$

$a = 8$, $\frac{b}{8} = \frac{3}{4}$, $b = 6$

$c^2 = a^2 + b^2 = 64 + 36 = 100$

$c = 10$

Focus: $(15, 2)$, $(-5, 2)$

The answer is (a).

17. $P$ is on a parabola with directrix $y = -4$ and focus $(3, 0)$.

Vertex: $(3, -2)$

$p = 2$

$(x-3)^2 = 8(y+2)$ or $y + 2 = \frac{1}{8}(x-3)^2$

The answer is (b).

18. $y^2 - x^2 = 4$ or $y^2 = x^2 + 4$

$8x^2 + 24y^2 = 192$

$8x^2 + 24(x^2 + 4) = 192$

$32x^2 = 96$

$x^2 = 3$

$x = \pm\sqrt{3}$

$y^2 = 3 + 4 = 7$

$y = \pm\sqrt{7}$

There are four solutions, so the answer is (d).

19. $y = \frac{1}{18}x^2$

$x^2 = 18y$

$4p = 18$

$p = 4.5$

The answer is (c).

**20.** $x^2 + y^2 = 97$

$3x + x - y + 3y = 44$ or $\quad 4x + 2y = 44$

$\hspace{9cm} 2x + y = 22$

The answer is (d).

# Chapter 11: Sequences and Series

## Exercises 11.1

**1.** $a_1 = 2 \cdot 1 - 1 = 1$
$a_2 = 2 \cdot 2 - 1 = 3$
$a_3 = 2 \cdot 3 - 1 = 5$
$a_4 = 2 \cdot 4 - 1 = 7$
$a_5 = 2 \cdot 5 - 1 = 9$
$1, 3, 5, 7, 9$

**3.** $a_1 = (-1)^1 = -1$
$a_2 = (-1)^2 = 1$
$a_3 = (-1)^3 = -1$
$a_4 = (-1)^4 = 1$
$a_5 = (-1)^5 = -1$
$-1, 1, -1, 1, -1$

**5.** $b_1 = 8\left(-\dfrac{1}{2}\right)^1 = -4$

$b_2 = 8\left(-\dfrac{1}{2}\right)^2 = +2$

$b_3 = 8\left(-\dfrac{1}{2}\right)^3 = -1$

$b_4 = 8\left(-\dfrac{1}{2}\right)^4 = \dfrac{1}{2}$

$b_5 = 8\left(-\dfrac{1}{2}\right)^5 = -\dfrac{1}{4}$

$-4, 2, -1, \dfrac{1}{2}, -\dfrac{1}{4}$

**7.** $c_1 = (-1)^1 1^2 = -1$
$c_2 = (-1)^2 2^2 = 4$
$c_3 = (-1)^3 3^2 = -9$
$c_4 = (-1)^4 4^2 = 16$
$-1, 4, -9, 16$

**9.** $c_1 = 3\left(\dfrac{1}{10}\right)^1 = \dfrac{3}{10}$

$c_2 = 3\left(\dfrac{1}{10}\right)^2 = \dfrac{3}{100}$

$c_3 = 3\left(\dfrac{1}{10}\right)^3 = \dfrac{3}{1000}$

$c_4 = 3\left(\dfrac{1}{10}\right)^4 = \dfrac{3}{10,000}$

$\dfrac{3}{10}, \dfrac{3}{100}, \dfrac{3}{1000}, \dfrac{3}{10,000}$

**11.** $a_1 = 3\left(\dfrac{1}{10}\right)^{2 \cdot 1} = \dfrac{3}{100}$

$a_2 = 3\left(\dfrac{1}{10}\right)^{2 \cdot 2} = \dfrac{3}{10,000}$

$a_3 = 3\left(\dfrac{1}{10}\right)^{2 \cdot 3} = \dfrac{3}{1,000,000}$

$a_4 = 3\left(\dfrac{1}{10}\right)^{2 \cdot 4} = \dfrac{3}{100,000,000}$

$\dfrac{3}{100}, \dfrac{3}{10,000}, \dfrac{3}{1,000,000}, \dfrac{3}{100,000,000}$

**13.** $c_1 = \dfrac{1}{1} - \dfrac{1}{2} = \dfrac{1}{2}$

$c_2 = \dfrac{1}{2} - \dfrac{1}{3} = \dfrac{1}{6}$

$c_3 = \dfrac{1}{3} - \dfrac{1}{4} = \dfrac{1}{12}$

$c_4 = \dfrac{1}{4} - \dfrac{1}{5} = \dfrac{1}{20}$

$\dfrac{1}{2}, \dfrac{1}{6}, \dfrac{1}{12}, \dfrac{1}{20}$

**15.** $a_1 = (2 \cdot 1 - 10)^2 = 64$
$a_2 = (2 \cdot 2 - 10)^2 = 36$
$a_3 = (2 \cdot 3 - 10)^2 = 16$
$a_4 = (2 \cdot 4 - 10)^2 = 4$
$64, 36, 16, 4$

**17.** $a_1 = -2 + (1 - 1)(3) = -2$
$a_2 = -2 + (2 - 1)(3) = 1$
$a_3 = -2 + (3 - 1)(3) = 4$
$a_4 = -2 + (4 - 1)(3) = 7$
$-2, 1, 4, 7$

**19.** $b_1 = \dfrac{1 - 1}{1 + 1} = 0$

$b_2 = \dfrac{2 - 1}{2 + 1} = \dfrac{1}{3}$

$$b_3 = \frac{3-1}{3+1} = \frac{2}{4} = \frac{1}{2}$$

$$b_4 = \frac{4-1}{4+1} = \frac{3}{5}$$

$$0, \frac{1}{3}, \frac{1}{2}, \frac{3}{5}$$

**21.** $b_1 = \left(1 + \frac{1}{1}\right)^{1-1} = 1$

$$b_2 = \left(1 + \frac{1}{2}\right)^{2-1} = \frac{3}{2}$$

$$b_3 = \left(1 + \frac{1}{3}\right)^{3-1} = \frac{16}{9}$$

$$b_4 = \left(1 + \frac{1}{4}\right)^{4-1} = \frac{125}{64}$$

$$1, \frac{3}{2}, \frac{16}{9}, \frac{125}{64}$$

**23.** $u_1 = -2\left(\frac{3}{4}\right)^{1-1} = -2$

$$u_2 = -2\left(\frac{3}{4}\right)^{2-1} = -\frac{3}{2}$$

$$u_3 = -2\left(\frac{3}{4}\right)^{3-1} = -\frac{9}{8}$$

$$u_4 = -2\left(\frac{3}{4}\right)^{4-1} = -\frac{27}{32}$$

$$-2, -\frac{3}{2}, -\frac{9}{8}, -\frac{27}{32}$$

**25.** $x_1 = \frac{1}{2^1} = \frac{1}{2}$

$$x_2 = \frac{2}{2^2} = \frac{1}{2}$$

$$x_3 = \frac{3}{2^3} = \frac{3}{8}$$

$$x_4 = \frac{4}{2^4} = \frac{1}{4}$$

$$\frac{1}{2}, \frac{1}{2}, \frac{3}{8}, \frac{1}{4}$$

**27.** $x_1 = \frac{1}{2} - \frac{2}{1} = -\frac{3}{2}$

$$x_2 = \frac{2}{3} - \frac{3}{2} = -\frac{5}{6}$$

$$x_3 = \frac{3}{4} - \frac{4}{3} = -\frac{7}{12}$$

$$x_4 = \frac{4}{5} - \frac{5}{4} = -\frac{9}{20}$$

$$-\frac{3}{2}, -\frac{5}{6}, -\frac{7}{12}, -\frac{9}{20}$$

**29.** $y_1 = 4$
$y_2 = 4$
$y_3 = 4$
$y_4 = 4$
$4, 4, 4, 4$

**31.** $\frac{1}{2}(1 + 3^{6-1}) = \frac{1}{2}(1 + 243) = 122$

**33.** $a_7 = 3(0.1)^{7-1} = 0.000003$

**35.** $a_{12} = 12$

**37.** $a_{12} = (1 - 12)^3 = (-11)^3 = -1331$

**39.** $4, 6, 8, 10$

**41.** $5, 10, 15, 20, 25$
$S_n = 5n$

**43.** $-5, 25, -125, 625, -3125$
$S_n = (-5)^n$

**45.** $a_5 = 10 + 5 = 15$
$a_6 = 15 + 6 = 21$
$a_7 = 21 + 7 = 28$
$15, 21, 28$

**47.** $a_1 = 12$

$$a_2 = -\frac{1}{2}(12) + 2 = -4$$

$$a_3 = -\frac{1}{2}(-4) + 2 = 4$$

$$a_4 = -\frac{1}{2}(4) + 2 = 0$$

$$a_5 = -\frac{1}{2}(0) + 2 = 2$$

$$a_6 = -\frac{1}{2}(2) + 2 = 1$$

$$a_7 = -\frac{1}{2}(1) + 2 = \frac{3}{2}$$

$$a_8 = -\frac{1}{2}\left(\frac{3}{2}\right) + 2 = \frac{5}{4}$$

$$12, -4, 4, 0, 2, 1, \frac{3}{2}, \frac{5}{4}$$

**49.** $a_1 = \dfrac{1 + (-1)^2}{2i^0} = 1$

$$a_2 = \frac{1 + (-1)^3}{2i^1} = 0$$

$$a_3 = \frac{1 + (-1)^4}{2i^2} = -1$$

$$a_4 = \frac{1 + (-1)^5}{2i^3} = 0$$

$$a_5 = \frac{1 + (-1)^6}{2i^4} = 1$$

$$a_6 = \frac{1 + (-1)^7}{2i^5} = 0$$

$$a_7 = \frac{1 + (-1)^8}{2i^6} = -1$$

$$a_8 = \frac{1 + (-1)^9}{2i^7} = 0$$

$$1, 0, -1, 0, 1, 0, -1, 0$$

**51.** $a_1 = \dfrac{3}{2}$

$$a_2 = \frac{3 \cdot 5}{2 \cdot 4} = \frac{15}{8}$$

$$a_3 = \frac{3 \cdot 5 \cdot 7}{2 \cdot 4 \cdot 6} = \frac{35}{16}$$

$$a_4 = \frac{3 \cdot 5 \cdot 7 \cdot 9}{2 \cdot 4 \cdot 6 \cdot 8} = \frac{315}{128}$$

$$\frac{3}{2}, \frac{15}{8}, \frac{35}{16}, \frac{315}{128}$$

**Challenge Problem**

Since the $n^{\text{th}}$ triangular number is the $n - 1^{\text{th}}$ triangular number plus $n$, we have that
$$a_1 = 1$$
$$a_2 = 3 = 1 + 2$$
$$a_3 = 6 = 1 + 2 + 3$$
$$\vdots$$

$$a_n = 1 + 2 + 3 + \ldots + n = \frac{n(n+1)}{2}$$

**Exercises 11.2**

**1.** $a_1 + a_2 + a_3 + a_4 + a_5$
$= 3 \cdot 1 + 3 \cdot 2 + 3 \cdot 3 + 3 \cdot 4 + 3 \cdot 5$
$= 3 + 6 + 9 + 12 + 15 = 45$

**3.** $a_1 + a_2 + a_3 + a_4 + a_5$
$= 1^2 + 2^2 + 3^2 + 4^2 + 5^2$
$= 1 + 4 + 9 + 16 + 25 = 55$

**5.** $b_1 + b_2 + b_3 + b_4 + b_5$
$$= \frac{3}{10^1} + \frac{3}{10^2} + \frac{3}{10^3} + \frac{3}{10^4} + \frac{3}{10^5}$$
$= 0.3 + 0.03 + 0.003 + 0.0003 + 0.00003$
$= 0.33333$

**7.** $\displaystyle\sum_{n=1}^{8} t_n = \sum_{n=1}^{8} 2^n$
$= 2^1 + 2^2 + 2^3 + 2^4 + 2^5 + 2^6 + 2^7 + 2^8$
$= 2 + 4 + 8 + 16 + 32 + 64 + 128 + 256$
$= 510$

**9.** $\displaystyle\sum_{k=1}^{20} y_k = \sum_{k=1}^{20} 3$
$= 3 + 3 + 3 + 3 + 3 + 3 + 3 + 3 + 3 + 3 + 3$
$\quad + 3 + 3 + 3 + 3 + 3 + 3 + 3 + 3 + 3$
$= 60$

**11.** $2 + 4 + 8 + 16 + 32 + 64 + 128 = 254$

**13.** $3 + \dfrac{3}{2} + \dfrac{3}{4} + \dfrac{3}{8} + \dfrac{3}{16} + \dfrac{3}{32} + \dfrac{3}{64} = \dfrac{381}{64}$

**15.** $5\left(\displaystyle\sum_{k=1}^{6} k\right) = 5(1 + 2 + 3 + 4 + 5 + 6) = 105$

**17.** $\displaystyle\sum_{n=1}^{4} n^2 + \sum_{n=1}^{4} n$

$= 1^2 + 2^2 + 3^2 + 4^2 + 1 + 2 + 3 + 4$
$= 40$

**19.** $\displaystyle\sum_{k=1}^{4} \frac{k}{2^k} = \frac{1}{2^1} + \frac{2}{2^2} + \frac{3}{2^3} + \frac{4}{2^4}$

$= \frac{1}{2} + \frac{1}{2} + \frac{3}{8} + \frac{1}{4} = \frac{13}{8}$

**21.** $\displaystyle\sum_{k=1}^{8} (-1)^k$

$= (-1)^1 + (-1)^2 + (-1)^3 + (-1)^4 + (-1)^5$
$\quad + (-1)^6 + (-1)^7 + (-1)^8$
$= -1 + 1 - 1 + 1 - 1 + 1 - 1 + 1 = 0$

**23.** $\displaystyle\sum_{j=1}^{6} [-3 + (j-1)5]$

$= [-3 + (1-1)5] + [-3 + (2-1)5]$
$\quad + [-3 + (3-1)5] + [-3 + (4-1)5]$
$\quad + [-3 + (5-1)5] + [-3 + (6-1)5]$
$= -3 + 2 + 7 + 12 + 17 + 22 = 57$

**25.** $\displaystyle\sum_{k=-3}^{3} \frac{1}{10^k}$

$= \frac{1}{10^{-3}} + \frac{1}{10^{-2}} + \frac{1}{10^{-1}} + \frac{1}{10^0} + \frac{1}{10^1}$

$\quad + \frac{1}{10^2} + \frac{1}{10^3}$

$= 1000 + 100 + 10 + 1 + 0.1 + 0.01$
$\quad + 0.001$
$= 1111.111$

**27.** $\displaystyle\sum_{i=1}^{4} (-1)^i 3^i$

$= (-1)^1 \cdot 3^1 + (-1)^2 \cdot 3^2 + (-1)^3 \cdot 3^3$
$\quad + (-1)^4 \cdot 3^4$
$= -3 + 9 - 27 + 81 = 60$

**29.** $\displaystyle\sum_{n=1}^{3} \frac{n+1}{n} - \sum_{n=1}^{3} \frac{n}{n+1} = \frac{2}{1} + \frac{3}{2} + \frac{4}{3} - \frac{1}{2} - \frac{2}{3} - \frac{3}{4}$

$= \frac{35}{12}$

**31.** $\displaystyle\sum_{k=1}^{3} (0.1)^{2k} = (0.1)^{2\cdot1} + (0.1)^{2\cdot2} + (0.1)^{2\cdot3}$

$= 0.01 + 0.0001 + 0.000001$
$= 0.010101$

**33.** $5 + 10 + 15 + 20 + \ldots + 50 = \displaystyle\sum_{n=1}^{10} 5n$

**35.** $-9 - 6 - 3 + 0 + 3 + \ldots + 24 = \displaystyle\sum_{k=-3}^{8} 3k$

**37.** **(a)** $\displaystyle\sum_{k=1}^{2} (2k-1) = 1 + 3 = 4$

$\displaystyle\sum_{k=1}^{3} (2k-1) = 1 + 3 + 5 = 9$

$\displaystyle\sum_{k=1}^{4} (2k-1) = 1 + 3 + 5 + 7 = 16$

$\displaystyle\sum_{k=1}^{5} (2k-1) = 1 + 3 + 5 + 7 + 9 = 25$

$\displaystyle\sum_{k=1}^{6} (2k-1) = 1 + 3 + 5 + 7 + 9 + 11 = 36$

**(b)** $\displaystyle\sum_{k=1}^{n} (2k-1) = n^2$

**39.** $a_n = a_1 + a_{n-1} = a_1 + (a_1 + a_{n-2})$
$\quad = \ldots = \underbrace{a_1 + a_1 + \ldots + a_1}_{n \text{ terms}}$
$\quad = 2 + 2 + \ldots + 2 = 2n$

**41.** $\displaystyle\sum_{k=1}^{n} s_k + \sum_{k=1}^{n} t_k$

$= (s_1 + s_2 + \ldots + s_n) + (t_1 + t_2 + \ldots + t_n)$

$= (s_1 + t_1) + (s_2 + t_2) + \ldots + (s_n + t_n)$

$= \displaystyle\sum_{k=1}^{n} (s_k + t_k)$

**43.** $\displaystyle\sum_{k=1}^{n} (s_k + c)$

$= (s_1 + c) + (s_2 + c) + \ldots + (s_n + c)$

$= (s_1 + s_2 + \ldots + s_n) + \underbrace{(c + c + \ldots + c)}_{n \text{ terms}}$

$= \displaystyle\sum_{k=1}^{n} s_k + nc$

**45. (a)** $\displaystyle\sum_{k=1}^{10} \frac{2}{k(k+2)} = \sum_{k=1}^{10} \left( \frac{1}{k} - \frac{1}{k+2} \right)$

$= \left(1 - \frac{1}{3}\right) + \left(\frac{1}{2} - \frac{1}{4}\right) + \left(\frac{1}{3} - \frac{1}{5}\right) + \left(\frac{1}{4} - \frac{1}{6}\right) + \left(\frac{1}{5} - \frac{1}{7}\right) + \left(\frac{1}{6} - \frac{1}{8}\right)$

$\quad + \left(\frac{1}{7} - \frac{1}{9}\right) + \left(\frac{1}{8} - \frac{1}{10}\right) + \left(\frac{1}{9} - \frac{1}{11}\right) + \left(\frac{1}{10} - \frac{1}{12}\right)$

$= 1 + \frac{1}{2} - \frac{1}{11} - \frac{1}{12} = \frac{175}{132}$

**(b)** $\displaystyle\sum_{k=1}^{n} \left( \frac{1}{k} - \frac{1}{k+2} \right)$

$= \left(1 - \frac{1}{3}\right) + \left(\frac{1}{2} - \frac{1}{4}\right) + \left(\frac{1}{3} - \frac{1}{5}\right) + \left(\frac{1}{4} - \frac{1}{6}\right) + \ldots + \left(\frac{1}{n-3} - \frac{1}{n-1}\right)$

$\quad + \left(\frac{1}{n-2} - \frac{1}{n}\right) + \left(\frac{1}{n-1} - \frac{1}{n+1}\right) + \left(\frac{1}{n} - \frac{1}{n+2}\right)$

$= 1 + \frac{1}{2} - \frac{1}{n+1} - \frac{1}{n+2} = \frac{2(n+1)(n+2) + (n+1)(n+2) - 2(n+2) - 2(n+1)}{2(n+1)(n+2)}$

$= \frac{2n^2 + 6n + 4 + n^2 + 3n + 2 - 2n - 4 - 2n - 2}{2(n+1)(n+2)} = \frac{3n^2 + 5n}{2(n+1)(n+2)} = \frac{n(3n+5)}{2(n+1)(n+2)}$

## Challenge Problem

$c_{ij} = a_{i1}b_{1j} + a_{i2}b_{2j} + \ldots + a_{in}b_{nj} = \displaystyle\sum_{k=1}^{n} a_{ik}b_{kj}$

## Exercises 11.3

1. $d = 3 - 1 = 2$
The next three terms are: 5, 7, 9
$a_n = 1 + (n-1) \cdot 2 = 2n - 1$
$S_{20} = \frac{20}{2}[2 \cdot 1 + 19 \cdot 2] = 10(40) = 400$

3. $d = -4 - 2 = -6$
The next three terms are: $-10, -16, -22$
$a_n = 2 + (n-1)(-6) = -6n + 8$
$S_{20} = \frac{20}{2}[2 \cdot 2 + 19 \cdot (-6)] = 10(-110)$
$= -1100$

5. $d = 8 - \frac{15}{2} = \frac{1}{2}$

    The next three terms are: $\frac{17}{2}, 9, \frac{19}{2}$

    $a_n = \frac{15}{2} + (n-1) \cdot \frac{1}{2} = \frac{1}{2}n + 7$

    $S_{20} = \frac{20}{2}\left[2 \cdot \frac{15}{2} + 19 \cdot \frac{1}{2}\right] = 10\left(\frac{49}{2}\right) = 245$

7. $d = -\frac{1}{5} - \frac{2}{5} = -\frac{3}{5}$

    The next three terms are: $-\frac{4}{5}, -\frac{7}{5}, -2$

    $a_n = \frac{2}{5} + (n-1)\left(-\frac{3}{5}\right) = -\frac{3}{5}n + 1$

    $S_{20} = \frac{20}{2}\left[2 \cdot \frac{2}{5} + 19\left(-\frac{3}{5}\right)\right]$

    $= 10\left(\frac{-53}{5}\right) = -106$

9. $d = 100 - 50 = 50$
The next three terms are: 150, 200, 250
$a_n = 50 + (n-1) \cdot 50 = 50n$
$S_{20} = \frac{20}{2}[2 \cdot 50 + (19) \cdot 50] = 10(1050)$
$= 10,500$

11. $d = 10 - (-10) = 20$
The next three terms are: 30, 50, 70
$a_n = -10 + (n-1) \cdot 20 = 20n - 30$
$S_{20} = \frac{20}{2}[2 \cdot (-10) + 19 \cdot 20] = 10(360)$
$= 3600$

13. $5 + 10 + 15 + 20 + 25 + 30 + 35 + 40 + 45$
$\quad + 50 + 55 + 60 + 65$
$= 455$
$= \sum_{k=1}^{13} 5k = \frac{13}{2}[2 \cdot 5 + 12 \cdot 5] = \frac{13}{2} \cdot 70 = 455$

15. $\frac{3}{4} + 1 + \frac{5}{4} + \frac{3}{2} + \frac{7}{4} + 2 + \frac{9}{4} + \frac{5}{2} + \frac{11}{4} = \frac{63}{4}$

    $= \sum_{k=1}^{9}\left(\frac{3}{4} + (k-1) \cdot \frac{1}{4}\right) = \frac{9}{2}\left[2 \cdot \frac{3}{4} + 8 \cdot \frac{1}{4}\right]$

    $= \frac{9}{2}\left(\frac{7}{2}\right) = \frac{63}{4}$

17. $a_1 = -30$
$a_{10} = 69 = -30 + 9 \cdot d$
$99 = 9d$
$d = 11$
$a_n = -30 + 11(n-1) = 11n - 41$
$a_{30} = 11 \cdot 30 - 41 = 289$

19. $S_{100} = \frac{100}{2}[2 \cdot 3 + 99 \cdot 3] = 50(303)$
$= 15,150$

21. $S_{100} = \frac{100}{2}[2(-91) + 99 \cdot 21) = 50(1897)$
$= 94,850$

23. $S_{100} = \frac{100}{2}\left[2 \cdot \frac{1}{7} + 99 \cdot 5\right] = 50\left(\frac{3467}{7}\right)$
$= \frac{173,350}{7}$

25. $S_{100} = \frac{100}{2}[2 \cdot 725 + 99 \cdot 100]$
$= 50(11,350) = 567,500$

27. $a_1 = -8, d = 16$
$S_{28} = \frac{28}{2}[2(-8) + 27 \cdot 16] = 14(416)$
$= 5824$

29. The first multiple of 12 is 12, and the 50th multiple is 600.
$S_{50} = \frac{50}{2}(12 + 600) = 25(612) = 15,300$

**31. (a)** $a_1 = 1, d = 2$

$$S_{100} = \frac{100}{2}(2 \cdot 1 + 99 \cdot 2) = 50(200)$$
$$= 10,000$$

**(b)** $S_n = \frac{n}{2}[2 \cdot 1 + (n-1) \cdot 2] = \frac{n}{2}(2n) = n^2$

**33.** $\sum_{k=1}^{9}\left[-6 + (k-1)\frac{1}{2}\right] = \frac{9}{2}\left[2(-6) + 8 \cdot \frac{1}{2}\right]$

$= \frac{9}{2}(-8) = -36$

**35.** $\sum_{k=1}^{30}(10k - 1)$

$a_1 = 9, d = 10$

$\sum_{k=1}^{30}(10k - 1) = \frac{30}{2}(2 \cdot 9 + 29 \cdot 10)$

$= 15(308) = 4620$

**37.** $a_1 = \frac{1}{4}, d = \frac{3}{4}$

$\sum_{k=1}^{49}\left(\frac{3}{4}k - \frac{1}{2}\right) = \frac{49}{2}\left[2 \cdot \frac{1}{4} + 48\left(\frac{3}{4}\right)\right]$

$= \frac{49}{2}\left(\frac{73}{2}\right) = \frac{3577}{4}$

**39.** $a_1 = 5, d = 5$

$\sum_{k=1}^{n}5k = \frac{n}{2}[2 \cdot 5 + (n-1) \cdot 5]$

$= \frac{n}{2}(5n + 5) = \frac{5}{2}n(n+1)$

**41.** $u = \dfrac{-7 + \frac{5}{2}}{2} = -\frac{9}{4}$

**43.** $a_1 = -\frac{2}{3}, d = -\frac{1}{5} + \frac{2}{3} = \frac{7}{15}$

$a_n = -\frac{2}{3} + (n-1)\left(\frac{7}{15}\right) = \frac{7}{15}n - \frac{17}{15}$

$a_{35} = \frac{7}{15}(35) - \frac{17}{15} = \frac{228}{15} = \frac{76}{5}$

**45.** $a_n = a_1 + (n-1)d$

$38 = 3 + 5d$
$5d = 35$
$d = 7$
$3, 10, 17, 24, 31, 38$

**47.** $a_n = a_1 + (n-1)d$

$48 = -8 + 7d$
$7d = 56$
$d = 8$
$-8, 0, 8, 16, 24, 32, 40, 48$

**49.** $a_n = a_1 + (n-1)d$

$-3 = -\frac{1}{5} + 7d$

$7d = -\frac{14}{5}$

$d = -\frac{2}{5}$

$-\frac{1}{5}, -\frac{3}{5}, -1, -\frac{7}{5}, -\frac{9}{5}, -\frac{11}{5}, -\frac{13}{5}, -3$

**51.** $a_1 = 10, d = 0.50$

$S_{52} = \frac{52}{2}[2 \cdot 10 + 51(0.50)] = 26(45.50)$

$= \$1183$

**53. (a)** The monthly balance is $12,000 - 1000(k - 1)$. So the monthly interest payment is $240 - 20(k - 1)$.
For $k = 1$, $\$240$
For $k = 2$, $\$220$
For $k = 3$, $\$200$

**(b)** Monthly loan balance is $12,000 - 1000(k - 1)$, monthly interest is $260 - 20k$.

**(c)** Total interest $\sum_{k=1}^{12}(260 - 20k)$

$= \frac{12}{2}[2 \cdot 240 + 11(-20)]$

$= 6(260) = \$1560$

Annual interest rate $= \dfrac{1560}{12,000} \times 100$

$= 13\%.$

**55.** $\displaystyle\sum_{n=6}^{20}(5n-3) = \sum_{n=1}^{20}(5n-3) - \sum_{n=1}^{5}(5n-3)$

$= \dfrac{20}{2}[2\cdot 2 + 19\cdot 5] - \dfrac{5}{2}[2\cdot 2 + 4\cdot 5]$

$= 10(99) - \dfrac{5}{2}(24) = 990 - 60 = 930$

**57.** $a_1 = 3\cdot 1 - 8 = -5$

$a_{80} = 3\cdot 80 - 8 = 232$

$S_{80} = \dfrac{80}{2}(-5 + 232) = 40(227) = 9080$

**59.** $a_1 = 33,\ a_n = 427,\ d = 2$

$427 = 33 + (n-1)\cdot 2$

$394 = 2(n-1)$

$197 = n - 1$

$n = 198$

$\text{Sum} = \dfrac{198}{2}(33 + 427) = 99(460) = 45{,}540$

**61.** $-5865 = \dfrac{30}{2}[2a_1 + 29d]$ or

$\qquad -5865 = 30a_1 + 435d$

$-2610 = \dfrac{20}{2}[2a_1 + 19d]$ or

$\qquad -2610 = 20a_1 + 190d$

$-2610 = 20a_1 + 190d$

$20a_1 = -2610 - 190d$

$a_1 = -\dfrac{261}{2} - \dfrac{19}{2}d$

$-5865 = 30\left(-\dfrac{261}{2} - \dfrac{19}{2}d\right) + 435d$

$-5865 = -3915 - 285d + 435d$

$-1950 = 150d$

$d = -13$

$a_1 = -\dfrac{261}{2} - \dfrac{19}{2}(-13) = -7$

**63.** $a_n = a_1 + (n-1)d$

$10 = 3 + 3d$

$7 = 3d$

$d = \dfrac{7}{3}$

$u = 3 + \dfrac{7}{3} = \dfrac{16}{3}$

$v = \dfrac{16}{3} + \dfrac{7}{3} = \dfrac{23}{3}$

**65.** $\displaystyle\sum_{k=1}^{16} f_k = \sum_{k=1}^{16}(3k+7) = \dfrac{16}{2}[2\cdot 10 + 15(3)]$

$= 8(65) = 520$

## Critical Thinking

**1.** (a) $a_n = 2$

(b) $a_n = 1 + (-1)^n$

(c) $a_n = [1 + (-1)^n]i^n$, where $i = \sqrt{-1}$

**3.** $\dfrac{4}{(k+1)(k+3)} = \dfrac{A}{k+1} + \dfrac{B}{k+3}$

$4 = A(k+3) + B(k+1)$

Let $k = -3$:  $4 = A\cdot 0 + B(-3+1)$

$\qquad\qquad\quad 4 = -2B$

$\qquad\qquad\quad B = -2$

Let $k = -1$:  $4 = A(-1+3) + B\cdot 0$

$\qquad\qquad\quad 4 = 2A$

$\qquad\qquad\quad A = 2$

$\dfrac{4}{(k+1)(k+3)} = \dfrac{2}{k+1} - \dfrac{2}{k+3}$

$\displaystyle\sum_{k=1}^{10} a_k = \sum_{k=1}^{10}\left(\dfrac{2}{k+1} - \dfrac{2}{k+3}\right)$

$= \left(\dfrac{2}{2} - \dfrac{2}{4}\right) + \left(\dfrac{2}{3} - \dfrac{2}{5}\right) + \left(\dfrac{2}{4} - \dfrac{2}{6}\right)$

$\quad + \left(\dfrac{2}{5} - \dfrac{2}{7}\right) + \left(\dfrac{2}{6} - \dfrac{2}{8}\right) + \left(\dfrac{2}{7} - \dfrac{2}{9}\right)$

$\quad + \left(\dfrac{2}{8} - \dfrac{2}{10}\right) + \left(\dfrac{2}{9} - \dfrac{2}{11}\right) + \left(\dfrac{2}{10} - \dfrac{2}{12}\right)$

$\quad + \left(\dfrac{2}{11} - \dfrac{2}{13}\right)$

$= \dfrac{2}{2} + \dfrac{2}{3} - \dfrac{2}{12} - \dfrac{2}{13} = 1 + \dfrac{2}{3} - \dfrac{1}{6} - \dfrac{2}{13}$

$= \dfrac{78}{78} + \dfrac{52}{78} - \dfrac{13}{78} - \dfrac{12}{78} = \dfrac{105}{78}$

## Exercises 11.4

**1.** $a_1 = 2,\ r = \dfrac{4}{2} = 2$

The next three terms are: 16, 32, 64

$a_n = 2\cdot 2^{n-1} = 2^n$

**3.** $a_1 = 1$, $r = \dfrac{3}{1} = 3$

The next three terms are: 27, 81, 243

$a_n = 1 \cdot 3^{n-1} = 3^{n-1}$

**5.** $a_1 = -3$, $r = \dfrac{1}{-3} = -\dfrac{1}{3}$

The next three terms are: $\dfrac{1}{9}, -\dfrac{1}{27}, \dfrac{1}{81}$

$a_n = -3\left(-\dfrac{1}{3}\right)^{n-1}$

**7.** $a_1 = -1$, $r = \dfrac{-5}{-1} = 5$

The next three terms are: $-125, -625,$
$-3125$

$a_n = -1 \cdot 5^{n-1} = -5^{n-1}$

**9.** $a_1 = -6$, $r = \dfrac{-4}{-6} = \dfrac{2}{3}$

The next three terms are: $-\dfrac{16}{9}, -\dfrac{32}{27}, -\dfrac{64}{81}$

$a_n = -6\left(\dfrac{2}{3}\right)^{n-1}$

**11.** $a_1 = \dfrac{1}{1000}$, $r = \dfrac{\frac{1}{10}}{\frac{1}{1000}} = 100$

The next three terms are: 1000, 100,000,
10,000,000

$a_n = \dfrac{1}{1000}(100)^{n-1}$

**13.** $2 + 4 + 8 + 16 + 32 + 64 = 126$

$S_6 = \displaystyle\sum_{k=1}^{6} 2^k = \dfrac{2(1-2^6)}{1-2} = \dfrac{2(-63)}{-1} = 126$

**15.** $-6 - 4 - \dfrac{8}{3} - \dfrac{16}{9} - \dfrac{32}{27} - \dfrac{64}{81} = -\dfrac{1330}{81}$

$S_6 = \displaystyle\sum_{k=1}^{6} -6\left(\dfrac{2}{3}\right)^{n-1} = \dfrac{-6\left(1-\left(\frac{2}{3}\right)^6\right)}{1-\frac{2}{3}}$

$= -18\left(1 - \dfrac{64}{729}\right) = -18\left(\dfrac{665}{729}\right)$

$= -\dfrac{1330}{81}$

**17.** $a_1 = \dfrac{1}{8}$, $r = \dfrac{\frac{1}{4}}{\frac{1}{8}} = 2$

$a_n = \dfrac{1}{8}(2)^{n-1}$

$a_{14} = \dfrac{1}{8} \cdot 2^{14-1} = \dfrac{1}{8} \cdot 2^{13} = 1024$

**19.** $a_n = 3(-1)^{n-1}$

$a_{101} = 3(-1)^{101-1} = 3(-1)^{100} = 3$

**21.** $a_6 = a_1 r^{6-1}$

$-\dfrac{5}{8} = 20r^5$

$r^5 = -\dfrac{1}{32}$

$r = \sqrt[5]{-\dfrac{1}{32}} = -\dfrac{1}{2}$

**23.** $\displaystyle\sum_{k=1}^{10} 2^{k-1} = \dfrac{1(1-2^{10})}{1-2} = \dfrac{-1023}{-1} = 1023$

**25.** $\displaystyle\sum_{k=1}^{n} 2^{k-1} = \dfrac{1(1-2^n)}{1-2} = -(1-2^n) = 2^n - 1$

**27.** $\displaystyle\sum_{k=1}^{5} 3^{k-4} = \sum_{k=1}^{5} 3^{-3} \cdot 3^{k-1} = \dfrac{3^{-3}(1-3^5)}{1-3}$

$= \dfrac{1}{27}\left(\dfrac{1-243}{-2}\right) = \dfrac{121}{27}$

**29.** $\displaystyle\sum_{j=1}^{5} \left(\dfrac{2}{3}\right)^{j-2} = \sum_{j=1}^{5} \left(\dfrac{2}{3}\right)^{-1}\left(\dfrac{2}{3}\right)^{j-1}$

$= \dfrac{\left(\frac{2}{3}\right)^{-1}\left(1-\left(\frac{2}{3}\right)^5\right)}{1-\frac{2}{3}} = \dfrac{3}{2} \cdot 3\left(1 - \dfrac{32}{243}\right)$

$= \dfrac{9}{2}\left(\dfrac{211}{243}\right) = \dfrac{211}{54}$

**31.** $\displaystyle\sum_{k=1}^{8}16\left(-\frac{1}{2}\right)^{k+2} = \sum_{k=1}^{8}16\left(-\frac{1}{2}\right)^{3}\left(-\frac{1}{2}\right)^{k-1}$

$\displaystyle = \sum_{k=1}^{8}(-2)\left(-\frac{1}{2}\right)^{k-1} = \frac{-2\left(1-\left(-\frac{1}{2}\right)^{8}\right)}{1-\left(-\frac{1}{2}\right)}$

$\displaystyle = \frac{-2\left(1-\frac{1}{256}\right)}{\frac{3}{2}} = -\frac{4}{3}\left(\frac{255}{256}\right) = -\frac{85}{64}$

**33.** $u^2 = \frac{1}{7}\cdot\frac{25}{63} = \frac{25}{441}$

$u = -\sqrt{\frac{25}{441}} = -\frac{5}{21}$

**35.** Geometric means $= \pm\sqrt{8\cdot 12} = \pm\sqrt{96}$
$= \pm 4\sqrt{6}$

**37.** $a_n = a_1 r^{n-1}$
$1536 = 6r^4$
$r^4 = 256$
$r = \pm 4$
6, 24, 96, 384, 1536 or
6, –24, 96, –384, 1536

**39.** $\displaystyle\sum_{k=1}^{30}2^{k-1} = \frac{1(1-2^{30})}{1-2} = 2^{30}-1$
$= 1,073,741,823$ cents or $10,737,418.23

**41.** The following answer assumes that we're answering the question "How many (total) are there after the 10th, $n$th day?"

$S_n = \dfrac{1000(1-2^n)}{1-2}$   $S_{10} = \dfrac{1000(1-2^{10})}{1-2}$

$= 1000(2^n - 1)$   $= 1000(2^{10} - 1)$
$= 1,023,000$

In response to the question, "How many are born on the 10th, $n$th day," the answer is $a_n = 1000(2)^{n-1}$
$a_{10} = 1000\cdot 2^9 = 512,000$

**43.** $a_1 = 14{,}280, \ r = \dfrac{9}{10}$

$a_6 = 14{,}280\left(\dfrac{9}{10}\right)^5 = \$8432.20$

**45.** **(a)** $a_1 = 800, \ r = 0.11$
$a_n = 800(1.11)^n$
**(b)** $a_5 = 800(1.11)^5 = \$1348.05$

**47.** $a_n = 1500(1.08)^n$
$a_5 = 1500(1.08)^5 = 2203.99$
Interest $= 2203.99 - 1500 = \$703.99$

**49.** **(a)** If the volume of the first container is

$V$, then $\dfrac{1}{2}V + \left(\dfrac{1}{2}\right)^2 V + \left(\dfrac{1}{2}\right)^3 V$

$+ \left(\dfrac{1}{2}\right)^4 V + \left(\dfrac{1}{2}\right)^5 V = \dfrac{31}{32}V$ is the sum
of the volumes of the other five. Since
$\dfrac{31}{32}V < V$, the answer is yes.

**(b)** $\displaystyle\sum_{k=1}^{5}\left(\dfrac{2}{3}\right)^k V = \dfrac{422}{243}V > V$; therefore no.

## Challenge Problem

You snap your fingers once in 0 minutes, twice in 1 minute, three times in $1 + 2 = 3$ minutes, etc. In general, you snap your fingers $n$ times in

$1 + 2 + 2^2 + 2^3 + \ldots + 2^{n-1} = \displaystyle\sum_{k=1}^{n}2^{k-1} = 2^n - 1$

minutes.
**(a)** It will take you $2^{10} - 1 = 1023$ minutes, or 17 hours and 3 minutes, to snap your fingers 10 times.
**(b)** It will take you 32,767 minutes, or 22 days, 18 hours, and 7 minutes, to snap your fingers 15 times.
**(c)** It will take you 1,048,575 minutes, or almost 2 years, to snap your fingers 20 times.

**1.** $2 + 1 + \dfrac{1}{2} + \ldots = \dfrac{2}{1 - \frac{1}{2}} = \dfrac{2}{\frac{1}{2}} = 4$

**3.** $25 + 5 + 1 + \ldots = \dfrac{25}{1 - \frac{1}{5}} = \dfrac{25}{\frac{4}{5}} = \dfrac{125}{4}$

**5.** $1 - \dfrac{1}{2} + \dfrac{1}{4} - \ldots = \dfrac{1}{1 + \frac{1}{2}} = \dfrac{1}{\frac{3}{2}} = \dfrac{2}{3}$

**7.** $1 + 0.1 + 0.01 + \ldots = \dfrac{1}{1 - \frac{1}{10}} = \dfrac{1}{\frac{9}{10}} = \dfrac{10}{9}$

**9.** $-2 - \dfrac{1}{4} - \dfrac{1}{32} - \ldots = \dfrac{-2}{1 - \frac{1}{8}} = -\dfrac{2}{\frac{7}{8}} = -\dfrac{16}{7}$

**11.** The numerator at the right should be $a_1 = \dfrac{1}{4}$, not 1.

**13.** The denominator at the right should be $1 - \left(-\dfrac{1}{3}\right)$ since $r = -\dfrac{1}{3}$, not $\dfrac{1}{3}$.

**15.** $\displaystyle\sum_{k=1}^{\infty} \left(\dfrac{1}{3}\right)^{k-1} = \dfrac{1}{1 - \frac{1}{3}} = \dfrac{1}{\frac{2}{3}} = \dfrac{3}{2}$

**17.** $\displaystyle\sum_{k=1}^{\infty} \left(\dfrac{1}{3}\right)^{k+1} = \dfrac{\frac{1}{9}}{1 - \frac{1}{3}} = \dfrac{\frac{1}{9}}{\frac{2}{3}} = \dfrac{1}{6}$

**19.** $\displaystyle\sum_{n=1}^{\infty} \dfrac{1}{2^{n-2}} = \dfrac{2}{1 - \frac{1}{2}} = \dfrac{2}{\frac{1}{2}} = 4$

**21.** $\displaystyle\sum_{k=1}^{\infty} 2(0.1)^{k-1} = \dfrac{2}{1 - 0.1} = \dfrac{2}{0.9} = \dfrac{20}{9}$

**23.** $\displaystyle\sum_{n=1}^{\infty} \left(\dfrac{3}{2}\right)^{n-1}$ has no finite sum.

**25.** $\displaystyle\sum_{k=1}^{\infty} (0.7)^{k-1} = \dfrac{1}{1 - 0.7} = \dfrac{1}{0.3} = \dfrac{10}{3}$

**27.** $\displaystyle\sum_{k=1}^{\infty} 5(1.01)^k$ has no finite sum.

**29.** $\displaystyle\sum_{k=1}^{\infty} 10\left(\dfrac{2}{3}\right)^{k-1} = \dfrac{10}{1 - \frac{2}{3}} = \dfrac{10}{\frac{1}{3}} = 30$

**31.** $\displaystyle\sum_{k=1}^{\infty} (0.45)^{k-1} = \dfrac{1}{1 - 0.45} = \dfrac{1}{0.55} = \dfrac{20}{11}$

**33.** $\displaystyle\sum_{n=1}^{\infty} 7\left(-\dfrac{3}{4}\right)^{n-1} = \dfrac{7}{1 + \frac{3}{4}} = \dfrac{7}{\frac{7}{4}} = 4$

**35.** $\displaystyle\sum_{k=1}^{\infty} \left(-\dfrac{2}{5}\right)^{2k} = \displaystyle\sum_{k=1}^{\infty} \left(\dfrac{4}{25}\right)^{k} = \dfrac{\frac{4}{25}}{1 - \frac{4}{25}}$

$= \dfrac{\frac{4}{25}}{\frac{21}{25}} = \dfrac{4}{21}$

**37.** $0.777\ldots = \dfrac{7}{10} + \dfrac{7}{100} + \dfrac{7}{1000} + \ldots$

$= \dfrac{7}{10^1} + \dfrac{7}{10^2} + \dfrac{7}{10^3} + \ldots$

$= \dfrac{7}{10} + \dfrac{7}{10}\left(\dfrac{1}{10}\right) + \dfrac{7}{10}\left(\dfrac{1}{10}\right)^2 + \ldots$

$= \displaystyle\sum_{k=1}^{\infty} \dfrac{7}{10}\left(\dfrac{1}{10}\right)^{k-1} = \dfrac{\frac{7}{10}}{1 - \frac{1}{10}} = \dfrac{\frac{7}{10}}{\frac{9}{10}} = \dfrac{7}{9}$

**39.** $0.131313\ldots$

$= \dfrac{13}{100} + \dfrac{13}{10,000} + \dfrac{13}{1,000,000} + \ldots$

$= \dfrac{13}{100} + \dfrac{13}{100^2} + \dfrac{13}{100^3} + \ldots$

$= \dfrac{13}{100} + \dfrac{13}{100}\left(\dfrac{1}{100}\right) + \dfrac{13}{100}\left(\dfrac{1}{100}\right)^2 + \ldots$

$= \displaystyle\sum_{k=1}^{\infty} \dfrac{13}{100}\left(\dfrac{1}{100}\right)^{k-1} = \dfrac{\frac{13}{100}}{1 - \frac{1}{100}} = \dfrac{\frac{13}{100}}{\frac{99}{100}} = \dfrac{13}{99}$

**41.** $0.0131313$

$$= \frac{13}{1000} + \frac{13}{100,000} + \frac{13}{10,000,000} + \dots$$

$$= \frac{13}{1000} + \frac{13}{1000}\left(\frac{1}{100}\right) + \frac{13}{1000}\left(\frac{1}{100}\right)^2 + \dots$$

$$= \sum_{k=1}^{\infty} \frac{13}{1000}\left(\frac{1}{100}\right)^{k-1} = \frac{\frac{13}{1000}}{1 - \frac{1}{100}}$$

$$= \frac{\frac{13}{1000}}{\frac{99}{100}} = \frac{13}{990}$$

**43.** $0.9999\dots = \frac{9}{10} + \frac{9}{100} + \frac{9}{1000} + \dots$

$$= \frac{9}{10^1} + \frac{9}{10^2} + \frac{9}{10^3} + \dots$$

$$= \frac{9}{10} + \frac{9}{10}\left(\frac{1}{10}\right) + \frac{9}{10}\left(\frac{1}{10}\right)^2 + \dots$$

$$= \sum_{k=1}^{\infty} \frac{9}{10}\left(\frac{1}{10}\right)^{k-1} = \frac{\frac{9}{10}}{1 - \frac{1}{10}} = \frac{\frac{9}{10}}{\frac{9}{10}} = 1$$

**45.** **(a)** Since time $= \dfrac{\text{distance}}{\text{rate}}$, the sequence is

$$\frac{4}{3}, \frac{4}{9}, \frac{4}{27}, \dots, \frac{4}{3n}, \dots$$

**(b)** $\displaystyle\sum_{n=1}^{\infty} \frac{4}{3^n} = \frac{\frac{4}{3}}{1 - \frac{1}{3}} = \frac{\frac{4}{3}}{\frac{2}{3}} = 2$

**47.** $a_1 = 4, r = \dfrac{1}{2}$

$$\sum_{k=1}^{\infty} a_1 r^{k-1} = \frac{4}{1 - \frac{1}{2}} = \frac{4}{\frac{1}{2}} = 8 \text{ hours}$$

**49.** The time for the last $\dfrac{1}{2}$ mile would have to

be $\displaystyle\sum_{n=1}^{\infty} \frac{2}{5}\left(\frac{10}{9}\right)^{n-1}$, which is not a finite

sum since $\dfrac{10}{9} > 1$.

**51.** **(a)** Area $= \dfrac{1}{2}(AC)(CB) = \dfrac{1}{2}(4)(4) = 8$

**(b)** $4 + 2 + 1 + \dfrac{1}{2} + \dots = \dfrac{4}{1 - \frac{1}{2}} = \dfrac{4}{\frac{1}{2}} = 8$

**(c)** For the odd-numbered triangles:

$$4 + 1 + \frac{1}{4} + \dots = \frac{4}{1 - \frac{1}{4}} = \frac{4}{\frac{3}{4}} = \frac{16}{3}$$

For the even-numbered triangles:

$$2 + \frac{1}{2} + \frac{1}{8} + \dots = \frac{2}{1 - \frac{1}{4}} = \frac{2}{\frac{3}{4}} = \frac{8}{3}$$

$$\frac{16}{3} + \frac{8}{3} = \frac{24}{3} = 8$$

**53.** Sum of areas

$$= \frac{1}{2}(9)(3) + \frac{1}{2}(6)(2) + \frac{1}{2}(4)\left(\frac{4}{3}\right) + \dots$$

$$= \frac{27}{2} + 6 + \frac{8}{3} + \dots$$

$$= \frac{\frac{27}{2}}{1 - \frac{4}{9}} = \frac{\frac{27}{2}}{\frac{5}{9}} = \frac{243}{10}$$

**55.** **(a)** $a_1 = 1200, r = \dfrac{6}{10}$

$$\sum_{k=1}^{\infty} 1200\left(\frac{6}{10}\right)^{k-1} = \frac{1200}{1 - \frac{6}{10}}$$

$$= \frac{1200}{\frac{4}{10}} = \$3000$$

**(b)** $a_1 = 800, r = \dfrac{6}{10}$

$$\sum_{k=1}^{\infty} 800\left(\frac{6}{10}\right)^{k-1} = \frac{800}{1 - \frac{6}{10}}$$

$$= \frac{800}{\frac{4}{10}} = \$2000$$

## Challenge Problem

$$S_n = \sum_{k=1}^{n} \frac{1}{k(k+1)} = \sum_{k=1}^{n}\left(\frac{1}{k} - \frac{1}{k+1}\right)$$

$$= \left(\frac{1}{1} - \frac{1}{2}\right) + \left(\frac{1}{2} - \frac{1}{3}\right) + \ldots + \left(\frac{1}{n-1} - \frac{1}{n}\right)$$

$$+ \left(\frac{1}{n} - \frac{1}{n+1}\right)$$

$$= 1 - \frac{1}{n+1}$$

As $n$ gets large, $\dfrac{1}{n+1}$ tends to 0, so $S_\infty = 1$.

## Critical Thinking

**1.** The sum of the first $n$ powers of 2, starting

with $2^0$ is $\displaystyle\sum_{k=1}^{n} 2^{k-1} = \frac{1-2^n}{1-2} = 2^n - 1$. Since

$2^n - 1 < 2^n$, $2^n$ is larger than the sum.

**3.**

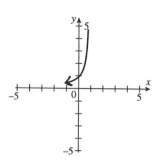

**5.** The series $\displaystyle\sum_{k=1}^{\infty}(3x-4)^{k-1}$ will have a

finite sum when

$$|3x - 4| < 1$$
$$-1 < 3x - 4 < 1$$
$$-1 < 3x < 5$$
$$3 < 3x < 5$$
$$1 < x < \frac{5}{3}$$

For example, if $k = \dfrac{3}{2}$,

$$\sum_{k=1}^{\infty}\left(3 \cdot \frac{3}{2} - 4\right)^{k-1} = \sum\left(\frac{1}{2}\right)^{k-1} = 2$$

## Exercises 11.6

**1.** Let $S_n$ be the statement

$$1 + 2 + 3 + \ldots + n = \frac{n(n+1)}{2}.$$

1. Since $1 = \dfrac{1(1+1)}{2}$, $S_1$ is true.

2. Assume $S_k$ and add $k + 1$ to obtain
$$1 + 2 + 3 + \ldots + k + (k+1)$$
$$= \frac{k(k+1)}{2} + (k+1)$$
$$= \frac{k(k+1)}{2} + \frac{2(k+1)}{2}$$
$$= \frac{(k+1)(k+2)}{2} = \frac{(k+1)[(k+1)+1]}{2}$$

Therefore, $S_{k+1}$ is true. Since $S_1$ is true and $S_k$ implies $S_{k+1}$, the principle of mathematical induction makes $S_n$ true for all integers $n \geq 1$.

**3.** Let $S_n$ be the statement $\displaystyle\sum_{i=1}^{n} 3i = \frac{3n(n+1)}{2}$.

1. Since $\displaystyle\sum_{i=1}^{1} 3i = 3 = \frac{3(1+1)}{2}$, $S_1$ is true.

2. Assume $S_k$ and add $3(k+1)$ to obtain
$$\sum_{i=1}^{k+1} 3i = \left(\sum_{i=1}^{k} 3i\right) + 3(k+1)$$
$$= \frac{3k(k+1)}{2} = 3(k+1)$$
$$= \frac{3k(k+1)}{2} + \frac{3 \cdot 2(k+1)}{2}$$
$$= \frac{3(k+1)(k+2)}{2}$$
$$= \frac{3(k+1)[(k+1)+1]}{2}$$

Therefore, $S_{k+1}$ is true. Since $S_1$ is true and $S_k$ implies $S_{k+1}$, the principle of mathematical induction makes $S_n$ true for all integers $n \geq 1$.

**5.** Let $S_n$ be the statement
$$\frac{5}{3}+\frac{4}{3}+1+\ldots+\left(-\frac{1}{3}n+2\right)=\frac{n(11-n)}{6}.$$

1. Since $\frac{5}{3}=\frac{1(11-1)}{6}$, $S_1$ is true.

2. Assume $S_k$ and add $-\frac{1}{3}(k+1)+2$ to obtain
$$\frac{5}{3}+\frac{4}{3}+1+\ldots+\left(-\frac{1}{3}k+2\right)$$
$$+\left[-\frac{1}{3}(k+1)+2\right]$$
$$=\frac{k(11-k)}{6}+\left[-\frac{1}{3}(k+1)+2\right]$$
$$=\frac{11k-k^2}{6}+\frac{-2k+10}{6}$$
$$=\frac{10+9k-k^2}{6}=\frac{(k+1)(10-k)}{6}$$
$$=\frac{(k+1)[11-(k+1)]}{6}$$
Therefore, $S_{k+1}$ is true. Since $S_1$ is true and $S_k$ implies $S_{k+1}$, the principle of mathematical induction makes $S_n$ true for all integers $n \geq 1$.

**7.** Let $S_n$ be the statement
$$\frac{1}{1\cdot 2}+\frac{1}{2\cdot 3}+\frac{1}{3\cdot 4}+\ldots+\frac{1}{n(n+1)}=\frac{n}{n+1}.$$

1. Since $\frac{1}{1\cdot 2}=\frac{1}{1+1}$, $S_1$ is true.

2. Assume $S_k$ and add $\frac{1}{(k+1)[(k+1)+1]}$ to obtain
$$\frac{1}{1\cdot 2}+\frac{1}{2\cdot 3}+\ldots+\frac{1}{k(k+1)}$$
$$+\frac{1}{(k+1)[(k+1)+1]}$$
$$=\frac{k}{k+1}+\frac{1}{(k+1)(k+2)}$$
$$=\frac{k^2+2\cdot k+1}{(k+1)(k+2)}=\frac{(k+1)^2}{(k+1)(k+2)}$$
$$=\frac{k+1}{k+2}=\frac{k+1}{(k+1)+1}$$

Therefore, $S_{k+1}$ is true. Since $S_1$ is true and $S_k$ implies $S_{k+1}$, the principle of mathematical induction makes $S_n$ true for all integers $n \geq 1$.

**9.** Let $S_n$ be the statement
$$-2-4-6-\ldots-2n=-n-n^2.$$

1. Since $-2=-1-1^2$, $S_1$ is true.
2. Assume $S_k$ and add $-2(k+1)$ to obtain
$$-2-4-6-\ldots-2k-2(k+1)$$
$$=-k-k^2-2(k+1)$$
$$=-(k+1)-(k^2+2k+1)$$
$$=-(k+1)-(k+1)^2$$
Therefore, $S_{k+1}$ is true. Since $S_1$ is true and $S_k$ implies $S_{k+1}$, the principle of mathematical induction makes $S_n$ true for all integers $n \geq 1$.

**11.** Let $S_n$ be the statement
$$1+\frac{2}{5}+\frac{4}{25}+\ldots+\left(\frac{2}{5}\right)^{n-1}=\frac{5}{3}\left[1-\left(\frac{2}{5}\right)^n\right].$$

1. Since $1=\frac{5}{3}\left[1-\frac{2}{5}\right]$, $S_1$ is true.

2. Assume $S_k$ and add $\left(\frac{2}{5}\right)^k$ to obtain
$$1+\frac{2}{5}+\frac{4}{25}+\ldots+\left(\frac{2}{5}\right)^{k-1}+\left(\frac{2}{5}\right)^k$$
$$=\frac{5}{3}\left[1-\left(\frac{2}{5}\right)^k\right]+\left(\frac{2}{5}\right)^k$$
$$=\frac{5}{3}\left[1-\left(\frac{2}{5}\right)^k+\frac{3}{5}\left(\frac{2}{5}\right)^k\right]$$
$$=\frac{5}{3}\left[1-\left(\frac{2}{5}\right)^k\left(1-\frac{3}{5}\right)\right]$$
$$=\frac{5}{3}\left[1-\left(\frac{2}{5}\right)^k\left(\frac{2}{5}\right)\right]=\frac{5}{3}\left[1-\left(\frac{2}{5}\right)^{k+1}\right]$$
Therefore, $S_{k+1}$ is true. Since $S_1$ is true and $S_k$ implies $S_{k+1}$, the principle of mathematical induction makes $S_n$ true for all integers $n \geq 1$.

**13.** Let $S_n$ be the statement

$$1^3 + 2^3 + 3^3 + \ldots + n^3 = \frac{n^2(n+1)^2}{4}.$$

1. Since $1^3 = \frac{1^2(1+1)^2}{4}$, $S_1$ is true.

2. Assume $S_k$ and add $(k+1)^3$ to obtain

$$1^3 + 2^3 + 3^3 + \ldots + k^3 + (k+1)^3$$

$$= \frac{k^2(k+1)^2}{4} + (k+1)^3$$

$$= \frac{k^2(k+1)^2 + 4(k+1)^3}{4}$$

$$= \frac{(k+1)^2[k^2 + 4k + 4]}{4}$$

$$= \frac{(k+1)^2(k+2)^2}{4}$$

$$= \frac{(k+1)^2[(k+1)+1]^2}{4}$$

Therefore, $S_{k+1}$ is true. Since $S_1$ is true and $S_k$ implies $S_{k+1}$, the principle of mathematical induction makes $S_n$ true for all integers $n \geq 1$.

**15.** Let $S_n$ be the statement

$$\sum_{i=1}^{n} ar^{i-1} = \frac{a(1-r^n)}{1-r}, \quad r \neq 1$$

1. $\sum_{i=1}^{1} ar^{i-1} = a = \frac{a(1-r)}{1-r}$, so $S_1$ is true.

2. Assume $S_k$ and add $ar^{(k+1)-1}$ to obtain

$$\sum_{i=1}^{k+1} ar^{i-1} = \frac{a(1-r^k)}{1-r} + ar^k$$

$$= \frac{a - ar^k + ar^k - ar^{k+1}}{1-r}$$

$$= \frac{a - ar^{k+1}}{1-r} = \frac{a(1-r^{k+1})}{1-r}$$

Therefore, $S_{k+1}$ is true. Since $S_1$ is true and $S_k$ implies $S_{k+1}$, the principle of mathematical induction makes $S_n$ true for all integers $n \geq 1$.

**17.** Let $S_n$ be the statement $2^n > 4n$.
1. $2^5 = 32 > 20 = 4 \cdot 5$, so $S_5$ is true.
2. Assume $S_k$ and multiply by two to obtain

$$2 \cdot 2^k > 2 \cdot 4k$$

$$2^{k+1} > 4k + 4k > 4k + 4 = 4(k+1)$$

Therefore $2^{k+1} > 4(k+1)$ and $s_{k+1}$ is true. Since $S_5$ is true and $S_k$ implies $S_{k+1}$, the principle of mathematical induction makes $S_n$ true for all integers $n \geq 5$.

**19.** Let $S_n$ be the statement $a^n < 1, 0 < a < 1$.
1. Since $a^1 = a < 1$ ($0 < a < 1$ is given), $S_1$ is true.
2. Assume $a^k < 1$. Since $a > 0$, $a^{k+1} < a$, but $a < 1$. Therefore $a^{k+1} < 1$ and $S_{k+1}$ is true. Since $S_1$ is true and $S_k$ implies $S_{k+1}$, the principle of mathematical induction makes $S_n$ true for all integers $n \geq 1$.

**21.** Let $S_n$ be the statement $(ab)^n = a^n b^n$.
1. Since $(ab)^1 = a^1 b^1$, $S_1$ is true.
2. Assume $S_k$ and multiply by $(ab)$ to obtain

$$(ab)^k(ab) = (a^k b^k)(ab)$$

$$(ab)^{k+1} = (a^k a)(b^k b) = a^{k+1} b^{k+1}$$

Therefore, $S_{k+1}$ is true. Since $S_1$ is true and $S_k$ implies $S_{k+1}$, the principle of mathematical induction makes $S_n$ true for all integers $n \geq 1$.

**23.** Let $S_k$ be the statement

$$|a_0 + a_1 + \ldots + a_n| \leq |a_0| + |a_1| + \ldots + |a_n|$$

1. Since $|a_0 + a_1| \leq |a_0| + |a_1|$, $S_1$ is true.
2. Assume $S_k$. Then,

$$|a_0 + a_1 + \ldots + a_k + a_{k+1}|$$

$$= |(a_0 + a_1 + \ldots + a_k) + a_{k+1}|$$

$$\leq |a_0 + a_1 + \ldots + a_k| + |a_{k+1}| \quad \text{(by } S_1\text{)}$$

$$\leq |a_0| + |a_1| + \ldots + |a_k| + |a_{k+1}| \quad \text{(by } S_k\text{)}$$

Therefore, $S_{k+1}$ is true. Since $S_1$ is true and $S_k$ implies $S_{k+1}$, the principle of mathematical induction makes $S_n$ true for all integers $n \geq 1$.

**25. (a)** Let $S_n$ be the statement

$$\frac{a^n - b^n}{a - b} = a^{n-1} + a^{n-2}b$$
$$+ \ldots + ab^{n-2} + b^{n-1}$$

1. Since $\dfrac{a^2 - b^2}{a - b} = a + b = a^{2-1} + b^{2-1}$,

   $S_2$ is true.

2. Assume $S_k$. Then,

$$\frac{a^{k+1} - b^{k+1}}{a - b} = \frac{a^k a - b^k b}{a - b}$$
$$= \frac{a^k a - b^k a + b^k a - b^k b}{a - b}$$
$$= \frac{a(a^k - b^k) + b^k(a - b)}{a - b}$$
$$= \frac{a(a^k - b^k)}{a - b} + b^k$$
$$= a[a^{k-1} + a^{k-2}b + \ldots + ab^{k-2}$$
$$+ b^{k-1}] + b^k$$
$$= a^k + a^{k-1}b + \ldots + a^2 b^{k-2}$$
$$+ ab^{k-1} + b^k$$

Therefore $S_{k+1}$ is true. Since $S_2$ is true and $S_k$ implies $S_{k+1}$, the principle of mathematical induction makes $S_n$ true for all integers $n \geq 2$.

**(b)** Since $\dfrac{a^n - b^n}{a - b}$

$$= a^{n-1} + a^{n-2}b + \ldots + ab^{n-2} + b^{n-1},$$

multiplying by $a - b$ gives

$$a^n - b^n = (a - b)(a^{n-1} + a^{n-2}b + \ldots + ab^{n-2} + b^{n-1}).$$

**27. (a)**
$$1 = 1$$
$$1 + 2 + 1 = 4$$
$$1 + 2 + 3 + 2 + 1 = 9$$
$$1 + 2 + 3 + 4 + 3 + 2 + 1 = 16$$
$$1 + 2 + 3 + 4 + 5 + 4 + 3 + 2 + 1 = 25$$

**(b)** $1 + 2 + 3 + \ldots + (n-1) + n + (n-1)$
$$+ \ldots + 3 + 2 + 1$$
$$= n^2$$

**(c)** Let $S_n$ be the statement in (b).

1. Since $1 = 1^2$, $S_1$ is true.

2. Assume $S_k$ and add $k + (k+1)$ to obtain

$$1 + 2 + 3 + \ldots + (k-1) + k$$
$$+ (k+1) + k + (k-1) + \ldots$$
$$+ 3 + 2 + 1$$
$$= k^2 + k + (k+1) = k^2 + 2k + 1$$
$$= (k+1)^2$$

Therefore, $S_{k+1}$ is true. Since $S_1$ is true and $S_k$ implies $S_{k+1}$, the principle of mathematical induction makes $S_n$ true for all integers $n \geq 1$.

**29.** Let $S_n$ be the statement $7^n - 1$ is divisible by 6.

1. Since $7^1 - 1 = 6$ is divisible by 6, $S_1$ is true.

2. Assume $S_k$. Then $7^k - 1 = 6b$ and $7^k = 6b + 1$. Multiply by 7 to get $7^{k+1} = 42b + 7$ and $7^{k+1} - 1 = 42b + 6$, which is divisible by 6. Therefore $S_{k+1}$ is true. Since $S_1$ is true and $S_k$ implies $S_{k+1}$, the principle of mathematical induction makes $S_n$ true for all integers $n \geq 1$.

**Challenge Problem**

Let $f(x) = x^n - 1$. Since $f(1) = 1^n - 1 = 0$, $(x - 1)$ is a factor of $x^n - 1$. So, $f(x) = (x - 1)g(x)$ for some polynomial $g(x)$. Now, let $x = 7$. Then $f(7) = 7^n - 1 = (7 - 1)g(7) = 6g(7)$. Hence $7^n - 1$ is divisible by 6 for $n \geq 1$.

**Chapter 11 Review Exercises**

**1.** A sequence is a function whose domain is a set of consecutive positive integers. An infinite sequence is a sequence in which the domain is infinite.

**3.** $\left(1-\dfrac{1}{1}\right)^1 = 0$

$\left(1-\dfrac{1}{2}\right)^2 = 0.250$

$\left(1-\dfrac{1}{3}\right)^3 = 0.296$

$\left(1-\dfrac{1}{4}\right)^4 = 0.316$

$\left(1-\dfrac{1}{5}\right)^5 = 0.328$

$\left(1-\dfrac{1}{6}\right)^6 = 0.335$

$\left(1-\dfrac{1}{7}\right)^7 = 0.340$

Differences: 0.250, 0.046, 0.020, 0.012, 0.007, 0.005

**5.** $b_1 = \dfrac{2}{1(1+2)} = \dfrac{2}{3}$

$b_2 = \dfrac{2}{2(2+2)} = \dfrac{1}{4}$

$b_3 = \dfrac{2}{3(3+2)} = \dfrac{2}{15}$

$b_4 = \dfrac{2}{4(4+2)} = \dfrac{1}{12}$

$b_5 = \dfrac{2}{5(5+2)} = \dfrac{2}{35}$

$\dfrac{2}{3}, \dfrac{1}{4}, \dfrac{2}{15}, \dfrac{1}{12}, \dfrac{2}{35}$

**7.** $a_1 = \dfrac{(-2)^1}{1(1+1)} = -1$

$a_2 = \dfrac{(-2)^2}{2(2+1)} = \dfrac{2}{3}$

$a_3 = \dfrac{(-2)^3}{3(3+1)} = -\dfrac{2}{3}$

$a_4 = \dfrac{(-2)^4}{4(4+1)} = \dfrac{4}{5}$

$a_5 = \dfrac{(-2)^5}{5(5+1)} = -\dfrac{16}{15}$

$-1, \dfrac{2}{3}, -\dfrac{2}{3}, \dfrac{4}{5}, -\dfrac{16}{15}$

**9.** $a_1 = 5$

$a_2 = 2 \cdot 5 = 10$

$a_3 = 2 \cdot 10 = 20$

$a_4 = 2 \cdot 20 = 40$

$a_5 = 2 \cdot 40 = 80$

5, 10, 20, 40, 80

**11.** $\displaystyle\sum_{i=1}^{n} a_i = a_1 + a_2 + \ldots + a_n$

**13.** $\displaystyle\sum_{k=1}^{6} (-1)^k (k+1)$

$= (-1)^1(1+1) + (-1)^2(2+1)$
$\quad + (-1)^3(3+1) + (-1)^4(4+1)$
$\quad + (-1)^5(5+1) + (-1)^6(6+1)$
$= -2 + 3 - 4 + 5 - 6 + 7 = 3$

**15.** $\displaystyle\sum_{k=1}^{4} (3k-1)$

$= (3 \cdot 1 - 1) + (3 \cdot 2 - 1) + (3 \cdot 3 - 1)$
$\quad + (3 \cdot 4 - 1)$
$= 2 + 5 + 8 + 11 = 26$

**17.** $4 + 8 + 12 + 16 + 20 + 24 = \displaystyle\sum_{k=1}^{6} 4k$

**19.** $a_n = a_1 + (n-1)d$

**21.** $a_1 = 12$

$a_5 = 12 + 4d = -8$

$4d = -20$

$d = -5$

$a_n = 12 + (n-1)(-5) = 17 - 5n$

$a_{20} = 17 - 5(20) = -83$

**23.** $S_{30} = \dfrac{30}{2}[2 \cdot 5 + 29 \cdot (-3)] = 15(-77)$

$= -1155$

**25. (a)** $\displaystyle\sum_{k=1}^{40}(-3k+5) = \frac{40}{2}[2\cdot 2 + 39(-3)]$

$= 20(-113) = -2260$

**(b)** $\displaystyle\sum_{k=1}^{40}\left(-\frac{1}{3}k+5\right)$

$= \frac{40}{2}\left[2\cdot\left(\frac{14}{3}\right) + 39\left(-\frac{1}{3}\right)\right]$

$= 20\left(-\frac{11}{3}\right) = -\frac{220}{3}$

**27.** The arithmetic mean of $a_1$ and $a_2$ is

$\dfrac{a_1+a_2}{2}$.

**29.** A sequence is said to be geometric if each term, after the first, is obtained by multiplying the preceding term by a common value.

**31.** $a_n = \dfrac{1}{3}\left(\dfrac{1}{3}\right)^{n-1} = \left(\dfrac{1}{3}\right)^{n}$

$a_{100} = \left(\dfrac{1}{3}\right)^{100} = \dfrac{1}{3^{100}}$

**33.** $a_n = 12\left(-\dfrac{2}{3}\right)^{n-1}$

$a_8 = 12\left(-\dfrac{2}{3}\right)^{7} = -\dfrac{512}{729}$

**35.** $a_5 = a_1 r^4$

$\dfrac{9}{4} = 36r^4$

$r^4 = \dfrac{1}{16}$

$r = \pm\dfrac{1}{2}$

**37. (a)** $\displaystyle\sum_{k=1}^{7}2\left(\frac{1}{10}\right)^{k+1} = \dfrac{\frac{2}{100}\left(1-\left(\frac{1}{10}\right)^{7}\right)}{1-\frac{1}{10}}$

$= \dfrac{\frac{2}{100}(0.9999999)}{0.9} = 0.02222222$

**(b)** $\displaystyle\sum_{k=1}^{7}2\left(-\frac{1}{10}\right)^{k+1} = \dfrac{\frac{2}{100}\left(1-\left(-\frac{1}{10}\right)^{7}\right)}{1+\frac{1}{10}}$

$= \dfrac{\frac{2}{100}(1.0000001)}{1.1} = 0.01818182$

**39.** The geometric means of $a$ and $b$ are $\pm\sqrt{ab}$

**41.** The $n^{\text{th}}$ partial sum of a geometric series is

$S_n = \displaystyle\sum_{k=1}^{n}a_1 r^{k-1}$. For example,

$\displaystyle\sum_{k=1}^{n}\left(\frac{1}{10}\right)^{k+1} = \dfrac{\frac{1}{100}\left[1-\left(\frac{1}{10}\right)^{n}\right]}{1-\frac{1}{10}}$.

$= \dfrac{\frac{1}{100}(10)}{9}\left(1-\left(\frac{1}{10}\right)^{n}\right)$

$= \dfrac{1}{90}\left[1-\left(\frac{1}{10}\right)^{n}\right]$

**43.** $36 + 24 + 16 + \ldots = \displaystyle\sum_{k=1}^{\infty}36\left(\frac{2}{3}\right)^{k-1}$

$= \dfrac{36}{1-\frac{2}{3}} = \dfrac{36}{\frac{1}{3}} = 108$

**45.** Since $r = \dfrac{5}{3} > 1$, there is no finite sum.

**47.** Since $|r| = 5 > 1$, there is no finite sum.

**49.** $\displaystyle\sum_{k=1}^{\infty}40\left(\frac{6}{10}\right)^{k-1} = \dfrac{40}{1-\frac{6}{10}} = \dfrac{40}{\frac{4}{10}} = 100$ inches

**51.** Let $S_n$ be the statement

$3 + 6 + 9 + \ldots + 3n = \dfrac{3}{2}n(n+1)$

1. Since $3 = \dfrac{3}{2}\cdot 1(1+1)$, $S_1$ is true.

2. Assume $S_k$ and add $3(k+1)$ to obtain

$3 + 6 + 9 + \ldots + 3k + 3(k+1)$

$= \dfrac{3}{2}k(k+1) + 3(k+1)$

$= 3(k+1)\left(\dfrac{1}{2}k + 1\right)$

$= \dfrac{3}{2}(k+1)(k+2)$

Therefore $S_{k+1}$ is true. Since $S_1$ is true and $S_k$ implies $S_{k+1}$, the principle of mathematical induction makes $S_n$ true for all integers $n \geq 1$.

**53.** Let $S_n$ be the statement

$\dfrac{1}{1 \cdot 3} + \dfrac{1}{3 \cdot 5} + \dfrac{1}{5 \cdot 7} + \ldots + \dfrac{1}{(2n-1)(2n+1)}$

$= \dfrac{n}{2n+1}$

1.  Since $\dfrac{1}{1 \cdot 3} = \dfrac{1}{2 \cdot 1 + 1}$, $S_1$ is true.
2.  Assume $S_k$ and add

$\dfrac{1}{[2(k+1)-1][2(k+1)+1]}$

$= \dfrac{1}{(2k+1)(2k+3)}$

$\dfrac{1}{1 \cdot 3} + \dfrac{1}{3 \cdot 5} + \ldots + \dfrac{1}{(2k-1)(2k+1)}$

$+ \dfrac{1}{(2k+1)(2k+3)}$

$= \dfrac{k}{2k+1} + \dfrac{1}{(2k+1)(2k+3)}$

$= \dfrac{k(2k+3)+1}{(2k+1)(2k+3)} = \dfrac{2k^2 + 3k + 1}{(2k+1)(2k+3)}$

$= \dfrac{(2k+1)(k+1)}{(2k+1)(2k+3)} = \dfrac{k+1}{2k+3} = \dfrac{k+1}{2(k+1)+1}$

Therefore, $S_{k+1}$ is true. Since $S_1$ is true and $S_k$ implies $S_{k+1}$, the principle of mathematical induction makes $S_n$ true for all integers $n \geq 1$.

**55.** Let $S_n$ be the statement

$3^n > 27n$

1.  Since $3^5 = 243 > 135 = 27(5)$, $S_5$ holds.

2.  Assume $3^k > 27k$. Multiply by 3 to get

$3^{k+1} > 3(27k)$

$= 27k + 54k > 27k + 27 = 27(k+1)$.

Since $3^{k+1} > 27(k+1)$, $S_{k+1}$ is true. Since $S_5$ is true and $S_k$ implies $S_{k+1}$, the principle of mathematical induction makes $S_n$ true for all integers $n \geq 1$.

## Chapter 11 Test: Standard Answer

**1.** $a_1 = \dfrac{1^2}{6 - 5 \cdot 1} = 1$

$a_2 = \dfrac{2^2}{6 - 5 \cdot 2} = -1$

$a_3 = \dfrac{3^2}{6 - 5 \cdot 3} = -1$

$a_4 = \dfrac{4^2}{6 - 5 \cdot 4} = -\dfrac{8}{7}$

$a_{40} = \dfrac{40^2}{6 - 5 \cdot 40} = -\dfrac{800}{97}$

**2.** $a_{100} = \dfrac{100 + 2}{3(100)^2 + 6(100)} = \dfrac{102}{30,600} = \dfrac{1}{300}$

**3.** $b_1 = (-1)^1 + 1 = 0$
$b_2 = (-1)^2 + 2 = 3$
$b_3 = (-1)^3 + 3 = 2$
$b_4 = (-1)^4 + 4 = 5$
$0, 3, 2, 5$

**4.** $u_{10} = \dfrac{(1-10)^4}{(-3)^{10-1}} = \dfrac{(-9)^4}{(-3)^9} = -\dfrac{1}{3}$

**5.** $a_2 = (-1)^2\left(-\dfrac{2}{3}\right) + \dfrac{4}{3} = \dfrac{2}{3}$

$a_3 = (-1)^3\left(\dfrac{2}{3}\right) + \dfrac{4}{3} = \dfrac{2}{3}$

$a_4 = (-1)^4\left(\dfrac{2}{3}\right) + \dfrac{4}{3} = \dfrac{6}{3} = 2$

$a_5 = (-1)^5(2) + \dfrac{4}{3} = -\dfrac{2}{3}$

**6.** $\displaystyle\sum_{n=1}^{5}\frac{(-2)^n}{n}=\frac{(-2)^1}{1}+\frac{(-2)^2}{2}+\frac{(-2)^3}{3}$

$\displaystyle +\frac{(-2)^4}{4}+\frac{(-2)^5}{5}$

$\displaystyle =-2+2-\frac{8}{3}+4-\frac{32}{5}=-\frac{76}{15}$

**7.** $a_{49}=-3+(48)\frac{1}{2}=21$

**8.** $S_{20}=\dfrac{20}{2}\left[2(-3)+19\cdot\dfrac{1}{2}\right]=10\left(\dfrac{7}{2}\right)=35$

**9.** $a_1=-768,\ r=\dfrac{192}{-768}=-\dfrac{1}{4}$

$a_4=12,\ a_5=-3,\ a_6=\dfrac{3}{4}$

$a_n=-768\left(-\dfrac{1}{4}\right)^{n-1}$

**10.** $a_1=7,\ d=7$

$\displaystyle\sum_{n=1}^{50}7n=\frac{50}{2}[2\cdot7+49\cdot7]=25(357)=8925$

**11.** $a_9=a_1+8d$

$10=-\dfrac{2}{3}+8d$

$8d=\dfrac{32}{3}$

$d=\dfrac{4}{3}$

$a_{21}=-\dfrac{2}{3}+20\cdot\dfrac{4}{3}=\dfrac{78}{3}=26$

**12.** $a_1=90,\ a_n=300$

$300=90+(n-1)\cdot3=3n+87$

$3n=213$ or $n=71$

$\text{Sum}=\dfrac{71}{2}(90+300)=13{,}845$

**13.** $\displaystyle\sum_{k=1}^{4}8\left(\frac{1}{2}\right)^k=\frac{4\left(1-\frac{1}{2^4}\right)}{1-\frac{1}{2}}=\frac{4\left(\frac{15}{16}\right)}{\frac{1}{2}}=\frac{15}{2}$

**14.** $\displaystyle\sum_{j=1}^{101}(4j-50)=\frac{101}{2}[2\cdot(-46)+100\cdot4]$

$\displaystyle =\frac{101}{2}(308)=15{,}554$

**15.** $\displaystyle\sum_{k=1}^{8}12\left(\frac{1}{2}\right)^{k-1}=\frac{12\left(1-\frac{1}{2^8}\right)}{1-\frac{1}{2}}=\frac{12\left(1-\frac{1}{2^8}\right)}{\frac{1}{2}}$

$\displaystyle =24\left(1-\frac{1}{2^8}\right)=\frac{765}{32}$

**16.** $a_6=a_1r^5$

$\dfrac{3}{4}=-24r^5$

$r^5=-\dfrac{1}{32}$

$r=-\dfrac{1}{2}$

**17.** $\displaystyle\sum_{k=1}^{\infty}8\left(\frac{3}{4}\right)^{k+1}=\frac{\frac{9}{2}}{1-\frac{3}{4}}=\frac{\frac{9}{2}}{\frac{1}{4}}=18$

**18.** $1+\dfrac{3}{2}+\dfrac{9}{4}+\ldots$ has no finite sum since

$r=\dfrac{3}{2}>1.$

**19.** $0.06-0.009+0.00135-\ldots$

$\displaystyle =\sum_{k=1}^{\infty}(0.06)\left(\frac{-3}{20}\right)^{k-1}=\frac{\frac{6}{100}}{1+\frac{3}{20}}$

$\displaystyle =\frac{\frac{6}{100}}{\frac{23}{20}}=\frac{6}{115}$

**20.** $0.363636\ldots$

$\displaystyle =\frac{36}{100}+\frac{36}{10{,}000}+\frac{36}{1{,}000{,}000}+\ldots$

$\displaystyle =\frac{36}{100}+\frac{36}{100}\left(\frac{1}{100}\right)+\frac{36}{100}\left(\frac{1}{100}\right)^2+\ldots$

$$= \sum_{k=1}^{\infty} \frac{36}{100}\left(\frac{1}{100}\right)^{k-1} = \frac{\frac{36}{100}}{1-\frac{1}{100}}$$

$$= \frac{\frac{36}{100}}{\frac{99}{100}} = \frac{4}{11}$$

**21.** Amount saved $= \sum_{k=1}^{52}\left[10+(k-1)\frac{1}{10}\right]$

$$= \frac{52}{2}\left[2\cdot10+51\cdot\frac{1}{10}\right]$$

$$= 26\left[\frac{251}{10}\right] = \$652.60$$

**22.** $\sum_{k=3}^{22} k = \sum_{k=1}^{20}(k+2) = \frac{20}{2}[2\cdot3+19\cdot1]$
$= 10(25) = 250$

**23.** $a_1 = 24,\ r = \frac{1}{3}$

Distance $= \dfrac{24}{1-\frac{1}{3}} = \dfrac{24}{\frac{2}{3}} = 36$ feet

**24.** Let $S_n$ be the statement

$$5+10+15+\ldots+5n = \frac{5n(n+1)}{2}$$

1.   Since $5 = \dfrac{5\cdot1(1+1)}{2}$, $S_1$ is true.

2.   Assume $S_k$ and add $5(k+1)$ to obtain
$5+10+15+\ldots+5k+5(k+1)$

$$= \frac{5k(k+1)}{2}+5(k+1)$$

$$= \frac{5k(k+1)+10(k+1)}{2}$$

$$= \frac{5(k+1)(k+2)}{2}$$

$$= \frac{5(k+1)[(k+1)+1]}{2}$$

Therefore $S_{k+1}$ is true. Since $S_1$ is true and $S_k$ implies $S_{k+1}$, the principle of mathematical induction makes $S_n$ true for all integers $n \geq 1$.

**25.** Let $S_n$ be the statement

$$\frac{1}{1\cdot4}+\frac{1}{2\cdot6}+\frac{1}{3\cdot8}+\ldots+\frac{1}{n(2n+2)}$$

$$= \frac{n}{2(n+1)}$$

1.   Since $\dfrac{1}{1\cdot4} = \dfrac{1}{2(1+1)}$, $S_1$ is true.

2.   Assume $S_k$ and add

$$\frac{1}{(k+1)(2k+2+2)} \text{ to obtain}$$

$$\frac{1}{1\cdot4}+\frac{1}{2\cdot6}+\frac{1}{3\cdot8}+\ldots+\frac{1}{k(2k+2)}$$

$$+\frac{1}{(k+1)(2k+4)}$$

$$= \frac{k}{2(k+1)}+\frac{1}{2(k+1)(k+2)}$$

$$= \frac{k(k+2)+1}{2(k+1)(k+2)} = \frac{k^2+2k+1}{2(k+1)(k+2)}$$

$$= \frac{(k+1)^2}{2(k+1)(k+2)} = \frac{k+1}{2(k+2)}$$

$$= \frac{k+1}{2[(k+1)+1]}$$

Therefore, $S_{k+1}$ is true. Since $S_1$ is true and $S_k$ implies $S_{k+1}$, the principle of mathematical induction makes $S_n$ true for all integers $n \geq 1$.

## Chapter 11 Test: Multiple Choice

**1.** $a_{10} = \dfrac{10+2}{2\cdot10^2+3\cdot10-2} = \dfrac{12}{228} = \dfrac{1}{19}$
The answer is (b).

**2.** $a_1 = 1+(-1)^1 = 0$
$a_2 = 1+(-1)^2 = 2$
$a_3 = 1+(-1)^3 = 0$
$a_4 = 1+(-1)^4 = 2$
The answer is (a).

**3.** $\sum_{n=1}^{4} 3^{n-1}(n-1) = 3^{1-1}(1-1) + 3^{2-1}(2-1)$

$\qquad + 3^{3-1}(3-1) + 3^{4-1}(4-1)$

$\qquad = 0 + 3 + 18 + 81 = 102$

The answer is (b).

**4.** $a_{50} = -2 + 49 \cdot 5 = 243$
The answer is (a).

**5.** $S_{20} = \dfrac{20}{2}[2 \cdot 2 + 19(-3)] = 10(-53) = -530$
The answer is (d).

**6.** $\sum_{k=1}^{50}(-2k+3) = \dfrac{50}{2}[2 \cdot 1 + 49(-2)]$

$\qquad = 25(-96) = -2400$
The answer is (d).

**7.** $a_n = 1 + \dfrac{1}{2}(n-1)$

$\sum_{n=1}^{60} a_n = \dfrac{60}{2}\left[2 \cdot 1 + 59\left(\dfrac{1}{2}\right)\right] = 30\left[\dfrac{63}{2}\right] = 945$

The answer is (c).

**8.** $a_n = -\dfrac{1}{2}\left(\dfrac{1}{2}\right)^{n-1} = -\left(\dfrac{1}{2}\right)^n = -\dfrac{1}{2^n}$

$a_{100} = -\dfrac{1}{2^{100}}$

The answer is (c).

**9.** $\sum_{k=1}^{100} 3\left(\dfrac{1}{3}\right)^k = \dfrac{1\left(1-\left(\frac{1}{3}\right)^{100}\right)}{1-\frac{1}{3}} = \dfrac{1-\frac{1}{3^{100}}}{\frac{2}{3}}$

$\qquad = \dfrac{3}{2}\left(1 - \dfrac{1}{3^{100}}\right)$

The answer is (a).

**10.** $a_1 = 512, \; r = \dfrac{1}{2}$

$\sum_{n=1}^{12} 512\left(\dfrac{1}{2}\right)^{n-1} = \dfrac{512\left(1-\left(\frac{1}{2}\right)^{12}\right)}{1-\frac{1}{2}}$

$\qquad = \dfrac{512\left(1-\frac{1}{2^{12}}\right)}{\frac{1}{2}} = \$1023.75$

The answer is (d).

**11.** $a_1 = -27, \; r = \dfrac{1}{3}$

$\sum_{k=1}^{\infty} a_1 r^{k-1} = \dfrac{-27}{1-\frac{1}{3}} = -\dfrac{27}{\frac{2}{3}} = -\dfrac{81}{2}$

The answer is (e).

**12.** The sum is $\sum_{n=1}^{10} 5n$, so

the answer is (a).

**13.** $A.M. = \dfrac{\frac{5}{2} + \frac{45}{2}}{2} = \dfrac{25}{2}$

$G.M. = \pm\sqrt{\dfrac{5}{2} \cdot \dfrac{45}{2}} = \pm\dfrac{15}{2}$

Since $\dfrac{15}{2} < \dfrac{25}{2}$,

the answer is (b).

**14.** $a_1 = \dfrac{16}{9}, \; a_5 = 144$

$144 = \dfrac{16}{9}r^4$

$r^4 = 81$

$r = \pm 3$

$\dfrac{16}{9}, \pm\dfrac{16}{3}, 16, \pm 48, 144$

The answer is (c).

**15.** $\displaystyle\sum_{n=1}^{\infty} 100\left(\frac{7}{100}\right)^{n+1}$

$\displaystyle = \sum_{n=1}^{\infty} 100\left(\frac{7}{100}\right)^{2}\left(\frac{7}{100}\right)^{n-1}$

$\displaystyle = \sum_{n=1}^{\infty} \frac{49}{100}\left(\frac{7}{100}\right)^{n-1}$

$\displaystyle = \frac{\frac{49}{100}}{1-\frac{7}{100}} = \frac{\frac{49}{100}}{\frac{93}{100}} = \frac{49}{93}$

The answer is (a).

**16.** $\displaystyle\sum_{n=1}^{\infty}(-1)^{n}2^{n}$ has no finite sum since $|r| > 1$.

The answer is (d).

**17.** Areas $= \dfrac{1}{2}(8)(16) + \dfrac{1}{2}(4)(8) + \dfrac{1}{2}(2)(4) + \ldots$

$= 64 + 16 + 4 + \ldots = \dfrac{64}{1-\frac{1}{4}} = \dfrac{64}{\frac{3}{4}} = \dfrac{256}{3}$

The answer is (b).

**18.** I and II are true, III is false.

The answer is (c).

**19.** Let $a_n$ be the value at the beginning of year $n$. Then $a_5 =$ value after 4 years (at beginning of 5th year).

$a_n = 6000(0.88)^{n-1}$, so $a_5 = 6000(0.88)^4$

The answer is (b).

**20.** $\displaystyle\sum_{n=21}^{40}(4n-11) = \frac{20}{2}[2\cdot 73 + (19)\cdot 4]$

$= 10(222) = 2220$

The answer is (c).

# Chapter 12: Permutations, Combinations, Probability

**1.** $\dfrac{7!}{6!} = \dfrac{7 \cdot 6!}{6!} = 7$

**3.** $\dfrac{12!}{2! \cdot 10!} = \dfrac{12 \cdot 11 \cdot 10!}{2! \cdot 10!} = \dfrac{12 \cdot 11}{2} = 66$

**5.** $_5P_4 = \dfrac{5!}{(5-4)!} = \dfrac{5!}{1!} = 120$

**7.** $_4P_1 = \dfrac{4!}{(4-1)!} = \dfrac{4!}{3!} = \dfrac{4 \cdot 3!}{3!} = 4$

**9.** $_nP_{n-3} = \dfrac{n!}{[n-(n-3)]!} = \dfrac{n!}{3!}$

**11.** $5 \cdot 3 = 15$

**13.** $_5P_3 = \dfrac{5!}{(5-3)!} = \dfrac{5!}{2!} = \dfrac{120}{2} = 60$

**15.** $_4P_2 = \dfrac{4!}{(4-2)!} = \dfrac{4!}{2!} = \dfrac{24}{2} = 12$

**17.** $3 \cdot 2 \cdot 1 = 6$

**19.** $3(4 \cdot 3) = 36$

**21.** $9 \cdot 8 \cdot 7 = 504$

**23.** There are four possibilities for the last digit, eight for the middle, and seven for the first. $4 \cdot 8 \cdot 7 = 224$

**25.** There are four possibilities for the first digit, eight for the middle, and seven for the last. $4 \cdot 8 \cdot 7 = 224$

**27.** (a) $5! = 120$
  (b) ST _ _ _, $3! = 6$
      TS _ _ _, $3! = 6$
      $6 + 6 = 12$
  (c) There are four possible places for S and T to be together, and two ways to arrange them. $(4 \cdot 2) \cdot 3! = 48$
  (d) $120 - 48 = 72$

**29.** $9! = 362,880$
       $8! = 40,320$

**31.** If one person's position is fixed, there are $6!$ ways for the remaining six people to be seated. $6! = 720$

**33.** (a) $26 \cdot 25 \cdot 10 \cdot 9 \cdot 8 = 468,000$
  (b) $26 \cdot 26 \cdot 10 \cdot 10 \cdot 10 = 676,000$

**35.** (a) $_nP_1 = 10$
  $$\dfrac{n!}{(n-1)!} = 10$$
  $$\dfrac{n \cdot (n-1)!}{(n-1)!} = 10$$
  $$n = 10$$
  (b) $_nP_2 = 12$
  $$\dfrac{n!}{(n-2)!} = 12$$
  $$\dfrac{n(n-1)(n-2)!}{(n-2)!} = 12$$
  $$n^2 - n = 12$$
  $$n^2 - n - 12 = 0$$
  $$(n-4)(n+3) = 0$$
  $$n - 4 = 0 \text{ or } n + 3 = 0$$
  $$n = 4 \text{ or } n = -3$$
  Since $n \geq 0$, $n = 4$.

**37.** $6 \cdot 6 = 36$
| (1, 1) | (2, 1) | (3, 1) | (4, 1) | (5, 1) | (6, 1) |
| (1, 2) | (2, 2) | (3, 2) | (4, 2) | (5, 2) | (6, 2) |
| (1, 3) | (2, 3) | (3, 3) | (4, 3) | (5, 3) | (6, 3) |
| (1, 4) | (2, 4) | (3, 4) | (4, 4) | (5, 4) | (6, 4) |
| (1, 5) | (2, 5) | (3, 5) | (4, 5) | (5, 5) | (6, 5) |
| (1, 6) | (2, 6) | (3, 6) | (4, 6) | (5, 6) | (6, 6) |

**39.** (a) $10^9 = 1,000,000,000$
  (b) $10 \cdot 9 \cdot 8 \cdot 7 \cdot 6 \cdot 5 \cdot 4 \cdot 3 \cdot 2 = {}_{10}P_9$
      $= 3,628,800$
  (c) $1,000,000,000 - 3,628,800$
      $= 996,371,200$

**41.** (a) $6! = 720$
  (b) $5! = 120$

**(c)** There are five ways in which the two books will be next to each other, and two ways in which they may be positioned with respect to each other.
$5 \cdot 2 \cdot 4! = 240$

**43.** $_{12}P_r = 8(_{12}P_{r-1})$

$$\frac{12!}{(12-r)!} = 8 \cdot \frac{12!}{(12-r+1)!}$$

$$\frac{(12-r+1)!}{(12-r)!} = 8$$

$$\frac{(12-r+1)(12-r)!}{(12-r)!} = 8$$

$13 - r = 8$

$-r = -5$

$r = 5$

**45. (a)** $\dfrac{8!}{3!1!1!1!1!1!} = \dfrac{8!}{3!} = \dfrac{40,320}{6} = 6720$

**(b)** $\dfrac{7!}{2!2!1!1!1!} = \dfrac{5040}{4} = 1260$

**(c)** $\dfrac{10!}{2!3!1!2!1!1!} = \dfrac{3,628,800}{24} = 151,200$

**Challenge Problem**

**(a)** $1 \cdot 2 \cdot 3 \cdot 4 \cdot 5 \cdot 6 \cdot 7 = 5040$

**(b)** $1 \cdot 2 \cdot 2 \cdot 2 \cdot 2 \cdot 2 \cdot 2 = 64$

**(c)** The number of paths from $P$ to any $A$ is the same as the number of paths from that $A$ to the middle $D$. There is exactly one path from $P$ to each of the outside $A$'s. There are a total of $1 \cdot 2 \cdot 2 \cdot 2 = 8$ paths from $P$ to the $A$'s, so there are six paths from $P$ to the middle $A$'s. By symmetry of the pyramid, there are three paths from $P$ to each middle $A$. Hence, the total number of paths from $P$ to the middle $D$ is $1^2 + 3^2 + 3^2 + 1^2 = 20$.

**Exercises 12.2**

**1.** $_5C_2 = \dfrac{5!}{2!(5-2)!} = \dfrac{5!}{2!3!} = \dfrac{120}{2 \cdot 6} = 10$

**3.** $_{10}C_0 = \dfrac{10!}{0!(10-0)!} = \dfrac{10!}{0!10!} = 1$

**5.** $\dbinom{15}{15} = \dfrac{15!}{15!(15-15)!} = \dfrac{15!}{15!0!} = 1$

**7.** $\dbinom{30}{27} = \dfrac{30!}{27!(30-27)!} = \dfrac{30 \cdot 29 \cdot 28 \cdot 27!}{27! \cdot 3!}$
$= 4060$

**9. (a)** Since order is not important,
$$\dbinom{20}{4} = \dfrac{20!}{4!(20-4)!}$$
$$= \dfrac{20 \cdot 19 \cdot 18 \cdot 17 \cdot 16!}{4! \cdot 16!} = 4845$$

**(b)** Since order is important,
$$_{20}P_4 = \dfrac{20!}{(20-4)!}$$
$$= \dfrac{20 \cdot 19 \cdot 18 \cdot 17 \cdot 16!}{16!} = 116,280$$

**11.** $\dbinom{6}{3} = \dfrac{6!}{3!(6-3)!} = \dfrac{6!}{3!3!} = \dfrac{720}{6 \cdot 6} = 20$

**13.** $\dbinom{15}{5} = \dfrac{15!}{5!(15-5)!} = \dfrac{15!}{5!10!}$
$$= \dfrac{15 \cdot 14 \cdot 13 \cdot 12 \cdot 11 \cdot 10!}{5!10!} = 3003$$

**15.** $\dbinom{20}{2} = \dfrac{20!}{2!(20-2)!} = \dfrac{20 \cdot 19 \cdot 18!}{2 \cdot 18!} = 190$

**17.** $\dbinom{12}{3}\dbinom{10}{4} = \dfrac{12!}{3!(12-3)!} \cdot \dfrac{10!}{4!(10-4)!}$
$$= \dfrac{12 \cdot 11 \cdot 10 \cdot 9!}{6 \cdot 9!} \cdot \dfrac{10 \cdot 9 \cdot 8 \cdot 7 \cdot 6!}{24 \cdot 6!} = 46,200$$

**19. (a)** $_nC_{n-1} = \dfrac{n!}{(n-1)![n-(n-1)]!}$
$$= \dfrac{n \cdot (n-1)!}{(n-1)! \cdot 1!} = n$$

**(b)** $_nC_{n-2} = \dfrac{n!}{(n-2)![n-(n-2)]!}$
$$= \dfrac{n(n-1)(n-2)!}{(n-2)!2!} = \dfrac{n(n-1)}{2}$$

**21.** $_nC_4 = \dfrac{_nP_4}{4!} = \dfrac{1680}{24} = 70$

**23. (a)** $\binom{2}{0}+\binom{2}{1}+\binom{2}{2}=\dfrac{2!}{0!2!}+\dfrac{2!}{1!1!}+\dfrac{2!}{2!0!}$

$= 1+2+1 = 4 = 2^2$

**(b)** $\binom{3}{0}+\binom{3}{1}+\binom{3}{2}+\binom{3}{3}$

$= \dfrac{3!}{0!3!}+\dfrac{3!}{1!2!}+\dfrac{3!}{2!1!}+\dfrac{3!}{0!3!}$

$= 1+3+3+1 = 8 = 2^3$

**(c)** $\binom{4}{0}+\binom{4}{1}+\binom{4}{2}+\binom{4}{3}+\binom{4}{4}$

$= \dfrac{4!}{0!4!}+\dfrac{4!}{1!3!}+\dfrac{4!}{2!2!}+\dfrac{4!}{3!1!}+\dfrac{4!}{4!0!}$

$= 1+4+6+4+1 = 16 = 2^4$

**(d)** $\binom{5}{0}+\binom{5}{1}+\binom{5}{2}+\binom{5}{3}+\binom{5}{4}+\binom{5}{5}$

$= \dfrac{5!}{0!5!}+\dfrac{5!}{1!4!}+\dfrac{5!}{2!3!}+\dfrac{5!}{3!2!}$

$\quad +\dfrac{5!}{4!1!}+\dfrac{5!}{5!0!}$

$= 1+5+10+10+5+1 = 32 = 2^5$

**25.** Each time a subset of $r$ elements is chosen out of $n$ elements, there are $n-r$ elements left over. Similarly, when $n-r$ elements are chosen out of $n$ elements there are $n-(n-r)=r$ elements left over. Therefore, there must be the same number of subsets of size $r$, $\binom{n}{r}$, as there are

subsets of size $n-r$, $\binom{n}{n-r}$.

**27.** A set of $n$ elements has a total of $2^n$ subsets of all possible sizes, including the empty set and the set itself.

**29.** $4\binom{13}{5}=4\cdot\dfrac{13!}{5!8!}=5148$

**31.** $13\cdot\binom{4}{2}\cdot 12\cdot\binom{4}{3}=13\cdot\dfrac{4!}{2!2!}\cdot 12\cdot\dfrac{4!}{3!1!}$

$= 13\cdot 6\cdot 12\cdot 4 = 3744$

**33. (a)** $\binom{25}{5}\binom{19}{4}=\dfrac{25!}{5!20!}\cdot\dfrac{19!}{4!15!}$

$= 205{,}931{,}880$

**(b)** $\binom{25}{4}\binom{19}{5}=\dfrac{25!}{4!21!}\cdot\dfrac{19!}{5!14!}$

$= 147{,}094{,}200$

**35.** $\binom{3}{1}\binom{5}{2}\binom{4}{2}=\dfrac{3!}{1!2!}\cdot\dfrac{5!}{2!3!}\cdot\dfrac{4!}{2!2!}$

$= 3\cdot 10\cdot 6 = 180$

**37. (a)** $\binom{8}{2}\binom{9}{2}=\dfrac{8!}{2!6!}\cdot\dfrac{9!}{2!7!}=1008$

**(b)** A committee can't contain all men, nor can it contain all women. There are $\binom{17}{4}$ possible committees, so

$\binom{17}{4}-\binom{9}{4}=\dfrac{17!}{4!\cdot 13!}-\dfrac{9!}{4!5!}$

$= 2380 - 126 = 2254$

**39.** $\binom{9}{5}+\binom{9}{6}+\binom{9}{7}+\binom{9}{8}+\binom{9}{9}$

$= \dfrac{9!}{4!5!}+\dfrac{9!}{6!3!}+\dfrac{9!}{7!2!}+\dfrac{9!}{8!1!}+\dfrac{9!}{9!0!}$

$= 126+84+36+9+1 = 256$

**41.** There are 210 ways the people may be divided. There are 5! ways they may be seated at the first table and 3! ways they may be seated at the second table. Therefore, $210\cdot 5!\cdot 3! = 151{,}200$ ways.

**Challenge Problem**

$\binom{n}{r-1}+\binom{n}{r}=\dfrac{n!}{(r-1)!(n-r+1)!}+\dfrac{n!}{r!(n-r)!}$

$= \dfrac{n!\cdot r}{r\cdot(r-1)!(n-r+1)!}+\dfrac{n!(n-r+1)}{r!(n-r)!(n-r+1)}$

$= \dfrac{n!r}{r!(n+1-r)!}+\dfrac{n!(n+1)-n!r}{r!(n+1-r)!}=\dfrac{n!(n+1)}{r!(n+1-r)!}$

$= \dfrac{(n+1)!}{r!(n+1-r)!}=\binom{n+1}{r}$

## Exercises 12.3

**1.** $(x+1)^5 = \binom{5}{0}x^5 1^0 + \binom{5}{1}x^4 1^1 + \binom{5}{2}x^3 1^2 + \binom{5}{3}x^2 1^3 + \binom{5}{4}x^1 1^4 + \binom{5}{5}x^0 1^5$

$= x^5 + 5x^4 + 10x^3 + 10x^2 + 5x + 1$

**3.** $(x+1)^7 = \binom{7}{0}x^7 1^0 + \binom{7}{1}x^6 1^1 + \binom{7}{2}x^5 1^2 + \binom{7}{3}x^4 1^3 + \binom{7}{4}x^3 1^4 + \binom{7}{5}x^2 1^5 + \binom{7}{6}x^1 1^6 + \binom{7}{7}x^0 1^7$

$= x^7 + 7x^6 + 21x^5 + 35x^4 + 35x^3 + 21x^2 + 7x + 1$

**5.** $(a-b)^4 = \binom{4}{0}a^4(-b)^0 + \binom{4}{1}a^3(-b)^1 + \binom{4}{2}a^2(-b)^2 + \binom{4}{3}a^1(-b)^3 + \binom{4}{4}a^0(-b)^4$

$= a^4 - 4a^3 b + 6a^2 b^2 - 4ab^3 + b^4$

**7.** $(3x-y)^5 = \binom{5}{0}(3x)^5(-y)^0 + \binom{5}{1}(3x)^4(-y)^1 + \binom{5}{2}(3x)^3(-y)^2 + \binom{5}{3}(3x)^2(-y)^3 + \binom{5}{4}(3x)^1(-y)^4$

$\quad + \binom{5}{5}(3x)^0(-y)^5$

$= 243x^5 - 405x^4 y + 270x^3 y^2 - 90x^2 y^3 + 15xy^4 - y^5$

**9.** $(a^2+1)^5 = \binom{5}{0}(a^2)^5 \cdot 1^0 + \binom{5}{1}(a^2)^4 \cdot 1^1 + \binom{5}{2}(a^2)^3 \cdot 1^2 + \binom{5}{3}(a^2)^2 \cdot 1^3$

$\quad + \binom{5}{4}(a^2)^1 \cdot 1^4 + \binom{5}{5}(a^2)^0 \cdot 1^5$

$= a^{10} + 5a^8 + 10a^6 + 10a^4 + 5a^2 + 1$

**11.** $(1-h)^{10} = \binom{10}{0}1^{10}(-h)^0 + \binom{10}{1}1^9(-h)^1 + \binom{10}{2}1^8(-h)^2 + \binom{10}{3}1^7(-h)^3$

$\quad + \binom{10}{4}1^6(-h)^4 + \binom{10}{5}1^5(-h)^5 + \binom{10}{6}1^4(-h)^6 + \binom{10}{7}1^3(-h)^7 + \binom{10}{8}1^2(-h)^8 + \binom{10}{9}1^1(-h)^9$

$\quad + \binom{10}{10}1^0(-h)^{10}$

$= 1 - 10h + 45h^2 - 120h^3 + 210h^4 - 252h^5 + 210h^6 - 120h^7 + 45h^8 - 10h^9 + h^{10}$

**13.** $\left(\dfrac{1}{2}-a\right)^4 = \binom{4}{0}\left(\dfrac{1}{2}\right)^4(-a)^0 + \binom{4}{1}\left(\dfrac{1}{2}\right)^3(-a)^1 + \binom{4}{2}\left(\dfrac{1}{2}\right)^2(-a)^2 + \binom{4}{3}\left(\dfrac{1}{2}\right)^1(-a)^3 + \binom{4}{4}\left(\dfrac{1}{2}\right)^0(-a)^4$

$= \dfrac{1}{16} - \dfrac{1}{2}a + \dfrac{3}{2}a^2 - 2a^3 + a^4$

**15.** $\left(\dfrac{1}{x}-x^2\right)^6 = \dbinom{6}{0}\left(\dfrac{1}{x}\right)^6(-x^2)^0 + \dbinom{6}{1}\left(\dfrac{1}{x}\right)^5(-x^2)^1 + \dbinom{6}{2}\left(\dfrac{1}{x}\right)^4(-x^2)^2$

$\qquad + \dbinom{6}{3}\left(\dfrac{1}{x}\right)^3(-x^2)^3 + \dbinom{6}{4}\left(\dfrac{1}{x}\right)^2(-x^2)^4 + \dbinom{6}{5}\left(\dfrac{1}{x}\right)^1(-x^2)^5 + \dbinom{6}{6}\left(\dfrac{1}{x}\right)^0(-x^2)^6$

$\qquad = \dfrac{1}{x^6} - \dfrac{6}{x^3} + 15 - 20x^3 + 15x^6 - 6x^9 + x^{12}$

**17.** $\left(\dfrac{x}{2}+4y\right)^6 = \dbinom{6}{0}\left(\dfrac{x}{2}\right)^6(4y)^0 + \dbinom{6}{1}\left(\dfrac{x}{2}\right)^5(4y)^1 + \dbinom{6}{2}\left(\dfrac{x}{2}\right)^4(4y)^2 + \dbinom{6}{3}\left(\dfrac{x}{2}\right)^3(4y)^3 + \dbinom{6}{4}\left(\dfrac{x}{2}\right)^2(4y)^4$

$\qquad + \dbinom{6}{5}\left(\dfrac{x}{2}\right)^1(4y)^5 + \dbinom{6}{6}\left(\dfrac{x}{2}\right)^0(4y)^6$

$\qquad = \dfrac{1}{64}x^6 + \dfrac{3}{4}x^5y + 15x^4y^2 + 160x^3y^3 + 960x^2y^4 + 3072xy^5 + 4096y^6$

**19.** $(2x-3)^3 = \dbinom{3}{0}(2x)^3(-3)^0 + \dbinom{3}{1}(2x)^2(-3)^1 + \dbinom{3}{2}(2x)^1(-3)^2 + \dbinom{3}{3}(2x)^0(-3)^3$

$\qquad = 8x^3 - 36x^2 + 54x - 27$

**21.**

|   |   |   |   |   |   |   |   |   |   |   |   |   |   |   |   |   |   |   |
|---|---|---|---|---|---|---|---|---|---|---|---|---|---|---|---|---|---|---|
|   |   |   | 1 |   | 7 |   | 21 |   | 35 |   | 35 |   | 21 |   | 7 |   | 1 |   |
|   |   | 1 |   | 8 |   | 28 |   | 56 |   | 70 |   | 56 |   | 28 |   | 8 |   | 1 |   |
|   | 1 |   | 9 |   | 36 |   | 84 |   | 126 |   | 126 |   | 84 |   | 36 |   | 9 |   | 1 |
| 1 |   | 10 |   | 45 |   | 120 |   | 210 |   | 252 |   | 210 |   | 120 |   | 45 |   | 10 |   | 1 |

**23.** $(x-h)^{10} = x^{10} - 10x^9h + 45x^8h^2 - 120x^7h^3 + 210x^6h^4 - 252x^5h^5 + 210x^4h^6 - 120x^3h^7$

$\qquad + 45x^2h^8 - 10xh^9 + h^{10}$

**25.** $(x+1)^{15} = \dbinom{15}{0}x^{15}1^0 + \dbinom{15}{1}x^{14}1^1 + \dbinom{15}{2}x^{13}1^2 + \dbinom{15}{3}x^{12}1^3 + \dbinom{15}{4}x^{11}1^4 + \ldots + \dbinom{15}{11}x^4 1^{11}$

$\qquad + \dbinom{15}{12}x^3 1^{12} + \dbinom{15}{13}x^2 1^{13} + \dbinom{15}{14}x^1 1^{14} + \dbinom{15}{15}x^0 1^{15}$

$\qquad = x^{15} + 15x^{14} + 105x^{13} + 455x^{12} + 1365x^{11} + \ldots + 1365x^4 + 445x^3 + 105x^2 + 15x + 1$

**27.** $(a-1)^{30} = \dbinom{30}{0}a^{30}(-1)^0 + \dbinom{30}{1}a^{29}(-1)^1 + \dbinom{30}{2}a^{28}(-1)^2 + \dbinom{30}{3}a^{27}(-1)^3 + \ldots + \dbinom{30}{27}a^3(-1)^{27}$

$\qquad + \dbinom{30}{28}a^2(-1)^{28} + \dbinom{30}{29}a^1(-1)^{29} + \dbinom{30}{30}a^0(-1)^{30}$

$\qquad = a^{30} - 30a^{29} + 435a^{28} - 4060a^{27} + \ldots - 4060a^3 + 435a^2 - 30a + 1$

**29.** $\dfrac{(3+h)^4 - 81}{h} = \dfrac{81 + 108h + 54h^2 + 12h^3 + h^4 - 81}{h} = \dfrac{108h + 54h^2 + 12h^3 + h^4}{h}$

$\qquad = 108 + 54h + 12h^2 + h^3$

**31.** $\dfrac{(x+h)^6 - x^6}{h} = \dfrac{x^6 + 6x^5h + 15x^4h^2 + 20x^3h^3 + 15x^2h^4 + 6xh^5 + h^6 - x^6}{h}$

$= \dfrac{6x^5h + 15x^4h^2 + 20x^3h^3 + 15x^2h^4 + 6xh^5 + h^6}{h} = 6x^5 + 15x^4h + 20x^3h^2 + 15x^2h^3 + 6xh^4 + h^5$

**33.** $\dfrac{\frac{1}{(2+h)^2} - \frac{1}{4}}{h} \cdot \dfrac{4(2+h)^2}{4(2+h)^2} = \dfrac{4 - (2+h)^2}{4h(2+h)^2} = \dfrac{4 - 4 - 4h - h^2}{4h(2+h)^2} = \dfrac{-4h - h^2}{4h(2+h)^2}$

$= \dfrac{-h(4+h)}{4h(2+h)^2} = -\dfrac{4+h}{4(2+h)^2}$

**35.** 6th term: $\dbinom{10}{5}(a)^5(2b)^5 = 8064a^5b^5$

**37.** 4th term: $\dbinom{7}{3}\left(\dfrac{1}{x}\right)^4\left(\sqrt{x}\right)^3 = 35\dfrac{1}{x^4} \cdot x^{3/2} = 35x^{-5/2}$

**39.** $\dbinom{8}{3}(2x)^5(3y)^3 = 56(32x^5)(27y^3) = 48{,}384x^5y^3$

**41.** Middle term: $\dbinom{10}{5}(3a)^5\left(-\dfrac{b}{2}\right)^5 = -\dfrac{15{,}309}{8}a^5b^5$

**43.** $(2.1)^4 = (2+0.1)^4 = 2^4 + 4 \cdot 2^3(0.1) + 6 \cdot 2^2(0.1)^2 + 4 \cdot 2(0.1)^3 + (0.1)^4$
$= 16 + 3.2 + 0.24 + 0.008 + 0.0001 = 19.4481$

**45.** $(3.98)^3 = (4 - 0.02)^3 = 4^3 + 3 \cdot 4^2(-0.02) + 3 \cdot 4(-0.02)^2 + (-0.02)^3$
$= 64 - 0.96 + 0.0048 - 0.000008 = 63.044792$

**47.** $\dbinom{n}{r-1}a^{n-r+1}b^{r-1}, \ \dbinom{n}{r-1}a^{n-r+1}(-b)^{r-1}$

**49.** $\dbinom{n}{r}a^{n-r}b^r$

## Challenge Problem

1. Since $\displaystyle\sum_{n=1}^{\infty} x^n = \frac{x}{1-x}$ for $|x| < 1$, $(1-x)^{-1} = \frac{1}{1-x} = \frac{1}{x}\sum_{n=1}^{\infty} x^n = \frac{1}{x}(x^1 + x^2 + x^3 + \dots)$

$= 1 + x + x^2 + x^3 + x^4 + \dots$ and $\displaystyle\sum_{n=1}^{\infty}(-x)^n = \sum_{n=1}^{\infty}(-x)(-x)^{n-1} = \frac{-x}{1-(-x)} = \frac{-x}{1+x}$ so

$(1+x)^{-1} = \frac{1}{1+x} = \left(-\frac{1}{x}\right)\sum_{n=1}^{\infty}(-x)^n = \left(-\frac{1}{x}\right)(-x^1 + x^2 - x^3 + x^4 - \dots) = 1 - x + x^2 - x^3 + \dots$

## Critical Thinking

1. True, since the formula was derived is Section 1 using the fundamental principle of counting.

3. The number of groups of six numbers is a combination since order doesn't matter. The number of groups is $\dbinom{40}{6} = 3{,}838{,}380$.

5. $\dbinom{n}{0} - \dbinom{n}{1} + \dbinom{n}{2} - \dbinom{n}{3} + \dots + (-1)^n \dbinom{n}{n} = (1-1)^n = 0$

So, the number of subsets formed from a set of $n$ elements that have an even number of elements is the same as the number of subsets that have an odd number of elements.

## Exercises 12.4

1. $E = \{HHH\}$

$P(E) = \dfrac{n(E)}{n(S)} = \dfrac{1}{8}$

3. $E = \{HTT, THT, TTH, HHT, HTH,$
$\quad THH, HHH\}$

$P(E) = \dfrac{n(E)}{n(S)} = \dfrac{7}{8}$

5. $E = \{TTT, HTT, THT, TTH\}$

$P(E) = \dfrac{4}{8} = \dfrac{1}{2}$

7. $P(\text{odd number}) = \dfrac{n(\text{odd number})}{n(S)} = \dfrac{8}{16} = \dfrac{1}{2}$

9. $P(\text{white}) = \dfrac{n(\text{white})}{n(S)} = \dfrac{9}{16}$

11. $P(\text{even and red}) = \dfrac{n(\text{even and red})}{n(S)}$

$= \dfrac{4}{16} = \dfrac{1}{4}$

13. $P(\text{prime}) = \dfrac{n(\text{prime})}{n(S)} = \dfrac{6}{16} = \dfrac{3}{8}$

15. $P(\text{same number}) = \dfrac{n(\text{same number})}{n(S)}$

$= \dfrac{6}{36} = \dfrac{1}{6}$

17. $P(\text{sum is 7}) = \dfrac{n(\text{sum is 7})}{n(S)} = \dfrac{6}{36} = \dfrac{1}{6}$

19. $P(\text{sum is 2, 3, or 12})$
$= \dfrac{n(\text{sum is 2, 3, or 12})}{n(S)} = \dfrac{4}{36} = \dfrac{1}{9}$

21. $P(\text{sum is odd}) = \dfrac{n(\text{sum is odd})}{n(S)} = \dfrac{18}{36} = \dfrac{1}{2}$

**23.** $\dfrac{26}{52} \cdot \dfrac{25}{51} = \dfrac{25}{102}$

**25.** $\dfrac{1}{52} \cdot \dfrac{0}{51} = 0$

**27.** $P(\text{both black}) = \dfrac{26}{52} \cdot \dfrac{26}{52} = \dfrac{1}{4}$

$P(\text{both red}) = \dfrac{26}{52} \cdot \dfrac{26}{52} = \dfrac{1}{4}$

$P(\text{both black or both red}) = \dfrac{1}{4} + \dfrac{1}{4} = \dfrac{1}{2}$

**29.** $\dfrac{1}{52} \cdot \dfrac{1}{52} = \dfrac{1}{2704}$

**31.** $\dfrac{48}{52} \cdot \dfrac{48}{52} = \dfrac{144}{169}$

**33.** $\dfrac{4}{52} \cdot \dfrac{4}{52} = \dfrac{1}{169}$

**35.** $P(\text{all red}) = \dfrac{\binom{8}{3}}{\binom{13}{3}} = \dfrac{\frac{8!}{3!5!}}{\frac{13!}{3!10!}} = \dfrac{8 \cdot 7 \cdot 6}{13 \cdot 12 \cdot 11}$

$= \dfrac{28}{143}$

**37.** $P(\text{2 red, 1 green}) = \dfrac{\binom{8}{2}\binom{5}{1}}{\binom{13}{3}} = \dfrac{28 \cdot 5}{\frac{13 \cdot 12 \cdot 11}{3!}}$

$= \dfrac{140}{286} = \dfrac{70}{143}$

**39.** **(a)** $P(100\%) = \dfrac{\binom{10}{10}}{2^{10}} = \dfrac{1}{1024}$

**(b)** $P(0\%) = \dfrac{\binom{10}{0}}{2^{10}} = \dfrac{1}{1024}$

**(c)** $P(\geq 80\%) = \dfrac{\binom{10}{8}}{2^{10}} + \dfrac{\binom{10}{9}}{2^{10}} + \dfrac{\binom{10}{10}}{2^{10}}$

$= \dfrac{45}{1024} + \dfrac{10}{1024} + \dfrac{1}{1024} = \dfrac{56}{1024} = \dfrac{7}{128}$

**41.** $P(\text{not 2 red cards}) = 1 - P(\text{2 red cards})$

$= 1 - \dfrac{25}{102}$      (by Exercise 23)

$= \dfrac{77}{102}$

**43.** **(a)** $\dfrac{48}{\binom{52}{5}} = 0.0000185$

**(b)** $13 \cdot \dfrac{48}{\binom{52}{5}} = 0.0002401$

**(c)** $4 \cdot \left[ \dfrac{\binom{13}{5}}{\binom{52}{5}} \right] = 0.0019808$

**(d)** $\dfrac{4}{\binom{52}{5}} = 0.0000015$

**(e)** $\dfrac{\binom{4}{2}\binom{48}{3} + \binom{4}{3}\binom{48}{2} + \binom{4}{4}\binom{48}{1}}{\binom{52}{5}}$

$= 0.0416844$

**45.** $P(A) = \dfrac{4}{52} = \dfrac{1}{13}, \ P(\text{not } A) = \dfrac{12}{13}$

Odds in favor: $\dfrac{\frac{1}{13}}{\frac{12}{13}} = \dfrac{1}{12}$ or 1 to 12

Odds against: 12 to 1

**47.** $P(\text{7 or 11}) = \dfrac{8}{36} = \dfrac{2}{9}$

Odds against: $\dfrac{\frac{7}{9}}{\frac{2}{9}} = \dfrac{7}{2}$ or 7 to 2

**49.** $P(\text{2 heads}) = \dfrac{3}{2^3} = \dfrac{3}{8}$

Odds for: $\dfrac{\frac{3}{8}}{\frac{5}{8}} = \dfrac{3}{5}$ or 3 to 5

**51.** Let $x = P(E)$. Then $P(\text{not } E) = 1 - x$ and the odds for $E$ are $\dfrac{a}{b} = \dfrac{x}{1-x}$. Then,

$$a(1 - x) = bx$$
$$a - ax = bx$$
$$a = ax + bx = (a + b)x$$
$$\frac{a}{a+b} = x \text{ or } P(E) = \frac{a}{a+b}$$

**53.** $P(2B, 2G) = \dfrac{\binom{4}{2}}{2^4} = \dfrac{6}{16} = \dfrac{3}{8}$

**55.** Probability $= \dfrac{1}{\binom{45}{5}} \cdot \dfrac{1}{45} = \dfrac{1}{1,221,759} \cdot \dfrac{1}{45}$

$= \dfrac{1}{54,979,155} \approx \dfrac{1}{55,000,000}$

**57.** $P(\text{straight only}) = \dfrac{10 \cdot 4 \cdot 4 \cdot 4 \cdot 4 \cdot 4 - 40}{\binom{52}{5}}$

$= \dfrac{10,200}{\binom{52}{5}} = 0.0039$

**59.** It is a fair game using sums since $P(\text{odd}) = P(\text{even}) = \dfrac{1}{2}$. It is not a fair game using products since

$P(\text{even product}) = \dfrac{27}{36} = \dfrac{3}{4}$ and

$P(\text{odd product}) = \dfrac{9}{36} = \dfrac{1}{4}$.

**Challenge Problem**

Let the three cards be $\dfrac{R_1}{R_2}, \dfrac{R_3}{G_1}, \dfrac{G_2}{G_3}$. Since a red side is showing, there are three possibilities, all equally likely: $\dfrac{R_1}{R_2}, \dfrac{R_2}{R_1}, \dfrac{R_3}{G_1}$. Two of these have a red side not showing, so the probability that the other side is also red is $\dfrac{2}{3}$.

**Chapter 12 Review Exercises**

**1.** Suppose that a first task can be completed in $m_1$ ways, a second task in $m_2$ ways, and so on, until we reach the $r^{\text{th}}$ task that can be done in $m_r$ ways; then the total number of ways in which these tasks can be completed together is the product $m_1 m_2 \ldots m_r$. For example, the total number of ways of rolling two dice is $6 \cdot 6 = 36$.

**3.** $_nP_r = n(n-1)(n-2) \ldots (n-r+1)$

$= \dfrac{n!}{(n-r)!}$

**5.** (a) $_9P_3 = \dfrac{9!}{(9-3)!} = \dfrac{9!}{6!} = \dfrac{9 \cdot 8 \cdot 7 \cdot 6!}{6!}$
$= 504$

(b) $_5P_5 = \dfrac{5!}{(5-5)!} = \dfrac{5!}{0!} = 120$

(c) $\dfrac{10!}{7!3!} = \dfrac{10 \cdot 9 \cdot 8 \cdot 7!}{7!3!} = \dfrac{720}{6} = 120$

**7.** $_{20}P_3 = \dfrac{20!}{(20-3)!} = \dfrac{20!}{17!} = \dfrac{20 \cdot 19 \cdot 18 \cdot 17!}{17!}$
$= 6840$

**9.** $26 \cdot 26 \cdot 26 \cdot 26 = 26^4 = 456,976$

**11.** A combination is a subset of $r$ elements out of $n$ distinct elements without regard to order.

**13.** (a) $_{25}C_5 = \dfrac{25!}{5!(25-5)!}$
$= \dfrac{25 \cdot 24 \cdot 23 \cdot 22 \cdot 21 \cdot 20!}{5!20!} = 53,130$

(b) $_{25}C_0 = \dfrac{25!}{0!(25-0)!} = \dfrac{25!}{0!25!} = 1$

(c) $\binom{12}{3} = \dfrac{12!}{3!(12-3)!} = \dfrac{12!}{3!9!}$
$= \dfrac{12 \cdot 11 \cdot 10 \cdot 9!}{6 \cdot 9!} = 220$

**15.** $\binom{6}{3} = \dfrac{6!}{3!3!} = \dfrac{720}{6 \cdot 6} = 20$

**17.** $\dbinom{4}{3} = \dfrac{4!}{3!(4-3)!} = \dfrac{4!}{3!1!} = \dfrac{24}{6} = 4$

**19.** $\dbinom{10}{6} = \dbinom{10}{4} = \dfrac{10!}{6!4!} = 210$

**21.** 1   8   28   56   70   56   28   8   1

**23.** $\dbinom{n}{r}a^{n-r}b^r$

**25.** $(x-3)^5 = \dbinom{5}{0}x^5(-3)^0 + \dbinom{5}{1}x^4(-3)^1$

$+\dbinom{5}{2}x^3(-3)^2 + \dbinom{5}{3}x^2(-3)^3$

$+\dbinom{5}{4}x^1(-3)^4 + \dbinom{5}{5}x^0(-3)^5$

$= x^5 - 15x^4 + 90x^3 - 270x^2 + 405x - 243$

**27.** $(2a-3b)^3 = \dbinom{3}{0}(2a)^3(-3b)^0$

$+\dbinom{3}{1}(2a)^2(-3b)^1 + \dbinom{3}{2}(2a)^1(-3b)^2$

$+\dbinom{3}{3}(2a)^0(-3b)^3$

$= 8a^3 - 36a^2b + 54ab^2 - 27b^3$

**29.** 7th term: $\dbinom{10}{6}a^4b^6 = 210a^4b^6$

**31.** $\{HHH, HHT, HTH, THH, HTT, THT,$
$THH, TTT\}$

**33.** $E = \{HHH, TTT\}$
$P(E) = \dfrac{n(E)}{n(S)} = \dfrac{2}{8} = \dfrac{1}{4}$

**35.** $P(\text{sum is 5}) = \dfrac{4}{36} = \dfrac{1}{9}$

**37.** $P(\text{sum of 2, 3, 12}) = \dfrac{4}{36} = \dfrac{1}{9}$

**39.** $P(\text{a pair with one number twice the other})$
$= \dfrac{6}{36} = \dfrac{1}{6}$

**41.** $P(\text{both red}) = \dfrac{26}{52} \cdot \dfrac{25}{51} = \dfrac{25}{102}$

**43.** $P(\text{both kings}) = \dfrac{4}{52} \cdot \dfrac{4}{52} = \dfrac{1}{13} \cdot \dfrac{1}{13} = \dfrac{1}{169}$

**45.** $P(AK) = \dfrac{4}{52} \cdot \dfrac{4}{52} = \dfrac{1}{169}$

**47.** $P(\text{at least three hearts})$

$= \dfrac{\dbinom{13}{3}\dbinom{39}{2} + \dbinom{13}{4}\dbinom{39}{1} + \dbinom{13}{5}\dbinom{39}{0}}{\dbinom{52}{5}}$

$= 0.0927671$

## Chapter 12 Test: Standard Answer

**1. (a)** $_{10}P_3 = \dfrac{10!}{(10-3)!} = \dfrac{10!}{7!} = \dfrac{10 \cdot 9 \cdot 8 \cdot 7!}{7!}$
$= 720$

   **(b)** $_{10}C_3 = \dfrac{_{10}P_3}{3!} = \dfrac{720}{6} = 120$

**2.** $5! = 120$

**3. (a)** $9 \cdot 10 \cdot 10 = 900$
   **(b)** $9 \cdot 9 \cdot 8 = 648$

**4.** $\dbinom{15}{3} = \dfrac{15!}{3!12!} = \dfrac{15 \cdot 14 \cdot 13 \cdot 12!}{6 \cdot 12!} = 455$

**5.** $9 \cdot 10 \cdot 5 = 450$
(if 0 cannot be the hundreds digit)

**6.** $_6P_4 = \dfrac{6!}{(6-4)!} = \dfrac{6!}{2!} = \dfrac{720}{2} = 360$

**7.** $\dfrac{7!}{3!2!1!1!} = \dfrac{5040}{6 \cdot 2} = 420$

**8.** $\dbinom{12}{2}\dbinom{10}{2} = \dfrac{12!}{2!10!} \cdot \dfrac{10!}{2!8!} = 2970$

**9.** $7\dbinom{n}{2} = 3\dbinom{n+1}{3}$

$7 \cdot \dfrac{n!}{2!(n-2)!} = 3 \cdot \dfrac{(n+1)!}{3!(n+1-3)!}$

$\dfrac{7}{2} \cdot \dfrac{n!}{(n-2)!} = \dfrac{1}{2} \cdot \dfrac{(n+1)!}{(n-2)!}$

$7(n!) = (n+1)!$

$7 \cdot n! = (n+1)n!$

$7 = n+1$

$n = 6$

**10. (a)** $2^5 = 32$

**(b)** $P(100\%) = \dfrac{\dbinom{5}{5}}{32} = \dfrac{1}{32}$

**(c)** $P(0\%) = \dfrac{\dbinom{5}{0}}{32} = \dfrac{1}{32}$

**11.** $P(\text{even or} < 3) = P(1, 2, 4, 6) = \dfrac{4}{6} = \dfrac{2}{3}$

**12. (a)** $P(\text{sum is }12) = \dfrac{1}{36}$

**(b)** $P(\text{sum not }12) = 1 - P(\text{sum is }12)$

$= 1 - \dfrac{1}{36} = \dfrac{35}{36}$

**(c)** $P(\text{sum} < 5) = P(\text{sum is }2, 3, 4) = \dfrac{6}{36}$

$= \dfrac{1}{6}$

**13. (a)** $P(\text{both aces or both kings})$

$= \dfrac{4}{52} \cdot \dfrac{4}{52} + \dfrac{4}{52} \cdot \dfrac{4}{52} = \dfrac{1}{169} + \dfrac{1}{169} = \dfrac{2}{169}$

**(b)** $P = \dfrac{4}{52} \cdot \dfrac{3}{51} + \dfrac{4}{52} \cdot \dfrac{3}{51} = \dfrac{1}{221} + \dfrac{1}{221}$

$= \dfrac{2}{221}$

**14.** $P(\text{4 picture cards}) = \dfrac{\dbinom{12}{4}\dbinom{40}{1}}{\dbinom{52}{5}}$

$= 0.0076184$

**15.** $P(\text{4 red chips}) = \dfrac{\dbinom{20}{4}\dbinom{15}{1}}{\dbinom{35}{5}} = 0.2238689$

**16. (a)** $P(\text{4 heads}) = \dfrac{\dbinom{4}{4}}{2^4} = \dfrac{1}{16}$

**(b)** $P(\text{0 heads}) = \dfrac{\dbinom{4}{0}}{2^4} = \dfrac{1}{16}$

**17.** $P(\text{both odd}) = \dfrac{5}{11} \cdot \dfrac{4}{10} = \dfrac{20}{110} = \dfrac{2}{11}$

**18.** $P(\text{both red}) = \dfrac{3}{8} \cdot \dfrac{6}{10} = \dfrac{18}{80} = \dfrac{9}{40}$

**19.** $P(\text{red}) = \dfrac{1}{2} \cdot \dfrac{3}{8} + \dfrac{1}{2} \cdot \dfrac{6}{10} = \dfrac{3}{16} + \dfrac{6}{20}$

$= \dfrac{15}{80} + \dfrac{24}{80} = \dfrac{39}{80}$

**20.** $P(\text{both not blue}) = \dfrac{14}{27} \cdot \dfrac{13}{26} = \dfrac{7}{27}$

**21.** $(x-2y)^5 = \dbinom{5}{0}x^5(-2y)^0 + \dbinom{5}{1}x^4(-2y)^1$

$+ \dbinom{5}{2}x^3(-2y)^2 + \dbinom{5}{3}x^2(-2y)^3$

$+ \dbinom{5}{4}x^1(-2y)^4 + \dbinom{5}{5}x^0(-2y)^5$

$= x^5 - 10x^4y + 40x^3y^2 - 80x^2y^3 + 80xy^4$

$- 32y^5$

**22.** $\dbinom{10}{0}\left(\dfrac{1}{2a}\right)^{10}(-b)^0 + \dbinom{10}{1}\left(\dfrac{1}{2a}\right)^9(-b)^1$

$+ \dbinom{10}{2}\left(\dfrac{1}{2a}\right)^8(-b)^2$

$+ \dbinom{10}{3}\left(\dfrac{1}{2a}\right)^7(-b)^3$

$= \dfrac{1}{1024a^{10}} - \dfrac{5b}{256a^9} + \dfrac{45b^2}{256a^8} - \dfrac{15b^3}{16a^7}$

**23.** $7^{\text{th}}$ term:

$$\binom{11}{6}(3a)^5 b^6 = 462(243a^5)b^6$$
$$= 112{,}166a^5 b^6$$

**24.** Middle term:

$$\binom{16}{8}(2x)^8(-y)^8 = (12{,}870)(256x^8)y^8$$
$$= 3{,}294{,}720x^8 y^8$$

**25.** $(3.1)^4 = (3+0.1)^4$
$= 3^4 + 4\cdot3^3(0.1) + 6\cdot3^2(0.1)^2$
$\qquad + 4\cdot3(0.1)^3 + (0.1)^4$
$= 81 + 10.8 + 0.54 + 0.012 + 0.0001$
$= 92.3521$

## Chapter 12 Test: Multiple Choice

**1.** $9\cdot9\cdot8\cdot7 = 4536$
The answer is (c).

**2.** All three are correct, so
the answer is (d).

**3.** 5! ways.
The answer is (c).

**4.** $\binom{5}{2}\binom{6}{3} = \dfrac{5!}{2!3!}\cdot\dfrac{6!}{3!3!} = 10\cdot20 = 200$
The answer is (b).

**5.** $\dfrac{8!}{1!2!2!2!1!} = \dfrac{8!}{8} = 5040$
The answer is (b).

**6.** I and II are true, but III is false.
The answer is (e).

**7.** $\binom{6}{3} = \dfrac{6!}{3!3!} = 20$
The answer is (d).

**8.** Coefficient of $x^3$:
$$\binom{5}{3}(2)^2 x^3 = 10\cdot4x^3 = 40x^3$$
The answer is (a).

**9.** $5^{\text{th}}$ term:
$$\binom{6}{4}(3a)^2(-b)^4 = 15\cdot9a^2 b^4 = 135a^2 b^4$$
The answer is (b).

**10.** Middle term:
$$\binom{6}{3}x^3(-2)^3 = 20x^3(-8) = -160x^3$$
The answer is (c).

**11.** $9^{\text{th}}$ term: $\binom{n}{8}a^{n-8}b^8$
The answer is (a).

**12.** $P(\text{not all same}) = \dfrac{6}{8} = \dfrac{3}{4}$
The answer is (d).

**13.** $P(\text{4 aces}) = \dfrac{(_4C_4)(_{48}C_1)}{_{52}C_5}$
The answer is (c).

**14.** $P(\text{no spades}) = \dfrac{39}{52}\cdot\dfrac{39}{52} = \dfrac{3}{4}\cdot\dfrac{3}{4} = \dfrac{9}{16}$
The answer is (a).

**15.** $P(\text{sum is 10}) = \dfrac{3}{36} = \dfrac{1}{12}$
The answer is (b).

**16.** The answer is (c).

**17.**
$$\begin{array}{ccccccc} 1 & 6 & 15 & 20 & 15 & 6 & 1 \\ 1 & 7 & 21 & 35 & 35 & 21 & 7 & 1 \end{array}$$
The answer is (d).

**18.** I is correct, but II and II are incorrect.
The answer is (a).

**19.** The term containing $a^r$ is: $\binom{n}{n-r}a^r b^{n-r}$
The answer is (b).

**20.** $P(\text{2 red, 2 blue}) = \dfrac{\binom{8}{2}\binom{5}{2}}{\binom{13}{4}} = \dfrac{(_8C_2)(_5C_2)}{_{13}C_4}$

The answer is (c).

# Graphing Calculator Exercises

## Notes About This Manual:

1. Most graphing calculators do not place solid or open dots on the graphs of number lines or piecewise-defined functions. These were added during production of this manual so that the students could see the calculator screen along with the proper solution.
2. Most graphing calculators will not label the graphed functions, as below. However, many graphing calculators have a small number in the upper right-hand corner when using the TRACE operation that shows which function is currently being traced.
3. The decimal answers given in this manual, especially the coordinates of points on the graphs, will vary slightly between calculators. The values in this manual were obtained using the TI-81 and/or the TI-82.
4. All the calculator screens in this manual were graphed on the Texas Instruments TI-82 Graphing Calculator.

## Graphing Calculator Exercises 1.3

**1.** $y = (3x + 1 \le 2x + 5)(1)$

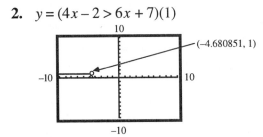

Algebraic solution: $x \le 4$

**2.** $y = (4x - 2 > 6x + 7)(1)$

Algebraic solution: $x < -4.5$

**3.** $y = (3 < 2x + 4)(2x + 4 < 15)(1)$

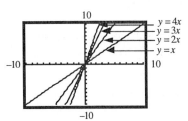

Algebraic solution: $-0.5 < x < 5.5$

## Graphing Calculator Exercises 2.1

**1.** (a) 4  (b) 13  (c) 1  (d) −1
   (e) 57.9358

**2.** (a) 0  (b) 24  (c) 1.414213562
   (d) 3.464101615  (e) 3.81587146
   [**c**, **d**, and **e** are calculator approximations.]

**3.** It gives an error message in both cases, since the values of $x$ are not in the domain of the function. It also gives the choice of quitting or going to the location of the error in the expression for $f(x)$. The domain consists of all real $x$ such that $x > -1$.

## Graphing Calculator Exercises 2.2

**1.**

Equations of the form $y = mx$ are all straight lines through the origin. In fact, the $m$'s are their slopes. All the slopes are positive, and all the lines slope upward from left to right. The larger the slope, the steeper the line.

**2.**

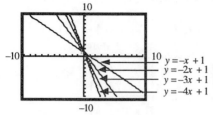

$$y = -x + 1$$
$$y = -2x + 1$$
$$y = -3x + 1$$
$$y = -4x + 1$$

Equations of the form $y = mx$ having negative $m$'s slope downward from left to right. As the absolute value of $m$ increases, the line gets steeper.

**3.**

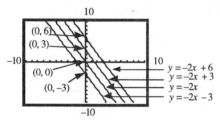

$$y = -2x + 6$$
$$y = -2x + 3$$
$$y = -2x$$
$$y = -2x - 3$$

All the equations $y = -2x + b$ intersect the $y$-axis at the points $(0, b)$. The $y$-intercepts of each pair of adjacent lines differ by 3 units.

**4.** All the lines in Exercise **3** have the same slope and they are parallel to one another.

**5.**

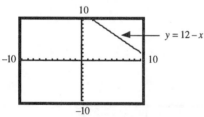

$$y = 12 - x$$

The figure with the standard window does not show the $x$- or $y$-intercepts of the line. A window of $0 \le X \le 12$, $0 \le Y \le 12$ with $x$ and $y$ scales equal to 1 shows both intercepts $(12, 0)$ and $(0, 12)$. These coordinates both satisfy the equation of the line.

**6. (a)**

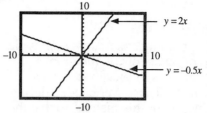

$$y = 2x$$
$$y = -0.5x$$

**(b)** The lines do not appear perpendicular.
**(c)** The scales on the axes are different. This distorts the figure.
**(d)** Use the ZoomSquare function on your calculator to get a screen with a proper aspect ratio.

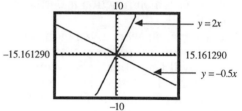

$$y = 2x$$
$$y = -0.5x$$

### Graphing Calculator Exercises 2.3

**1.** $y = 3x - 13$

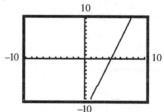

**2.** $y = 2x + 5$

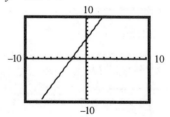

**3.** $y = -2x + 3$

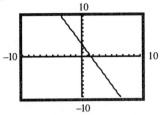

**4.** $y = 2x + 6$

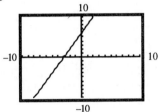

**5.** $y = -\dfrac{2}{5}x + 2$

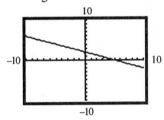

**6.** (4.105623, −0.6842105) on the TI-81. Other calculators may differ slightly.

**7.** (0.10525316, 1.9578947) and (4.9473684, 0.2105262) on the TI-81. Other calculators may differ slightly.

**8.** A friendly window lets you trace right on the point.

**9.** The suggested window lets you TRACE on the first point but not the second (on the TI-81). Doubling the RANGE variables lets you trace the second point as well.

## Graphing Calculator Exercises 2.4

**1.**

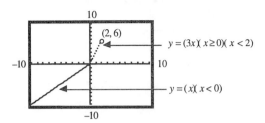

**2.**

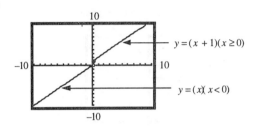

**3.**

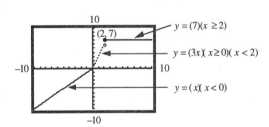

**4.** The last point of the second interval is incorrectly connected to the first point of the third interval in CONNECTED mode.

**5.** (a)

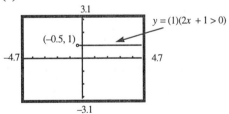

Algebraic solution: $x > -0.5$

**(b)**

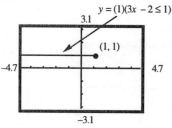

$$y = (1)(3x - 2 \le 1)$$

Algebraic solution: $x \le 1$

**(c)**

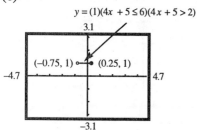

$$y = (1)(4x + 5 \le 6)(4x + 5 > 2)$$

Algebraic solution: $-0.75 < x \le 0.25$

## Graphing Calculator Exercises 2.5

1. The solution is $x = 2$, $y = 1$.

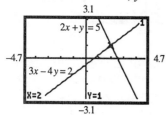

2. No.

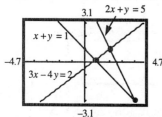

3. $x = 1.48$, $y = -0.13$. The exact coordinates are $x = \dfrac{67}{45}$ and $y = -\dfrac{6}{45}$. Using a calculator, the decimal approximations are $x = 1.488888889$ and $y = -0.133333333$.

## Graphing Calculator Exercises 2.6

1. The added or subtracted constant translates the graph of $y = x^2$ up or down respectively.

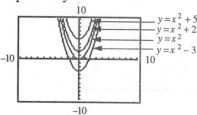

2. The coefficent 2 stretches the parabola in (a). The coefficients $-2$ and $-3$ stretch and reflect it across the $x$-axis.

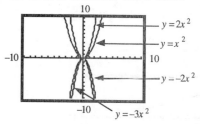

3. The $-3$ in (b) translates the parabola in (a) three units to the right. The $+3$ in (c) translates the parabola in (a) three units to the left.

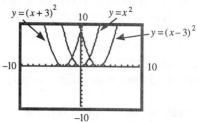

4. The subtracted 2 translates (a) to the right two units. The multiplier 0.5 contracts (b). The added 3 translates (c) up three units.

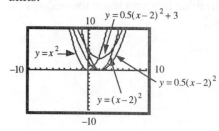

**5.**

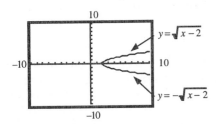

## Graphing Calculator Exercises 2.7

**1.** The numbers $a$ and $b$ are the $x$-intercepts of the parabolas. For Exercise **93**, they are $\pm 3$; for Exercise **94**, they are $\pm 1$; for Exercise **95**, they are $-2$ and $3$.

**2.** See the answer to Exercise **96**.

**3.** Define $y = (4)(x^2 - 10x + 21 \le 15)$.

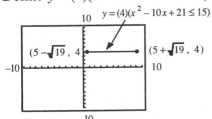

Algebraic solution: $5 - \sqrt{19} \le x \le 5 + \sqrt{19}$

**4.** $y = \dfrac{-4 \pm \sqrt{4^2 - 4 \times 0.5 \times (-(2 + x))}}{2 \times 0.5}$

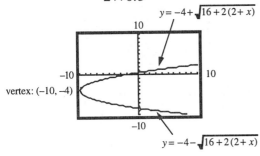

## Graphing Calculator Exercises 2.8

**1.**

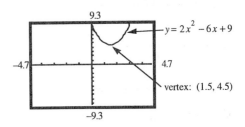

**2.** The function has a minimum value of $-1$ at $x = \pm 1$, but no maximum.

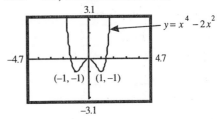

**3.** This function has neither a maximum nor a minimum value.

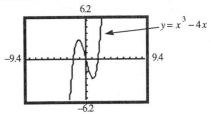

## Graphing Calculator Exercises 3.1

**1.** Window used: $-10 \le x \le 10$, $-2744 \le y \le 10$

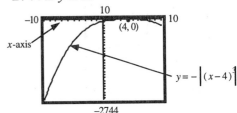

**2.** Window used: $-10 \le x \le 10$,
$-10 \le y \le 1280$

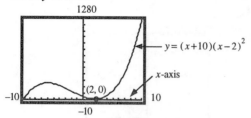

**3.** Window used: $-10 \le x \le 10$,
$-10 \le y \le 860$

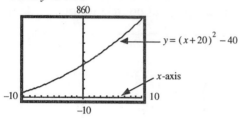

**4.** ZOOM out twice: $-40 \le x \le 40$,
$-40 \le y \le 40$

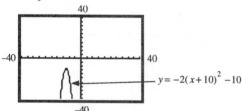

**5.** ZOOM out once: $-20 \le x \le 20$,
$-20 \le y \le 20$

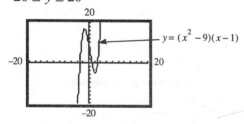

**6.** ZOOM out once: $-20 \le x \le 20$,
$-20 \le y \le 20$

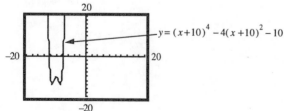

**7.** No.

**8.** No.

**9.** The following, using a window of
$-20 \le x \le 0$ and $-20 \le y \le 10$, is one of
many possible answers. It shows that the
graph is symmetric with respect to the line
$x = -10$, for example.

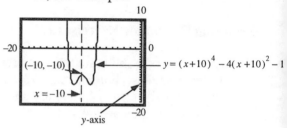

## Graphing Calculator Exercises 3.2

**1. (a)** **Note:** The graph below includes
the vertical asymptote $x = 3$,
since the calculator is in
CONNECTED mode. To
correctly sketch the solution,
you should draw the
asymptote and show that the
function follows it, not
simply copy the screen
below onto your page.

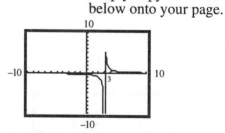

**(b)**

**(c)** (2.9787234, −46.99999) and
(3.1914894, 5.2222223) on the
TI-82.

**(d)** "$X = 3$, $Y = $ " (The $y$-coordinate on
the screen is blank.)

**2. (a)**

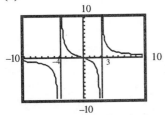

**(b)**

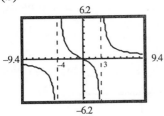

**(c)** Answers will vary depending on your calculator. On the TI-81 the values are:

For $x = -4$:
$(-4.10563, -32.93333)$ and $(-3.894737, 32.198473)$

For $x = 3$:
$(2.8421053, -15.78462)$ and $(3.0526316, 49.343284)$

**(d)** Should display "$X = -4$, $Y=$ " at $x = -4$ asymptote and "$X = 3$, $Y =$ " at $x = 3$ asymptote.

**3. (a)**

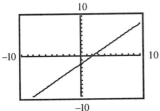

**(b)**

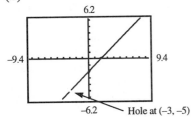

Hole at $(-3, -5)$

**(c)** No.

**(d)** The friendly window shows the absence of the point $(-3, -5)$, as it should.

## Graphing Calculator Exercises 3.3

**1.** $x$-intercepts at $x = 1, -2, 3$, and $-4$. No asymptotes.

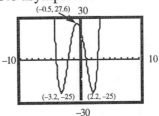

**2.** $x$-intercepts at $x = -1, 0$, and $1$. Turning points at about $(-0.67, 0.53)$ and $(0.67, -0.53)$. No asymptotes.

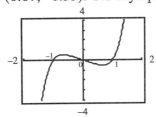

**3.** $x$-intercept at $x = 1$. Turning point at $x = -1$. $x = 0$ is a vertical asymptote.

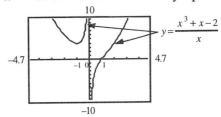

$$y = \frac{x^3 + x - 2}{x}$$

**4. and 5.** $x$-intercept at $x = 1$. Turning point at about $x = -1.8$. $x = 0$ is a vertical asymptote. $y = x$ is an oblique asymptote for the function in Exercise **4**. There are no oblique asymptotes in Exercises **1 - 3**.

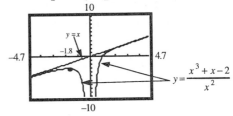

$$y = x$$

$$y = \frac{x^3 + x - 2}{x^2}$$

# Graphing Calculator Exercises 3.4

**1. (a)**  $y = \dfrac{2}{x}$ and $y = -2x + 5$

**(b) and (c)**

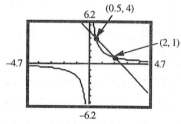

**2. (a)**  $y = \dfrac{9}{x}$ and $y = 5x + 12$

**(b) and (c)**

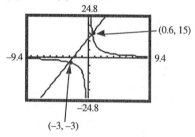

**3.**  $y = (x - 2)/(x + 2)$ and $y \geq 0$ yield the following graph and intervals on the $x$-axis:

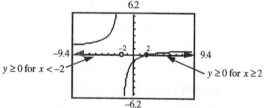

**Note:**  $x = -2$ is a vertical asymptote.

**4.**  $y = (6 - x)(3 + x)/(x + 1)$ and $y \leq 0$ yields the following graph and intervals on the $x$-axis:

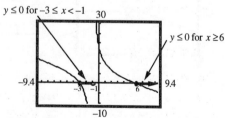

**Note:**  $x = -1$ is a vertical asymptote.

**5.**  Graphing $y = ((x - 2)/(x + 2) \geq 0)(2)$ yields:

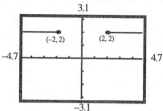

**Note:**  The graph was obtained using Dot Mode.

**6.**  Graphing
$y = ((6 - x)(3 + x)/(x + 1) \leq 0)(2)$ yields

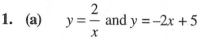

# Graphing Calculator Exercises 3.6

**1.**  $x^3 - 3x^2 + 4x + 1 = ((1x - 3)x + 4)x + 1$

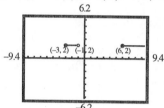

**Note:**  Two screens are shown here since all the entries did not fit on one. Your screen will simply scroll up as more space is needed. Hence, you may wish to record the coefficients as they appear for polynomials of degree greater than 3.

Quotient $= x^2 + 2x + 14$
Remainder $= 71$

**2.** $x^4 + 2x - 7 = x^4 + 0x^3 + 0x^2 + 2x - 7$
$$= (((1x + 0)x + 0)x + 2)x - 7$$

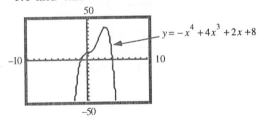

See note on Problem **1**.

Quotient $= x^3 - 6x^2 + 36x - 214$
Remainder $= 1277$

**3.** By the factor theorem $f(2) = 0$. Then
$f(2) = 2^3 + k^3 2^2 + k2 - k^4 = 0$. To find
$k$ graphically, replace $k$ by $x$ and graph
$y = 2^3 + 2^2 x^3 + 2x - x^4$. The $k$-values
(i.e., the $x$-intercepts) are approximately
$-1.1$ and $4.2$.

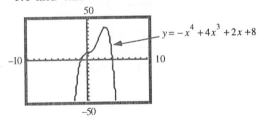

## Graphing Calculator Exercises 3.7

**1.** TRACE and ZOOM just continue to show
that the root is between 1.4 and 1.5 again
and again, without showing how to round
up or down. It is better to start ZOOMing
with a window of about
$1.445 \le X \le 1.455,\ -0.005 \le Y \le 0.005$.
TRACE now shows the root is closer to
1.4 than to 1.5.

**2.** $p(x) = x^5 - x^4 - 3x^3 + 3x^2 - 4x + 4 = 0$

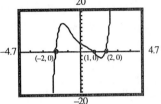

The graph confirms the real roots $-2$, 1,
and 2. Then the factor theorem implies
that $x + 2$, $x - 1$, and $x - 2$ are factors of
$p(x)$. Use synthetic division to obtain the
remaining factor $x^2 + 1$ which has
imaginary roots. Factoring by grouping
will also confirm these results.

**3.** $f(x) = 2x^4 + 3x^2 + 8$

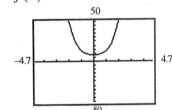

The graph shows that there are no $x$-
intercepts, which means that $p(x) = 0$ has
no real roots. Another way to conclude
that the graph will always be above the $x$-
axis is to observe that $2x^4 + 3x^2 + 8 \ge 8$
since $x^4$ and $x^2$ are $\ge 0$ for all $x$. (The
lowest point on the curve is $(0, 8)$.)

**4.** $p(x) = x^5 + x^4 - x - 1$

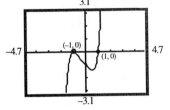

The graph shows that $-1$ is an even root
and $+1$ is an odd root. Dividing $p(x)$ by
$(x + 1)^2 (x - 1)$ yields $x^2 + 1$, which has
imaginary roots. Hence,

$p(x) = (x+1)^2 (x-1)(x^2+1)$. Another way to verify the graphical observations of the roots is to factor $p(x)$ by using the method of factoring by grouping.

## Graphing Calculator Exercises 4.1

1. **(a)** $y = \pm\sqrt{9-x^2}$; No.

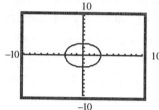

   **(b)**

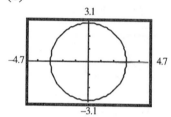

2. **(a)** $y = -2 \pm \sqrt{2-(x-1)^2}$; No.

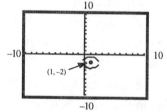

   **(b)**

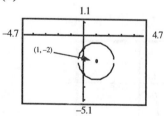

3. Head: $x^2 + y^2 = 9$

   Left eye: $(x+1.5)^2 + (y-1)^2 = 0.25$

   Right eye: $(x-1.5)^2 + (y-1)^2 = 0.25$

   Nose: $x^2 + y^2 = 0.25$

   Mouth: bottom semicircle of
   $$x^2 + (y+1.5)^2 = 1$$

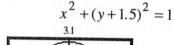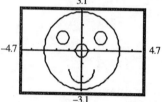

   This was graphed on the TI-82. Each circle requires two functions—one for the top semicircle, one for the bottom semicircle. Thus the picture above requires 9 functions to be graphed. The TI-81 only allows 4 functions to be plotted, so you must use the DRAW facility to place the other circles.

## Graphing Calculator Exercises 4.2

1. The student will obtain the straight line $y = x/2$. The 1/2 should have been enclosed within parentheses: $y = x\wedge(1/2)$.

2.

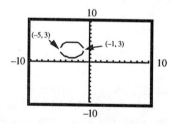

   **(a)** $(x+3)^2 + (y-3)^2 = 4$
   **(b)** There is no pixel corresponding to the points $(-1, 3)$ and $(-5, 3)$, since the $x$-interval, for example, of $[-10, 10]$ has been divided equally into 96 (TI-81) or 95 (TI-82 or Casio) points. Hence, few points with integer coordinates will be plotted. (See the graphing calculator appendix for more on this.)

**(c)** The graph doesn't look circular because the scales on the $x$ and $y$ axes are unequal.

**(d)** $-9.6 \le X \le 9.4, -6.4 \le Y \le 6.2$ on the TI-81 or $-9.4 \le X \le 9.4$, $-6.2 \le Y \le 6.2$ on the TI-82 or Casio.

## Graphing Calculator Exercises 4.3

$$f(x) = (x^2 - 1) / (2\sqrt{(x-1)}) + 2x\sqrt{(x-1)}$$

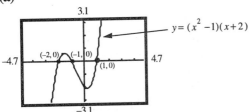

**Notes:** ZOOMing out repeatedly gives a similar graph. The point $(1, 1)$ is not on the graph.

## Graphing Calculator Exercises 4.4

**1. (a)**

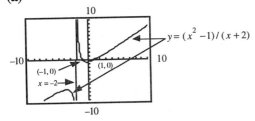

**(b)** The zeros of $f(x)g(x)$ consist of all the zeros of both $f(x)$ and $g(x)$.

**2. (a)**

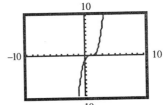

**(b)** The zeros of $\dfrac{f(x)}{g(x)}$ are the same as those of $f(x)$. The graph

has a vertical asymptote at the zero of $g(x)$.

**Notes:** $y = x - 2$ is an oblique asymptote of the quotient. The calculator plots the vertical asymptote since it is in CONNECTED mode.

**3. (a)** $y = f(g(x)) = |x - 2|$ is a translation two units to the right of $y = |x|$.

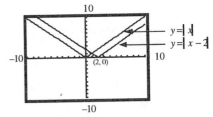

**(b)** $y = g(f(x)) = |x| - 2$ is a translation two units downward of $y = |x|$.

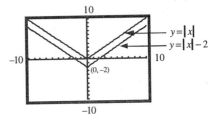

**4.**

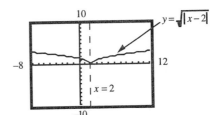

## Graphing Calculator Exercises 4.5

**1.** Yes

**2.** No

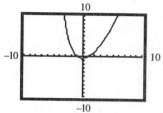

**Note:** $x = -4$ is a vertical asymptote.

**3.** $[a, b] = [-0.1, 2.1]$. The corresponding domain of $f^{-1}(x)$ is $[-3.1, 3.1]$. The first screen shown is the function $f(x) = x^3 - 3x^2 - x + 3$ graphed in a friendly window in CONNECTED mode. From this you can determine the one-to-one section to be the interval listed above.

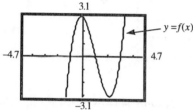

To achieve the second screen, restrict the function to the desired interval, plot the function in Dot mode, then select **DrawInv** from the **DRAW** menu on your calculator. Find **Y₁** from the **Y-VARS** menu and hit enter. The relevant screens for the TI-82 are below.

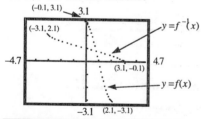

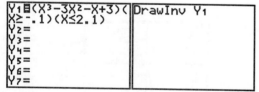

## Graphing Calculator Exercises 5.1

**1.** Intersections at about $(1.37, 2.59)$ and $(9.94, 981.97)$

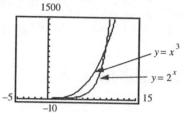

**2.** The student probably left out the parentheses and used multiplication instead of exponentiation after the 3. That is, the student keyed in $y = 3 * 2x$ instead of $y = 3 \wedge (2x)$.

**3.** The graph is symmetric with respect to the $y$-axis. It passes through the origin and intersects the graph of $y = x^2$ at about the points $(4.43, 19.66)$ and $(-4.43, 19.66)$. The calculator screen below shows the two graphs, their three points of intersection, and the intersection point as determined by the TI-82. From the graph, we see that $y = x^2 < y = 2^x + 2^{-x} - 2$ for approximately $|x| > 4.43$.

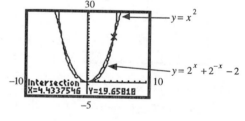

4. The graph has no intercepts (it is only graphed for $x > 0$, so it cannot hit the $y$-axis). It does have a horizontal asymptote of about $y = 2.7$. (The correct number is actually $2.718281828...$ and is denoted by the letter $e$, as we will see in later sections.)

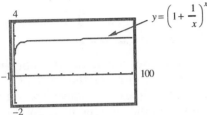

5. It will take about 299 minutes. The screen shows the function with the TRACE function showing the point $(299, 1000611.8)$

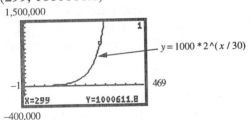

## Graphing Calculator Exercises 5.2

1. The $y$-coordinates of (**b**) are twice the $y$-coordinates of (**a**) for corresponding $x$ values. Both are undefined at $x = 0$, concave downward for all $x$, increasing where $x > 0$. The $y$-axis is a vertical asymptote for both.

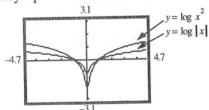

2. The domain is all $x \neq 0$. There are no asymptotes. The $x$-intercepts are at $x = \pm 1$. The function is symmetric about the origin, since $f(-x) = -f(x)$.

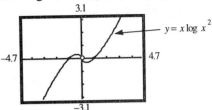

3. The inverse is $y = 2^{x-2}$. First graph this function, then use the **DrawInverse** function of your calculator to graph the desired function.

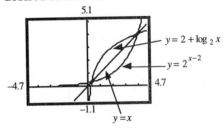

4. The calculator shows that as $x$ approaches zero, the graph is simply a horizontal line at $y = 10$. The graph only exists for $x > 0$ since that is the domain of $y = \log x$. Notice that as $x$ approaches zero, $1/\log x$ approaches zero. So it seems that as $x$ approaches zero we get $0^0 = 10$. So it seems there is no reason to define $0^0 = 1$!

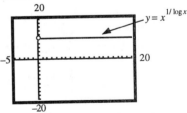

# Graphing Calculator Exercises 5.3

1. **(a)** $1 \text{ m}^2 = 10,000 \text{ cm}^2 = 10^4 \text{ cm}^2$
   converts $10^{-16} \text{ cm}^2$ to $10^{-20} \text{ m}^2$.
   Hence, we obtain
   $$D = 10\log(I/10^{-12})$$
   $$= 10(\log I - \log 10^{-12})$$
   $$= 10\log I + 120$$

   **(b)** The graph is linear in log $I$ and, hence, easier to comprehend. To do this, you simply replace log $I$ in the formula with $x$ when you enter it into the graphing calculator. Then the $x$-axis becomes the (log $I$)-axis.

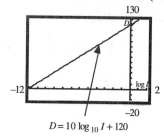

$$D = 10\log_{10} I + 120$$

2. Use $\log_2 x = \log x / \log 2$

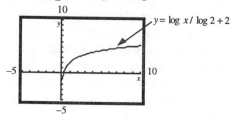

3. We take the log of both sides and apply rules of logarithms and simple algebra to obtain
   $$\log Y = (3\log D + \log 0.001)/2$$
   $$= 1.5\log D - 1.5$$
   Then, on the graphing calculator, replace log $Y$ with $y$ and log $D$ with $x$ and graph it.

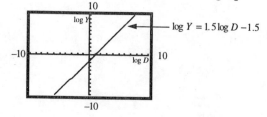

4. Domain $= \{x: x < -2 \text{ or } x > 2\}$.
   Asymptotes: $x = \pm 2$.
   $x$-intercepts: $x = \pm\sqrt{5}$.

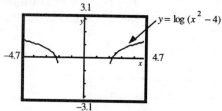

   Notice that the calculator does not show the function following the vertical asymptotes. This is because it is simply connecting dots, and the next dot does not exist, so it stops. When you write up your solution, you should sketch in the vertical asymptotes and show that the function follows these asymptotes.

5. Domain = all real numbers.
   No asymptotes and no $x$-intercepts.
   $y$-intercept: $y = \log 4$ or approximately $y = 0.60$.

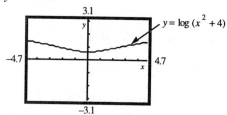

# Graphing Calculator Exercises 5.4

1.

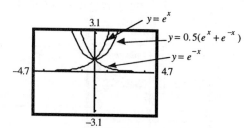

**2.** Use the Intersection function on your calculator (if you have one) to assist you. First intersection is (1.86, 6.41).

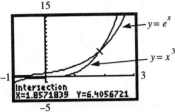

Second intersection is (4.54, 93.35).

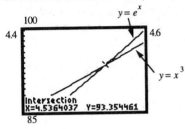

**3.** max $= 0.40 = 1 / \sqrt{2\pi}$ at $x = 0$

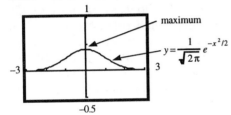

**4.** $y$-intercept at $(0, 1)$. $y = 0$ and $y = 2$ are asymptotes.

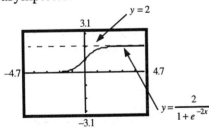

## Graphing Calculator Exercises 6.2

**1. (a)** We get *most* of the unit circle, but there is a small piece missing. By using trace, we see that the $T$-range has been truncated to $0 \le T \le 6.2$.

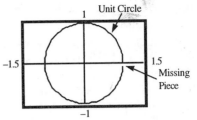

**(b)** $X$ and $Y$ are cos $T$ and sin $T$, respectively, where $T$ is the radian measure of the angle defining the sine and cosine functions with the unit circle. Since $T$ is in radians, it would be better to use $T$step $= 0.1*(\pi/2)$

**2.** We get *all* of the unit circle, since $360 = 6 * 60$. The rest is the same as in Exercise **1**, except there is no need to change the $T$step.

**3.** In Parametric and Radian modes, set $-3 \le X \le 3, -2 \le Y \le 2, 0 \le T \le 2\pi$, $T$step $= 0.1*(\pi/2)$, and graph $X_1(T) = 2 \cos T$ and $Y_1(T) = 2 \sin T$.

**4.** Corresponding points on the circle and the tangent line $x = 1$ are always collinear with the origin. That is, the cursor oscillates between the circle and the tangent line along a line which passes through the origin, as shown in the following graph. This line makes an angle of $T$ radians with the $x$-axis, and the height of the point on the line $x = 1$ is tan $T$, as claimed in Exercise 83 of this section.

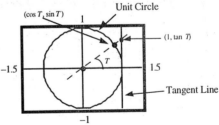

**Note:** Exercises **5** and **6** were obtained on a TI-82 by following Exercise **1** and using the "improved" $T$step $= 0.1*(\pi/2)$.

**5. (a)** We only get the half-line in the third quadrant, where both $X$ and $Y$ are negative, since $T$ is positive.

**(b)** $T = 3.6128316$ radians, $X = -.8910065 = \cos T$, and $Y = -.4539905 = \sin T$.

**(c)** The exact answer is $X = -\dfrac{\sqrt{3}}{2}$, $Y = -0.5$, and $T = 7\pi/6$. The two answers are fairly close, but certainly not exactly the same.

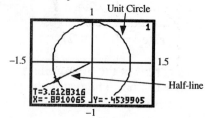

**6. (a)** $T = 4.2411501$, $X = \cos T = -.4539505$ $Y = \sin T = -.8910065$. The graph is similar to that in Exercise **5c**.

**(b)** $T = 4.3982297$, $X = \cos T = -.30901$ $Y = \sin T = -.9510565$. The graph is similar to that in Exercise **5c**.

## Graphing Calculator Exercises 6.3

**Note:** None of Exercises **1** to **4** is defined at $x = 0$.

**1.** The graph gets closer and closer to $y = 1$.

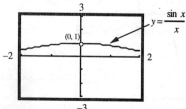

**2.** The graph approaches $y = 0$.

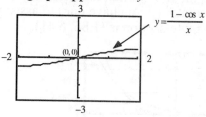

**3.** The graph gets higher and higher, without bound.

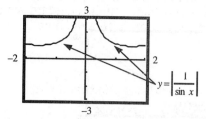

**4.** The graph oscillates infinitely often between $y = -1$ and $y = 1$. The calculator screen only shows a few of these, since it is simply connecting dots, and can only plot a finite number of dots to connect. To show the oscillation, you can change the $X$min and $X$max to smaller and smaller values, and the graph will continually oscillate.

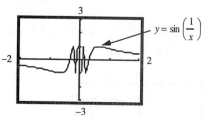

**5.** $f(x)$ resembles $g(x) = \cos x$.

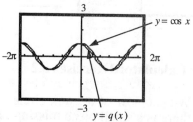

**6.** $f(x)$ resembles $g(x) = -\sin x$.

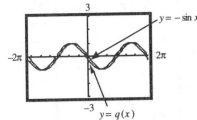

**7.** $g(x)$ and $q(x)$ become virtually indistinguishable on the standard window, although Zooming In will show the difference.

## Graphing Calculator Exercises 6.4

**1.** All the graphs agree with the text, except for the vertical lines included by some calculators. These are the lines through the highest and lowest adjacent pixels on either side of values, such as $x = \pi/2$ for $y = \sec x$, where the function is not defined.

**2.** $x = 0.67$ and $x = 2.48$

**3.** $x = 0.95$ and $x = 2.70$

**4.** $x = 0.90$

**5.**

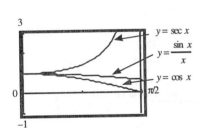

## Graphing Calculator Exercises 6.5

**1.** They are not the same function, as can be seen from the graphs. The domain of $\tan(\arctan x)$ is all real numbers, and the range is also all real numbers. The domain of $\arctan(\tan x)$ is all real numbers except $x = \pi/2 + k\pi$, and the range is $(-\pi/2, \pi/2)$.

$y = \tan(\tan^{-1} x)$

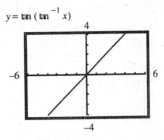

$y = \tan^{-1}(\tan x)$

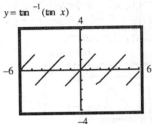

**2.** The domain of $\sin(\arcsin x)$ is $[-1, 1]$; range is $[-1, 1]$. The domain of $\arcsin(\sin x)$ is all real numbers; range is $[-\pi/2, \pi/2]$.

$y = \sin(\sin^{-1} x)$

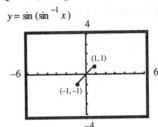

$y = \sin^{-1}(\sin x)$

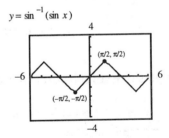

**3.** Domain of cos(arccos $x$) is [–1, 1]; range is [–1, 1]. The domain of arccos(cos $x$) is all real numbers; range is [0, $\pi$]

$y = \cos(\cos^{-1} x)$

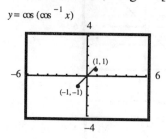

$y = \cos^{-1}(\cos x)$

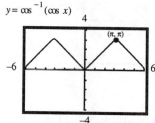

**4.** Maximum about 0.91 when $x = 5.83$, to the nearest hundredth, as determined using the **CALC maximum** function on the TI-82.

$\alpha = \tan^{-1}\left(\dfrac{17}{x}\right) - \tan^{-1}\left(\dfrac{2}{x}\right)$

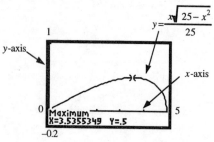

## Graphing Calculator Exercises 7.1

**1.** The maximum of 0.50 is achieved at an $x$ value of about 3.54 and $\theta = 45°$.

$y = \dfrac{x\sqrt{25 - x^2}}{25}$

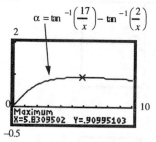

**2.** The maximum of 0.50 is achieved an $x$ value of about 2.83 and $\theta = 45°$.

$y = \dfrac{x\sqrt{16 - x^2}}{16}$

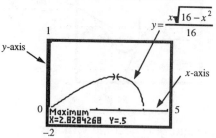

## Graphing Calculator Exercises 7.2

$\text{Area} = \dfrac{bh}{2}$, but $\sin\theta = \dfrac{b}{2}$ and $\cos\theta = h$.

Hence, Area = sin $\theta$ cos $\theta$.

Next, the graph of $y = \sin x \cos x$ has a maximum of 0.5 when $x = 45°$. Hence, angle $\theta$ should be $45°$ to get the maximum area.

$y = \sin x \cos x$

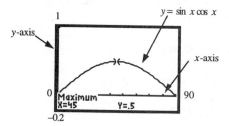

## Graphing Calculator Exercises 7.3

**1. (a)** The graphs, and therefore the functions, are different.

$y = \sin 2x$

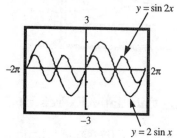

$y = 2 \sin x$

**(b)** These are identical. See Example 4, and take reciprocals of both sides.

$$y = \frac{\sin x}{1 - \cos x} = \frac{1 + \cos x}{\sin x}$$

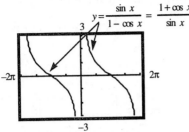

**(c)** These are not identical.

$$y = \sin x + \sin 1$$

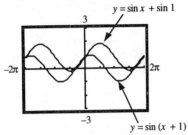

$$y = \sin(x + 1)$$

**(d)** The graphs are identical.

$$\frac{\tan x \csc^2 x}{\tan^2 x + 1} = \frac{\tan x \csc^2 x}{\sec^2 x}$$

$$= \tan x \cdot \frac{\csc^2 x}{\sec^2 x}$$

$$= \tan x \cdot \frac{\frac{1}{\sin^2 x}}{\frac{1}{\cos^2 x}}$$

$$= \tan x \cdot \frac{\cos^2 x}{\sin^2 x}$$

$$= \tan x \cdot \cot^2 x$$

$$= \cot x$$

$$y = \frac{\tan x \csc^2 x}{\tan^2 x + 1} = \cot x$$

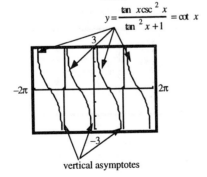

vertical asymptotes

**2.** The graph is just a horizontal line, $g(x) = 1$.

$$y = f(x) = \sin^4 x + 2\sin^2 x \cos^2 x + \cos^4 x$$

—looks like $g(x) = 1$

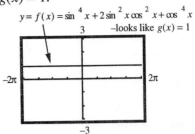

To prove that $f(x) = g(x)$, notice that $f(x)$ is a perfect square trinomial in $\sin^2 x$ and $\cos^2 x$:

$$f(x) = \sin^4 x + 2\sin^2 x \cos^2 x + \cos^4 x$$

$$= \left(\sin^2 x\right)^2 + 2\sin^2 x \cos^2 x + \left(\cos^2 x\right)^2$$

$$= \left(\sin^2 x + \cos^2 x\right)^2 = 1^2 = 1 = g(x)$$

**3.** The graph looks like $g(x) = \cos x$, except that it jumps down to $y = 0$ when $x = \pm 2\pi$. This happens since these values are not in the domain of $f(x)$.

$$y = f(x) = \frac{\cos x - \cos^3 x}{\sin^4 x + \sin^2 x \cos^2 x}$$

—looks like $g(x) = \cos x$

X=6.2831853   Y=0

To prove that $f(x) = g(x)$, factor $\cos x$ out of the numerator and $\sin^2 x$ out of the denominator, simplify and reduce:

$$f(x) = \frac{\cos x - \cos^3 x}{\sin^4 x + \sin^2 x \cos^2 x}$$

$$= \frac{\cos x (1 - \cos^2 x)}{\sin^2 x (\sin^2 x + \cos^2 x)}$$

$$= \frac{\cos x (\sin^2 x)}{\sin^2 x (1)} = \cos x = g(x)$$

## Graphing Calculator Exercises 7.4

**1.**

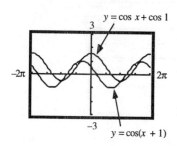

**2.**

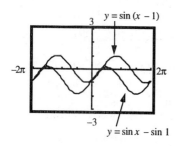

**3.** The graph looks like $s(x) = 0.5$.

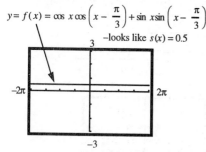

To prove that $f(x) = s(x)$, use the difference formulas for sine and cosine, find the appropriate values using special triangles, and simplify:

$$f(x) = \cos x \cos\left(x - \frac{\pi}{3}\right) + \sin x \sin\left(x - \frac{\pi}{3}\right)$$

$$= \cos x \left( \cos x \cos\frac{\pi}{3} + \sin x \sin\frac{\pi}{3} \right)$$

$$+ \sin x \left( \sin x \cos\frac{\pi}{3} - \cos x \sin\frac{\pi}{3} \right)$$

$$= \cos x \left( \cos x \cdot \frac{1}{2} + \sin x \cdot \frac{\sqrt{3}}{2} \right)$$

$$+ \sin x \left( \sin x \cdot \frac{1}{2} - \cos x \cdot \frac{\sqrt{3}}{2} \right)$$

$$= \frac{1}{2}\cos^2 x + \frac{\sqrt{3}}{2}\cos x \sin x$$

$$+ \frac{1}{2}\sin^2 x - \frac{\sqrt{3}}{2}\sin x \cos x$$

$$= \frac{1}{2}\cos^2 x + \frac{1}{2}\sin^2 x$$

$$= \frac{1}{2}\left( \cos^2 x + \sin^2 x \right)$$

$$= \frac{1}{2}(1) = \frac{1}{2} = s(x)$$

## Graphing Calculator Exercises 7.5

**1.** The graph is a sine wave with an amplitude of 1. It completes three periods between 0 and $2\pi$, so the period is $2\pi/3$. The graph starts at the origin by sloping downward, so it must have a coefficient of $-1$. The graph must be $g(x) = -\sin 3x$.

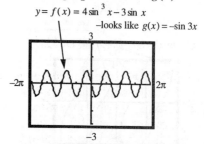

To prove that $f(x) = g(x)$, we will start with $g(x)$ and make it look like $f(x)$:

$$g(x) = -\sin 3x$$
$$= -\sin(2x + x)$$
$$= -(\sin 2x \cos x + \cos 2x \sin x)$$
$$= -\sin 2x \cos x - \cos 2x \sin x$$
$$= -(2\sin x \cos x)\cos x - (1 - 2\sin^2 x)\sin x$$
$$= -2\sin x \cos^2 x - \sin x + 2\sin^3 x$$
$$= -2\sin x(1 - \sin^2 x) - \sin x + 2\sin^3 x$$
$$= -2\sin x + 2\sin^3 x - \sin x + 2\sin^3 x$$
$$= 4\sin^3 x - 3\sin x = f(x)$$

**2.** The graph looks like $g(x) = 1$.

$$y = f(x) = \frac{\cos^4 x - \sin^4 x}{1 - 2\sin^2 x}$$

—looks like $g(x) = 1$

To prove that $f(x) = g(x)$, we notice that the numerator is a difference of squares, factor it, and simplify:

$$f(x) = \frac{\cos^4 x - \sin^4 x}{1 - 2\sin^2 x}$$
$$= \frac{(\cos^2 x - \sin^2 x)(\cos^2 x + \sin^2 x)}{\cos 2x}$$
$$= \frac{(\cos 2x)(1)}{\cos 2x} = 1 = g(x)$$

## Graphing Calculator Exercises 7.6

**Note:** Recall that the vertical lines on some of the following screens are representative of the vertical asymptotes. You should not consider places where a graph crosses these asymptotes to be solutions.

**1.** $x = 1.01$ and $x = 2.13$

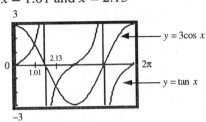

**2.** $x = 3.74$ and $x = 5.68$

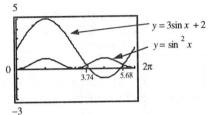

**3.** $x = 0.00$, $x = 2.09$, and $x = 6.28$

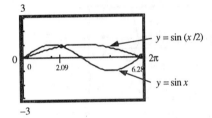

**4.** $x = 7.29$ and $x = 8.41$

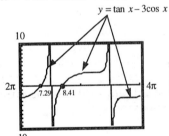

**5.** $x = 10.02$ and $x = 11.96$

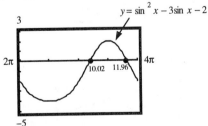

**6.** $x = 6.29$ (since $6.28 < 2\pi$), $x = 10.47$, and $x = 12.56$ (since $12.57 > 4\pi$)

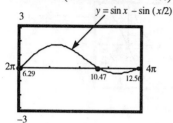

**7.** For Exercise **1**: $x = 1.01 + 2k\pi$, and
$\quad\quad x = 2.13 + 2k\pi$, $k = 0, \pm1, \pm2, ...$
For Exercise **2**: $x = 3.74 + 2k\pi$, and
$\quad\quad x = 5.68 + 2k\pi$, $k = 0, \pm1, \pm2, ...$
For Exercise **3**: $x = 2k\pi$, $x = 2.09 + 4k\pi$,
$\quad\quad$ and $x = 6.28 + 4k\pi$, $k = 0, \pm1, \pm2, ...$

**8.** $x = -0.46$ and $x = 1.26$

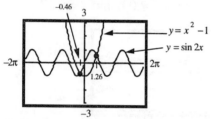

There is no general solution in the usual sense because $x^2 - 1$ is not a periodic function.

**9.** $x$ in the intervals $[1.01, \pi/2]$ and $[2.13, 3\pi/2]$. See calculator screen in Exercise **1**.

**10.** $x$ in the interval $[-0.46, 1.26]$. See calculator screen in Exercise **2**.

## Graphing Calculator Exercises 8.1

**1. (a)** $c(x) = \sqrt{25 - 24\cos x}$
**(b)**

10

$(\pi, 7) = \text{maximum}$

$y = c(x)$

0
$\quad\quad 0 \quad (0, 1) = \text{minimum} \quad\quad \pi$

**(c)** maximum at $c(\pi) = 7$

**(d)** minimum at $c(0) = 1$
**Note:** The maximum and minimum values, 1 and 7, for the function $c(x)$ are, in reality, not sides of triangles because when $x = 0$ or $x = \pi$, the "triangles" are straight lines. So, we may regard $c(0) = 1$ and $c(\pi) = 7$ as limiting values for side $c$. For example, since the function $c(x)$ is increasing from left to right on the interval $0 \le x \le \pi$, the side $c$ increases toward 7 as the angle gets larger and decreases toward 1 as the angle gets smaller.

**2. (a)**
$$K(x) = \sqrt{\left(\frac{x+2}{2}\right)\left(\frac{x+2}{2} - 1\right)\left(\frac{x+2}{2} - 1\right)\left(\frac{x+2}{2} - x\right)}$$
$$= \sqrt{\frac{(x+2)(x^2)(2-x)}{2^4}}$$
$$= \sqrt{\frac{4x^2 - x^4}{16}}$$

**(b)**

1

$y = K(x)$

$\underset{\text{X=1.4142132 \ Y=.5}}{\text{Maximum}}$
$0 \quad\quad\quad\quad\quad\quad\quad 2$
$\quad\quad 0$

**(c)** $K = 0.500$ when $x$ is about $1.414$.

## Graphing Calculator Exercises 8.4

**1.** $a = $ 8th roots of unity
$b = $ 4th roots of unity
$c = $ square roots of unity

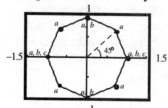

**2.** $T$step $= 60°$
$a = $ 6th roots of unity
$b = $ 3rd roots of unity

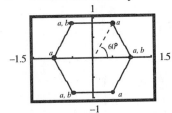

**3.** $T$step $= 120°$, $X_1(T) = (2 \wedge (1/3) \cos T)$,
$Y_1(T) = (2 \wedge (1/3) \sin T)$

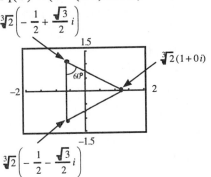

## Graphing Calculator Exercises 8.5

**1.**

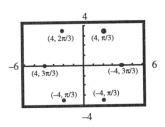

**2.**

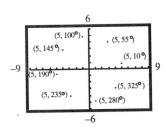

## Graphing Calculator Exercises 8.6

**1.** Symmetric with respect to the $x$-axis, the $y$-axis, and the origin.

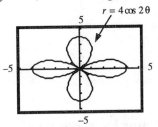

**2.** Symmetric only with respect to the $x$-axis.

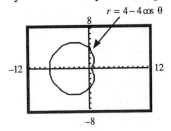

**3.** Symmetric only with respect to the $y$-axis.

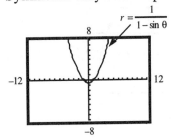

**4.**

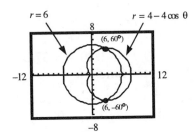

**5.**

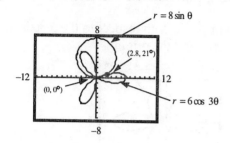

## Graphing Calculator Exercises 9.1

**1.** $x = y = z = w = u = 1$

**2. (a)**

$4a + 2b + c = 1$
$9a + 3b + c = 2$
$16a + 4b + c = -1$
yields $a = -2$, $b = 11$, $c = -13$ and the
function $y = -2x^2 + 11x - 13$

**(b)**

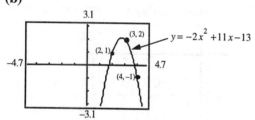

**3. (a)**

$$
\begin{array}{rcrcrcrcr}
8a & + & 4b & + & 2c & + & d & = & 1 \\
27a & + & 9b & + & 3c & + & d & = & 2 \\
-1a & + & b & - & c & + & d & = & -1 \\
-8a & + & 4b & - & 2c & + & d & = & -2
\end{array}
$$

yields $a = 1/30$, $b = -1/20$, $c = 37/60$, and
$d = -3/10$ and the function

$$y = f(x) = \frac{1}{30}x^3 - \frac{1}{20}x^2 + \frac{37}{60}x - \frac{3}{10}$$

**(b)**

## Graphing Calculator Exercises 9.2

**(a)** $A^2 = \begin{bmatrix} 2 & 0 & 2 & 0 \\ 0 & 2 & 1 & 1 \\ 1 & 1 & 2 & 0 \\ 0 & 2 & 1 & 1 \end{bmatrix}$

**(b)** $A^3 = \begin{bmatrix} 0 & 4 & 2 & 2 \\ 3 & 1 & 4 & 0 \\ 1 & 3 & 3 & 1 \\ 3 & 1 & 4 & 0 \end{bmatrix}$

**(c)** $A^4 = \begin{bmatrix} 6 & 2 & 8 & 0 \\ 1 & 7 & 5 & 3 \\ 4 & 4 & 7 & 1 \\ 1 & 7 & 5 & 3 \end{bmatrix}$

**(d)** $A^5 = \begin{bmatrix} 2 & 14 & 10 & 6 \\ 10 & 6 & 15 & 1 \\ 5 & 11 & 12 & 4 \\ 10 & 6 & 15 & 1 \end{bmatrix}$

## Graphing Calculator Exercises 9.3

**1. Note:** Your calculator will give decimal
approximations of the following
fractional matrix entries.

$$A^{-1} = \begin{bmatrix} -\frac{17}{14} & \frac{9}{7} & -\frac{11}{7} & \frac{1}{14} & \frac{13}{28} \\ -\frac{4}{7} & \frac{3}{7} & \frac{1}{7} & -\frac{1}{7} & \frac{1}{14} \\ -\frac{5}{14} & \frac{1}{7} & -\frac{2}{7} & \frac{11}{14} & \frac{3}{28} \\ \frac{5}{7} & -\frac{2}{7} & \frac{4}{7} & -\frac{4}{7} & -\frac{3}{14} \\ \frac{11}{14} & -\frac{5}{7} & \frac{3}{7} & \frac{1}{14} & -\frac{1}{28} \end{bmatrix}$$

**2. Note:** Your calculator will give values such as 1E–13 in place of the zeros in the matrix below. This is expected since the calculator is rounding off the decimal representations of the numbers in each fraction in $A^{-1}$. However, these values are so small as to be essentially zero.

$$AA^{-1} = A^{-1}A = I = \begin{bmatrix} 1 & 0 & 0 & 0 & 0 \\ 0 & 1 & 0 & 0 & 0 \\ 0 & 0 & 1 & 0 & 0 \\ 0 & 0 & 0 & 1 & 0 \\ 0 & 0 & 0 & 0 & 1 \end{bmatrix}$$

**3.** We solve $AX = C$ where

$$C = \begin{bmatrix} 4 \\ 5 \\ 6 \\ 7 \\ 1 \end{bmatrix} \text{ by multiplying both sides by } A^{-1}$$

to get

$$X = \begin{bmatrix} x \\ y \\ z \\ w \\ v \end{bmatrix} = A^{-1}C = A^{-1}\begin{bmatrix} 4 \\ 5 \\ 6 \\ 7 \\ 1 \end{bmatrix} = \begin{bmatrix} -\frac{193}{28} \\ -\frac{3}{14} \\ \frac{89}{28} \\ \frac{9}{14} \\ \frac{73}{28} \end{bmatrix}$$

## Graphing Calculator Exercises 9.5

**1.**

$$x = \frac{\begin{vmatrix} 5 & 1 & 2 & 1 \\ 3 & -1 & 3 & -1 \\ -2 & 1 & -1 & -1 \\ 7 & 1 & 2 & 1 \end{vmatrix}}{\begin{vmatrix} 1 & 1 & 2 & 1 \\ 2 & -1 & 3 & -1 \\ -1 & 1 & -1 & -1 \\ 3 & 1 & 2 & 1 \end{vmatrix}}$$

$$y = \frac{\begin{vmatrix} 1 & 5 & 2 & 1 \\ 2 & 3 & 3 & -1 \\ -1 & -2 & -1 & -1 \\ 3 & 7 & 2 & 1 \end{vmatrix}}{\begin{vmatrix} 1 & 1 & 2 & 1 \\ 2 & -1 & 3 & -1 \\ -1 & 1 & -1 & -1 \\ 3 & 1 & 2 & 1 \end{vmatrix}}$$

$$z = \frac{\begin{vmatrix} 1 & 1 & 5 & 1 \\ 2 & -1 & 3 & -1 \\ -1 & 1 & -2 & -1 \\ 3 & 1 & 7 & 1 \end{vmatrix}}{\begin{vmatrix} 1 & 1 & 2 & 1 \\ 2 & -1 & 3 & -1 \\ -1 & 1 & -1 & -1 \\ 3 & 1 & 2 & 1 \end{vmatrix}}$$

$$w = \frac{\begin{vmatrix} 1 & 1 & 2 & 5 \\ 2 & -1 & 3 & 3 \\ -1 & 1 & -1 & -2 \\ 3 & 1 & 2 & 7 \end{vmatrix}}{\begin{vmatrix} 1 & 1 & 2 & 1 \\ 2 & -1 & 3 & -1 \\ -1 & 1 & -1 & -1 \\ 3 & 1 & 2 & 1 \end{vmatrix}}$$

**2.** $x = y = z = w = 20/20 = 1$

## Graphing Calculator Exercises 9.6

**1.** The solution of the system is the shaded triangle as shown.

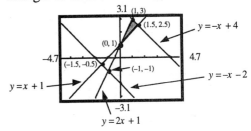

**2. (a)** $(0, 1)$, $(1, 3)$, and $(1.5, 2.5)$ are the vertices of the solution set.

**(b)** The vertices $(-1.5, -0.5)$ and $(-1, -1)$ are not in the solution set. $(-1, -1)$ does not satisfy $y \geq x + 1$, since $-1 < -1 + 1 = 0$, and $(-1.5, -0.5)$ does not satisfy $y \leq 2x + 1$, since $-0.5 > 2(-1.5) + 1 = -2$.

## Graphing Calculator Exercises 9.7

**1. (a) Note:** The solution set (the triangle) was drawn using the DRAW feature of the calculator.

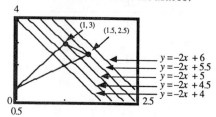

**(b)** Two of the given lines intersect the solution set at feasible points (vertices). The highest of these is the function $y = -2x + 5.5$.

**(c)** The maximum value of the objective function is $P = 5.5$.

**2.** 1521 of the first species and none of the second yields a maximum of about 2737.97 pounds of fish. Note that 1521.6797 and 0.0268037 is feasible, but rounding the first number up yields the point $(1522, 0)$, which is not in the region of feasible points.

## Graphing Calculator Exercises 10.1

**1.** $y = \pm \sqrt{\dfrac{1 - (x+1)^2}{2}}$

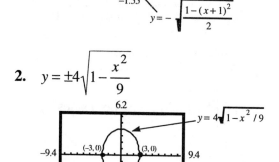

**2.** $y = \pm 4\sqrt{1 - \dfrac{x^2}{9}}$

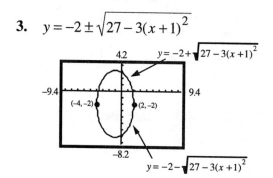

**3.** $y = -2 \pm \sqrt{27 - 3(x+1)^2}$

**Graphing Calculator Exercises 10.2**

1. $y = \pm 4\sqrt{\dfrac{x^2}{9} - 1}$

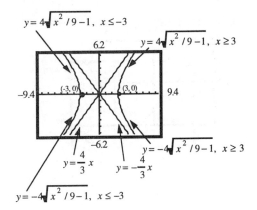

$y = 4\sqrt{x^2/9-1}, \; x \le -3$

$y = 4\sqrt{x^2/9-1}, \; x \ge 3$

$y = -4\sqrt{x^2/9-1}, \; x \ge 3$

$y = \dfrac{4}{3}x$

$y = -\dfrac{4}{3}x$

$y = -4\sqrt{x^2/9-1}, \; x \le -3$

2. $y = -2 \pm \sqrt{(x-1)^2 - 16}$

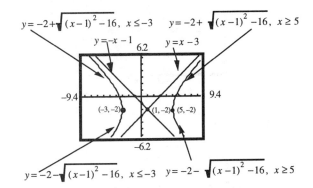

$y = -2 + \sqrt{(x-1)^2 - 16}, \; x \le -3$     $y = -2 + \sqrt{(x-1)^2 - 16}, \; x \ge 5$

$y = -x - 1$     $y = x - 3$

$y = -2 - \sqrt{(x-1)^2 - 16}, \; x \le -3$     $y = -2 - \sqrt{(x-1)^2 - 16}, \; x \ge 5$

**Graphing Calculator Exercises 10.3**

$y = 2 \pm \sqrt{x+3}$

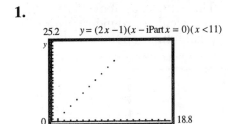

$y = 2 + \sqrt{x+3}$

$y = 2$

$y = 2 - \sqrt{x+3}$

**Graphing Calculator Exercises 10.4**

1 - 4.   See answers to Exercises **1** to **4**.

5 - 8.   See answers to Exercises **29** to **33**.

**Graphing Calculator Exercises 11.1**

**Note:**   The values for $X$min and $X$max below will work on the TI-82 or the Casio. Use $X$max = 19 on the TI-81.

1.

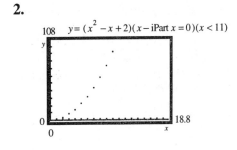

$y = (2x-1)(x - \text{iPart}\,x = 0)(x < 11)$

2.

$y = (x^2 - x + 2)(x - \text{iPart}\,x = 0)(x < 11)$

3. **Note:**   Although the dots appear to bounce up and down a bit, this sequence is steadily increasing toward the limiting value of $y = 2$. The bounce of the dots is due to roundoff error by the calculator and the pixel limitations of the calculator screen.

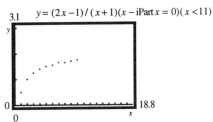

$y = (2x-1)/(x+1)(x - \text{iPart}\,x = 0)(x < 11)$

**4.**

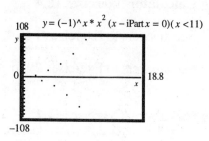

$y = (-1)^\wedge x * x^2 (x - \text{iPart } x = 0)(x < 11)$

**5. Note:** As in Exercise **3**, this is an increasing sequence, this one with limiting value $e = 2.718281828....$ The bounce, as above, is due to the calculator roundoff error and pixel limitations.

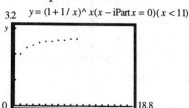

$y = (1 + 1/x)^\wedge x(x - \text{iPart } x = 0)(x < 11)$

**6. (a)** 1, 1, 2, 3, 5, 8, 13, 21, 34, 55, 89, 144, 233, 377, 610, 987, 1597, 2584, 4181, 6765

**(b)** For example, with $n = 20$,

$$a_{20} = \left(\frac{1}{\sqrt{5}}\right)\left(\frac{1+\sqrt{5}}{2}\right)^{20} - \left(\frac{1}{\sqrt{5}}\right)\left(\frac{1-\sqrt{5}}{2}\right)^{20}$$
$$= 6765$$

**(c)** 1, 2, 1.5, 1.6667, 1.6, 1.625, 1.6154, 1.6190, 1.6176, 1.6182. The golden mean is approximately 1.618, to the nearest thousandth. (Also, see the Challenge on page 209.)

## Graphing Calculator Exercises 11.5

**1.** Each time another term is added, the corresponding curve is closer to the graph of $f(x) = \dfrac{1}{1-x}$, for $-1 < x < 1$, than the preceeding curve.

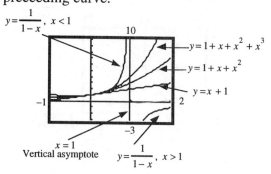

**2.** The observation in Exercise **1** is confirmed.

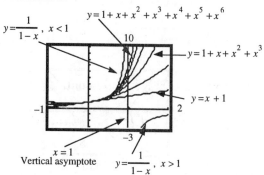

**3.** Each time another term is added, the corresponding curve is closer to the graph of $f(x) = \dfrac{1}{1+x}$, for $-1 < x < 1$, than the preceeding curve.

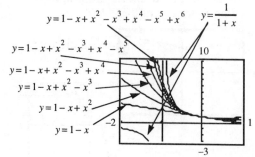

**4.** A suitable power series is
$$y = 1 - 2x + 4x^2 - 8x^3$$ which is valid for
$-0.5 < x < 0.5$.

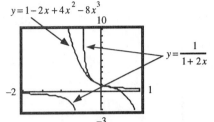

## Graphing Calculator Exercises 12.3

**1.** $f(x)$ and $g(x)$ are equivalent.

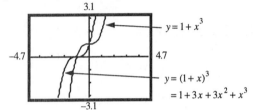

**2.** $f(x)$ and $g(x)$ are equivalent.

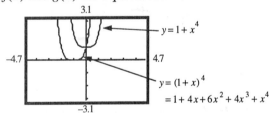

**Note:** In the solutions to Exercises **3** to **5** below, the values of $Y$min and $Y$max were changed from those in the instructions to make the graphs easier to distinguish.

**3.** As $h$ gets smaller, the lines get closer to $y = d(x)$.

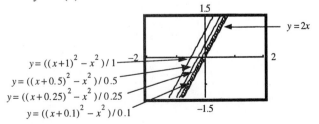

**4.** As $h$ gets smaller, the curves get closer to $y = d(x)$.

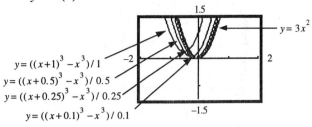

**5.** As $h$ gets smaller, the curves get closer to $y = d(x)$.

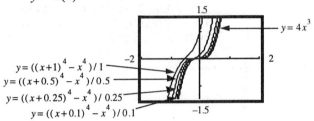

**6.** An appropriate curve is $d(x) = 5x^4$.

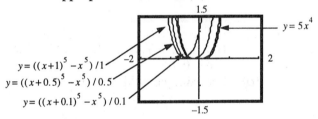

## Graphing Calculator Exercises 12.4

**Note:** The dots on the following graphs were quickly generated on the TI-82 by writing the following simple program and running it after graphing each function. Then you can simply count the number of dots above and below the graph. Refer to your calculator manual to learn how to write programs on your calculator.

```
PROGRAM:PLOT
:0→I
:Lbl 1
:While I<20
:Pt-On(rand,rand)
:I+1→I
:End
:Stop
```

1. Expect 10 points above the line; in this figure there are 14. Answers will vary.

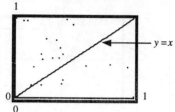

2. Expect 5 = 20/4; in this figure there are 4. Answers will vary.

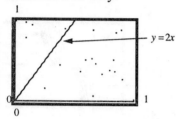

3. Expect $1 - \pi/4 \approx 4$; in this figure there are 3. Answers will vary.

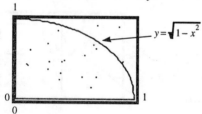